# *Paleozoic and Triassic Paleogeography and Tectonics of Western Nevada and Northern California*


Edited by
M.J. Soreghan
Department of Geosciences
University of Arizona
Tucson, AZ 85721
USA

and

G.E. Gehrels
Department of Geosciences
University of Arizona
Tucson, AZ 85721
USA

**347**


Geological Society of America
3300 Penrose Place
P.O. Box 9140
Boulder, CO 80301-9140
USA
**2000**

Published by The Geological Society of America, Inc.
3300 Penrose Place, P.O. Box 9140, Boulder, Colorado 80301

Printed in U.S.A.

GSA Books Science Editor Abhijit Basu
GSA Books Managing Editor Rebecca Herr
Cover design by Diane Lorenz

**Library of Congress Cataloging-in-Publication Data**

Paleozoic and Triassic paleogeography and tectonics of western Nevada and northern California / edited by Michael J. Soreghan and George E. Gehrels.
p. cm. — (Special paper ; 347)
Includes bibliographical references and index.
ISBN 0-8137-2347-7
1. Geology—Nevada. 2. Geology—California. 3. Geology, Stratigraphic. 4. Geology, Structural. I. Soreghan, Michael J., 1963- . II. Gehrels, George E. III. Series: Special papers (Geological Society of America) ; 347.
QE137.P35 2000
557.93—dc21
00-037199

**Cover:** (Front cover) Upper: The Vinini Formation of the Roberts Mountains allochthon, taken in the Roberts Mountains. (Photo by Stanley C. Finney.) Lower left: Students collecting U-Pb geochronology sample from the Pine Forest Range, northern Nevada. Shown, from left to right, are Angela Smith, Dani Montague-Judd, Brook Riley, Brian Darby, Jeff Manuszak, and Paul Kapp. (Photo by George E. Gehrels.) Lower right: Canyon Creek in the northern Sierra Nevada. (Photo by Gary H. Girty.)

(Back cover) Group of students and geologists during a field trip to collect detrital zircon samples in the Sierra Nevada. Front row, left to right: Jeff Manuszak, Dr. Joe Satterfield, Angela Smith, Amy Snapp, Dr. Meghan Miller, Brook Riley, Shannon Mack. Back row, left to right: Paul Kapp, Dr. Sandra Wyld, Dani Montague-Judd, Dr. Bill Dickinson, Brian Darby, Dr. Dave Harwood, Dr. George Gehrels.

10 9 8 7 6 5 4 3 2 1

# Contents

Geological Society of America
Special Paper 347
2000

# *Introduction to detrital zircon studies of Paleozoic and Triassic strata in western Nevada and northern California*

**George E. Gehrels**
*Department of Geosciences, University of Arizona, Tucson, Arizona 85721, USA*

## ABSTRACT

**U-Pb geochonologic analyses have been conducted on 648 individual detrital zircon grains from Paleozoic and Triassic strata of western Nevada and northern California. These strata belong to several distinct terranes that are fault bounded and potentially displaced from their sites of origin. The analyses have been conducted in an attempt to provide additional constraints on where these terranes may have formed in relation to each other and to western North America.**

**This paper provides background information that is essential to each of the accompanying chapters presenting detrital zircon data. Main components include (1) an outline of the general tectonic evolution of the region, (2) an overview of the detrital zircon reference for western North America, with a new quantitative analysis of the data that make up this reference, (3) a detailed discussion of the analytical methods used in analyzing detrital zircon grains in our laboratories, and (4) an assessment of the various biases and interpretations involved in this type of study.**

## INTRODUCTION

This volume contains several papers that present the results of a detrital zircon geochronologic study of terranes in western Nevada and northern California (Fig. 1). The main objective of this study was to place tighter constraints on the displacement history of terranes outboard of the Cordilleran miogeocline in this part of the Cordillera (Fig. 2). As described in the following, there are few direct constraints on where many of the terranes in the western United States formed and evolved, and when and where they were accreted to North America. End-member possibilities are that the terranes formed and evolved (1) near their present position along the North American margin, (2) elsewhere along the Cordilleran margin, (3) in the paleo-Pacific or other ocean basins, far from a continent, or (4) near a continent other than North America.

Detrital zircon geochronology is used to evaluate these options by comparing the ages of zircons present in the terranes with the ages of grains that accumulated along the western margin of the North American continent. Comparisons with zircon ages along the Cordilleran margin are facilitated by a study of detrital zircon ages in Cambrian to Triassic miogeoclinal strata (Gehrels et al., 1995). Unfortunately, such data sets do not exist for other continents, so it remains difficult to test the option that a terrane formed or evolved near some other continent.

The detrital zircon data presented in this volume were generated in large part by six undergraduate students at the University of Arizona, each of whom was "in charge" of one of the terranes in western Nevada–northern California. With help from a researcher familiar with their terrane, the students decided which units should be sampled and where the strata are best known and exposed; they collected the samples during a group field trip in May–June 1995, extracted and analyzed the zircons in the U-Pb laboratory at the University of Arizona, and interpreted the data in light of other types of information and other detrital zircon data sets. Papers prepared by students present data from the Black rock terrane (Darby et al., this volume, Chapter 5), lower Paleozoic strata of the northern Sierra terrane (Harding et al., Chapter 2), Triassic strata of western Nevada (Manuszak et al., Chapter 8), the Golconda allochthon (Riley et al., Chapter 4), and upper Paleozoic and lower Mesozoic strata of the northern Sierra

Gehrels, G.E., 2000, Introduction to detrital zircon studies of Paleozoic and Triassic strata in western Nevada and northern California, *in* Soreghan, M.J., and Gehrels, G.E., eds., Paleozoic and Triassic paleogeography and tectonics of western Nevada and northern California: Boulder, Colorado, Geological Society of America Special Paper 347, p. 1–17.

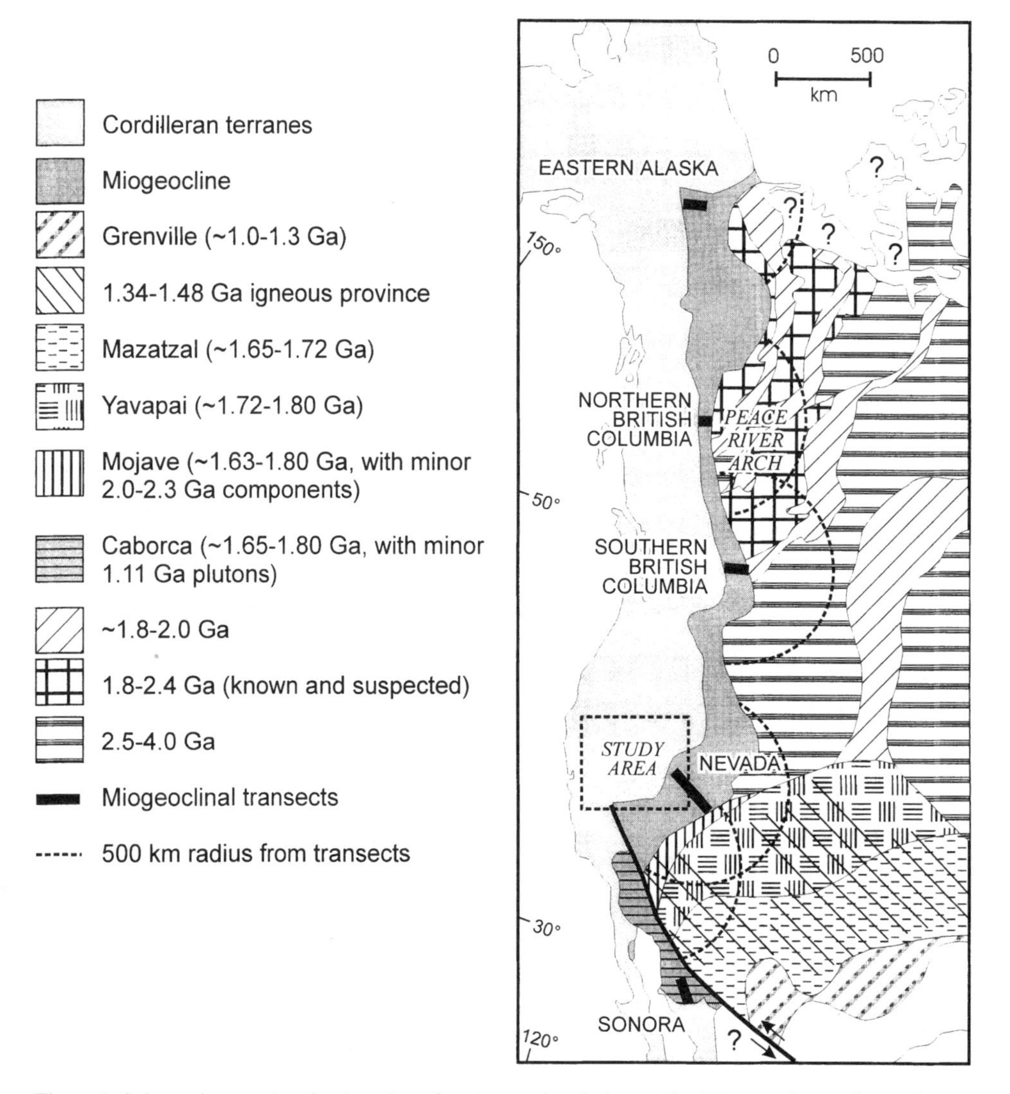


Figure 1. Schematic map showing location of study area in relation to Cordilleran miogeocline and outboard terranes, and first-order basement provinces of western North America. Also shown are locations of five transects that yield miogeoclinal detrital zircon reference (Gehrels et al., 1995), and areas within 500 km of these transects. Basement provinces are simplified from Hoffman (1989) for cratonal interior, from Ross (1991) and Villeneuve et al. (1993) for western Canadian shield, from Van Schmus et al. (1993) for southwestern United States, and from Stewart et al. (1990) for northwestern Mexico.

terrane (Spurlin et al., Chapter 6). Detrital zircon papers prepared by faculty focus on data from the Roberts Mountains allochthon (Gehrels et al., this volume, Chapter 1), the eastern Klamath terrane (Gehrels and Miller, Chapter 7), strata overlying the Roberts Mountains allochthon (Gehrels and Dickinson, Chapter 3), and the Yreka terrane (Wallin et al., Chapter 9). The sedimentary petrology of all the detrital zircon samples analyzed is described by Dickinson and Gehrels (Chapter 11). The large-scale implications of our study are described in a separate chapter (Gehrels et al., Chapter 10), co-written by all participants in our study.

Each of the papers that present geochronologic data contains sections that describe: (1) previous geologic work in the region studied, (2) the stratigraphy of the terrane with an emphasis on the units sampled, (3) characteristics of zircons that were extracted from the samples, (4) the resulting U-Pb ages, (5) how the ages compare with ages of zircons in other regions, and (6) the provenance and tectonic implications of these comparisons. To avoid repetition, this introductory chapter presents information that is relevant for all of the detrital zircon papers, including an overview of the geologic framework and tectonic evolution of western Nevada and northern California, a discussion of the detrital zircon reference for western North America, a summary of the analytical methods used in our geochronologic studies, and a discussion of potential biases in detrital zircon data.

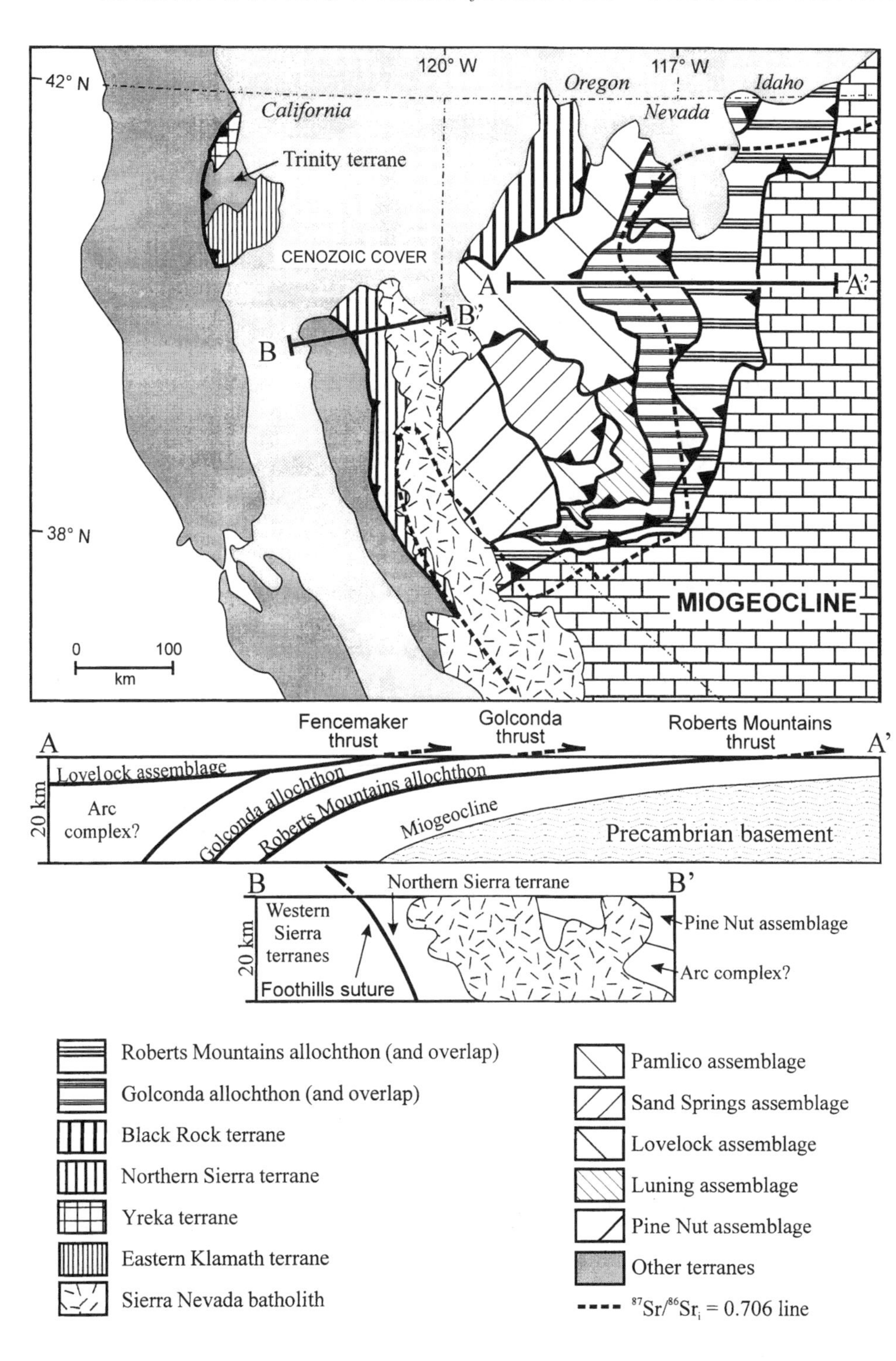


Figure 2. Schematic map and cross sections of main terranes and assemblages and their bounding faults in western Nevada and northern California. Map configuration of terranes and assemblages is adapted primarily from Oldow (1984), Silberling (1991), and Silberling et al. (1987), and cross sections are highly simplified from Blake et al. (1985) and Saleeby (1986). Location of $^{87}Sr/^{86}Sr_i$ = 0.706 line is from Kistler and Peterman (1973) and Kistler (1991). Note that widespread Cretaceous and younger rocks and structures are omitted in map and sections in effort to emphasize configuration of pre-Cretaceous features.

## GEOLOGIC FRAMEWORK AND TECTONIC EVOLUTION OF WESTERN NEVADA AND NORTHERN CALIFORNIA

Following the advent of plate tectonics in the late 1960s and early 1970s, it was realized that much of western North America is underlain by terranes, or crustal fragments, that were tectonically added to the North American continent (Wilson, 1968; Moores, 1970; Monger and Ross, 1971; Burchfiel and Davis, 1972). These early syntheses framed a debate that continues today; i.e., whether the accreted terranes formed in close proximity to their present positions or are far traveled. The first-order observation made by these workers, and by others before them (e.g., Schuchert, 1923; Kay, 1951), is that the terranes are currently situated outboard of the Cordilleran miogeocline (Fig. 1), which marks the western margin of the continent during much of Paleozoic and Mesozoic time. Most of these terranes are also outboard of the $^{87}Sr/^{86}Sr_i$ = 0.706 line (Fig. 2), which is interpreted to generally delineate the western edge of North American Precambrian basement (Kistler and Peterman, 1973; Kistler, 1991). Because tectonic mobility of crustal elements is inherent along active continental margins, terranes outboard of the miogeocline were recognized to be potentially far traveled (Helwig, 1974) or suspect (Coney et al., 1980).

### *Neoproterozoic–early Paleozoic history of the Miogeocline*

The Cordilleran miogeocline, which serves as the backstop for the accreted terranes (Churkin et al., 1980), originated during a phase of rifting in Neoproterozoic to Cambrian time (Stewart, 1972; Bond and Kominz, 1984). Continents that may have been removed from the Cordilleran margin during this rifting event include Siberia (Sears and Price, 1978), Antarctica and Australia (Hoffman, 1991; Dalziel, 1991; Moores, 1991), and South China (Li et al., 1995). During much of Paleozoic and early Mesozoic time, miogeoclinal strata accumulated in shallow-marine environments, forming a westward-thickening and fining sequence of mainly sandstone, shale, and limestone. Sandstones in the sequence provide an opportunity to reconstruct the ages of detrital zircons that were accumulating along the western margin of North America from Neoproterozoic through mid-Mesozoic time. This set of ages comprises a detrital zircon reference for the Cordilleran margin, as described in a separate section herein.

### *Antler orogeny*

Passive-margin sedimentation along the western margin of the continent was interrupted during two important tectonic events. The first event occurred in Late Devonian–Mississippian time, when ocean-floor or eugeoclinal assemblages were tectonically emplaced onto the continental margin in the western United States (Roberts et al., 1958) and perhaps western Canada (Smith et al., 1993). The overthrust eugeoclinal rocks were uplifted into an orogenic highland in Nevada, and a clastic wedge was formed from sediment shed eastward toward the craton. This thrusting event is referred to as the Antler orogeny (Nilsen and Stewart, 1980). The allochthonous rocks in Nevada belong to the Roberts Mountains allochthon (Fig. 2) and consist of Cambrian through Devonian sandstone, chert, shale, and subordinate basalt and limestone (Roberts et al., 1958; Poole et al., 1992). Most workers envision these rocks as having accumulated in slope, rise, and perhaps basinal settings along the Cordilleran margin, near their present position, because of faunal and stratigraphic ties with the miogeocline (Ketner, 1966; Palmer, 1971; Miller and Larue, 1983; Finney and Perry, 1991; Finney et al., 1993). The long-standing problem with this simple paleogeography is the presence in the allochthon of feldspathic sands deposited during times when sands deposited on the shelf were either mature arenites or were lacking altogether. The feldspathic sands may accordingly have been shed from promontories along the continental margin, such as the Salmon River arch (Rowell et al., 1979; Stewart and Suczek, 1977), or perhaps from an outboard region which has been referred to as the Harmony platform (Madrid, 1987).

The Antler orogeny remains enigmatic because the tectonic setting and driving force of allochthon emplacement are uncertain (Nilsen and Stewart, 1980). Many workers have considered scenarios involving arc-continent collision, but these are problematic because it is difficult to identify an appropriate volcanic arc outboard of the allochthon. Lower Paleozoic rocks of the northern Sierra and eastern Klamath terranes (Fig. 2) are likely candidates because they contain strata that are generally similar to rocks of the Roberts Mountains allochthon as well as arc-type(?) volcanic rocks and plutons (Burchfiel and Davis, 1972, 1975; Davis et al., 1978; Schweickert and Snyder, 1981; Miller and Harwood, 1990; Poole et al., 1992).

Various workers have also proposed that the collisional arc system is no longer exposed in the Nevada-California region. Dickinson (1977) suggested that any Antler-related arc system was removed from the margin prior to the Permian-Triassic Sonoma orogeny. The Alexander terrane of Alaska and coastal British Columbia could be this arc (Schweickert and Snyder, 1981), but this is unlikely because Middle Devonian through Mississippian strata in the Alexander terrane record tectonic quiescence following Silurian–Early Devonian deformation (Gehrels et al., 1987). Speed and Sleep (1982) and Burchfiel and Royden (1991) offered the possibility that the colliding arc subsided during or soon after the Antler orogeny and was either buried or overridden by strata of outboard terranes such as the Golconda allochthon (Fig. 2).

We have attempted to shed light on the early Paleozoic paleogeography and tectonic setting of the Cordilleran margin by determining detrital zircon ages for lower Paleozoic strata of the Roberts Mountains allochthon (Gehrels et al., this volume, Chapter 1), the northern Sierra terrane (Harding et al., this volume, Chapter 2), and the Yreka terrane (Wallin et al., this volume, Chapter 9) (Fig. 2). Comparison of the detrital zircon ages in these terranes provides a test of their proximity to each other, and comparisons with the miogeocline can reveal if they were located along the Cordilleran margin.

### *Late Paleozoic history*

Following the Antler orogeny, passive margin sedimentation resumed along the Cordilleran margin, and the Antler highlands were progressively eroded and buried beneath an overlap assemblage. By Permian time the continental margin had stepped outboard (westward) of the allochthon and its overlap assemblage.

### *Sonoma orogeny*

The second major accretionary event along the Cordilleran margin, the Sonoma orogeny, occurred during Late Permian–Triassic time (Gabrielse et al., 1983). This event, like the Antler orogeny, involved the emplacement of ocean-floor assemblages onto shelf strata of the continental margin (the Antler overlap assemblage). Little clastic material was shed eastward, but a Middle and Upper Triassic overlap assemblage accumulated on the allochthon following emplacement and erosion.

Rocks of the Golconda allochthon, which consists of highly dismembered and imbricated chert, shale, sandstone, basalt, and limestone (Fig. 2), were emplaced during the Sonoma orogeny. Most strata are Carboniferous in age, but rocks as old as Late Devonian and as young as Early Triassic have been recognized (Stewart et al., 1986; Miller et al., 1984, 1992). There is little direct evidence bearing on the displacement history of these rocks. Most scenarios portray strata of the allochthon as accumulating in deep-water basins along the continental margin, but the lack of evidence of the Antler orogeny in Upper Devonian–Mississippian strata indicates a degree of separation from inboard rocks.

There is also considerable debate about the tectonic setting of the Sonoma orogeny. Like the Antler orogeny, upper Paleozoic–lower Mesozoic volcanic rocks in the northern Sierra and eastern Klamath terranes are commonly interpreted as parts of an outboard volcanic arc (or arcs) that collided with the margin during allochthon emplacement (Burchfiel and Davis, 1972, 1975; Davis et al., 1978; Speed, 1979; Schweickert and Snyder, 1981; Gabrielse et al., 1983; Miller et al., 1992; Burchfiel et al., 1992).

We have attempted to resolve primary relations between the Golconda allochthon, the northern Sierra and eastern Klamath terranes, and the craton margin by analyzing detrital zircons from upper Paleozoic strata in each assemblage. Results from the Golconda allochthon are described by Riley et al. (this volume, Chapter 4); Spurlin et al. (this volume, Chapter 6) present data from the northern Sierra, and Gehrels and Miller (this volume, Chapter 7) describe the results of analyses of zircons from the eastern Klamath terrane. The Black Rock terrane of northwestern Nevada (Fig. 2), an additional arc-type assemblage, was studied by Darby et al. (this volume, Chapter 5).

### *Triassic and later history*

Beginning in Middle Triassic time and continuing through the Early Jurassic, shallow-marine clastic strata accumulated on rocks of the Golconda allochthon. Upper Triassic strata in the sequence are probably the outboard equivalents of fluvial deposits of the Chinle Formation, which drained much of the southwestern United States (Lupe and Silberling, 1985; Gehrels and Dickinson, 1995; Riggs et al., 1996). Several assemblages of Triassic and Jurassic clastic strata and subordinate volcanic and carbonate rocks that underlie much of westernmost Nevada and a portion of northeastern California (Fig. 2) are outboard of the shallow-marine strata. These generally deeper water rocks are structurally juxtaposed over inboard shelf facies strata along the Luning-Fencemaker thrust system, which moved during Middle Jurassic time (Oldow, 1984; Oldow et al., 1989). The outboard strata have been divided into several different assemblages on the basis of known and inferred faults separating distinct stratigraphic sequences (Speed, 1978; Oldow, 1984; Silberling, 1991). The westernmost of these assemblages is characterized by abundant mafic to intermediate volcanic rocks, and these rocks are bounded to the west by the northern Sierra terrane, which also contains abundant lower Mesozoic volcanic rocks and related plutons (Fig. 2).

Most previous syntheses portray the deeper water strata as having accumulated within a marginal basin separating a west-facing magmatic arc in the Sierra-Klamath (and perhaps Black Rock) region on the west from the continental margin to the east (Burchfiel and Davis, 1972, 1975; Speed, 1978; Oldow, 1984; Oldow et al., 1989). The marginal basin apparently narrowed to the south, where the Triassic magmatic arc is developed on Precambrian basement of the North American craton, and opened to the north into a basin represented by the Slide Mountain terrane of British Columbia (Saleeby and Busby-Spera, 1992). The basinal system is interpreted to have closed during Middle Jurassic thrusting and imbrication within the basinal assemblages. We have attempted to reconstruct the pre-Middle Jurassic paleogeography of the basinal and arc-type terranes by analyzing detrital zircons from three of the main basinal assemblages (Manuszak et al., this volume, Chapter 8) and from the northern Sierra (Spurlin et al., this volume, Chapter 6) and Black Rock (Darby et al., this volume, Chapter 5) arc-type terranes.

Middle Jurassic emplacement of the basinal assemblages and the outboard Sierra–Klamath–Black Rock terranes marks the end of marine deposition in a large tract of the western United States, as the continental margin stepped from central Nevada to the west side of the Sierra-Klamath region in California (Speed, 1979; Oldow, 1984). Since Middle Jurassic time, the Cordilleran margin in California has been largely convergent, which has resulted in further imbrication and strike-slip displacement of the terranes in western Nevada and northern California, transport of all of these rocks eastward on thrust faults of the fold and thrust belt, and emplacement of various suites of igneous rocks. The final phase in the tectonic evolution of the region is Tertiary extension, which has created the basin and range topography seen in much of Nevada and eastern California.

## DETRITAL ZIRCON REFERENCE FOR WESTERN NORTH AMERICA

One of the primary objectives of our research is to determine where the terranes of western Nevada and northern California were located prior to their accretion to western North America. To test the possibility that the terranes were in proximity to North America, either near their present position or elsewhere along the margin, our detrital zircon ages can be compared to the ages of grains extracted from miogeoclinal sandstones that accumulated along the western margin of the continent (Gehrels et al., 1995). This is presumably more accurate than comparisons with the ages of crystalline basement rocks in western North America because of the possibilities that (1) grains from distant regions of the craton were brought to the margin by large river systems, (2) longshore transport brought grains from regions far to the north or south along the margin, and (3) the detrital zircon suites in some areas could be dominated by igneous suites that have not been recognized or accurately dated. In contrast, the miogeoclinal strata presumably contain grains of various ages that were shed off of the western margin of the continent, regardless of where the grains originated and how they were transported to their final position along the margin.

To construct the detrital zircon reference, miogeoclinal sandstones of Neoproterozoic or Cambrian through Triassic age were collected from Sonora, Nevada, southern British Columbia, and northern British Columbia, and Cambrian and Devonian strata were collected from eastern Alaska (Fig. 2). More than 600 single zircon grains were analyzed from strata in these five areas, and the data were presented by Gehrels and Dickinson (1995) for Nevada, by Gehrels and Ross (1998) and Ross et al. (1997) for British Columbia, by Gehrels et al. (1999) for Alaska, and by Gehrels and Stewart (1998) for Sonora. The laboratory methods used in these analyses are described in detail in the following sections.

The data for all five transects were displayed by Gehrels et al. (1995) on a series of histograms. The histograms were constructed using a 50 m.y. time interval, and the dominant populations were highlighted for comparison with age data from outboard assemblages. This general approach has been criticized because a 50 m.y. interval obscures much of the age information available, because the "dominant" populations for each transect were selected in a qualitative fashion, and because rigorous comparisons with this type of reference are not possible.

The miogeoclinal data are accordingly shown on relative age-probability plots in Figures 3–8, and the main age groups and their proportions are listed in Table 1. The curves in Figures 3–8 represent the sum of the probability distributions of all well-constrained grains present in each sample and along each transect. Each curve is normalized to the number of grains plotted, so that the area under the curve has a value of 1.0. In subsequent chapters, ages of grains in outboard terranes are compared with these curves by constructing an age-probability curve for a sample or a set of samples that also has a value of 1.0, and then conducting statistical tests of the degree of similarity of the two curves.

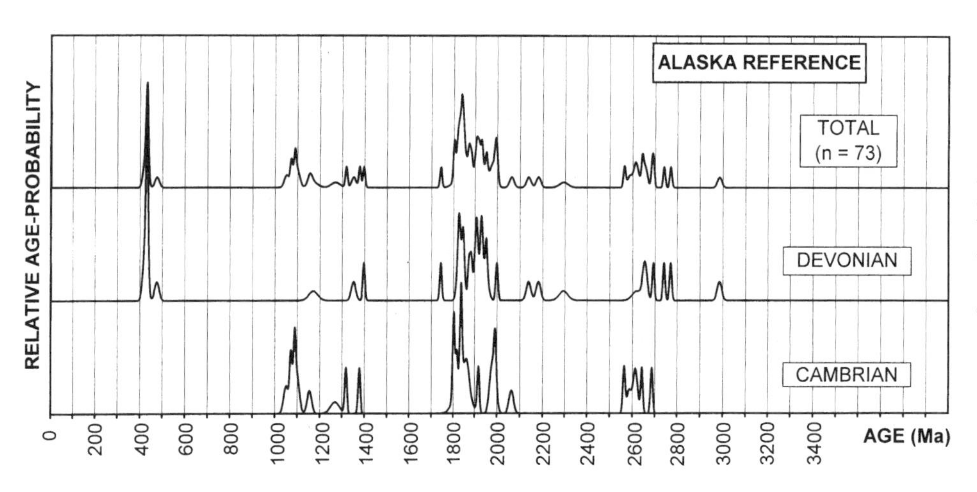


Figure 3. Relative age-probability plot of detrital zircon ages determined for Cambrian and Devonian miogeoclinal strata in east-central Alaska. Plot has been constructed with ages reported in Gehrels et al. (1999).

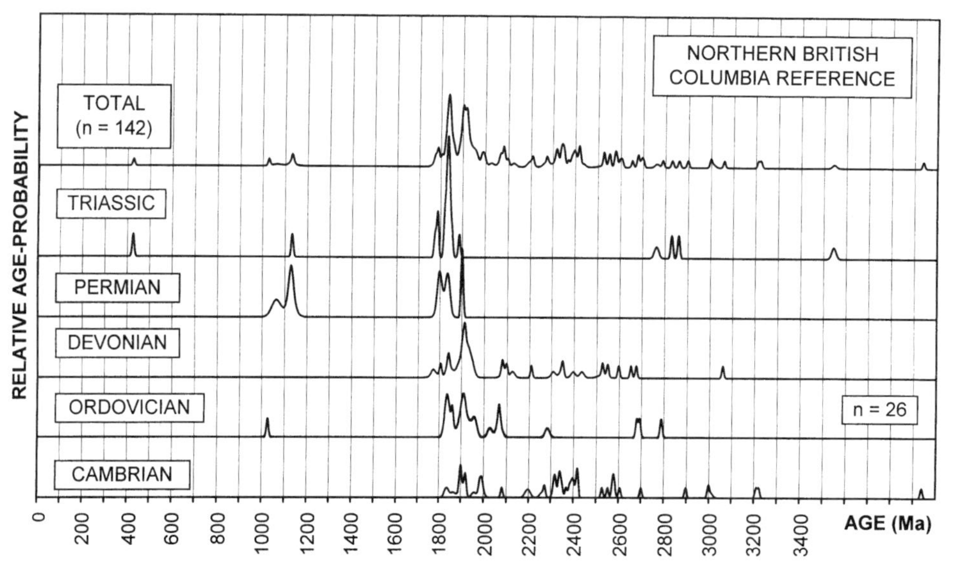


Figure 4. Relative age-probability plot of detrital zircon ages determined for Cambrian, Ordovician, Devonian, Permian, and Triassic miogeoclinal strata in northern British Columbia. Plot has been constructed with ages reported in Ross et al. (1997) and Gehrels and Ross (1998).

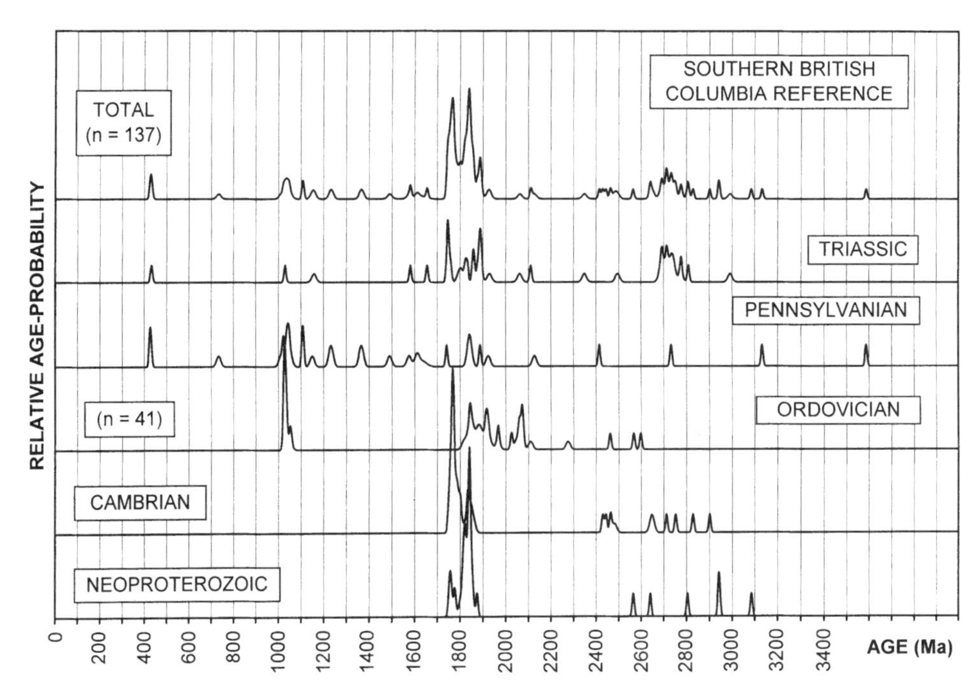


Figure 5. Relative age-probability plot of detrital zircon ages determined for Neoproterozoic, Cambrian, Ordovician, Devonian, Pennsylvanian, and Triassic miogeoclinal strata in southeastern British Columbia–southwestern Alberta. Plot has been constructed with ages reported in Ross et al. (1997) and Gehrels and Ross (1998). Note that ages of grains in Ordovician strata are not included in uppermost combined curve.

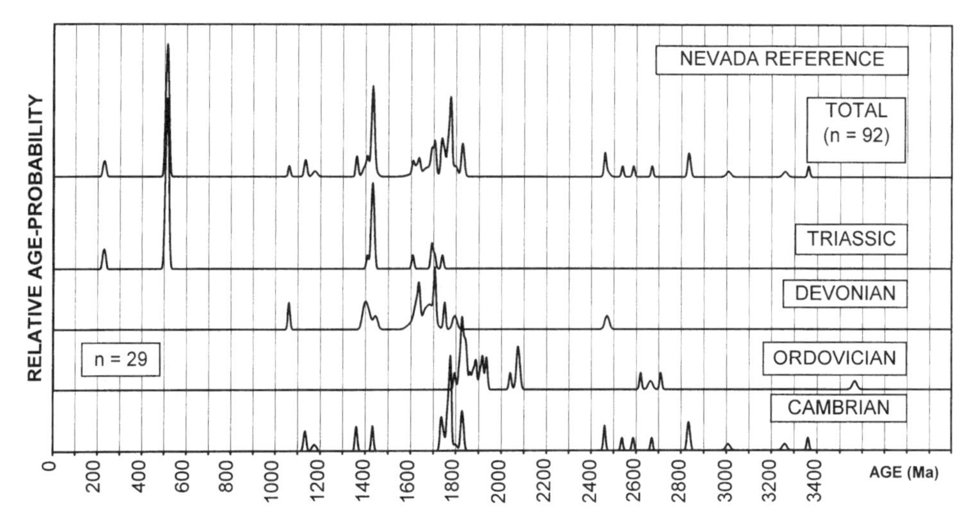


Figure 6. Relative age-probability plot of detrital zircon ages determined for Cambrian, Ordovician, Devonian, and Triassic miogeoclinal strata in central Nevada. Plot has been constructed mainly with ages reported in Gehrels and Dickinson (1995). Note that ages of grains in Ordovician strata are not included in uppermost combined curve. Ages of grains that accumulated along continental margin in this region during late Paleozoic time are preserved in strata of Antler overlap assemblage, as described by Gehrels and Dickinson (this volume, Chapter 3).

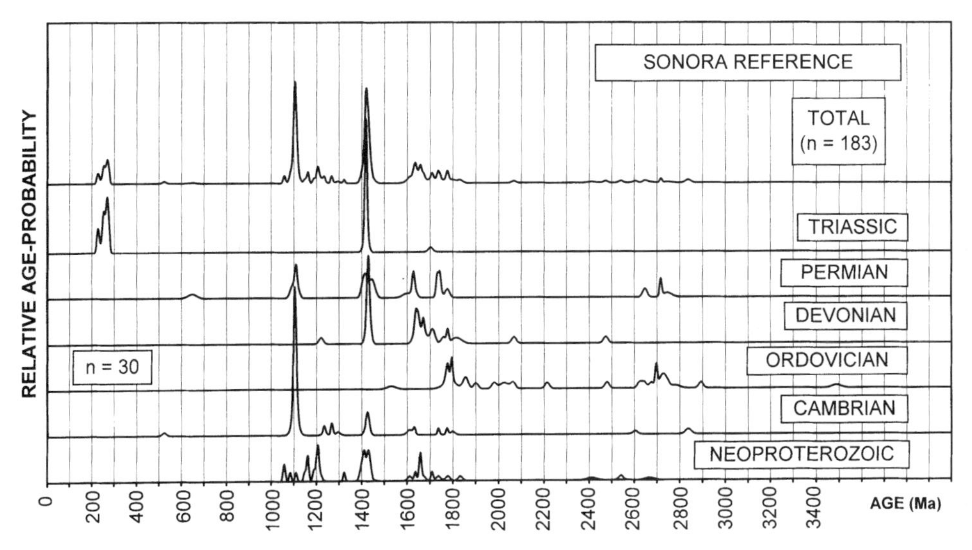


Figure 7. Relative age-probability plot of detrital zircon ages determined for Neoproterozoic, Cambrian, Ordovician, Devonian, and Permian miogeoclinal strata and for Triassic basinal strata in northern Sonora. Plot has been constructed mainly with ages reported in Gehrels and Stewart (1998). Note that ages of grains in Ordovician strata are not included in uppermost combined curve.

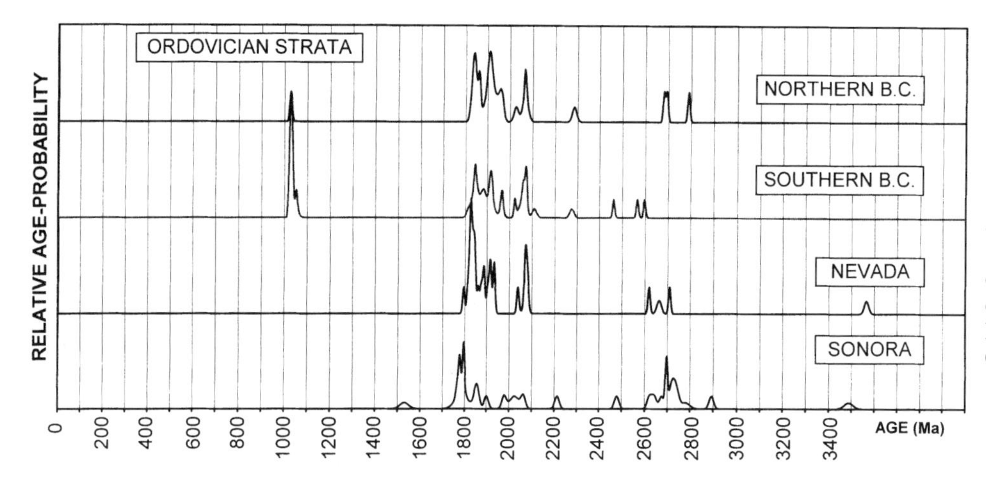


Figure 8. Relative age-probability plot of detrital zircon ages determined for Ordovician strata in northern and southern British Columbia, central Nevada, and northern Sonora. General similarity of these zircon ages all along continental margin supports suggestions of Ketner (1968) that much of detritus in Ordovician quartzites of Cordilleran miogeocline was derived from Peace River arch region of northwestern Canada (Gehrels et al., 1995).

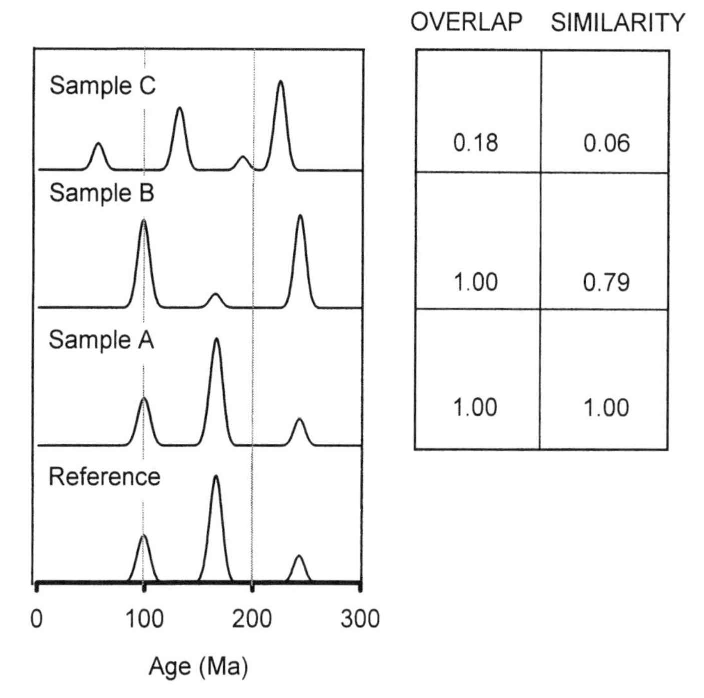


Figure 9. Simple example of two means of comparing detrital zircon age spectra used in this study. Each curve is age-probability plot containing 15 ages with uncertainties of ±10 Ma (95% confidence level). Curves for three samples are compared to reference curve using degree of overlap (measure of whether two curves contain overlapping ages) and degree of similarity (measure of whether overlapping ages have similar proportions). Sample A is identical to reference curve, sample B has different proportions of similar ages, and sample C has different ages.

Two statistical comparisons are used. The first is a measure of the degree of overlap of two curves, which indicates the degree to which grain ages in the sample(s) overlap with the ages in a region of the miogeocline. A value of 1.0 indicates a perfect match (that all grains could have been derived from the same source as the miogeoclinal strata) and 0.0 indicates that no ages match. The second is a measure of whether the proportions of overlapping ages are similar. This measure, referred to as the degree of similarity, is calculated by summing, over the time period of interest, the square root of the product of each pair of probabilities. It yields a value of 1.0 for curves that have similar proportions of perfectly overlapping ages, and low values for curves that do not overlap or have very different proportions of overlapping ages. Examples of these two measures are shown in Figure 9. Sample A contains the same ages, in the same proportions, as the reference curve, and accordingly yields values of 1.0 for both degree of overlap and degree of similarity. Sample B has similar sets of ages as the reference curve, but in different proportions. This yields 1.0 for the degree of overlap and 0.79 for the degree of similarity. Sample C yields low values for both overlap and similarity because there is little overlap of ages.

The miogeocline samples are divided by stratigraphic age in Figures 3–8 because Ordovician samples yield a different pattern of ages than the Neoproterozoic–Cambrian and Devonian to Triassic samples. As summarized by Gehrels et al. (1995), Neoproterozoic–Cambrian and Devonian to Triassic strata yield a consistent set of ages that are generally similar to the ages of basement rocks nearby (Fig. 1). This conclusion is quantified in Table 2 and shown graphically in Figure 10, which shows the degree of overlap of ages in each sample with the age ranges of basement rocks within 500 km of the sample site. A value of 1.0 indicates that all grains in a sample yield ages that are within the reported range of ages of nearby basement rocks, whereas a value of 0.0 indicates that none of the grains yield ages that match the ages of nearby rocks. It is important to note that these comparisons help resolve the original igneous sources from which the grains may have come, but do not reveal the history of sedimentary recycling that most miogeoclinal zircons have undergone.

Of significance for this study is the latitudinal variation in the dominant ages of zircons in these strata (Figs. 3–7), as follows. (1) Grains having ages of ~1.45 Ga and 1.6–1.8 Ga are common in Nevada and Sonora but not farther north; (2) the Triassic sample in Nevada contains ~520 Ma grains that are not present to the north or south; (3) Neoproterozoic and Cambrian sandstones in Sonora include abundant 1.11 Ga grains, whereas the Triassic sample is dominated by 250–280 Ma grains; and (4) grains having ages of 430 Ma, 1.80–1.95 Ga, 2.05–2.45 Ga, and 2.5–3.0 Ga are common in samples from western Canada and Alaska but not from the United States or Mexico.

**TABLE 1. DOMINANT DETRITAL ZIRCON AGE GROUPS IN NEOPROTEROZOIC-CAMBRIAN AND DEVONIAN THROUGH TRIASSIC MIOGEOCLINAL STRATA OF ALASKA, BRITISH COLUMBIA, NEVADA, AND SONORA**

| Age range (Ma) | Peak ages (Ma) | Relative abundance |
|---|---|---|
| Alaska | | |
| 410–445 | 433 | 0.092 |
| 1040–1115 | 1073 | 0.078 |
| 1140–1185 | 1157 | 0.022 |
| 1310–1405 | 1319, 1379, 1398 | 0.054 |
| 1795–2010 | 1838, 1911, 1996 | 0.460 |
| 2558–2070 | 2645, 2691 | 0.130 |
| Northern British Columbia | | |
| 1116–1143 | 1133 | 0.017 |
| 1765–1796 | 1788 | 0.034 |
| 1818–1865 | 1838 | 0.171 |
| 1885–1940 | 1904, 1916 | 0.200 |
| 2050–2105 | 2074 | 0.049 |
| 2306–2360 | 2319, 2344 | 0.068 |
| 2380–2427 | 2398, 2418 | 0.055 |
| 2514–2614 | 2527, 2551, 2578 | 0.068 |
| 2647–2708 | 2679 | 0.034 |
| Southern British Columbia | | |
| 420–441 | 431 | 0.021 |
| 1010–1060 | 1037 | 0.047 |
| 1570–1662 | 1581, 1612, 1656 | 0.033 |
| 1737–1784 | 1767 | 0.188 |
| 1815–1859 | 1840 | 0.197 |
| 1878–1894 | 1888 | 0.034 |
| 2412–2505 | 2416, 2431, 2445, 2466 | 0.041 |
| 2635–2830 | 2642, 2713, 2734, 2807 | 0.148 |
| Nevada | | |
| 495–530 | 515 | 0.204 |
| 1388–1440 | 1432 | 0.150 |
| 1600–1647 | 1616, 1636 | 0.050 |
| 1660–1715 | 1705 | 0.088 |
| 1725–1805 | 1738, 1776 | 0.202 |
| 1818–1842 | 1828 | 0.041 |
| 2450–2480 | 2461 | 0.030 |
| Sonora | | |
| 215–285 | 229, 255, 270 | 0.092 |
| 1045–1255 | 1059, 1106, 1162, 1207, 1234 | 0.316 |
| 1390–1460 | 1421 | 0.274 |
| 1585–1790 | 1635, 1657, 1710, 1737, 1777 | 0.196 |

*Note:* These ages and relative abundances refer to the relative age-probability plots shown in Figures 3–8.

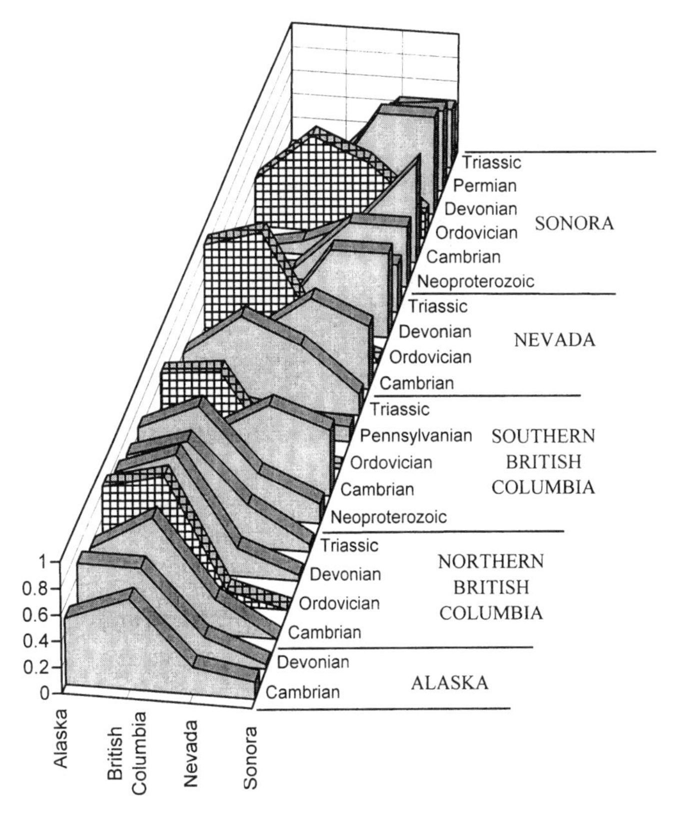


Figure 10. Curves showing that most detrital zircons in miogeoclinal strata were probably derived from nearby basement sources, except Ordovician strata, which were apparently shed from sources in the northern Canadian shield. This is graphical display of data in Table 2, which compares age spectra for each of miogeocline samples in Figures 3–7 with ages of basement rocks within 500 km of each transect through miogeocline (Fig. 1). Value of 1.0 indicates high degree of overlap, whereas value of 0.0 indicates no overlap of detrital ages and basement ages. Ordovician strata are shown with cross pattern to emphasize their difference from older and younger strata.

In contrast to the Neoproterozoic–Cambrian and Devonian to Triassic rocks, all of the Ordovician strata analyzed yield generally similar detrital zircon ages (Fig. 8). The source of the grains in these strata is most likely the Peace River arch region of northwestern Canada (Fig. 1), which contains crystalline rocks of the appropriate age and is the only region where Precambrian basement was exposed along the Cordilleran margin during Ordovician time (Gehrels et al., 1995; Gehrels and Ross, 1998). Ketner (1968) originally proposed that the ultramature quartz sand in these miogeoclinal units was shed from the Peace River arch region on the basis of a southward decrease in grain size, and on the paleogeography of western North America during Ordovician time. South-directed longshore transport was apparently responsible for distributing these grains along the Cordilleran margin. The lack of overlap of the detrital zircon ages in these samples with the ages of basement rocks south of the Peace River arch is indicated by the low values of overlap shown in Table 2 and Figure 10.

For this study, the miogeoclinal reference can be used to help resolve the original position of terranes along the North American margin if data are acquired from Cambrian or Devonian through Triassic strata. The ages of grains in a sample of a certain stratigraphic age from a terrane can be compared to the detrital zircon ages in miogeocline samples of the nearest stratigraphic age, or to the set of all grain ages from each of the five transects (Figs. 3–7). The positions of terranes along the margin during Ordovician time are more difficult to determine, however, because there are only minor variations in detrital zircon ages in miogeoclinal strata along the continental margin (Fig. 8).

TABLE 2. COMPARISON OF DETRITAL ZIRCON AGES IN MIOGEOCLINAL STRATA WITH THE AGES OF BASEMENT PROVINCES IN WESTERN NORTH AMERICA

| Transect | Age of strata | Alaska | British Columbia | Nevada | Sonora |
|---|---|---|---|---|---|
| Alaska | Cambrian | 0.613 | **0.690** | 0.207 | 0.130 |
| Alaska | Devonian | 0.694 | **0.698** | 0.207 | 0.036 |
| Northern British Columbia | Cambrian | 0.587 | **0.857** | 0.270 | 0 |
| Northern British Columbia | Ordovician | 0.844 | **0.959** | 0.115 | 0 |
| Northern British Columbia | Devonian | 0.738 | **0.910** | 0.217 | 0.045 |
| Northern British Columbia | Triassic | 0.654 | **0.797** | 0.332 | 0.058 |
| Southern British Columbia | Neoproterozoic | 0.651 | **0.858** | 0.362 | 0.156 |
| Southern British Columbia | Cambrian | 0.262 | 0.426 | **0.685** | 0.522 |
| Southern British Columbia | Ordovician | 0.657 | **0.706** | 0.050 | 0.002 |
| Southern British Columbia | Pennsylvanian | **0.257** | 0.24 | 0.149 | 0.121 |
| Southern British Columbia | Triassic | 0.393 | **0.726** | 0.528 | 0.171 |
| Nevada | Cambrian | 0.127 | 0.363 | **0.698** | 0.508 |
| Nevada | Ordovician | 0.856 | **0.993** | 0.168 | 0.030 |
| Nevada | Devonian | 0.039 | 0.039 | 0.681 | **0.685** |
| Nevada | Triassic | 0 | 0 | **0.387** | **0.387** |
| Sonora | Neoproterozoic | 0.029 | 0.077 | 0.498 | **0.499** |
| Sonora | Cambrian | 0.024 | 0.110 | 0.299 | **0.831** |
| Sonora | Ordovician | 0.394 | **0.795** | 0.606 | 0.206 |
| Sonora | Devonian | 0.097 | 0.097 | **0.817** | **0.817** |
| Sonora | Permian | 0.004 | 0.147 | 0.634 | **0.679** |
| Sonora | Triassic | 0 | 0 | **0.493** | **0.493** |

*Note:* The numerical values represent the degree to which the detrital zircon ages in a particular sample match the ages of known igneous rocks within a 500 km radius of each transect. A value of 1.0 indicates that all grains could have been sourced from nearby rocks, whereas a value of 0.0 indicates no overlap of detrital ages and basement ages. Bold values indicate the highest degree of overlap. The southern and northern British Columbia transects are combined because the known igneous rocks within 500 km of the two transects are of similar age. B.C.—British Columbia.

## ANALYTICAL METHODS

The laboratory techniques used in analyzing detrital zircons at the University of Arizona have evolved somewhat since they were described by Gehrels et al. (1991). The following accordingly describes the analytical techniques used in this study and in the miogeoclinal study of Gehrels et al. (1995).

### *Zircon extraction and separation*

Detrital zircons were generally separated from a 10–25 kg sample collected from an ~1-m-thick interval of sandstone at a single locality. In general, we tried to collect the coarsest sand available from a particular unit, and attempted to high-grade heavy-mineral-rich layers where possible. The samples were processed with a jaw crusher, roller mill, Wilfley table, Frantz LB-1 magnetic separator, and heavy liquid (diodomethane). The magnetic separator was operated with a steep side slope (20°) in an effort to retain all zircon grains, even those with high magnetic susceptibility. For most samples, several hundred milligrams to several grams of zircon were recovered, and few zircon grains >30 μm in diameter were lost or excluded in the separation process. The available grains were then sieved with disposable nylon screen into size fractions and examined in alcohol under a binocular microscope.

### *Grain selection and processing*

Our procedure at this point was to separate several hundred of the largest grains into populations based on optical characteristics such as color, shape, degree of rounding, and transparency. Smaller size fractions were also examined in an attempt to identify additional populations, and any distinctive grains >63 μm in sieve size were added to the pool to be analyzed. Grains with abundant inclusions or fractures were removed, and then representatives of each population were selected for analysis. This method of grain selection is discussed in the following section on potential biases and uncertainties.

All grains larger than 80 μm were abraded to ~70% of their original diameter (outer 30% removed) in an air abrasion device (Krogh, 1982) using abundant pyrite to prevent grain breakage. The pyrite was then removed and the abraded grains were cleaned in warm 3N $HNO_3$ for 20 min.

### *Dissolution and chemical separation*

After weighing, each grain was dissolved in HF>>$HNO_3$ in 0.1 ml microcapsules enclosed within 125 ml dissolution vessels (as described by Parrish, 1987). The resulting solutions were evaporated, spiked with a $^{205}$Pb-$^{233}$U-$^{235}$U mixed spike (Parrish and Krogh, 1987), and redissolved in 3.1N HCl in the dissolution vessels. Pb and U were purified and isolated using ion exchange resin in 0.1 ml columns (following Krogh [1982] and Parrish et al. [1987]), collected in separate beakers for Pb and U, and evaporated with a drop of ultrapure $H_3PO_4$. Organic molecules in the sample beads were removed by dissolution in hot $HNO_3$, and then the solutions were evaporated to dryness.

### *Mass spectrometry*

Pb and U were loaded in silica gel on two separate parts of a single Re filament, with a small kink separating the two beads. This allowed Pb and U to be run from the same filament, but produced higher ion intensities than if the Pb and U were mixed together in the same sample bead. Ion intensities generally yielded ~1 millivolt per picogram (mV/pg) for Pb and ~0.1 mV/pg for $UO_2$ (using $10^{-11}$ ohm resistors). All of these procedures contributed ~5 pg of common Pb and <1 pg of U to each analysis.

Mass spectrometry was conducted with a VG-354, operating in dynamic multicollector mode using a Daly detector system simultaneously with five Faraday collectors. The gain of the Daly detector system was determined continuously during every Pb run by comparing $^{206}Pb_{(Faraday)}/^{205}Pb_{(Faraday)}$ with $^{206}Pb_{(Faraday)}/^{205}Pb_{(Daly)}$ as well as $^{206}Pb_{(Faraday)}/^{207}Pb_{(Faraday)}$ with $^{206}Pb_{(Faraday)}/^{207}Pb_{(Daly)}$. The more stable of the two gain determinations was used to correct the $^{206}Pb_{(Faraday)}/^{204}Pb_{(Daly)}$. This measuring technique yielded data of moderate to high precision on samples with as little as 25 pg of total Pb. $UO_2$ analyses were conducted immediately following each Pb analysis using three Faraday detectors in static mode. Each pair of Pb and $UO_2$ analyses required ~35 min, and most analyses were conducted in fully automatic (computer-driven) mode.

### *Data reduction and interpretation*

The resulting isotopic data were processed and plotted following Ludwig (1991a, 1991b) and Mattinson (1987). The raw isotopic data were corrected for mass-dependent factors (mainly fractionation) based on analyses of NBS 981, 982, 983, and U-500. Resulting correction factors are 0.14 ± 0.06 per mass unit for Pb and 0.04 ± 0.06 for $UO_2$. Common Pb corrections included a Pb blank of 1–10 pg and U blank of 1 pg, and the remaining common Pb was interpreted to be initial Pb. Blank Pb composition was measured to be 18.6 ± 2.0 ($^{206}Pb/^{204}Pb$), 15.6 ± 0.3 ($^{207}Pb/^{204}Pb$), and 38.0 ± 2.0 ($^{208}Pb/^{204}Pb$), and initial Pb composition was interpreted from Stacey and Kramers (1975) with uncertainties of 1.0 for $^{206}Pb/^{204}Pb$, 0.3 for $^{207}Pb/^{204}Pb$, and 2.0 for $^{208}Pb/^{204}Pb$, and a correlation coefficient of 0.6.

Most of the grains analyzed yielded concordant to slightly discordant ages, where concordance is determined by overlap of a 95% level error ellipse with concordia. The discordant ages were divided into a highly discordant group, where the $^{206}Pb^{*}/^{238}U$ age is <85% of the $^{207}Pb^{*}/^{206}Pb^{*}$ age, and a group of slightly to moderately discordant grains with <15% difference in ages. Data tables in each paper divide the analyses into these three categories. Of the 648 analyses reported in this study, 214 are concordant and of moderate to high precision, 398 are slightly to moderately discordant and/or of low precision, and 36 are highly discordant.

In each of the tables, for grains that are apparently concordant and of moderate to high precision, the interpreted ages are $^{206}Pb^{*}/^{238}U$ ages for <1.0 Ga grains and $^{207}Pb^{*}/^{206}Pb^{*}$ ages for >1.0 Ga grains. This change at 1.0 Ga results from the fundamentally different way that $^{206}Pb^{*}/^{238}U$ and $^{207}Pb^{*}/^{206}Pb^{*}$ ages are determined. The former are calculated from measurements of the concentration of Pb and U (by comparison with the abundance of added $^{205}Pb$ and $^{233-235}U$), whereas the latter are calculated directly from the isotopic composition of radiogenic Pb. Concentration measurements (and calculated ages) have an uncertainty of ≥0.3% (at the 95% confidence level), regardless of age, due largely to uncertainties in the concentration and isotope composition of Pb and U in standard reference materials. In contrast, the uncertainty of ages calculated from $^{207}Pb^{*}/^{206}Pb^{*}$ actually decreases (in both percentage and in millions of years) with increasing age. The crossover, above which composition ages are more precise and below which concentration ages are more precise, is at ~1.0 Ga for most samples (Fig. 11).

Discordant grains that lie along obvious discordia lines are plotted as concordia upper intercept ages. Other discordant grains are plotted as upper intercept ages projected from 100 Ma, which is a reasonable approximate age of disturbance for all five regions. These ages are generally a few million years older than $^{207}Pb^{*}/^{206}Pb^{*}$ ages, and are probably more accurate indicators of age because there is little evidence for isotopic disturbance in the recent geologic past in most regions of the miogeocline. However, as discussed by Gehrels and Dickinson (1995), these projected ages are only accurate if discordance results from a single phase of Pb loss at about 100 Ma, and if no inherited components are present. Because the error resulting from this assumption increases as the degree of discordance increases, the uncertainty assigned to discordant ages increases with increasing degree of discordance.

The uncertainties assigned to each of the ages used in constructing Figures 3–8 depend primarily on analytical reproducibility for concordant grains (following Mattinson, 1987) and on the degree of discordance for the discordant grains, rather than on the analytical precision of the individual measurements. Because of the difficulty of measuring isotopic ratios on the small amounts of Pb and U present in individual zircon grains, particularly if they are small and/or young, analytical reproducibility is generally ~5 m.y. (at the 95% confidence level) for both concentration and composition ages that are >200 Ma. This is used as the minimum uncertainty for concordant ages. Because of the assumptions involved in interpreting the age of discordant grains, they are assigned a larger uncertainty, which decreases their significance in the plots of Figures 3–8. The minimum uncertainty assigned to grains that are slightly (<10%) discordant is ±20 m.y. (95% confidence level), and at least ± 40 m.y. is used for grains that are >10% discordant.

An additional means of discounting grains with complex discordance histories comes in the interpretation of the curves shown

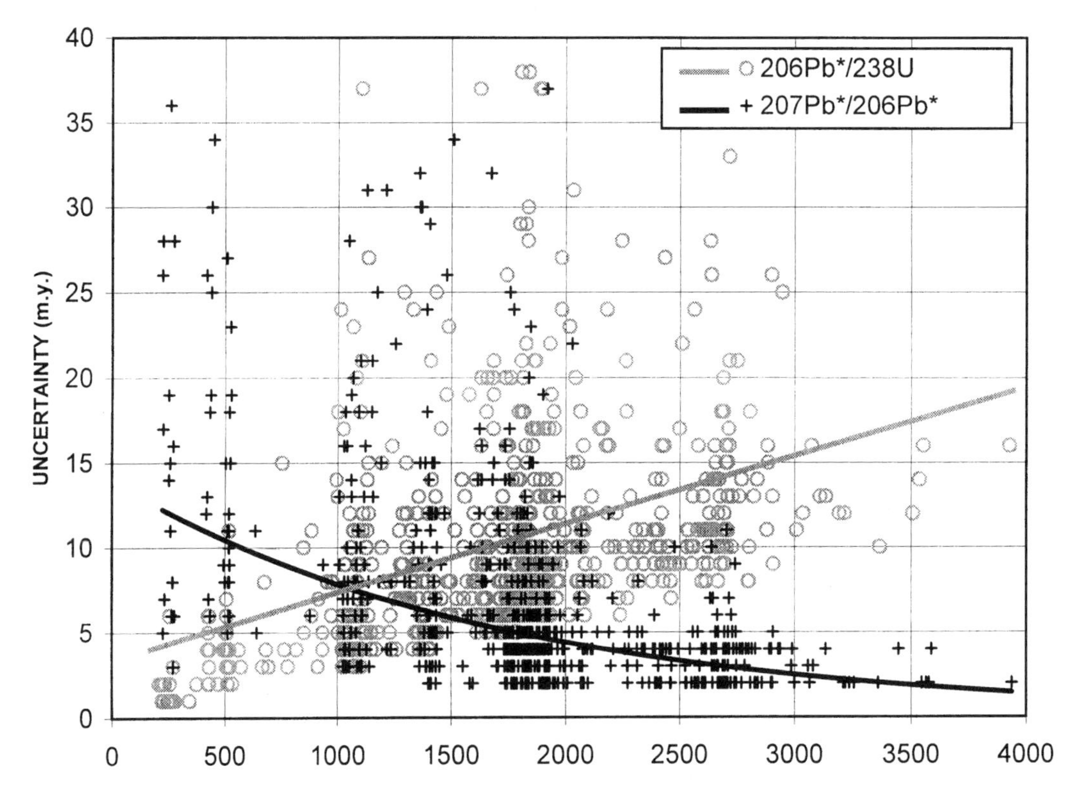


Figure 11. Plot showing how observed analytical uncertainties vary as function of age for $^{206}Pb^{*}/^{238}U$ and $^{207}Pb^{*}/^{206}Pb^{*}$ ages. Gray line is linear regression through uncertainties in $^{206}Pb^{*}/^{238}U$ ages, and solid black line is power law regression through uncertainties in $^{207}Pb^{*}/^{206}Pb^{*}$ ages. Cross-over of two lines indicates that $^{206}Pb^{*}/^{238}U$ ages are generally more precise for ages < ~1.0 Ga, whereas $^{207}Pb^{*}/^{206}Pb^{*}$ ages are generally more precise for older grains. Data points are for 707 analyses of single zircon grains in miogeoclinal strata reported in Gehrels et al. (1995). Uncertainties are at 95% confidence level.

in Figures 3–8. In general, little significance is attached to the age of a grain if there are no other grains in the sample or the region that are of the same age. Clusters of ages are considered a robust indicator of the age of detritus because grains with complex discordance histories would lie in such clusters only by coincidence.

## POTENTIAL BIASES AND UNCERTAINTIES IN OUR DETRITAL ZIRCON DATA

There are several aspects of geologic processes and analytical procedures that affect the results of a detrital zircon study. Because some of these processes create a bias in the determined ages, it is important to realize that the ages measured for detrital grains in a sandstone may not be accurate reflections of the ages of the dominant rocks in the source region. Following is a discussion of some of these biases.

### *Geologic processes*

In terms of geologic processes, an obvious bias in establishing provenance links results from the fact that most detrital zircons are derived from felsic to intermediate igneous rocks and, to a lesser degree, moderate- to high-grade metamorphic rocks, and from clastic sediments derived from these sources. However, most of the detritus in a sandstone could instead be derived from rocks such as basalt, chert, and carbonate, which would not be represented in detrital zircon populations.

A second bias results from the large size of zircons analyzed, generally >100 μm. Grains of this size are most common in coarse-grained granitoids because: (1) zircon abundance generally increases with silica content in an igneous rock, (2) plutons generally contain a higher concentration of zircon grains than volcanic rocks, and (3) the size of zircon grains commonly increases with grain size of the rock-forming minerals. For example, most of the large zircon grains in a sandstone may have been derived from a pluton that occupies a small region of the source area, if the pluton was the only coarse-grained granitoid in the area.

A third source of geological bias is that zircon grains with high U concentration and/or greater age will tend to be mechanically abraded faster than low-U, younger zircons during sedimentary transport. This is because resistance to abrasion is controlled largely by lattice damage due to radioactive decay, which increases with both U concentration and age. U concentration is probably not an important source of bias in age because there is no reason to suspect that the original U concentration of zircons has changed significantly through geologic time. In contrast, detrital grains in a sample should be biased toward younger ages, because the increase in lattice damage with age leads to preferential disintegration of older grains. In addition, older grains would have more opportunities for sedimentary transport and recycling, hence disintegration, than younger grains.

A fourth potential bias results from processes operating during sediment maturation, such as weathering and diagenesis (Basu et al., 1998). These processes are know to have biased the distribution of different zircon morphologies in Tertiary strata, suggesting that such processes may also have been important in our data set.

A complexity, rather than a bias, in interpreting provenance history results from the overall resistance of zircons to chemical weathering and mechanical abrasion. It is accordingly difficult to use degree of rounding of zircon grains as an indicator of distance of transport. An excellent example of this is the presence of multifaceted euhedral zircons in the Chinle and Osobb Formations of central Nevada that were transported more than 2000 km in a Triassic river system from Cambrian granites in the Texas-Oklahoma region (Gehrels and Dickinson, 1995; Riggs et al., 1996).

Resistance to weathering also creates the potential for recycling of zircons from older strata, rather than directly from crystalline rocks, which could lead to erroneous conclusions about paleodispersal systems. Because of this effect, it is important to note that most discussions of zircon provenance in following chapters refer to the original source of the grains, rather than where they resided prior to being incorporated into the sampled unit. An example is provided by the presence of 1.0–1.3 Ga detrital grains in upper Paleozoic miogeoclinal strata of western Canada (Gehrels and Ross, 1998). Many of these grains may have originated to the southeast in the Grenville orogen of eastern North America, but they probably traveled across the craton in Neoproterozoic time and then resided in platformal strata of northwestern Canada during latest Proterozoic and early to mid-Paleozoic time (Rainbird et al., 1997). Transport of these Grenville-derived grains during late Paleozoic time may accordingly have been from the north (rather than southeast), or perhaps from erosion of locally exposed Neoproterozoic or Paleozoic strata.

### *Laboratory procedures*

Several aspects of sample processing techniques also affect our results. One factor is the size of zircons that can be analyzed with conventional mass spectrometry. In general, the precision of an isotopic analysis decreases as the amount of Pb (and to a lesser extent U) decreases. Below 50 pg of Pb, our results are not sufficiently accurate to yield a well-determined age, and measurements are rarely possible below 25 pg of Pb. These minimum threshold values require grains of at least ~45 μm if they are Precambrian in age, and at least ~100 μm if they are Mesozoic. Because of these size cut-offs, sources that yield only small zircon grains (volcanic rocks, fine-grained granitoids, metamorphic rocks) are commonly not recognized if there are other sources present that yield coarser zircon grains. If a sample contains only small grains, our results would also be biased toward older ages, given the lower Pb content of young grains. This bias would be detectable, however, from the pattern of a Pb analysis with low signal intensity coupled with a U analysis of normal intensity.

A second potential bias comes from the process of mechanical abrasion of the grains in the lab. One might expect that grains with higher U content would be selectively destroyed during abrasion because of their increased lattice damage from radioactive decay. This effect is pronounced if the zircons are abraded by themselves, but experience has shown that abrasion with a significant volume of pyrite (~1000 grains of pyrite for each zircon grain) prevents grain destruction. Because all grains survive the abrasion process in most cases, this is not a significant source of bias in our data.

Another potential bias, as noted in the previous section, comes from our procedure for selecting the grains to be analyzed. Our methodology of analyzing grains from every color, shape, morphology, and/or transparency group present was adopted in an effort to identify as many different age groups as possible. In some samples, there is a general correlation between age and color, morphology, shape, and/or transparency, and the different populations have unequal abundances. By analyzing grains from all of the different populations present, we maximize our chance of identifying all of the main age groups present. Graphically, this tends to flatten and broaden the age distribution of grains on both histograms and relative age-probability plots. Other methods, such as selecting grains at random or according to population abundance, would have a greater likelihood of missing age groups, particularly if most grains in a sample have similar characteristics. Likewise, selecting grains on the basis of optical clarity or low color intensity might yield more concordant data, but a smaller number of age groups would be identified.

A comparison of selecting representatives of all populations versus random selection of grains is afforded by a study of detrital zircons in the Alexander terrane of southeast Alaska (Gehrels et al., 1996). Ordovician, Devonian, Permian, and Triassic sandstones from the terrane yield abundant zircon grains ranging to 250 μm in sieve size. Most of the grains present are euhedral (with little sign of rounding) and colorless to very light pink in color. Of these grains, 61 were analyzed, all of which yielded ages between 400 and 500 Ma. In two Devonian samples, however, smaller size fractions also contain a small proportion of well-rounded, pinkish grains. Of these grains, 40 were analyzed, 36 of which yielded ages between 1.0 and 3.0 Ga. Because the terrane contains widespread rocks having ages of 400–550 Ma, but none >600 Ma, the older grains provided strong evidence of proximity to a continental landmass during Devonian time. Our selection procedure yielded 36 out of 101 of these important grains, whereas selection of grains at random would have yielded few if any of the critical >1.0 Ga ages. Using one of the Devonian samples as a specific example, the average abundance of these pinkish, rounded grains is 0.3% of the total population (increasing to 0.5% for 30–45 μm grains and decreasing to 0% for >125 grains). The probability of analyzing one such grain in a random selection of 100 >30 μm grains is 25%, and the probability of analyzing a more useful number of Precambrian grains, such as 10, is negligible.

Nonrandom selection of grains has the disadvantage that the final results are not necessarily representative of the entire population of zircons from a sample. In certain circumstances, it would be important to have an accurate representation of the total population of grains. An example would be an attempt to resolve detailed changes in provenance within a single unit by comparing samples collected vertically through a sequence or horizontally along strike. However, for resolving potential source terranes, knowing the relative proportion of ages in a sample is not particularly useful given the many geologic biases that control the final distribution of grains determined in a sample. In our experience, identifying a particular source region can be done with greater certainty if a sample yields a larger number of age groups. An extreme example is provided by two samples of Triassic strata from western Nevada (Manuszak et al., this volume, Chapter 8). A sample from the Pine Nut assemblage yields only grains that are 225–235 Ma, whereas the adjacent Luning assemblage yields grains of 220–230, ~270, ~1037, ~1440, 1660–1800, and ~2320 Ma. A unique source for grains in the Pine Nut assemblage cannot be identified because Triassic igneous rocks are widespread in the Cordillera, whereas the only region of western North America that could yield the grains in the Luning sample is the southwestern United States.

The optimal situation would be to determine high-precision ages for a large number of grains (several hundred) from each sample, and thereby identify the relative abundance of most of the different age groups present. Unfortunately, the tremendous cost and effort involved in conducting high-precision U-Pb analyses makes this unreasonable with existing technology. An ion probe can produce detrital zircon analyses much more readily than conventional analyses (cf. Ireland, 1992), but the precision of the analyses is generally such that only general age comparisons can be made. Improving the precision of the analyses requires much longer counting times, which makes the analyses prohibitively expensive for routine analyses. In the future, an ion probe or ICPMS equipped with sensitivity-stable multiple detectors might allow large numbers of high-precision detrital zircon analyses to be determined efficiently and economically.

An important question in detrital zircon studies concerns the number of grains that needs to be analyzed to characterize the age distribution of a sample. If grains are selected at random, the number of grains that needs to be analyzed (N) to have a 95% probability of detecting an age group of given proportion (P) is determined by the following (Dodson et al., 1988):

$$N = \text{Ln}\,(0.05) / (1 - P).$$

Figure 12 shows this relationship in graphical form.

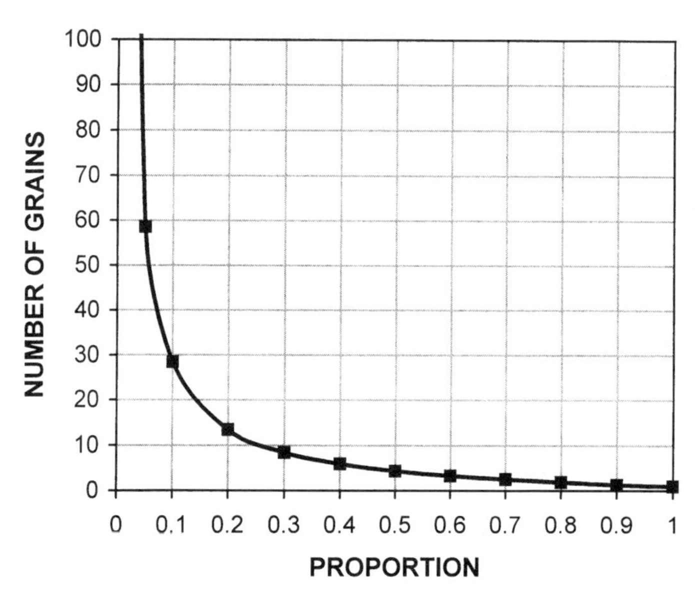


Figure 12. Plot showing number of randomly selected grains that need to be analyzed in order to have 95% probability of detecting grain from age group that makes up given proportion of total population of grains. Curve is plotted from relationship presented by Dodson et al. (1988).

**TABLE 3. ANALYSIS OF THE NUMBER OF GRAINS ANALYZED FROM A SAMPLE**

| | Atan Group (n = 48)* | Mount Wilson Formation (n = 41)* | Osobb Formation (n = 40)† |
|---|---|---|---|
| Number of grains in each age group | 1830-1870 Ma (n = 10) | 1020-1060Ma (n = 11) | ~230 Ma (n = 1) |
| | 1890-1930 Ma (n = 7) | 1810-1970 Ma (n = 18) | 260-420 Ma (n = 4) |
| | 1960-2000 Ma (n = 5) | 2020-2280 Ma (n = 9) | 500-530 Ma (n = 6) |
| | 2080-2280 Ma (n = 5) | 2460-2600 Ma (n = 3) | 960-1180 Ma (n = 15) |
| | 2310-2430 Ma (n = 15) | | 1400-1450 Ma (n = 10) |
| | 2520-3940 Ma (n = 13) | | 1700-1730 Ma (n = 3) |
| Number of color groups | 3 | 3 | 3 |
| Number of analyses required to identify ≥1 in every age group | 12.2 | 9.9 | 19.5 |
| Number of analyses required to identify ≥2 in every age group | 17.2 | 18.1 | >40 |

*Note:* Age groups are defined on the basis of matches with basement provinces in western North America. Number of analyses required is based on the averages of 10 random selections from the 40–48 grains that were actually analyzed.

* From Gehrels and Ross (1998).

† From Gehrels and Stewart (1998).

The situation is obviously more complex if grains in a sample exhibit a correlation between age and optically visible characteristics (such as color, shape, morphology, transparency), and representatives are analyzed from each group. We can address this empirically with analyzed detrital samples that contain several different age groups. In general, grains are processed in our lab in groups of 11 (an even division of the 22 microcapsules that fit in each dissolution chamber). We commonly find that many of the main age groups in a sample are recognized in the first batch of 11 grains, provided that each of the different color, shape, and/or morphology groups is represented. The second batch of 11 generally yields the remainder of the main age groups, and new age groups are rarely discovered in grains 23–33 and 34–44. Our general strategy is therefore to analyze 22 grains from each sample, and continue with additional grains only if specific questions need to be addressed.

This can be quantified using several samples from which 40 or more grains have been analyzed. The three samples presented in Table 3 contain between 4 and 7 different age groups (where each age group is defined as matching a particular province in western North America), and the abundance of grains in each age group ranges from 2.5% (1 out of 40 grains) to 44% (18 out of 41 grains). To recognize one grain from each age group, the total number of grains that need to be analyzed is 9, 12, and 19 (averages of 10 random selections of grains from each of the samples). To analyze at least 2 grains from each age group, the number of grains that need to be analyzed rises to 17, 18, and >40 (for the sample that contains only one grain in an age group). This simple analysis supports the conclusion that analysis of 22 grains, selected from the available color, shape, morphology, and/or transparency groups, should be sufficient to recognize each of the main age groups present.

A summary of the main geologic and analytical factors that control the set of ages determined for a detrital zircon sample is as follows.

1. The detrital zircon signature determined for a sandstone will be controlled largely by coarse-grained granitoids—detrital contributions from most other rocks will not be recognized.

2. Detrital zircon ages yield information primarily about the original igneous sources of the grains. Because zircons are so resistant to chemical weathering and mechanical abrasion, most grains (particularly in mature quartz sands) are multicyclic and may have had a long and complex transport history.

3. Sedimentary processes bias a zircon signature toward younger grains due to the removal of grains with greater lattice damage, which accumulates with age.

4. Sample processing techniques are designed to not bias the age distribution of a sample.

5. Grains are selected from all available color, shape, morphology, and transparency groups in an effort to resolve the maximum number of age groups in a sample. Most grains analyzed are from larger (>80 µm) size fractions, but smaller grains are analyzed if their characteristics are distinct from the larger grains.

6. Maximizing the number of age groups recognized in a sample is useful in identifying potential source regions. In general, a single age group is not particularly useful in identifying a unique source region, whereas the combination of several age groups can generally be tied to a specific region with confidence.

7. Nonrandom selection of grains yields an age spectrum that is not necessarily representative of the entire sample, which disallows some types of analyses. However, this selection procedure should not add a consistent bias to our results.

8. Experience shows that analyzing ~10 grains from a sample should be sufficient to identify at least one grain from each of the main age groups present (~18 grains are appropriate for at least 2 grains per group). Our standard procedure of analyzing 22 grains per sample should therefore allow recognition of the main age groups in most samples.

## ACKNOWLEDGMENTS

I thank Mark Brandon, Ralph Haugerud, and John Garver for encouragement to quantify the statistical treatment of detrital zircon ages. Jon Patchett provided invaluable assistance in developing the measurement routines used by our mass spectrometer. This research was supported by National Science Foundation grants EAR-9116000 and EAR-9416933. Reviewed by Richard Schweickert, James Mattinson, and Troy Rasbury.

## REFERENCES CITED

Basu, A., Molinaroli, E., and Andersson, S., 1998, Diversity of physical and chemical properties of detrital zircons and garnets in Cenozoic sediments of southwestern Montana: The Mountain Geologist, v. 35, p. 23–29.

Blake, M.C., Jr., Bruhn, R.L., Miller, E.L., Moores, E.M., Smithson, S.B., and Speed, R.C., 1985, C-1 Menodocino triple junction to North American craton: Geological Society of America Centennial Continent-Ocean Transect 12, 1:500,000 scale, 30 p.

Bond, G., and Kominz, M.A., 1984, Construction of tectonic subsidence curves for the early Paleozoic miogeocline, southern Rocky Mountains: Implications for subsidence mechanisms, age of breakup, and crustal thinning: Geological Society of America Bulletin, v. 95, p. 155–173.

Burchfiel, B.C., and Davis, G.A., 1972, Structural framework and evolution of the southern part of the Cordilleran orogen, western United States: American Journal of Science, v. 272, p. 97–118.

Burchfiel, B.C., and Davis, G.A., 1975, Nature and controls of Cordilleran orogenesis, western United States: Extensions of an earlier synthesis: American Journal of Science, v. 275-A, p. 363–396.

Burchfiel, B.C., and Royden, L.H., 1991, Antler orogeny: A Mediterranean-type orogeny: Geology, v. 19, p. 66–69.

Burchfiel, B.C., Cowan, D.S., and Davis, G.A., 1992, Tectonic overview of the Cordilleran orogen in the western United States, *in* Burchfiel, B.C., et al., eds., The Cordilleran orogen: Conterminous U.S.: Boulder, Colorado, Geological Society of America, Geology of North America, v. G-3, p. 407–480.

Churkin, M., Jr., Carter, C., and Trexler, J.H., 1980, Collision-deformed Paleozoic margin of Alaska—Foundation for microplate accretion: Geological Society of America Bulletin, v. 91, p. 648–654.

Coney, P.J., Jones, D.L., and Monger, J.W.H., 1980, Cordilleran suspect terranes: Nature, v. 288, p. 329–333.

Dalziel, I.W.D., 1991, Pacific margins of Laurentia and East Antarctica–Australia as a conjugate rift pair: Evidence and implications for an Eocambrian supercontinent: Geology, v. 19, p. 598–601.

Davis, G.A., Monger, J.W.H., and Burchfiel, B.C., 1978, Mesozoic construction of the Cordilleran collage, central British Columbia to central California, *in* Howell, D.G., and McDougall, K.A., eds., Mesozoic paleogeography of the western United States: Society of Economic Paleontologists and Mineralogists Pacific Coast Paleogeography Symposium 2, p. 1–32.

Dickinson, W.R., 1977, Paleozoic plate tectonics and the evolution of the Cordilleran continental margin, *in* Stewart, J.H., et al., eds., Paleozoic paleogeography of the western United States: Society of Economic Paleontologists and Mineralogists Pacific Coast Paleogeography Symposium I, p. 137–155.

Dodson, M.H., Compston, W., Williams, I.S., and Wilson, J.F., 1988, A search for ancient detrital zircons in Zimbabwean sediments: Geological Society of London Journal, v. 145, p. 977–983.

Finney, S.C., and Perry, B.D., 1991, Depositional setting and paleogeography of Ordovician Vinini Formation, central Nevada, *in* Cooper, J.D., and Stevens, C.H., eds., Paleozoic paleography of the western United States—II, Volume 2: Los Angeles, California, Pacific Section, Society of Economic Paleontologists and Mineralogists book 67, p. 747–766.

Finney, S.C., Perry, B.D., Emsbo, P., and Madrid, R.J., 1993, Stratigraphy of the Roberts Mountains allochthon, Roberts Mountains and Shoshone Range, Nevada, *in* Lahren, M.M., et al., eds., Crustal evolution of the Great Basin and Sierra Nevada: Cordilleran-Rocky Mountain Section, Geological Society of America Guidebook: Reno, University of Nevada, p. 197–230.

Gabrielse, H., Snyder, W.S., and Stewart, J.H., 1983, Sonoma orogeny and Permian to Triassic tectonism in western North America (Penrose Conference Report): Geology, v. 11, p. 484–486.

Gehrels, G.E., and Dickinson, W.R., 1995, Detrital zircon provenance of Cambrian to Triassic miogeoclinal and eugeoclinal strata in Nevada: American Journal of Science, v. 295, p. 18–48.

Gehrels, G.E., and Ross, G.M., 1998, Detrital zircon geochronology of Neoproterozoic to Triassic miogeoclinal strata in British Columbia and Alberta: Canadian Journal of Earth Sciences, v. 35, p. 1380–1401.

Gehrels, G.E., and Stewart, J.H, 1998, Detrital zircon geochronology of Cambrian to Triassic miogeoclinal and eugeoclinal strata of Sonora, Mexico: Journal of Geophysical Research, v. 103, p. 2471–2487.

Gehrels, G.E., Saleeby, J.B., and Berg, H.C., 1987, Geology of Annette, Gravina, and Duke Islands, southeastern Alaska: Canadian Journal of Earth Sciences, v. 24, p. 866–881.

Gehrels, G.E., McClelland, W.C., Samson, S.D., and Patchett, P.J., 1991, U-Pb geochronology of detrital zircons from a continental margin assemblage in the northern Coast Mountains, southeastern Alaska: Canadian Journal of Earth Sciences, v. 28, p. 1285–1300.

Gehrels, G.E., Dickinson, W.R., Ross, G.M., Stewart, J.H., and Howell, D.G., 1995, Detrital zircon reference for Cambrian to Triassic miogeoclinal strata of western North America: Geology, v. 23, p. 831–834.

Gehrels, G.E., Butler, R.F., and Bazard, D.R., 1996, Detrital zircon geochronology of the Alexander terrane, southeastern Alaska: Geological Society of America Bulletin, v. 108, p. 722–734.

Gehrels, G.E., Johnsson, M.J., and Howell, D.G., 1999, Detrital zircon geochronology of the Adams Argillite and Nation River Formation, east-central Alaska: Journal of Sedimentary Research, v. 69, p. 147–156.

Helwig, J., 1974, Eugeosynclinal basement and a collage concept of orogenic belts, *in* Dott, R.H., and Shaver, R.H., eds., Modern and ancient geosynclinal sedimentation: Society of Economic Paleontologists and Mineralogists Special Publication 19, p. 359–376.

Hoffman, P.F., 1989, Precambrian geology and tectonic history of North America, *in* Bally, A.W., and Palmer, A.R., eds., The geology of North America—An overview: Boulder, Colorado, Geological Society of America, Geology of North America, v. A, p. 447–512.

Hoffman, P.F., 1991, Did the breakout of Laurentia turn Gondwanaland inside-out?: Science, v. 252, p. 1409–1411.

Ireland, T.R., 1992, Crustal evolution of New Zealand: Evidence from age distributions of detrital zircons in Western Province paragneisses and Torlesse greywacke: Geochimica et Cosmochimica Acta, v. 56, p. 911–920.

Kay, M., 1951, North American geosynclines: Geological Society of America Memoir 48, 143 p.

Ketner, K.B., 1966, Comparison of Ordovician eugeosynclinal and miogeosynclinal quartzites of the Cordilleran geosyncline: U.S. Geological Survey Professional Paper 550-C, p. C54–C60.

Ketner, K.B., 1968, Origin of Ordovician Quartzite in the Cordilleran miogeosyncline: U.S. Geological Survey Professional Paper 600-B, p. 169–177.

Kistler, R.W., 1991, Chemical and isotopic characteristics of plutons in the Great Basin, *in* Raines, G.L., et al., eds., Geology and ore deposits of the Great Basin: Reno, Geological Society of Nevada, p. 107–109.

Kistler, R.W., and Peterman, Z.E., 1973, Variations in Sr, Rb, K, Na, and initial $^{87}Sr/^{86}Sr$ in Mesozoic granitic rocks and intruded wall rocks in central California: Geological Society of America Bulletin, v. 84, p. 3489–3512.

Krogh, T.E., 1982, Improved accuracy of U-Pb zircon ages by the creation of more concordant systems using an air abrasion technique: Geochimica et Cosmochimica Acta, v. 46, p. 637–649.

Li, Z., Zhang, L., and Powell, C.M., 1995, South China in Rodinia: Part of the missing link between Australia–east Antarctica and Laurentia?: Geology, v. 23, p. 407–410.

Ludwig, K.R., 1991a, A computer program for processing Pb-U-Th isotopic data: U.S. Geological Survey Open-File Report 88-542.

Ludwig, K.R., 1991b, A plotting and regression program for radiogenic-isotopic data: U.S. Geological Survey Open-File Report 91-445.

Lupe, R., and Silberling, N.J., 1985, Genetic relationship between lower Mesozoic continental strata of the Colorado plateau and marine strata of the western Great Basin: Significance for accretionary history of Cordilleran lithotectonic terranes, *in* Howell, D.G., ed., Tectonostratigraphic terranes of the Circum-Pacific region: Houston, Texas, Circum-Pacific Council for Energy and Mineral Resources, p. 263–271.

Madrid, R.J., 1987, Stratigraphy of the Roberts Mountains allochthon in north-central Nevada [Ph.D. thesis]: Stanford, California, Stanford University, 336 p.

Mattinson, J.M., 1987, U-Pb ages of zircons: A basic examination of error propagation: Chemical Geology, v. 66, p. 151–162.

Miller, E.L., and Larue, D.K, 1983, Ordovician quartzite in the Roberts Mountains allochthon, Nevada: Deep-sea fan deposits derived from cratonal North America, *in* Stevens, C.H., ed., Pre-Jurassic rocks in western North American suspect terranes: Pacific Section, Society of Economic Paleontologists and Mineralogists, p. 91–102.

Miller, E.L., Holdsworth, B.K., Whiteford, W.B., and Rodgers, D., 1984, Stratigraphy and structure of the Schoonover sequence, northeastern Nevada: Implications for Paleozoic plate-margin tectonics: Geological Society of America Bulletin, v. 95, p. 1063–1076.

Miller, E.L., Miller, M.M., Stevens, C.H., Wright, J.E., and Madrid, R., 1992, Late Paleozoic paleographic and tectonic evolution of the western U.S. Cordillera, *in* Burchfiel, B.C., Lipman, P.W., and Zoback, M.L., eds., The Cordilleran orogen: Conterminous U.S.: Boulder, Colorado, Geological Society of America, Geology of North America, v. G-3, p. 57–106.

Miller, M.M., and Harwood, D.S., 1990, Paleogeographic setting of upper Paleozoic rocks in the northern Sierra and eastern Klamath terranes, northern California, *in* Harwood, D.S., and Miller, M.M., eds., Paleozoic and early Mesozoic paleogeographic relations; Sierra Nevada, Klamath Mountains, and related terranes: Geological Society of America Special Paper 255, p. 175–192.

Monger, J.W.H., and Ross, C.A., 1971, Distribution of Fusulinaceans in the western Canadian Cordillera: Canadian Journal of Earth Sciences, v. 8, p. 259–278.

Moores, E.M., 1970, Ultramafics and orogeny, with models of the U.S. Cordillera and the Tethys: Nature, v. 228, p. 837–842.

Moores, E.M., 1991, Southwest U.S.–East Antarctica (SWEAT) connection: A hypothesis: Geology, v. 19, p. 425–428.

Nilsen, T.H., and Stewart, J.H., 1980, The Antler orogeny—Mid-Paleozoic tectonism in western North America (Penrose Conference Report): Geology, v. 8, p. 298–302.

Oldow, J.S., 1984, Evolution of a late Mesozoic back-arc fold and thrust belt, northwestern Great Basin, U.S.A.: Tectonophysics, v. 102, p. 245–274.

Oldow, J.S., Bally, A.W., Ave Lallement, H.G., and Leeman, W.P., 1989, Phanerozoic evolution of the North American Cordillera; United States and Canada, *in* Bally, A.W., and Palmer, A.R., eds., The Geology of North America—An overview: Boulder, Colorado, Geological Society of America, Geology of North America, v. A, p. 139–232.

Palmer, A.R., 1971, Cambrian of the Great Basin and adjacent areas, western United States, *in* Holland, C.H., ed., Cambrian of the new world: New York, John Wiley Interscience, p. 1–78.

Parrish, R.R., 1987, An improved microcapsule for zircon dissolution in U-Pb geochronology: Chemical Geology, v. 66, p. 99–102.

Parrish, R.R., and Krogh, T.E., 1987, Synthesis and purification of $^{205}Pb$ for U-Pb geochronology: Chemical Geology, v. 66, p. 103–110.

Parrish, R.R., Roddick, J.C., Loveridge, W.D., and Sullivan, R.W., 1987, Uranium-lead analytical techniques at the geochronology laboratory, *in* Radiogenic age and isotopic studies, Report 1, Geological Survey of Canada. Geological Survey of Canada Paper 87–2, p. 3–7.

Poole, F.G., Stewart, J.H., Palmer, A.R., Sandberg, C.A., Madrid, R.J., Ross, R.J., Hintze, L.F., Miller, M.M., and Wrucke, C.T., 1992, Latest Precambrian to latest Devonian time; development of a continental margin, *in* Burchfiel, B.C., et al., eds., The Cordilleran orogen: Conterminous U.S.: Boulder, Colorado, Geological Society of America, Geology of North America, v. G-3, p. 9–56.

Rainbird, R.H., McNicoll, V.J., Theriault, R.J., Heaman, L.M., Abbott, J.G., Long, D.G.F., and Thorkelson, D.J., 1997, Pan-continental river system draining Grenville orogen recorded by U-Pb and Sm-Nd geochronology of Neoproterozoic quartz arenites and mudrocks, northwestern Canada: Journal of Geology, v. 105, p. 1–17.

Riggs, N.R., Lehman, T.M., Gehrels, G.E., and Dickinson, W.R., 1996, Detrital zircon link between headwaters and terminus of the Upper Triassic Chinle-Dockum paleoriver system: Science, v. 273, p. 97–100.

Roberts, R.J., Hotz, P.E., Gilluly, J., and Ferguson, H.G., 1958, Paleozoic rocks of north-central Nevada: American Association of Petroleum Geologists Bulletin, v. 42, p. 2813–2857.

Ross, G.M., 1991, Precambrian basement in the Canadian Cordillera: An introduction: Canadian Journal of Earth Sciences, v. 28, p. 1133–1139.

Ross, G.M., Gehrels, G.E., and Patchett, P.J., 1997, Provenance of Triassic strata in the Cordilleran miogeocline, western Canada: Bulletin of the Canadian Society of Petroleum Geologists, v. 45, p. 461–473.

Rowell, A.J., Rees, M.N., and Suczek, C.A., 1979, Margin of the North American continent in Nevada during Late Cambrian time: American Journal of Science, v. 279, p. 1–18.

Saleeby, J.B., 1986, C-2 Central California offshore to Colorado Plateau: Geological Society of America Centennial Continent-Ocean Transect 10, 1:500,000 scale, 63 p.

Saleeby, J.B., and Busby-Spera, C., 1992, Early Mesozoic tectonic evolution of the western U.S. Cordillera, *in* Burchfiel, B.C., et al., eds., The Cordilleran orogen: Conterminous U.S.: Boulder, Colorado, Geological Society of America, Geology of North America, v. G-3, p. 107–168.

Schuchert, C., 1923, Sites and nature of the North American geosynclines: Geological Society of America Bulletin, v. 34, p. 151–230.

Schweickert, R.A., and Snyder, W.S., 1981, Paleozoic plate tectonics of the Sierra Nevada and adjacent regions, *in* Ernst, W.G., ed., The geotectonic development of California: Englewood Cliffs, New Jersey, Prentice Hall, p. 183–201.

Sears, J.W., and Price, R.A., 1978, The Siberian connection: A case for Precambrian separation of the North American and Siberian cratons: Geology, v. 6, p. 267–270.

Silberling, N.J., 1991, Allochthonous terranes of western Nevada: Current status, *in* Raines, G.L., et al., eds., Geology and ore deposits of the Great Basin: Reno, Geological Society of Nevada, p. 101–102.

Silberling, N.J., Jones, D.L., Blake, M.C., Jr., and Howell, D.G., 1987, Lithotectonic terrane map of the western conterminous United States: U.S. Geological Survey Miscellaneous Field Studies Map MF-1874-C, 1:2,500,000 scale.

Smith, M.T., Dickinson, W.R., and Gehrels, G.E., 1993, Contractional nature of Devonian-Mississippian tectonism along the North American continental margin: Geology, v. 21, p. 21–24.

Speed, R.C., 1978, Paleogeographic and plate tectonic evolution of the early Mesozoic marine province of the western Great Basin, *in* Howell D.G., and McDougall, K.A., eds., Mesozoic paleogeography of the western United States: Society of Economic Paleontologists and Mineralogists Pacific Coast Paleogeography Symposium 2, p. 253–270.

Speed, R.C., 1979, Collided Paleozoic microplate in the western United States: Journal of Geology, v. 87, p. 279–292.

Speed, R.C., and Sleep, N.H., 1982, Antler orogeny and foreland basin: A model: Geological Society of America Bulletin, v. 93, p. 815–828.

Stacey, J.S., and Kramers, J.D., 1975, Approximation of terrestrial lead isotope evolution by a two-stage model: Earth and Planetary Science Letters, v. 26, p. 207–221.

Stewart, J.H., 1972, Initial deposits in the Cordilleran geosyncline: Evidence of a Late Precambrian (<850 m.y.) continental separation: Geological Society of America Bulletin, v. 83, p. 1345–1360.

Stewart, J.H., and Suczek, C.A., 1977, Cambrian and latest Precambrian paleogeography and tectonics in the western United States, *in* Stewart, J.H., et al., eds., Paleozoic paleogeography of the western United States: Society of Economic Paleontologists and Mineralogists Pacific Coast Paleogeography Symposium I: p. 1–17.

Stewart, J.H., Murchey, B., Jones, D.L., and Wardlaw, B.R., 1986, Paleontologic evidence for complex tectonic interlayering of Mississippian to Permian deep-water rocks rocks of the Golconda allochthon in Tobin Range, north-central Nevada: Geological Society of America Bulletin, v. 97, p. 1122–1132.

Stewart, J.H., Poole, F.G., Ketner, K.B., Madrid, R.J., Roldan-Quintana, J., and Amaya-Martinez, R., 1990, Tectonics and stratigraphy of the Paleozoic and Triassic southern margin of North America, Sonora, Mexico, *in* Gehrels, G.E., and Spencer, J.E., eds., Geologic excursions through the Sonoran Desert region, Arizona and Sonora: Arizona Geological Survey Special Paper 7, p. 183–202.

Van Schmus, W.R., and 24 others, 1993, Transcontinental Proterozoic provinces, *in* Reed, J.C., et al., eds., Precambrian: Conterminous U.S.: Boulder, Colorado, Geological Society of America, Geology of North America, v. C-2, p. 171-334.

Villeneuve, M.E., Ross, G.M., Theriault, R.J., Miles, W., Parrish, R.R., and Broome, J., 1993, Tectonic subdivision and U-Pb geochronology of the crystalline basement of the Alberta basin, western Canada: Geological Survey of Canada Bulletin 447, 86 p.

Wilson, J.T., 1968, Static or mobile earth, the current scientific revolution: American Philosophical Society Proceedings, v. 112, p. 309–320.

MANUSCRIPT ACCEPTED BY THE SOCIETY JANUARY 24, 2000.

Printed in U.S.A.

Geological Society of America
Special Paper 347
2000

# *Detrital zircon geochronology of the Roberts Mountains allochthon, Nevada*

**George E. Gehrels, William R. Dickinson, Brook C.D. Riley***
*Department of Geosciences, University of Arizona,
Tucson, Arizona 85721, USA*

**Stanley C. Finney**
*Department of Geological Sciences, California State University,
Long Beach, California 90840, USA*

**Moira T. Smith**
*Tech Exploration Ltd., #350, 272 Victoria Street,
Kamloops, British Columbia V2C 2A2, Canada*

## ABSTRACT

**U-Pb geochronologic analyses have been conducted on 205 individual detrital zircon grains from Cambrian through Devonian sandstones of the Roberts Mountains allochthon in central Nevada. These strata were tectonically emplaced onto the Cordilleran margin during the mid-Paleozoic Antler orogeny, but their original depositional settings and provenance have been controversial. Our data, combined with previous detrital zircon studies, define four different age signatures for the eugeoclinal strata: (1) 690–715 and 1065–1350 Ma grains in a minor group of sandstones in the Upper Cambrian(?) Harmony Formation, (2) 1745–1790, 1820–1860, and 2595–2700 Ma grains for most of the Harmony Formation, (3) 1410–1445, 1665–1690, and 1705–1740 Ma grains for lower Middle Ordovician sandstones of the Vinini Formation, and (4) 1020–1045, 1815–1860, 1905–1940, and 2645–2740 Ma ages for lower Upper Ordovician sandstones in the Vinini, Valmy, Snow Canyon, and McAfee Formations, for the Silurian Elder Sandstone, and for the Devonian Slaven Chert.**

**Comparison of these data with the detrital zircon reference for the Cordilleran miogeocline and with ages of basement provinces in cratonal North America indicates that sandstones in the lower Vinini and parts of the Harmony Formations were derived from 1.0–1.3, ~1.43, and 1.6–1.8 Ga provinces of southwestern North America. In contrast, most older and younger units contain few grains of the appropriate ages to have come from the southwestern part of North America, and instead have strong similarities with the Peace River arch region of western Canada.**

**We propose that detritus in most of the Harmony Formation was shed from off-shelf basement rocks exposed along the Canadian continental margin, perhaps as a western continuation of the Peace River arch or as extensional fault blocks. In contrast, detritus**

*Current address: Department of Geological Sciences, University of Texas, Austin, Texas 78712, USA

Gehrels, G.E., et al., 2000, Detrital zircon geochronology of the Roberts Mountains allochthon, Nevada, *in* Soreghan, M.J., and Gehrels, G.E., eds., Paleozoic and Triassic paleogeography and tectonics of western Nevada and northern California: Boulder, Colorado, Geological Society of America Special Paper 347, p. 19–42.

**in the lower Upper Ordovician through Devonian strata is interpreted to have been recycled from platformal strata exposed along the flanks of the Peace River arch. Transport of the detritus is interpreted to have been largely via turbidity currents flowing in off-shelf basins or trenches, rather than by longshore currents on the shelf. These provenance links provide new insights into the paleodispersal history along the Cordilleran margin, and indicate that sandstones of the Roberts Mountains allochthon received detritus from, and therefore accumulated near, western North America.**

## INTRODUCTION

The Roberts Mountains allochthon consists of Cambrian through Devonian rocks that accumulated in deep-marine (eugeoclinal) settings outboard of coeval shelf facies (miogeoclinal) strata of the Cordilleran passive margin sequence (Figs. 1 and 2) (Schuchert, 1923; Kay, 1951; Roberts et al., 1958; Madrid, 1987; Miller et al., 1992; Burchfiel et al., 1992). Dominant rock types in the allochthon include chert and argillite, with less voluminous sandstone and/or quartzite, limestone, and mafic volcanic rocks. In general, chert and argillite predominate in southern and eastern exposures of the allochthon, and sandstone, quartzite, and mafic volcanic rocks are more common to the west and north (Poole et al., 1992). In most regions, these strata are complexly deformed, highly disrupted, and imbricated on older over younger thrusts. Metamorphic grade is generally greenschist facies or lower.

The allochthon is as thick as 5 km and structurally overlies miogeoclinal strata (Roberts et al., 1958; Poole et al., 1992). The present-day structural overlap is ~100 km (Figs. 2 and 3), and total displacement is estimated to have been 75–200 km (Poole et al., 1992). The allochthon was initially emplaced along the Roberts Mountains thrust during Late Devonian to Early Mississippian time (Roberts et al., 1958), although, as emphasized by Ketner (1984, 1993), repeated deformation after mid-Paleozoic time modified the original structural boundary. The timing of emplacement is recorded by the accumulation of thick (locally >4.5 km) and widespread foreland basin deposits on the shelf and craton to the east (Poole, 1974). Most of the deformation within the allochthon probably occurred during mid-Paleozoic time, because the eugeoclinal strata are overlain in many places by less-deformed Mississippian through Permian fluvial and shallow-marine strata (Dickinson et al., 1983; Little, 1987). Emplacement of the allochthon, deformation of the eugeoclinal rocks, and accumulation of the foreland basin deposits are all ascribed to the Antler orogeny (Roberts et al., 1958; Nilsen and Stewart, 1980; Miller et al., 1992).

## STRATIGRAPHY

Strata within the allochthon are divided into several formations based in large part on rock types, but also in part on age (Fig. 4). The oldest stratigraphic unit recognized is the Middle Cambrian Schwin Formation, which consists of shale, limestone, basalt, and chert. The Comus Formation, of possible Cambrian age, consists of limestone, shale, and chert that occur both as thrust slices within the allochthon and in depositional contact with Cambrian miogeoclinal strata (Madrid, 1987). These strata were originally assigned an age of Late Cambrian–Early Ordovician, but an analysis of age constraints by one of us (Finney) suggests that it is more likely Middle or Late Ordovician in age. The Schwin and Comus Formations are commonly among the lower thrust slices within the Roberts Mountains allochthon (Madrid, 1987).

### *Harmony Formation*

The Harmony Formation (Figs. 3 and 4) is a distinctive sequence of poorly sorted feldspathic and micaceous sandstone and siltstone, with subordinate limestone and shale, that accumulated in a submarine fan setting (Suczek, 1977; Rowell et al., 1979). Sandstones consisting of monocrystalline and polycrystalline quartz, white mica and biotite, and both potassium feldspar (mainly perthite and microcline) and plagioclase (mainly albite) make up most of the unit (Suczek, 1977; Madrid, 1987). During the course of our study, two distinctive petrofacies of the Harmony Formation were identified. As described by Dickinson and Gehrels (this volume, Chapter 11) the dominant petrofacies is a medium- to coarse-grained subarkose. A second petrofacies, apparently minor in volume and identified only in the Little Cottonwood Canyon area of the Galena Range, is generally finer grained and more quartzose. In the Little Cottonwood Canyon area, the subarkosic sandstones overlie the quartzose sandstones along a concordant but sharp depositional contact (Fig. 3B).

Paleocurrent studies of the Harmony Formation in several mountain ranges did not yield a consistent transport direction (Suczek, 1977).

The age of the Harmony Foundation is currently uncertain. Late Cambrian trilobites have been reported from several localities (Hotz and Willden, 1964; Palmer, 1971), and Late Cambrian-Early Ordovician microfossils have been recovered from Harmony Canyon (Madden-McGuire et al., 1991). In contrast, Jones (1997) reported that conodonts recovered from a limestone layer within the Harmony Formation are Late Devonian in age. Pending confirmation of the latter findings, the age of the Harmony Formation is accordingly assumed to be Late Cambrian in the following discussion, but it should be noted that this age assignment may need to be revised.

Rocks of the Harmony Formation generally occur in westernmost and structurally highest exposures of the Roberts Mountains allochthon (Fig. 3) (Suczek, 1977; Rowell et al., 1979; Schweickert and Snyder, 1981; Madrid, 1987).

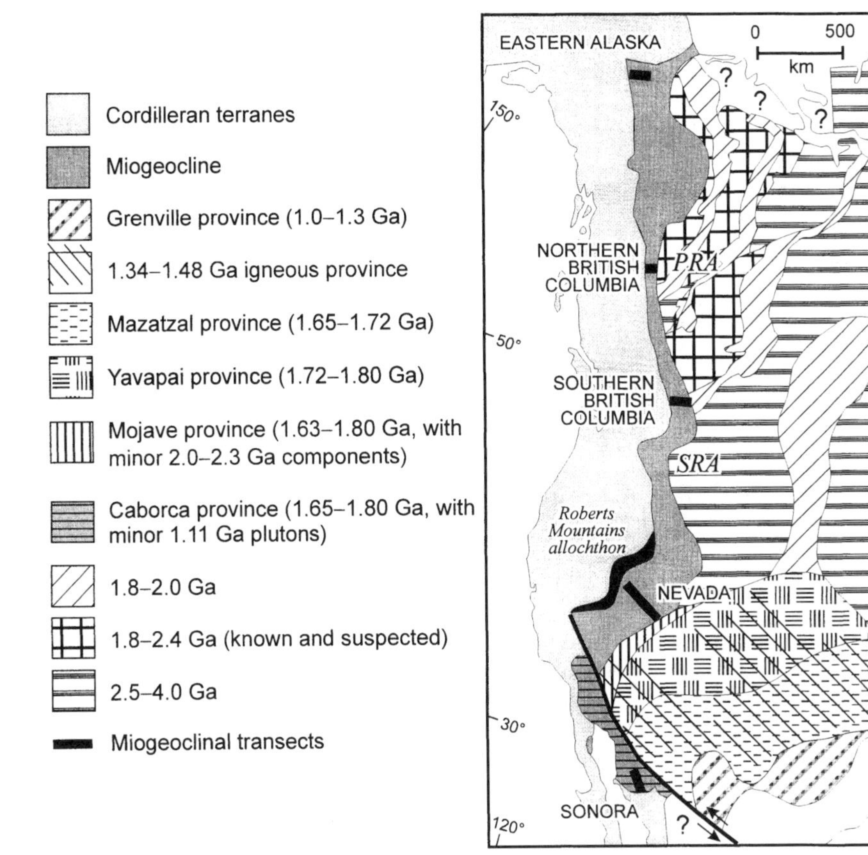


Figure 1. Schematic map showing location of study area in relation to Cordilleran miogeocline and first-order basement provinces of western North America. Also shown are locations of five transects that yield miogeoclinal detrital zircon reference (Gehrels et al., 1995; Gehrels, Introduction, this volume). Basement provinces are simplified from Hoffman (1989) for cratonal interior, from Ross (1991) and Villeneuve et al. (1993) for western Canadian shield, from Van Schmus et al. (1993) for southwestern United States, and from Stewart et al. (1990) for northwestern Mexico. PRA—Peace River arch, SRA—Salmon River arch.

### *Valmy, Vinini, Snow Canyon, and McAfee formations*

Ordovician eugeoclinal strata crop out in many ranges of central and northern Nevada and make up the dominant units of the Roberts Mountains allochthon (Fig. 3). In most regions, the Ordovician rocks occupy intermediate structural levels of the allochthon (Madrid, 1987). In southern and eastern parts of the allochthon, these strata are referred to the Vinini Formation, which consists mainly of chert and argillite. In northern and western parts of the allochthon, the Valmy Formation (or Valmy Group) contains thicker and more abundant quartzite beds, a greater proportion of greenstone, and less abundant chert and argillite.

The Valmy Formation generally consists of (1) a lower unit of chert and argillite with subordinate sandstone, greenstone, and limestone, (2) a middle unit of sandstone and siltstone with subordinate chert, argillite, and greenstone, and (3) an upper unit of black argillite with subordinate mudstone, siltstone, and limestone (Roberts, 1964; Miller and Larue, 1983; Madrid, 1987). In the northern Independence Range, these three subunits are referred to, respectively, as the Snow Canyon Formation, McAfee Quartzite, and Jacks Peak Formation (Churkin and Kay, 1967).

Separate upper and lower sandstone units have been recognized (Fig. 4). A lower Middle Ordovician sandstone occurs in the Snow Canyon Formation. In the northern Independence Range, the McAfee Quartzite contains lower Upper Ordovician sandstone that is apparently coeval with the Eureka Quartzite of the miogeocline (Ketner, 1966; Miller and Larue, 1983).

Valmy sandstones in the Shoshone Range and near Antler Peak can be divided into two different types that are commonly interbedded (Madrid, 1987). One type consists of well-sorted, medium- to fine-grained quartz arenite that yields sparse west-directed paleocurrents. The second type is poorly sorted and rich in detrital feldspar and mica. Sparse paleocurrent measurements from the latter sandstones yield an average transport direction toward the east. The ages of the sandstones are Middle and Late Ordovician, but limited biostratigraphic control hinders precise correlation.

The Vinini Formation, which has abundant biostratigraphic control, includes two members, each containing a quartz sandstone package (Finney and Perry, 1991; Finney et al., 1997). The lower member consists, in ascending order, of greenstone, limestone, greenstone, black shale, and a thick turbiditic quartz sandstone unit (>1 km thick in the Roberts Mountains). The quartz sandstone is early Middle Ordovician (early Whiterockian) age. The upper member consists largely of shale, argillite, and chert, capped by latest Ordovician limestone. An early Late Ordovician (early Mohawkian) quartz sandstone occurs within the upper unit, stratigraphically well above its base. The latter sandstone is coeval with the McAfee Quartzite of the Valmy Group in the northern Independence Range and probably correlates with lower Upper Ordovician sandstones in the Valmy Formation elsewhere in Nevada.

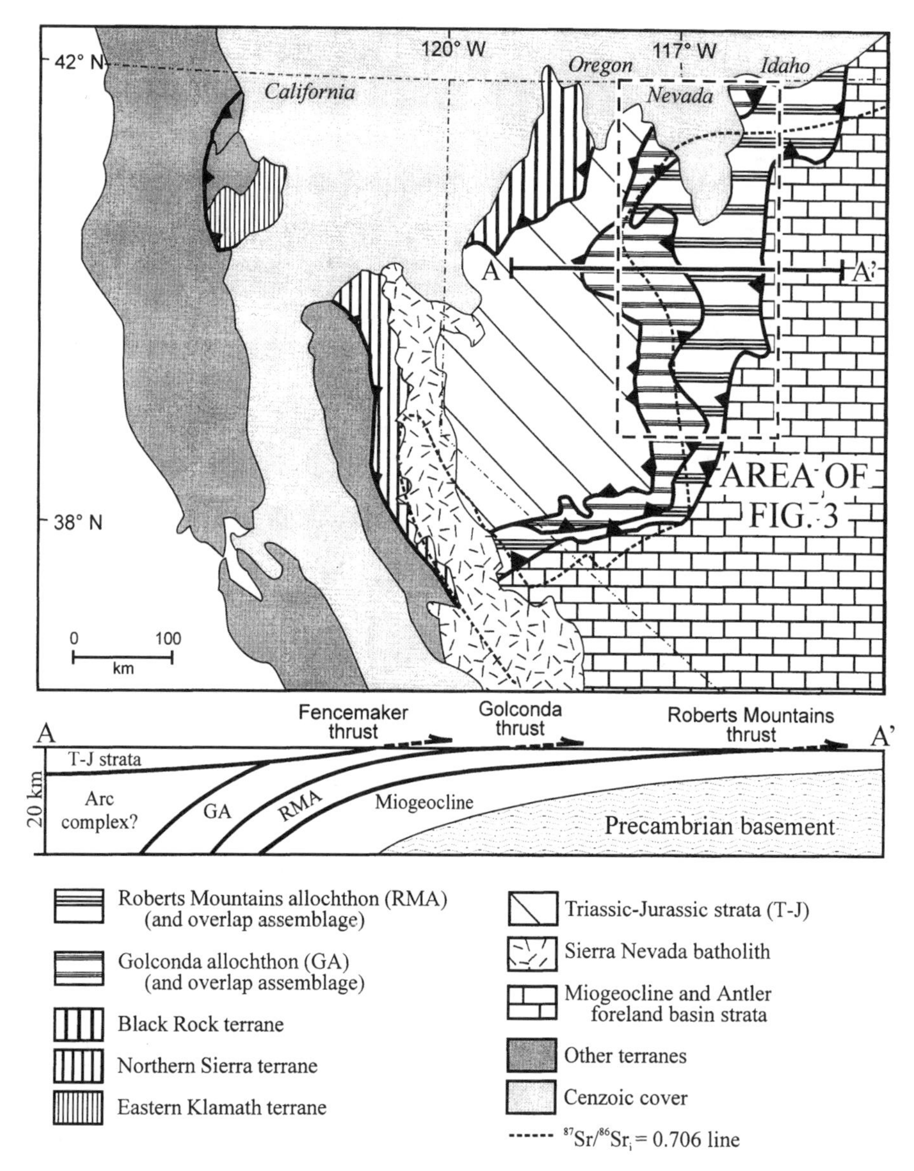


Figure 2. Schematic map and cross section of main terranes and assemblages and their bounding faults in western Nevada and northern California. Map configuration of terranes and assemblages is adapted primarily from Oldow (1984), Silberling (1991), and Silberling et al. (1987); cross section is highly simplified from Blake et al. (1985) and Saleeby (1986). Location of $^{87}Sr/^{86}Sr_i$ = 0.0706 line is from Kistler and Peterman (1973) and Kistler (1991). Note that widespread Cretaceous and younger rocks and structures are ignored in map and sections in effort to emphasize configuration of pre-Cretaceous features.

### *Elder Sandstone*

The Silurian Elder Sandstone consists mainly of orange-brown–weathering sandstone and siltstone with subordinate shale, chert, limestone, and greenstone (Gilluly and Gates, 1965; Girty et al., 1985; Madrid, 1987). Most of the sandstones are poorly sorted and arkosic, and include framework grains of monocrystalline and polycrystalline quartz, potassium feldspar, plagioclase, white mica, and biotite. Madrid (1987) reported that near the top of the section in the Shoshone Range, the uppermost arkosic sandstones are interbedded with well-sorted quartz arenite. Ketner (1977) reported that the average grain size in the Elder Sandstone generally decreases eastward.

Graptolite collections scattered stratigraphically through the Elder Sandstone among several ranges (Gilluly and Gates, 1965; Cluer et al., 1997; Finney et al., 1997) demonstrate that the sandstone ranges in age from near the beginning to the end of the Silurian System. Coeval strata of the miogeocline include the Lone Mountain Dolomite and the Roberts Mountain Formation, which lack quartz sandstone. The Elder Sandstone commonly occupies lower structural levels of the Roberts Mountains allochthon (Madrid, 1987).

### *Slaven Chert*

Devonian strata are usually assigned to the Slaven Chert, which comprises mainly chert and shale with subordinate limestone, quartzite, greenstone, and bedded barite (Gilluly and Gates, 1965; Madrid, 1987). As in the Valmy and Elder formations, both quartzose and feldspathic sandstones are present in the Slaven chert (Madrid, 1987). Most strata in the Slaven Chert are Middle and Late Devonian, the youngest fossils being late Famennian age (Gilluly and Gates, 1965; Madrid, 1987; Poole et al., 1992). The Slaven Chert commonly occurs in lower structural levels of the allochthon (Madrid, 1987).

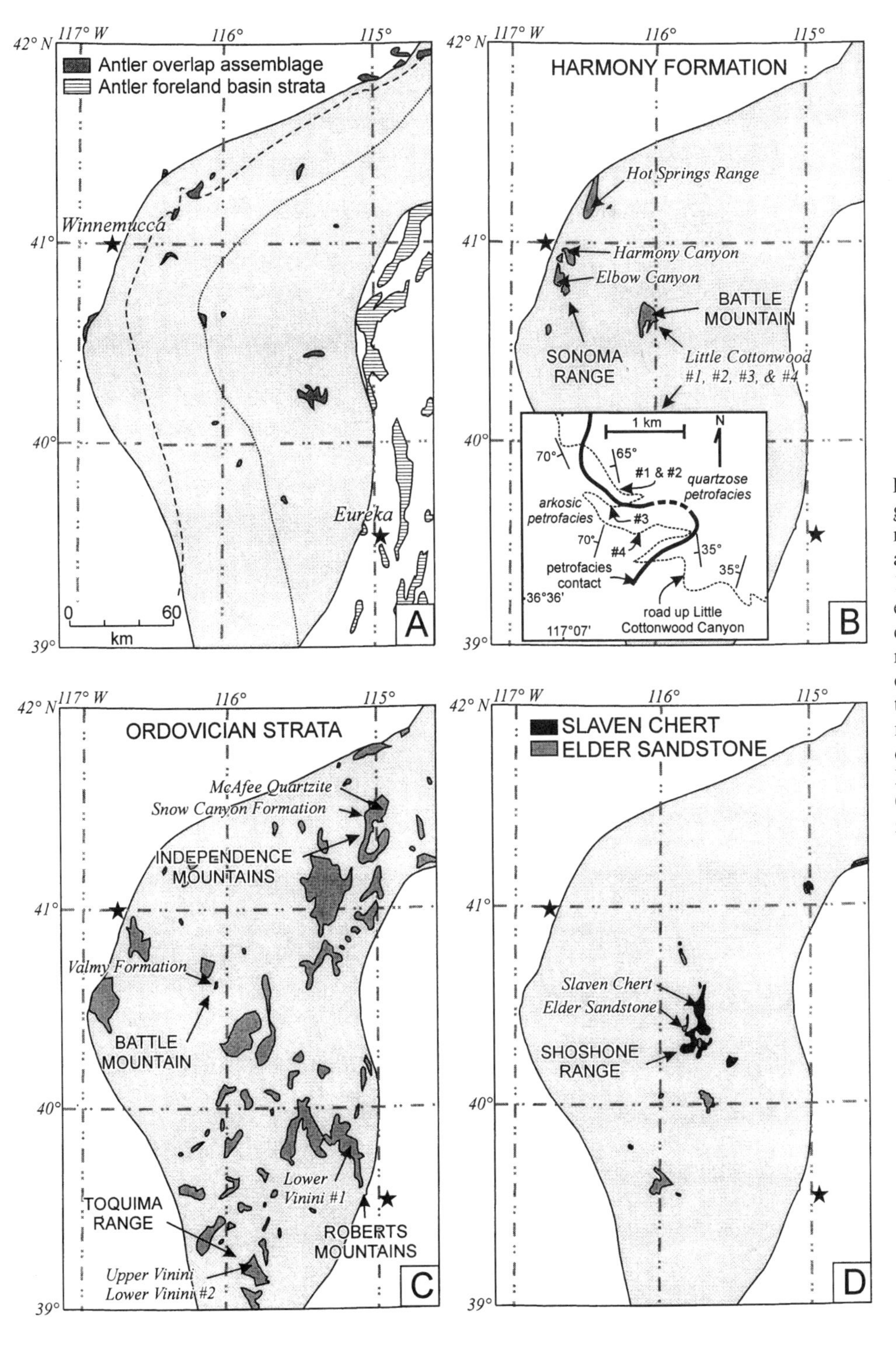


Figure 3. Map of central Nevada showing general outcrop patterns of several of main units of Roberts Mountains allochthon (from Stewart and Carlson, 1978). Also shown are western limit of exposed miogeoclinal strata (western dashed line in A), eastern limit of exposed rocks of Golconda allochthon (eastern dashed line in A), and approximate locations of our detrital zircon samples. Inset map shows stratigraphic relations of quartzose and subarkosic sandstones of Harmony Formation in Little Cottonwood Canyon area, and location of four detrital zircon samples from this region.

## PREVIOUS PROVENANCE STUDIES

The provenance of sandstones within the Roberts Mountains allochthon has been a subject of considerable controversy. The following is an outline of the various lines of evidence previously used to constrain the provenance of each of the main units of the allochthon.

### *Harmony Formation*

The provenance of the Harmony Formation has been particularly problematic because of (1) the immature composition of the Harmony Formation (in contrast to the presence of mature lower Paleozoic quartz arenites on the shelf), (2) the deposition mainly of carbonates along the Cordilleran margin during Late

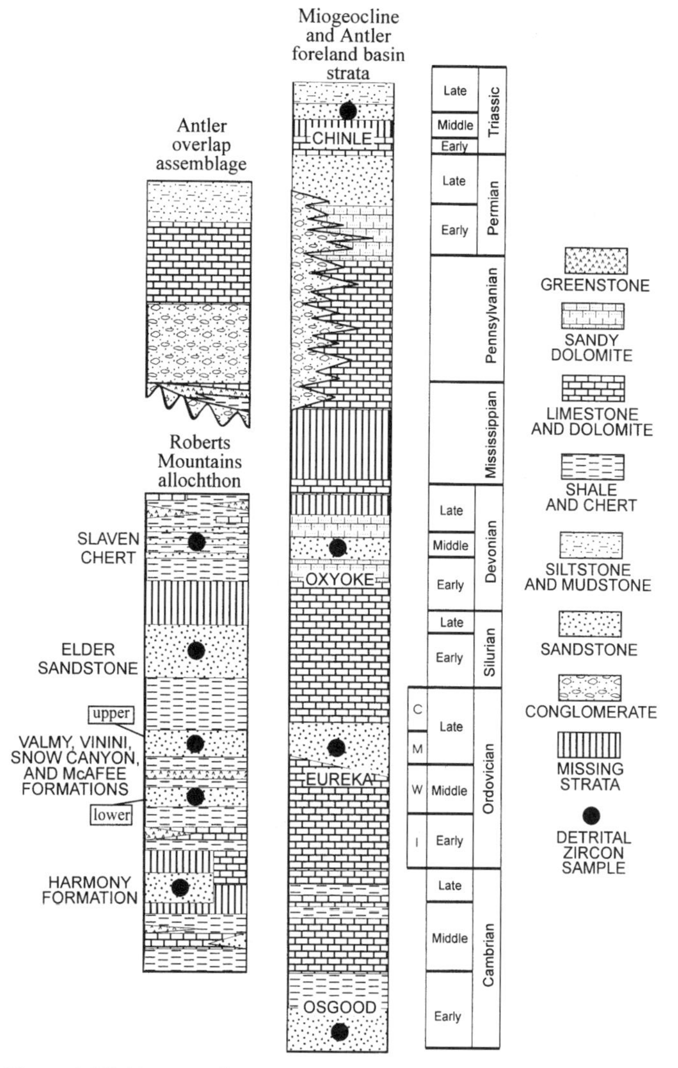


Figure 4. Highly generalized stratigraphic column of Roberts Mountains allochthon (compiled mainly from Madrid [1987] and Miller et al. [1992]). North American Ordovician series are Ibexian—I, Whiterockian—W, Mohawkian—M, and Cincinnatian—C. Global series are according to Webby (1998).

Cambrian time, and (3) the outboard position of the Harmony Formation when middle Paleozoic thrust displacements are restored (Suczek, 1977; Ketner, 1977; Stewart and Suczek, 1977; Madrid, 1987). These factors have led many workers to consider the possibility that the detritus was shed from an outboard granitic source region, perhaps a separate microcontinental sliver or an uplifted region of the in situ rifted continental margin (Suczek, 1977; Ketner, 1977; Miller et al., 1992).

Alternatively, detritus in the Harmony Formation may have been derived from basement rocks exposed along the continental margin. Suczek (1977), Stewart and Suczek (1977), Rowell et al. (1979), and Schweickert and Snyder (1981) proposed that detritus in the Harmony Formation originated in the Salmon River arch region of Idaho (Fig. 1), because Precambrian basement in that region was apparently exposed and was a source of feldspathic detritus during Cambrian time. The sands were interpreted to have been carried by turbidity currents from the continental margin in western Idaho out onto a deep sea fan in western Nevada.

Wallin (1990) analyzed multigrain detrital zircon fractions and a few single grains from Harmony sandstones in the Sonoma Range in an attempt to resolve their source. Multigrain fractions yielded $^{207}Pb^{*}/^{206}Pb^{*}$ ages ranging from ca. 1077 to 1836 Ma, but these ages are of little use for identifying specific sources because they are most likely averages of grains of disparate ages. Four single grains, however, yielded moderately discordant analyses with $^{207}Pb^{*}/^{206}Pb^{*}$ ages of about 691, 1336, 1752, and 1772 Ma. Because these ages do not match the dominant ca. 1370 and 1576 Ma ages of basement rocks in the Salmon River arch region (Evans and Fischer, 1986; Evans and Zartman, 1988), and instead overlap with the ages of basement rocks in southwestern North America, Wallin (1990) suggested that sandstones of the Harmony Formation may have accumulated south of their present position along the margin. Northward displacement would have occurred after deposition of the Harmony strata, and prior to Devonian-Mississippian emplacement of the Roberts Mountains allochthon.

Smith and Gehrels (1994) analyzed 27 detrital zircon grains from a sample of Harmony Formation in Little Cottonwood Canyon (at Battle Mountain), and obtained dominant ages of ca. 690–710 Ma and 1015–1330 Ma. These ages support Wallin's (1990) conclusions that the Salmon River arch is an unlikely source for Harmony Formation detritus, and that a southern source is possible. However, because these ages are common in detrital zircons from off-shelf sandstones all along the Cordilleran margin, Smith and Gehrels (1994) raised the possibility that Harmony sandstones accumulated near their present position, in proximity to ca. 690–710 Ma and 1015–1330 Ma crystalline rocks that may have been widespread along the Cordilleran margin.

### *Valmy, Vinini, Snow Canyon, and McAfee Formations*

Several alternatives for the provenance of detritus in the Ordovician units have been proposed. From observations that quartz grains in the miogeoclinal sandstones are consistently smaller and better sorted than grains in the eugeoclinal strata, Ketner (1966) concluded that detritus in the eugeoclinal strata was shed from a western source. In contrast, Churkin (1974), Miller and Larue (1983), and Finney and Perry (1991) argued that differences in grain size and sorting could be explained by sediment bypassing and channeling within the shelf, and concluded that the eugeoclinal strata are most likely outboard equivalents of the shelf-facies sandstones of the miogeocline. Finney and Perry (1991) linked the two main sandstone units of the Vinini Formation (Fig. 4) to patterns of deposition and sea-level change on the shelf, and thereby established a depositional link between the eugeoclinal strata and the miogeocline to the east. The strongest tie is provided by the presence of carbonate clasts and conodont and graptolite specimens in lower Vinini sandstones that were almost certainly derived from the coeval Antelope Valley Limestone of the miogeocline in Nevada (Finney and Perry, 1991). As outlined by Finney and Perry (1991), this probably occurred in response to a dramatic lowering of sea level that exposed basement rocks and platformal strata of the southwest and allowed sand-carrying channels to be carved through the Middle Ordovician and older shelf facies strata.

Madrid (1987) reported a significant feldspathic component in some of the Ordovician sandstones, and presented east-directed paleocurrent measurements from these same units. On the basis of these relations, he suggested that the feldspathic sands were derived from the west, perhaps reworked from sandstones of the Harmony Formation, whereas the quartzose sands were derived from the shelf to the east.

Wallin (1990) analyzed primarily multigrain detrital zircon fractions from Middle-Upper Ordovician sandstones of the Vinini Formation (from the Tuscarora Mountains) and Eureka Quartzite (from Lone Mountain), and a Lower Ordovician sandstone from the Palmetto Formation of southern Nevada. Because the various units yielded broad and generally similar ranges of ages, and each age is potentially an average of older and younger components, the multigrain ages are not particularly helpful in identifying specific sources for the various units.

Smith and Gehrels (1994) analyzed 27 individual detrital zircon grains from a Valmy quartzite in Galena Canyon, at Battle Mountain. They obtained dominant ages of 1830–1960 Ma, which are very similar to the ages of detrital zircon grains in the coeval Eureka Quartzite of the miogeocline (Gehrels and Dickinson, 1995). On the basis of this age similarity, and Ketner's (1968) conclusion that much of the quartz-rich detritus in Ordovician miogeoclinal strata along the Cordilleran margin was shed from the Peace River arch region of northwestern Canada (Fig. 1), Smith and Gehrels (1994) suggested that Valmy sandstones were derived largely from the Peace River arch area. This conclusion supported Churkin's (1974) and Miller and Larue's (1983) interpretation that Valmy and/or McAfee sandstones are the outboard equivalents of the Eureka Quartzite.

### *Elder Sandstone and Slaven Chert*

To explain the feldspathic composition and westward coarsening of sandstones in the Elder Sandstone, Ketner (1977) proposed that the detritus was derived primarily from a western source. Madrid (1987) proposed more specifically that feldspathic detritus in both Silurian and Devonian units was recycled from the Harmony Formation or related sandstones to the west, but that quartz-rich detritus in the two units was derived from the shelf to the east.

Girty et al. (1985) reported U-Pb ages from multigrain detrital zircon fractions from the Elder Sandstone in the Shoshone Range. The resulting data were highly discordant, with $^{207}Pb^{*}/^{206}Pb^{*}$ ages ranging from ca. 1.7 to 2.0 Ga, and an apparent discordia that yielded an upper intercept of ca. 2.2 Ga. These data confirmed a continental (rather than magmatic arc) source of the detritus, but were not sufficient to identify a specific source region.

## TECTONIC SETTING

There are several existing models for the origin and emplacement history of the Roberts Mountains allochthon (as summarized by Nilsen and Stewart, 1980). The interpretation that detritus was derived exclusively from the east has led to a traditional view in which the eugeoclinal strata accumulated in a slope-rise-basinal setting directly outboard of the Cordilleran miogeocline (Burchfiel and Davis, 1972, 1975; Churkin, 1974; and most subsequent syntheses). Building on Ketner's (1966, 1977) evidence for a western source, Madrid (1987) envisioned the eugeocline as a two-sided basin undergoing active extension during Ordovician through Devonian time, with detritus derived from both the miogeocline to the east and the Harmony platform to the west.

In most models, early Paleozoic island arcs along the Cordilleran margin are inferred to have been located a great enough distance to the west of the rifted margin that they contributed little or no detritus to the Devonian and older eugeoclinal strata. In most syntheses, however, emplacement of the Roberts Mountains allochthon during the Antler orogeny is attributed to collision or convergence with an outboard island arc. Most workers envision the arc system as facing westward, with the eugeoclinal strata of the Roberts Mountains allochthon deposited in a backarc or marginal basin (Burchfiel and Davis, 1972, 1975; Churkin, 1974; Miller et al., 1984; Burchfiel and Royden, 1991; Burchfiel et al., 1992). Thrust imbrication and emplacement of the Roberts Mountains allochthon are generally ascribed to changes in plate motion, leading to collapse of the backarc basin.

Others, however, propose that the early Paleozoic island-arc system faced eastward, with the Roberts Mountains allochthon forming as eugeoclinal strata that were imbricated in a forearc accretionary complex (Dickinson, 1977; Schweickert and Snyder, 1981; Speed and Sleep, 1982; Dickinson, this volume, Chapter 14). In this case, emplacement of the allochthon occurred when the arc and its accretionary complex collided with the continental margin.

The polarity of the inferred early Paleozoic arc system is difficult to determine because no arc system is preserved along the western margin of the Roberts Mountains allochthon today (Fig. 1), and because rocks of the Golconda allochthon, which are immediately west of and structurally above the Roberts Mountains allochthon, record continuous marine deposition through Late Devonian–Early Mississippian time (Miller et al., 1984). The lack of an early Paleozoic arc has been explained in most previous syntheses by rifting away of the arc soon after the Antler orogeny (Burchfiel and Davis, 1972, 1975; Churkin, 1974; Dickinson, 1977; Schweickert and Snyder, 1981; Miller et al., 1992). Alternatively, the arc may have subsided due to thermal contraction after the Antler orogeny (Speed and Sleep, 1982), and may be concealed beneath the Golconda allochthon.

Burchfiel and Royden (1991) offered a noncollisional model for the Antler orogeny that involves slab rollback and trench retreat within a short-lived, east-facing arc behind a generally west-facing arc. Based on Mediterranean analogues, they proposed that the east-facing arc underwent extension and subsidence due to slab rollback at the same time that the accretionary prism for the arc was being emplaced as the Roberts Mountains allochthon.

In addition to these two-dimensional models, it is possible that some or all of the eugeoclinal strata were transported southward along the Cordilleran margin along strike-slip faults (Eisbacher, 1983; Wallin, 1993; Wallin et al., this volume, Chapter 9).

By referring to the Roberts Mountains allochthon as a suspect terrane, Coney et al. (1980) emphasized the possibility that the eugeoclinal strata in the Roberts Mountains allochthon did not accumulate near their present position in Nevada and may therefore be far traveled.

## PRESENT STUDY

This study is an attempt to determine the provenance of detritus in the Roberts Mountains allochthon. We have analyzed 205 single detrital zircon grains from several of the sandstone-bearing units within the allochthon, including the Harmony Formation, Valmy, Vinini, and Snow Canyon Formations, McAfee Quartzite, Elder Sandstone, and Slaven Chert. The analyses were conducted using analytical techniques outlined by Gehrels (this volume, Introduction), and the results are presented in Table 1 and on concordia diagrams in Figures 5–7. The ages are shown on relative age-probability plots in Figure 8 and are compared to each other statistically in Table 2, and the main groups of ages and their abundances are listed in Table 3.

The ages of detrital zircons in our samples of eugeoclinal strata are compared with the ages of grains in miogeoclinal strata in Figure 9, and the results of statistical comparisons with the miogeoclinal strata are presented in Figure 10. Our conclusions based on these comparisons are presented in the following and in Figure 11, and in the synthesis chapter of this volume (Gehrels et al., this volume, Chapter 10). Dickinson and Gehrels (this volume, Chapter 11) present and discuss the petrography of the various age samples.

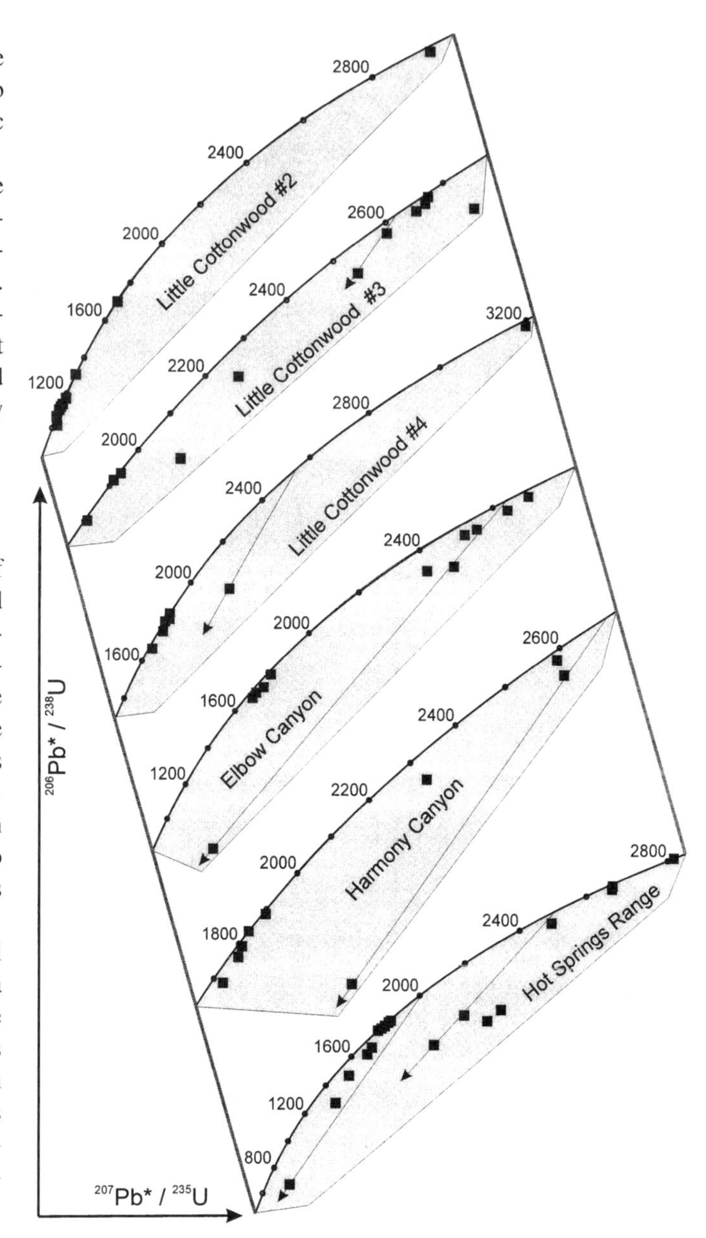


Figure 5. U-Pb concordia diagrams showing single-grain detrital zircon analyses from Upper Cambrian(?) strata of Harmony Formation. Discordia lines projected to 100 Ma are shown for reference. All data reduction and concordia plots are from programs of Ludwig (1991a, 1991b).

## RESULTS

### *Harmony Formation (Late Cambrian)*

We analyzed 73 individual zircon grains from 6 samples of the Harmony Formation (Fig. 3). Three of the six samples were collected from Little Cottonwood Canyon, near Battle Mountain, in the same region as the sample analyzed by Smith and Gehrels (1994). The latter sample is herein referred to as Little Cottonwood #1. As shown in Figure 3B, two of the samples (Little Cottonwood #1 and #2) were analyzed from a quartzose petrofacies of the Harmony Formation, and two (Little Cottonwood #3 and #4) were collected from overlying subarkosic sandstones. The other three samples were collected from the Hot Springs Range and the Sonoma Range. The following is a discussion of each of the samples analyzed in this study.

### *Little Cottonwood #2*

We collected ~8 kg of medium-grained quartzose sandstone from the 2172 m level of the road up Little Cottonwood Canyon at Battle Mountain (Fig. 3B). This sample was collected from within several meters of the same locality sampled by Smith and Gehrels (1994) and described herein as Little Cottonwood #1 (Figs. 3B and 8). Little Cottonwood #2 is a medium-grained quartzose wacke with moderately developed rounding and sorting, which is typical of the quartzose petrofacies (Dickinson and Gehrels, this volume, Chapter 11).

### TABLE 1. U-Pb ISOTOPIC DATA AND AGES

| Grain type | Grain wt. (µg) | $Pb_c$ (pg) | U (ppm) | $^{206}Pb_m/^{204}Pb$ | $^{206}Pb_c/^{208}Pb$ | Apparent ages (Ma) $^{206}Pb^*/^{238}U$ | $^{207}Pb^*/^{235}U$ | $^{207}Pb^*/^{206}Pb^*$ | Interpreted age (Ma) |
|---|---|---|---|---|---|---|---|---|---|
| Harmony Formation (Little Cottonwood #2: 40°36'21"N, 117°06'43"W) | | | | | | | | | |
| **LE** | **26** | **8** | **90** | **1590** | **5.4** | **1068 ± 10** | **1070 ± 12** | **1073 ± 11** | **1073 ± 10** |
| **CmR** | **14** | **14** | **75** | **865** | **7.2** | **1086 ± 10** | **1088 ± 14** | **1092 ± 17** | **1092 ± 10** |
| **CmR** | **19** | **7** | **66** | **1020** | **2.2** | **1088 ± 15** | **1089 ± 17** | **1092 ± 15** | **1092 ± 20** |
| **LE** | **31** | **12** | **98** | **1460** | **6.4** | **1108 ± 8** | **1109 ± 11** | **1110 ± 11** | **1110 ± 10** |
| **CmR** | **21** | **21** | **68** | **793** | **5.1** | **1124 ± 8** | **1124 ± 14** | **1123 ± 18** | **1123 ± 10** |
| LmR | 16 | 7 | 49 | 1295 | 8.7 | 1139 ± 13 | 1147 ± 15 | 1163 ± 11 | 1164 ± 10 |
| LmR | 32 | 21 | 422 | 7900 | 9.9 | 1172 ± 9 | 1191 ± 10 | 1226 ± 6 | 1229 ± 10 |
| GOE | 18 | 45 | 193 | 820 | 5.5 | 1016 ± 7 | 1072 ± 12 | 1188 ± 17 | 1203 ± 20 |
| **CmR** | **18** | **11** | **67** | **760** | **4.9** | **1306 ± 15** | **1308 ± 19** | **1311 ± 16** | **1312 ± 10** |
| **CmR** | **24** | **5** | **62** | **5300** | **14.0** | **1701 ± 9** | **1705 ± 12** | **1711 ± 7** | **1711 ± 10** |
| LmR | 21 | 5 | 54 | 3800 | 6.2 | 2916 ± 20 | 2941 ± 21 | 2959 ± 4 | 2959 ± 20 |
| Harmony Formation (Little Cottonwood #3: 40°36'18"N, 117°06'46"W) | | | | | | | | | |
| CE | 39 | 7 | 105 | 7300 | 4.9 | 1801 ± 11 | 1813 ± 12 | 1828 ± 5 | 1828 ± 10 |
| CE | 47 | 6 | 57 | 9300 | 4.9 | 1916 ± 13 | 1925 ± 14 | 1935 ± 6 | 1935 ± 10 |
| CsR | 30 | 8 | 41 | 3150 | 4.9 | 1935 ± 13 | 1941 ± 15 | 1947 ± 7 | 1947 ± 10 |
| MsR | 24 | 21 | 240 | 5700 | 19.1 | 1977 ± 18 | 2135 ± 15 | 2291 ± 6 | 2299 ± 20 |
| MOE | 36 | 485 | 92 | 173 | 3.1 | 2204 ± 21 | 2291 ± 55 | 2369 ± 34 | 2373 ± 40 |
| MOE | 42 | 8 | 56 | 8300 | 4.8 | 2471 ± 23 | 2551 ± 24 | 2615 ± 4 | 2618 ± 20 |
| MsR | 31 | 9 | 91 | 9450 | 6.2 | 2573 ± 18 | 2603 ± 19 | 2627 ± 4 | 2628 ± 10 |
| MsR | 28 | 18 | 96 | 4300 | 4.9 | 2630 ± 17 | 2655 ± 19 | 2675 ± 4 | 2676 ± 10 |
| CsR | 32 | 7 | 123 | 1810 | 1.2 | 2664 ± 26 | 2675 ± 29 | 2682 ± 6 | 2683 ± 10 |
| CsR | 18 | 7 | 49 | 4100 | 4.6 | 2648 ± 22 | 2670 ± 24 | 2686 ± 6 | 2687 ± 10 |
| CsR | 26 | 7 | 53 | 5900 | 8.4 | 2636 ± 24 | 2751 ± 27 | 2836 ± 7 | 2840 ± 20 |
| Harmony Formation (Little Cottonwood #4: 40°36'20"N, 117°06'39"W) | | | | | | | | | |
| LmR | 51 | 44 | 161 | 3080 | 4.8 | 1657 ± 9 | 1706 ± 10 | 1767 ± 6 | 1772 ± 20 |
| **LmR** | **33** | **5** | **172** | **20100** | **4.5** | **1804 ± 9** | **1809 ± 10** | **1814 ± 6** | **1815 ± 10** |
| **CmR** | **38** | **11** | **95** | **6460** | **7.5** | **1846 ± 8** | **1847 ± 11** | **1848 ± 7** | **1848 ± 10** |
| LE | 45 | 19 | 97 | 4050 | 9.8 | 1747 ± 7 | 1804 ± 9 | 1869 ± 6 | 1873 ± 20 |
| CE | 36 | 10 | 46 | 3230 | 18.2 | 1818 ± 9 | 1845 ± 11 | 1877 ± 5 | 1879 ± 10 |
| *LE* | *26* | *69* | *316* | *2405* | *8.5* | *1966 ± 9* | *2244 ± 13* | *2509 ± 6* | *2523 ± 40* |
| CmR | 42 | 12 | 144 | 18700 | 3.8 | 3174 ± 12 | 3199 ± 13 | 3214 ± 3 | 3215 ± 20 |
| Harmony Formation (Elbow Canyon: 40°46'44"N, 117°41'56"W) | | | | | | | | | |
| ME | 17 | 10 | 261 | 8310 | 8.4 | 1668 ± 7 | 1712 ± 8 | 1766 ± 4 | 1770 ± 10 |
| MsR | 9 | 12 | 222 | 2980 | 5.5 | 1696 ± 8 | 1731 ± 9 | 1775 ± 4 | 1778 ± 10 |
| LsR | 12 | 17 | 151 | 1987 | 8.9 | 1724 ± 8 | 1774 ± 9 | 1833 ± 5 | 1838 ± 10 |
| LE | 26 | 9 | 55 | 3159 | 3.9 | 1790 ± 10 | 1814 ± 11 | 1841 ± 4 | 1843 ± 10 |
| ME | 9 | 36 | 419 | 2530 | 6.6 | 2301 ± 9 | 2423 ± 11 | 2527 ± 4 | 2532 ± 20 |
| MsR | 20 | 30 | 281 | 4830 | 8.0 | 2472 ± 9 | 2528 ± 11 | 2574 ± 4 | 2576 ± 20 |
| LsR | 16 | 15 | 122 | 3510 | 4.7 | 2498 ± 10 | 2560 ± 12 | 2610 ± 4 | 2612 ± 20 |
| *ME* | *31* | *165* | *711* | *1027* | *3.9* | *825 ± 3* | *1444 ± 7* | *2528 ± 5* | *2636 ± 40* |
| MsR | 7 | 175 | 317 | 3172 | 5.5 | 2323 ± 9 | 2500 ± 11 | 2646 ± 4 | 2652 ± 20 |
| CsR | 12 | 5 | 56 | 2095 | 2.7 | 2588 ± 27 | 2636 ± 30 | 2674 ± 5 | 2675 ± 20 |
| LsR | 24 | 15 | 188 | 8690 | 13.1 | 2651 ± 10 | 2685 ± 12 | 2712 ± 4 | 2713 ± 10 |
| Harmony Formation (Harmony Canyon: 40°56'49"N, 117°38'20"W) | | | | | | | | | |
| CsR | 40 | 20 | 199 | 6290 | 7.7 | 1590 ± 6 | 1665 ± 8 | 1761 ± 5 | 1768 ± 20 |
| LsR | 36 | 30 | 382 | 7670 | 18.6 | 1687 ± 6 | 1737 ± 8 | 1798 ± 5 | 1802 ± 10 |
| CsR | 38 | 16 | 45 | 1972 | 11.6 | 1793 ± 9 | 1811 ± 12 | 1833 ± 7 | 1834 ± 10 |
| **CE** | **44** | **21** | **56** | **1138** | **2.6** | **1836 ± 9** | **1836 ± 12** | **1837 ± 13** | **1837 ± 10** |
| **CsR** | **30** | **17** | **96** | **3315** | **10.7** | **1837 ± 8** | **1838 ± 9** | **1838 ± 5** | **1838 ± 10** |
| CsR | 16 | 8 | 96 | 3540 | 12.8 | 1762 ± 9 | 1798 ± 12 | 1840 ± 7 | 1842 ± 10 |
| CsR | 19 | 7 | 174 | 9320 | 5.6 | 1886 ± 9 | 1897 ± 10 | 1909 ± 4 | 1910 ± 10 |
| LE | 14 | 23 | 502 | 7450 | 6.9 | 2256 ± 9 | 2338 ± 10 | 2410 ± 4 | 2414 ± 20 |
| CE | 18 | 11 | 71 | 3200 | 3.5 | 2569 ± 12 | 2596 ± 13 | 2617 ± 4 | 2618 ± 10 |
| *ME* | *10* | *12* | *261* | *4265* | *7.3* | *1683 ± 7* | *2157 ± 11* | *2644 ± 5* | *2672 ± 40* |
| LsR | 22 | 12 | 174 | 9110 | 4.6 | 2530 ± 10 | 2608 ± 12 | 2670 ± 4 | 2672 ± 20 |

**TABLE 1. U-Pb ISOTOPIC DATA AND AGES (continued)**

| Grain type | Grain wt. (µg) | $Pb_c$ (pg) | U (ppm) | $^{206}Pb_m$ / $^{204}Pb$ | $^{206}Pb_c$ / $^{208}Pb$ | Apparent ages (Ma) $^{206}Pb^*$ / $^{238}U$ | $^{207}Pb^*$ / $^{235}U$ | $^{207}Pb^*$ / $^{206}Pb^*$ | Interpreted age (Ma) |
|---|---|---|---|---|---|---|---|---|---|
| Harmony Formation (Hot Springs Range: 41°09'25"N, 117°28'17"W) | | | | | | | | | |
| *MsR* | *13* | *15* | *804* | *10340* | *14.5* | *1468 ± 7* | *1584 ± 9* | *1741 ± 5* | *1753 ± 40* |
| **CE** | **23** | **46** | **58** | **527** | **5.6** | **1772 ± 9** | **1773 ± 16** | **1775 ± 14** | **1775 ± 10** |
| LsR | 47 | 51 | 65 | 1100 | 4.3 | 1786 ± 7 | 1794 ± 12 | 1803 ± 10 | 1804 ± 10 |
| MsR | 21 | 16 | 262 | 4380 | 5.3 | 1277 ± 5 | 1481 ± 8 | 1788 ± 8 | *1813 ± 40* |
| ME | 13 | 15 | 208 | 3390 | 7.0 | 1804 ± 8 | 1816 ± 10 | 1829 ± 7 | 1829 ± 10 |
| ME | 10 | 10 | 183 | 3050 | 5.6 | 1615 ± 8 | 1708 ± 11 | 1825 ± 7 | 1833 ± 20 |
| LsR | 26 | 18 | 263 | 6270 | 11.6 | 1660 ± 9 | 1736 ± 11 | 1830 ± 7 | 1836 ± 20 |
| **LsR** | **31** | **10** | **149** | **8620** | **7.4** | **1835 ± 8** | **1838 ± 10** | **1841 ± 6** | **1842 ± 10** |
| LsR | 21 | 5 | 138 | 11780 | 8.7 | 1828 ± 8 | 1836 ± 10 | 1844 ± 6 | 1845 ± 10 |
| LsR | 26 | 33 | 129 | 1930 | 8.0 | 1836 ± 7 | 1840 ± 10 | 1846 ± 6 | 1846 ± 10 |
| LsR | 16 | 3 | 293 | 3150 | 9.6 | 1796 ± 11 | 1822 ± 15 | 1852 ± 9 | 1853 ± 10 |
| ME | 13 | 10 | 163 | 4110 | 8.2 | 1818 ± 8 | 1837 ± 12 | 1858 ± 8 | 1859 ± 10 |
| LsR | 24 | 12 | 155 | 5650 | 11.1 | 1838 ± 9 | 1849 ± 12 | 1862 ± 7 | 1863 ± 10 |
| DE | 12 | 15 | 455 | 2350 | 5.6 | 667 ± 3 | 1023 ± 6 | 1889 ± 7 | *2008 ± 40* |
| *MsR* | *13* | *11* | *173* | *3660* | *6.9* | *1676 ± 7* | *2067 ± 12* | *2482 ± 6* | *2507 ± 40* |
| *MsR* | *21* | *10* | *358* | *14810* | *7.2* | *1872 ± 8* | *2197 ± 12* | *2516 ± 4* | *2534 ± 40* |
| ME | 17 | 10 | 301 | 13430 | 9.7 | 2442 ± 12 | 2501 ± 15 | 2549 ± 5 | 2551 ± 20 |
| CE | 24 | 11 | 59 | 3940 | 2.2 | 2657 ± 13 | 2670 ± 15 | 2679 ± 5 | 2680 ± 10 |
| CE | 13 | 9 | 103 | 4600 | 5.1 | 2640 ± 13 | 2668 ± 15 | 2690 ± 4 | 2691 ± 10 |
| *MsR* | *13* | *13* | *177* | *3350* | *7.3* | *1835 ± 8* | *2286 ± 12* | *2719 ± 5* | *2742 ± 40* |
| *MsR* | *18* | *39* | *197* | *1740* | *6.8* | *1908 ± 8* | *2336 ± 12* | *2734 ± 5* | *2755 ± 40* |
| **LsR** | **66** | **485** | **121** | **515** | **3.4** | **2813 ± 20** | **2814 ± 27** | **2815 ± 9** | **2815 ± 10** |
| Lower Vinini #1 (39°52'57"N, 116°15'40"W) | | | | | | | | | |
| **LwR** | **25** | **9** | **227** | **2750** | **3.5** | **484 ± 3** | **181 ± 4** | **486 ± 12** | **485 ± 10** |
| **CwR** | **27** | **12** | **146** | **3515** | **6.9** | **1064 ± 4** | **1065 ± 6** | **1066 ± 7** | **1065 ± 10** |
| **CwR** | **28** | **10** | **233** | **8280** | **5.7** | **1274 ± 9** | **1277 ± 10** | **1283 ± 6** | **1283 ± 10** |
| LwR | 20 | 5 | 212 | 2220 | 7.2 | 1368 ± 17 | 1388 ± 20 | 1418 ± 10 | 1420 ± 20 |
| CwR | 27 | 11 | 114 | 3930 | 4.9 | 1396 ± 9 | 1409 ± 10 | 1427 ± 6 | 1428 ± 10 |
| CwR | 27 | 38 | 110 | 1140 | 4.4 | 1421 ± 6 | 1425 ± 10 | 1430 ± 10 | 1430 ± 10 |
| LwR | 28 | 25 | 128 | 2550 | 9.1 | 1699 ± 7 | 1705 ± 9 | 1711 ± 7 | 1712 ± 10 |
| LwR | 32 | 12 | 156 | 7270 | 5.3 | 1678 ± 11 | 1694 ± 12 | 1716 ± 5 | 1717 ± 10 |
| LwR | 17 | 3 | 37 | 4220 | 11.0 | 1666 ± 14 | 1689 ± 16 | 1717 ± 8 | 1719 ± 10 |
| LwR | 14 | 4 | 77 | 4680 | 11.6 | 1666 ± 12 | 1689 ± 14 | 1717 ± 8 | 1719 ± 10 |
| LwR | 26 | 49 | 260 | 2420 | 5.3 | 1697 ± 10 | 1707 ± 12 | 1718 ± 7 | 1719 ± 10 |
| LwR | 27 | 10 | 136 | 6910 | 10.6 | 1690 ± 9 | 1704 ± 10 | 1720 ± 5 | 1721 ± 10 |
| YwR | 20 | 74 | 1185 | 5040 | 19.3 | 1565 ± 16 | 1632 ± 18 | 1719 ± 6 | 1725 ± 20 |
| **CwR** | **29** | **10** | **179** | **9840** | **5.7** | **1730 ± 11** | **1729 ± 12** | **1728 ± 3** | **1730 ± 10** |
| LwR | 29 | 7 | 81 | 6060 | 4.1 | 1707 ± 8 | 1718 ± 10 | 1730 ± 6 | 1731 ± 10 |
| **CwR** | **21** | **6** | **153** | **9970** | **6.9** | **1731 ± 9** | **1734 ± 11** | **1737 ± 5** | **1737 ± 10** |
| LwR | 30 | 12 | 68 | 3190 | 3.8 | 1730 ± 8 | 1739 ± 10 | 1750 ± 7 | 1751 ± 10 |
| YwR | 16 | 13 | 105 | 2150 | 3.5 | 1633 ± 12 | 1684 ± 14 | 1748 ± 7 | 1752 ± 20 |
| LwR | 18 | 53 | 99 | 585 | 4.7 | 1663 ± 12 | 1707 ± 20 | 1761 ± 16 | 1765 ± 20 |
| LwR | 14 | 6 | 107 | 7540 | 3.6 | 1742 ± 11 | 1751 ± 13 | 1764 ± 7 | 1765 ± 10 |
| LwR | 27 | 9 | 206 | 10800 | 10.4 | 1568 ± 6 | 1661 ± 8 | 1780 ± 4 | 1789 ± 20 |
| CwR | 23 | 12 | 218 | 5870 | 6.3 | 1352 ± 5 | 1524 ± 8 | 1771 ± 6 | 1791 ± 40 |
| LwR | 16 | 4 | 56 | 3820 | 9.8 | 1711 ± 18 | 1748 ± 22 | 1791 ± 11 | 1794 ± 10 |
| LwR | 12 | 5 | 118 | 4690 | 8.3 | 1636 ± 10 | 1708 ± 13 | 1797 ± 8 | 1804 ± 20 |
| MwR | 11 | 3 | 32 | 2530 | 14.6 | 1805 ± 16 | 1814 ± 19 | 1824 ± 11 | 1824 ± 10 |
| **CwR** | **25** | **8** | **73** | **4610** | **13.5** | **1827 ± 8** | **1826 ± 11** | **1825 ± 6** | **1825 ± 10** |
| LwR | 8 | 7 | 104 | 2320 | 5.1 | 1767 ± 15 | 1797 ± 19 | 1832 ± 11 | 1835 ± 10 |
| LwR | 27 | 37 | 139 | 2030 | 6.5 | 1925 ± 8 | 1960 ± 11 | 1996 ± 6 | 1999 ± 10 |
| LwR | 31 | 13 | 189 | 12180 | 9.4 | 2381 ± 15 | 2421 ± 15 | 2456 ± 3 | 2457 ± 10 |
| Lower Vinini #2 (39°10'59"N, 116°47'26"W) | | | | | | | | | |
| CwR | 20 | 70 | 130 | 360 | 5.2 | 892 ± 34 | 919 ± 39 | 984 ± 33 | 992 ± 40 |
| LwR | 16 | 7 | 41 | 990 | 8.5 | 1069 ± 14 | 1094 ± 17 | 1142 ± 14 | 1147 ± 20 |
| CwR | 19 | 4 | 19 | 1340 | 5.5 | 1291 ± 14 | 1342 ± 16 | 1425 ± 10 | 1432 ± 20 |
| LwR | 23 | 4 | 116 | 10750 | 9.2 | 1634 ± 7 | 1650 ± 9 | 1670 ± 6 | 1672 ± 10 |
| MwR | 19 | 5 | 230 | 16100 | 17.4 | 1482 ± 8 | 1561 ± 9 | 1670 ± 4 | 1679 ± 20 |
| CwR | 18 | 6 | 52 | 2620 | 13.7 | 1562 ± 13 | 1613 ± 15 | 1681 ± 8 | 1686 ± 10 |
| CwR | 14 | 9 | 65 | 1600 | 6.3 | 1532 ± 13 | 1597 ± 15 | 1682 ± 8 | 1688 ± 20 |

**TABLE 1. U-Pb ISOTOPIC DATA AND AGES (continued)**

| Grain type | Grain wt. (μg) | $Pb_c$ (pg) | U (ppm) | $^{206}Pb_m/^{204}Pb$ | $^{206}Pb_c/^{208}Pb$ | Apparent ages (Ma) $^{206}Pb^*/^{238}U$ | | Apparent ages (Ma) $^{207}Pb^*/^{235}U$ | | Apparent ages (Ma) $^{207}Pb^*/^{206}Pb^*$ | | Interpreted age (Ma) | |
|---|---|---|---|---|---|---|---|---|---|---|---|---|---|
| MwR | 22 | 3 | 81 | 8200 | 4.9 | 1586 | ± 9 | 1640 | ± 11 | 1710 | ± 6 | 1715 | ± 20 |
| LwR | 24 | 5 | 51 | 1800 | 3.5 | 1407 | ± 16 | 1534 | ± 19 | 1715 | ± 8 | 1729 | ± 20 |
| LwR | 18 | 7 | 90 | 3250 | 8.7 | 1659 | ± 11 | 1709 | ± 13 | 1772 | ± 7 | 1776 | ± 10 |
| **CwR** | **18** | **7** | **59** | **3230** | **4.8** | **1827** | **± 13** | **1831** | **± 15** | **1837** | **± 7** | **1838** | **± 10** |
| **LwR** | **32** | **5** | **134** | **18100** | **3.3** | **1858** | **± 11** | **1857** | **± 13** | **1856** | **± 6** | **1856** | **± 10** |
| Upper Vinini (39°10'59"N, 116°47'26"W) | | | | | | | | | | | | | |
| **LwR** | **30** | **5** | **82** | **2650** | **2.9** | **1025** | **± 9** | **1024** | **± 10** | **1021** | **± 10** | **1025** | **± 10** |
| **LwR** | **28** | **3** | **96** | **4930** | **2.5** | **1034** | **± 7** | **1037** | **± 9** | **1041** | **± 10** | **1042** | **± 10** |
| CwR | 30 | 3 | 143 | 14500 | 6.9 | 1768 | ± 9 | 1792 | ± 11 | 1821 | ± 6 | 1823 | ± 10 |
| **LwR** | **38** | **9** | **185** | **16200** | **4.3** | **1824** | **± 9** | **1825** | **± 11** | **1825** | **± 6** | **1825** | **± 10** |
| **LwR** | **26** | **8** | **94** | **6400** | **1.1** | **1824** | **± 10** | **1826** | **± 13** | **1828** | **± 8** | **1828** | **± 10** |
| **CwR** | **25** | **4** | **203** | **28100** | **2.1** | **1841** | **± 11** | **1846** | **± 12** | **1851** | **± 4** | **1851** | **± 10** |
| CwR | 31 | 6 | 49 | 5320 | 6.8 | 1846 | ± 11 | 1853 | ± 13 | 1860 | ± 7 | 1861 | ± 10 |
| **LwR** | **31** | **8** | **148** | **9900** | **5.9** | **1909** | **± 10** | **1913** | **± 12** | **1919** | **± 6** | **1919** | **± 10** |
| MwR | 36 | 6 | 172 | 21800 | 1.9 | 1890 | ± 11 | 1906 | ± 12 | 1924 | ± 4 | 1925 | ± 10 |
| MwR | 30 | 6 | 209 | 24100 | 4.6 | 1885 | ± 11 | 1906 | ± 12 | 1930 | ± 4 | 1932 | ± 10 |
| **LwR** | **32** | **6** | **84** | **10710** | **4.1** | **1930** | **± 12** | **1936** | **± 14** | **1943** | **± 6** | **1945** | **± 10** |
| DwR | 30 | 9 | 508 | 33000 | 4.1 | 1896 | ± 10 | 1921 | ± 12 | 1947 | ± 6 | 1949 | ± 10 |
| **DwR** | **36** | **9** | **243** | **28500** | **7.3** | **2669** | **± 17** | **2669** | **± 19** | **2668** | **± 5** | **2668** | **± 10** |
| **CwR** | **28** | **4** | **39** | **8400** | **7.4** | **2747** | **± 13** | **2750** | **± 14** | **2753** | **± 4** | **2753** | **± 10** |
| DwR | 34 | 7 | 164 | 24900 | 3.7 | 2751 | ± 16 | 2796 | ± 18 | 2829 | ± 5 | 2831 | ± 20 |
| McAfee Quartzite (41°32'34"N, 115°54'48"W) | | | | | | | | | | | | | |
| **LwR** | **19** | **13** | **134** | **2040** | **3.5** | **999** | **± 5** | **1000** | **± 8** | **1004** | **± 12** | **1004** | **± 10** |
| **LwR** | **23** | **17** | **503** | **6770** | **7.8** | **1026** | **± 4** | **1027** | **± 5** | **1027** | **± 7** | **1027** | **± 10** |
| *YwR* | *19* | *31* | *1321* | *11440* | *5.1* | *1390* | *± 8* | *1541* | *± 10* | *1756* | *± 5* | *1773* | *± 40* |
| **YwR** | **22** | **22** | **514** | **9820** | **36.1** | **1793** | **± 7** | **1794** | **± 9** | **1794** | **± 6** | **1794** | **± 10** |
| YwR | 16 | 13 | 179 | 4300 | 2.4 | 1790 | ± 7 | 1796 | ± 10 | 1804 | ± 7 | 1805 | ± 10 |
| CwR | 62 | 49 | 315 | 7300 | 11.7 | 1804 | ± 7 | 1809 | ± 10 | 1815 | ± 6 | 1816 | ± 10 |
| CwR | 50 | 19 | 121 | 5580 | 4.3 | 1766 | ± 9 | 1790 | ± 11 | 1819 | ± 6 | 1820 | ± 10 |
| CwR | 44 | 13 | 445 | 29200 | 9.5 | 1797 | ± 7 | 1807 | ± 9 | 1819 | ± 6 | 1820 | ± 10 |
| *CwR* | *16* | *31* | *437* | *3170* | *5.6* | *1410* | *± 7* | *1582* | *± 10* | *1819* | *± 5* | *1837* | *± 40* |
| CwR | 14 | 26 | 454 | 4730 | 6.9 | 1823 | ± 9 | 1839 | ± 10 | 1859 | ± 5 | 1860 | ± 10 |
| CwR | 34 | 15 | 389 | 16800 | 4.8 | 1846 | ± 7 | 1853 | ± 9 | 1861 | ± 6 | 1862 | ± 10 |
| LwR | 9 | 14 | 482 | 6400 | 12.8 | 1866 | ± 8 | 1876 | ± 11 | 1887 | ± 7 | 1889 | ± 10 |
| CwR | 38 | 8 | 139 | 13110 | 1.3 | 1884 | ± 10 | 1901 | ± 11 | 1919 | ± 4 | 1921 | ± 10 |
| MwR | 18 | 11 | 81 | 2950 | 2.8 | 1953 | ± 12 | 1967 | ± 14 | 1981 | ± 7 | 1982 | ± 10 |
| CwR | 44 | 9 | 485 | 56890 | 8.5 | 2081 | ± 13 | 2089 | ± 14 | 2095 | ± 4 | 2096 | ± 10 |
| YwR | 23 | 13 | 195 | 9100 | 7.7 | 2351 | ± 12 | 2377 | ± 14 | 2401 | ± 4 | 2402 | ± 10 |
| CwR | 35 | 15 | 273 | 17810 | 1.9 | 2461 | ± 13 | 2503 | ± 15 | 2536 | ± 5 | 2539 | ± 20 |
| CwR | 13 | 10 | 863 | 32050 | 13.9 | 2542 | ± 15 | 2563 | ± 17 | 2580 | ± 5 | 2581 | ± 10 |
| LwR | 14 | 13 | 380 | 11800 | 3.9 | 2535 | ± 15 | 2607 | ± 17 | 2663 | ± 4 | 2665 | ± 20 |
| MwR | 18 | 67 | 343 | 2570 | 3.5 | 2527 | ± 10 | 2612 | ± 14 | 2678 | ± 6 | 2681 | ± 20 |
| CwR | 42 | 16 | 157 | 12070 | 4.2 | 2638 | ± 14 | 2685 | ± 16 | 2721 | ± 5 | 2723 | ± 20 |
| LwR | 34 | 34 | 238 | 6650 | 7.2 | 2518 | ± 13 | 2962 | ± 18 | 3279 | ± 5 | *3292* | *± 40* |
| Snow Canyon Formation (41°25'08"N, 116°02'44"W) | | | | | | | | | | | | | |
| **LwR** | **22** | **5** | **193** | **7650** | **4.4** | **1028** | **± 5** | **1030** | **± 7** | **1032** | **± 8** | **1032** | **± 10** |
| **GEO** | **22** | **4** | **396** | **20050** | **7.3** | **1041** | **± 6** | **1040** | **± 7** | **1037** | **± 7** | **1040** | **± 10** |
| **CwR** | **35** | **7** | **50** | **5350** | **5.3** | **1839** | **± 11** | **1839** | **± 14** | **1839** | **± 7** | **1839** | **± 10** |
| **CwR** | **30** | **6** | **97** | **9600** | **7.0** | **1835** | **± 12** | **1832** | **± 14** | **1841** | **± 6** | **1842** | **± 10** |
| **LwR** | **28** | **5** | **29** | **3205** | **8.4** | **1861** | **± 15** | **1963** | **± 17** | **1866** | **± 8** | **1866** | **± 10** |
| **CwR** | **41** | **8** | **38** | **4050** | **3.9** | **1881** | **± 11** | **1882** | **± 15** | **1882** | **± 9** | **1882** | **± 10** |
| MwR | 36 | 11 | 129 | 8220 | 7.2 | 1849 | ± 12 | 1877 | ± 13 | 1907 | ± 5 | 1909 | ± 10 |
| LwR | 26 | 10 | 256 | 12300 | 2.9 | 1777 | ± 10 | 1849 | ± 12 | 1931 | ± 6 | 1937 | ± 20 |
| CwR | 30 | 10 | 48 | 2920 | 2.5 | 1947 | ± 10 | 1958 | ± 13 | 1969 | ± 7 | 1970 | ± 10 |
| **CwR** | **36** | **5** | **175** | **26100** | **16.2** | **1998** | **± 12** | **1999** | **± 14** | **2000** | **± 6** | **2000** | **± 10** |
| LwR | 24 | 10 | 295 | 14900 | 5.2 | 1970 | ± 13 | 2016 | ± 15 | 2062 | ± 6 | 2065 | ± 10 |
| CwR | 32 | 6 | 45 | 5100 | 4.2 | 2058 | ± 10 | 2064 | ± 13 | 2070 | ± 7 | 2070 | ± 10 |
| GEO | 18 | 45 | 756 | 6060 | 21.6 | 1913 | ± 12 | 2134 | ± 16 | 2354 | ± 6 | 2360 | ± 40 |
| CwR | 47 | 13 | 104 | 11030 | 7.1 | 2528 | ± 16 | 2543 | ± 18 | 2555 | ± 5 | 2560 | ± 10 |
| CwR | 34 | 13 | 220 | 17600 | 4.3 | 2659 | ± 17 | 2665 | ± 18 | 2670 | ± 4 | **2670** | **± 10** |

**TABLE 1. U-Pb ISOTOPIC DATA AND AGES (continued)**

| Grain type | Grain wt. (μg) | $Pb_c$ (pg) | U (ppm) | $^{206}Pb_m/^{204}Pb$ | $^{206}Pb_c/^{208}Pb$ | Apparent ages (Ma) $^{206}Pb^*/^{238}U$ | $^{207}Pb^*/^{235}U$ | $^{207}Pb^*/^{206}Pb^*$ | Interpreted age (Ma) |
|---|---|---|---|---|---|---|---|---|---|
| Elder Sandstone (40°17'34"N, 116°48'34"W) | | | | | | | | | |
| CwR | 23 | 9 | 122 | 6030 | 8.7 | 1796 ± 10 | 1803 ± 11 | 1812 ± 5 | 1813 ± 10 |
| CwR | 20 | 12 | 84 | 2580 | 11.6 | 1784 ± 8 | 1803 ± 11 | 1825 ± 6 | 1826 ± 10 |
| CwR | 15 | 8 | 110 | 4030 | 5.4 | 1810 ± 9 | 1819 ± 11 | 1829 ± 7 | 1830 ± 10 |
| **CwR** | **18** | **11** | **119** | **4040** | **9.1** | **1827 ± 9** | **1829 ± 12** | **1830 ± 7** | **1830 ± 10** |
| CwR | 24 | 64 | 170 | 1190 | 11.2 | 1793 ± 10 | 1809 ± 13 | 1829 ± 8 | 1830 ± 10 |
| **LwR** | **16** | **7** | **81** | **3570** | **3.5** | **1846 ± 10** | **1849 ± 14** | **1853 ± 7** | **1854 ± 10** |
| CwR | 13 | 8 | 87 | 2890 | 3.9 | 1839 ± 10 | 1847 ± 12 | 1855 ± 6 | 1856 ± 10 |
| LwR | 27 | 23 | 357 | 8040 | 3.0 | 1837 ± 11 | 1846 ± 12 | 1857 ± 5 | 1858 ± 10 |
| LwR | 28 | 16 | 141 | 4840 | 8.1 | 1843 ± 9 | 1853 ± 11 | 1864 ± 6 | 1865 ± 10 |
| **LwR** | **27** | **9** | **125** | **7310** | **7.2** | **1866 ± 10** | **1868 ± 11** | **1871 ± 5** | **1871 ± 10** |
| CwR | 13 | 9 | 125 | 3840 | 8.4 | 1864 ± 9 | 1870 ± 12 | 1876 ± 7 | 1877 ± 10 |
| **CwR** | **16** | **16** | **416** | **8610** | **6.1** | **1887 ± 10** | **1891 ± 11** | **1896 ± 5** | **1896 ± 10** |
| CwR | 15 | 7 | 59 | 2860 | 2.2 | 1932 ± 12 | 1944 ± 14 | 1957 ± 5 | 1958 ± 10 |
| **CwR** | **13** | **8** | **43** | **1580** | **1.6** | **2060 ± 17** | **2063 ± 20** | **2066 ± 8** | **2066 ± 20** |
| **CwR** | **11** | **12** | **321** | **6550** | **7.4** | **2070 ± 10** | **2071 ± 12** | **2075 ± 5** | **2075 ± 10** |
| CwR | 16 | 6 | 222 | 12950 | 3.9 | 2050 ± 10 | 2064 ± 12 | 2076 ± 4 | 2077 ± 10 |
| MwR | 23 | 4 | 58 | 6880 | 4.7 | 2186 ± 9 | 2197 ± 10 | 2207 ± 5 | 2208 ± 10 |
| CwR | 16 | 25 | 232 | 4480 | 4.5 | 2673 ± 14 | 2686 ± 16 | 2697 ± 4 | 2698 ± 10 |
| MwR | 19 | 9 | 221 | 14990 | 2.0 | 2625 ± 13 | 2667 ± 14 | 2698 ± 4 | 2700 ± 10 |
| MwR | 19 | 8 | 158 | 12610 | 7.6 | 2850 ± 14 | 2858 ± 16 | 2864 ± 4 | 2866 ± 10 |
| MwR | 14 | 45 | 322 | 3420 | 3.2 | 2947 ± 18 | 2996 ± 19 | 3030 ± 4 | 3031 ± 20 |
| Slaven Chert (40°27'51"N, 116°46'15"W) | | | | | | | | | |
| **LmR** | **27** | **12** | **106** | **2620** | **7.6** | **1027 ± 7** | **1027 ± 9** | **1026 ± 10** | **1027 ± 10** |
| TOmR | 17 | 23 | 817 | 9660 | 6.8 | 1524 ± 11 | 1638 ± 12 | 1789 ± 5 | 1800 ± 40 |
| **CmR** | **26** | **24** | **42** | **895** | **4.0** | **1797 ± 11** | **1799 ± 16** | **1802 ± 11** | **1803 ± 10** |
| LmR | 11 | 7 | 490 | 16800 | 8.3 | 1806 ± 9 | 1814 ± 10 | 1822 ± 4 | 1823 ± 10 |
| CE | 15 | 34 | 192 | 1620 | 11.9 | 1815 ± 10 | 1821 ± 12 | 1827 ± 6 | 1827 ± 10 |
| CmR | 25 | 39 | 131 | 1620 | 4.4 | 1801 ± 8 | 1818 ± 12 | 1836 ± 8 | 1837 ± 10 |
| **CmR** | **20** | **16** | **77** | **1850** | **5.6** | **1829 ± 11** | **1833 ± 14** | **1836 ± 9** | **1837 ± 10** |
| **CE** | **20** | **13** | **87** | **2620** | **5.3** | **1840 ± 12** | **1839 ± 14** | **1839 ± 7** | **1839 ± 10** |
| CE | 22 | 22 | 83 | 4100 | 4.1 | 1835 ± 9 | 1839 ± 11 | 1842 ± 6 | 1843 ± 10 |
| CmR | 18 | 8 | 446 | 21850 | 7.4 | 1843 ± 11 | 1848 ± 12 | 1853 ± 4 | 1854 ± 10 |
| CE | 21 | 9 | 123 | 5730 | 4.5 | 1837 ± 8 | 1846 ± 10 | 1857 ± 5 | 1857 ± 10 |
| CmR | 18 | 9 | 239 | 10100 | 8.0 | 1874 ± 10 | 1888 ± 11 | 1904 ± 4 | 1905 ± 10 |
| CE | 17 | 10 | 80 | 2920 | 3.7 | 1882 ± 10 | 1898 ± 12 | 1915 ± 5 | 1916 ± 10 |
| CmR | 19 | 8 | 102 | 4860 | 4.1 | 1910 ± 8 | 1916 ± 11 | 1923 ± 6 | 1923 ± 10 |
| LE | 9 | 5 | 333 | 11970 | 15.2 | 1787 ± 10 | 1878 ± 11 | 1981 ± 4 | 1988 ± 20 |
| LE | 10 | 7 | 66 | 2740 | 3.6 | 2423 ± 15 | 2454 ± 16 | 2482 ± 3 | 2483 ± 10 |
| LmR | 22 | 11 | 47 | 2710 | 5.9 | 2460 ± 12 | 2487 ± 14 | 2508 ± 4 | 2509 ± 10 |
| LE | 11 | 7 | 203 | 8490 | 4.5 | 2412 ± 13 | 2503 ± 15 | 2576 ± 4 | 2580 ± 20 |

*Note:*
Grain with $^{206}Pb^*/^{238}U$ age not within 15% of $^{207}Pb^*/^{206}Pb^*$ age in italics
Grain with $^{206}Pb^*/^{238}U$ age within 15% OF $^{207}Pb^*/^{206}Pb^*$ age
Grain with concordant age and moderate to high precision analysis in bold.
* radiogenic Pb
Grain type: L—light pink; M—medium pink; D—dark pink; C—colorless; Y—yellow; T—tan; G—gray
sR—slightly rounded; mR—moderately rounded; wR—well rounded; E—euhedral; O—cloudy
Unless otherwise noted, all grains are clear or translucent and abraded with an air abrasion device.
$^{206}Pb/^{204}Pb$ is measured ratio, uncorrected for blank, spike, or fractionation.
$^{206}Pb/^{208}Pb$ is corrected for blank, spike, and fractionation.
Most concentrations have an uncertainty of up to 25% due to uncertainty in weight of grain.
Constants used: $^{238}U/^{235}U$ = 137.88. Decay constant for $^{235}U = 9.8485 \times 10^{-10}$. Decay constant for $^{238}U = 1.55125 \times 10^{-10}$.
All uncertainties are at the 95% confidence level.
Pb blank ranged from 2 to 10 pg. U blank was consistently <1 pg.
Interpreted ages for concordant grains are $^{206}Pb^*/^{238}U$ ages if <1.0 Ga and $^{207}Pb^*/^{206}Pb^*$ ages if >1.0 Ga.
Interpreted ages for discordant grains are projected from 100 Ma.
All analyses conducted using conventional isotope dilution and thermal ionization mass spectrometry, as described by Gehrels (this volume, Introduction).

This sample yielded 160 mg of zircon, with the largest grains reaching ~145 μm. Most of these grains are colorless; there are subordinate light pinkish grains. Most of the grains in the sample are moderately rounded and slightly spherical; euhedral grains are subordinate. In contrast, zircons from most other Harmony Formation samples display only minor degrees of rounding.

Of the 11 grains analyzed, most are concordant or slightly discordant (Fig. 5). Most grains yield ages between 1073 and 1312 Ma, with a peak in age probability at 1092 Ma (Fig. 8), and there are additional ages of ca. 1711 and 2959 Ma (Table 1). The ages from this sample are similar to one of the two main age populations reported from Little Cottonwood #1 by Smith and Gehrels (1994). The other population obtained from Little Cottonwood #1, 690–710 Ma, was not recognized in our sample.

### *Little Cottonwood #3*

We collected ~8 kg of coarse-grained feldspathic wacke from the ~2172 m level of the road up Little Cottonwood Canyon (Fig. 3B). The sample is from a thick bed of coarse-grained sandstone with large bluish quartz pebbles and large angular feldspars. We recovered ~140 mg of zircons, many of which are >175 μm in sieve size. Most are pinkish to purplish, with subordinate colorless grains, and only a slight degree of rounding on crystal edges and tips is apparent. Most grains are euhedral.

We analyzed 11 grains, most of which yielded slightly to moderately discordant ages (Fig. 5). Three apparent age groups of 1828–1947 (n = 3), 2299–2373 (n = 2), and 2618–2840 (n = 6) Ma (Table 1) are present. Age-probability peaks occur at 1936, 2627, and 2683 Ma (Fig. 8).

### *Little Cottonwood #4*

We collected ~15 kg of coarse feldspathic wacke from the ~2340m level of the road up Little Cottonwood Canyon (Fig. 3B), ~300 m from the location of the Little Cottonwood #3 sample described herein. The sample belongs to the subarkosic petrofacies (Dickinson and Gehrels, this volume, Chapter 11) and consists of moderately rounded quartz and feldspar grains to 5 mm in diameter, with rare bluish quartz grains and feldspar grains that are up to 1 cm in diameter.

The zircon yield was 380 mg, with a large number of grains >175 μm in sieve size. Most are slightly to moderately rounded, and some are euhedral. Pinkish grains are considerably more abundant than colorless grains.

Seven grains were analyzed, six of which are concordant to slightly discordant, and one of which is highly discordant (Fig. 5). Most of the grains are between 1772 and 1879 Ma, with two additional ages of ca. 2523 and 3215 Ma (Table 1).

### *Elbow Canyon*

This sample was collected from a prominent outcrop on the north side of Elbow Canyon in the Sonoma Range (Fig. 3B). We collected ~15 kg of sandstone from a 10-cm-thick quartzose layer at the top of a graded bed of more typical coarse-grained wacke. The sample consisted of well-rounded quartz and subordinate feldspar grains to 3 mm in diameter.

This sample yielded 1.03 g of zircon, with a few >175 μm grains and abundant grains between 145 and 175 μm. Most grains are euhedral, ~25% showing slight rounding of crystal edges and tips. A small proportion of the grains in the sample are colorless; most are light to medium pink. Most grains yield slightly to moderately discordant analyses, with one highly discordant Archean grain (Table 1; Fig. 5).

We analyzed 11 grains, with resulting ages belonging to two groups; 1770–1843 Ma (n = 4) and 2532–2713 Ma (n = 7). Age-probability peaks are at 1774, 1840, 2629, and 2667 Ma (Fig. 8).

### *Harmony Canyon*

Our sample consisted of ~15 kg of coarse-grained sandstone collected from a prominent 2-m-thick sandstone bed on the north side of Harmony Canyon in the Sonoma Range (Fig. 3). The sandstone consists of slightly to moderately rounded quartz and feldspar with an average diameter of 2 mm, but bluish quartz pebbles to 8 mm in diameter are also present. The sample yielded 740 mg of zircons, with abundant >175 μm grains. Most of the grains are euhedral or are broken along ragged fractures; a slight amount of rounding is apparent on ~25% of the grains. Colorless and light- to medium-pinkish–purple grains are subequal in abundance.

There were 11 grains that yielded a range of concordant to slightly discordant analyses, with one highly discordant grain (Table 1; Fig. 5). The resulting ages are mainly between 1768 and 1910 Ma (n = 7); four other grains are between 2414 and 2672 Ma (Table 1). The only age-probability peak defined by more than one grain is at 1838 Ma (Fig. 8).

This sample was collected in the same vicinity as the sample analyzed by Wallin (1990), who reported single-grain ages of ca. 691, 1336, 1752, and 1772 Ma. The older of these ages are consistent with our analyses, but the younger two grains were derived from a source not represented in our sample.

### *Hot Springs Range*

We collected ~20 kg of coarse-grained subarkose from a 2-m-thick bed near the microwave tower at the crest of the Hot Springs Range (Fig. 2). Quartz grains in the sample are as much as 6 mm in diameter, and some feldspars are 1 cm. Very little rounding and sorting is apparent. Most other sandstone beds in the area are similar in thickness and grading, but somewhat finer in grain size.

The sample yielded 360 mg of zircon; there were abundant grains >175 μm in sieve size. Most grains are pinkish and purplish; there are subordinate colorless grains. Many of the grains show a slight amount of rounding of crystal edges and tips, but crystal faces are well preserved. Others are perfectly euhedral, with no sign of rounding or abrasion. We analyzed 22 grains including each of the color and/or morphology groups. Most of the analyses are concordant to moderately discordant and a few are highly discordant (Table 1; Fig. 5).

The resulting ages define two distinct groups, 1753–1863 Ma (n = 13) and 2507–2815 Ma (n = 8) (Table 1). As shown in Figure 8, there are peaks in age probability at 1775, 1804, 1845, 2549, 2689, and 2738 Ma.

### *Vinini, Valmy, Snow Canyon, and McAfee formations (Ordovician)*

We analyzed 91 individual zircon grains from 6 samples of the Valmy Formation, Vinini Formation, Snow Canyon Formation, and McAfee Quartzite. These ages complement the previously reported single-grain ages from a sample of the Valmy Formation (Smith and Gehrels, 1994), which are summarized in Figure 8. The following is a discussion of each of the samples analyzed in this study.

#### *Lower Vinini #1*

We collected ~20 kg of mature quartz arenite from a 1-m-thick sandstone bed in the Vinini Creek area of the Roberts Mountains (Fig. 3). This sandstone occurs near the top of an ~2-km-thick sequence of predominantly sandstone in the lower part of the Vinini Formation. The horizon sampled is known to be early Whiterockian (early Middle Ordovician) on the basis of graptolites in interbedded shale (Finney and Perry, 1991; Finney et al., 1997. Petrographically, the sandstone is nearly pure quartz, with moderately rounded grains to 1.5 mm in diameter.

We recovered ~430 mg of zircon. Zircons in the sample are generally <125 µm in diameter, and all grains are moderately to well rounded and moderately spherical. The sample consists of approximately subequal colorless and light pink populations, and yellowish and darker pink grains are subordinate.

We analyzed 29 grains from this sample. Most analyses are concordant to slightly discordant (Fig. 6). Resulting ages are mainly between 1712 and 1999 Ma (n = 22), with three additional grains between 1420 and 1430 Ma, and single grains of ca. 485, 1065, and 1283 Ma (Table 1). Peaks in age probability are present at 1428, 1719, 1765, 1794, and 1825 Ma (Fig. 8).

#### *Lower Vinini #2*

We collected ~15 kg of mature quartz arenite from a site ~300 m east of Petes Summit in the Toquima Range (Fig. 3). The sample was collected from an ~20-cm-thick bed of quartz arenite in a sequence of interbedded sandstone, siltstone, and shale. The age of the unit is early Whiterockian (early Middle Ordovician) from graptolites in interbedded sediments.

Most of the zircon grains in this sample are <145 µm, and a total of 300 mg of zircon was recovered. The grains are predominantly light pink, and colorless, medium pink, and yellowish populations are subordinate. All grains recovered are moderately to well rounded, moderately spherical, and the surfaces of the grains are slightly to moderately polished.

We analyzed 12 zircon grains, and most yield slightly to moderately discordant ages (Fig. 6). The main age group is 1672–1856 Ma (n = 9), and individual grains have ages of ca. 992, 1147, and 1432 Ma. Age-probability peaks are present at 1656, 1672, and 1722 Ma (Fig. 8).

#### *Upper Vinini*

We collected 15 kg of mature quartz arenite from the main quartzite layer that holds up the ridge at Petes Summit in the Toquima Range (Fig. 3). The layer is ~10 m thick and consists of very well sorted, medium-grained quartzite. Its age is constrained by graptolites as uppermost Whiterockian to lowest Mohawkian (lowest Upper Ordovician).

The sample yielded a total of 260 mg of zircon, with abun-

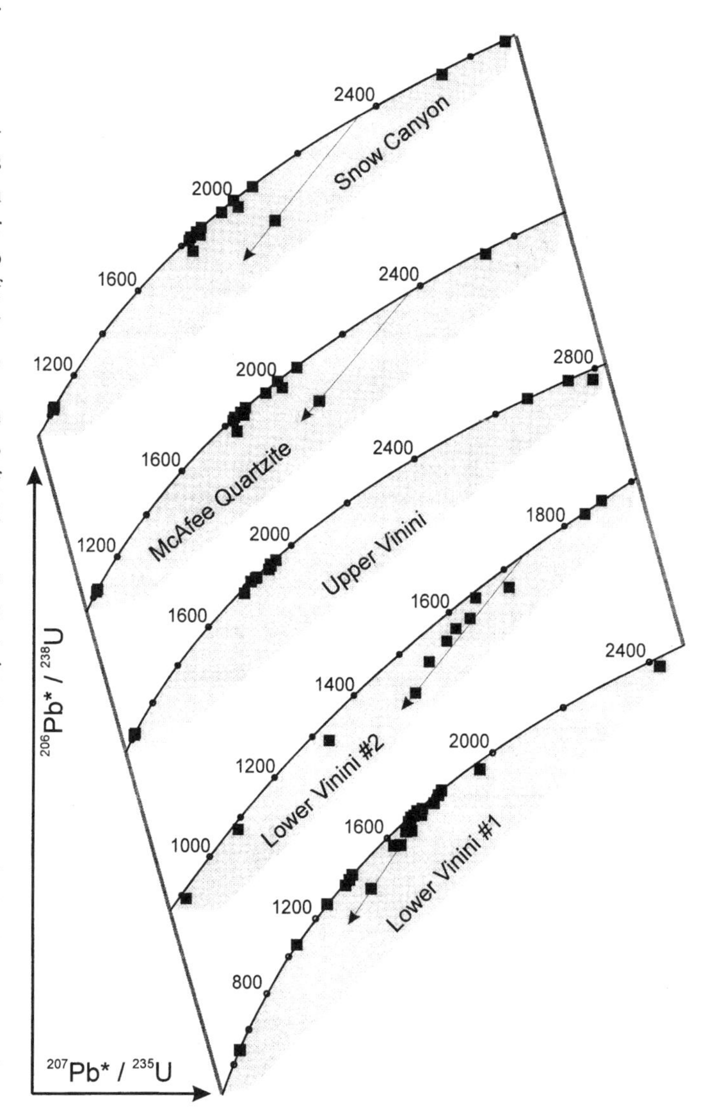


Figure 6. U-Pb concordia diagrams showing single-grain detrital zircon analyses from Ordovician strata of Valmy, Vinini, Snow Canyon, and McAfee Formations. Discordia lines projected to 100 Ma are shown for reference.

dant grains between 145 and 175 µm and a few grains >175 µm in sieve size. Most grains are colorless, but subordinate grains are yellowish and pinkish. All grains are well rounded, highly polished, and highly spherical.

There were 15 grains analyzed from this sample that yielded concordant to slightly discordant analyses (Fig. 6). The grains define three age groups, 1025–1042 (n = 2), 1823–1949 (n = 10), and 2668–2831 (n = 3) Ma. Age-probability peaks occur at 1825, 1924, and 1947 Ma (Fig. 8).

### *McAfee Quartzite*

We collected ~20 kg of quartz arenite from the northeasternmost exposure of a rib of quartz arenite along the east flank of the northern Independence Mountains (Fig. 3). The bed sampled is massive and at least 8 in thick. The sandstone is an ultramature quartz arenite, with sutured quartz grains to 1.5 mm in diameter. Early Mohawkian (early Late Ordovician) graptolites have been recovered from shales bounding the massive quartz arenite layer.

We recovered ~130 mg of zircon from the sample. Only a few grains are >145 µm, but 125–145 µm grains are abundant. Most grains are colorless, a small number are light pink, and a few grains are grayish and cloudy. All grains are extremely well rounded, well polished, and highly spherical, and no crystal faces are preserved on any grains.

Of 22 grains analyzed, most yielded concordant to slightly discordant ages (Fig. 6). Most ages are in three groups: 1004–1027 (n = 2), 1773–1982 (n = 12), and 2402–3292 (n = 7) Ma (Table 1). Age-probability peaks defined by two or more grains occur at 1819 and 1861 Ma (Fig. 8).

### *Snow Canyon Formation*

About 15 kg of mature quartz arenite was collected from a quartzite bed in the upper part of the Snow Canyon Formation exposed along the north side of the entrance to Snow Canyon on the west side of the northern Independence Mountains (Fig. 3). The quartzite is ~200 m stratigraphically below massive quartzite layers of the McAfee Quartzite. Mainly chert and argillite separate the two units. The age of this unit is not well constrained, but on the basis of stratigraphic position and regional stratigraphic patterns, it is considered to be correlative with sandstones in the lower part of the Vinini Formation in the Roberts Mountains and Toquima Range.

This sample yielded 350 mg of zircon, and only a few grains were greater than 145 µm in sieve size. Most grains are colorless, and a few are yellowish and light pink. All grains are well rounded, highly polished, and highly spherical.

Most of the 15 grains analyzed are concordant to slightly discordant (Fig. 6). Four general age groups are present, 1032–1040 (n = 2), 1839–2000 (n = 8), 2065–2070 (n = 2), and 2360–2670 (n = 3 ) Ma. Peaks in age probability occur at 1036, 1840, and 2067 Ma (Fig. 8).

### *Elder Sandstone (Silurian)*

One sample of the Elder Sandstone, from the Elder Creek area of the Shoshone Range (Fig. 3), was collected from a 50-cm-thick brownish sandstone bed in an area characterized by uniform lithology but disrupted and discontinuous bedding. We processed 20 kg of quartzose wacke, comprising well-rounded quartz grains to 3 mm in diameter, for zircons; 220 mg of zircon were recovered, with abundant grains in the 145–175 µm size range and a few dozen grains >175 µm. All of the grains are well rounded, highly polished, and moderately spherical. Colorless and light pink grains are subequal in abundance, and a small proportion of darker pink grains is present.

Of 21 grains analyzed, most are concordant to slightly discordant (Fig. 7). The resulting ages range from 1813 to 1958 Ma (n = 13), 2066 to 2077 Ma (n = 3), and 2698 to 3031 Ma (n = 4). Peaks in age probability are present at 1829, 1857, 2075, and 2699 Ma (Fig. 8).

### *Slaven Chert (Devonian)*

One sample of sandstone in the Slaven Chert was collected from the west side of Slaven Canyon in the Shoshone Range (Fig. 3). It was collected from a 20-cm-thick sandstone lens interbedded with layered black chert that is characteristic of much of the unit in the region (Gilluly and Gates, 1965). We

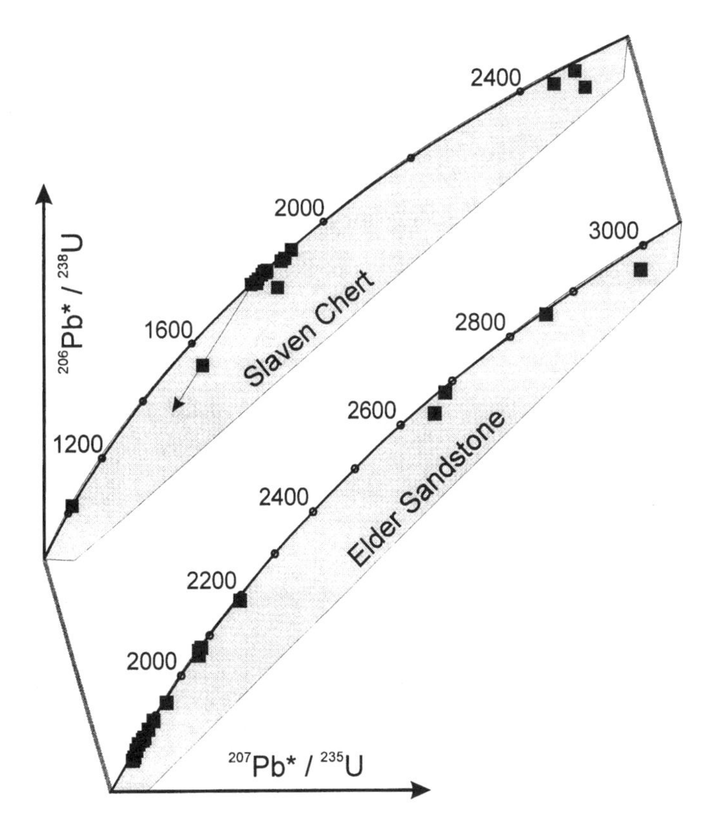


Figure 7. U-Pb concordia diagrams showing single-grain detrital zircon analyses from Silurian strata of Elder Sandstone and Devonian strata of Slaven Chert. Discordia line projected to 100 Ma is shown for reference.

recovered 470 mg of zircon from 20 kg of medium-grained, pure quartz wacke. Quartz grains in the sandstone are all very well rounded and to 3 mm in diameter, and most of the zircons are also very well rounded, highly polished, and highly spherical. A small number of grains are euhedral, with well-preserved crystal faces. Few grains >145 μm in sieve size were recovered.

The 18 grains analyzed from this sample yielded concordant to slightly discordant analyses (Fig. 7). Resulting ages are 1800–1857 Ma (n = 10) and 1905–1988 Ma (n = 4), with additional ages of ca. 1027, 2483, 2509, and 2580 Ma (Table 1). Age-probability peaks are at 1838, 1855, and 1919 Ma (Fig. 8).

## COMPARISON OF DETRITAL ZIRCON AGES

The detrital zircon ages reported above are compared to each other and to the single-grain ages available from other detrital zircon studies in Figure 8. The samples are divided in Figures 8 and 9 and in Tables 2 and 3 into four groups, based on their main sets of ages, as follows.

1. Harmony A comprises two quartzose Harmony sandstones (Little Cottonwood #1 and #2) that yield dominant ages of 690–715 and 1065–1130 Ma, with subordinate groups of 1155–1240 and 1305–1340 Ma. None of the other samples contains a significant proportion of grains of these ages.

2. Harmony B comprises five mainly subarkosic Harmony sandstones from three ranges that yield primarily 1820–1860 and 2595–2700 Ma ages, with subordinate groups of 1745–1790, 1900–1945, and 2480–2585 Ma. The 1745–1790 Ma ages in these samples are distinctive.

3. Lower Vinini comprises both samples of lower sandstone from the Vinini Formation (early Whiterockian), which yield dominant and distinctive age groups of 1410–1445, 1665–1690, and 1705–1740 Ma.

4. Upper Vinini–Valmy–Snow Canyon–McAfee–Elder–

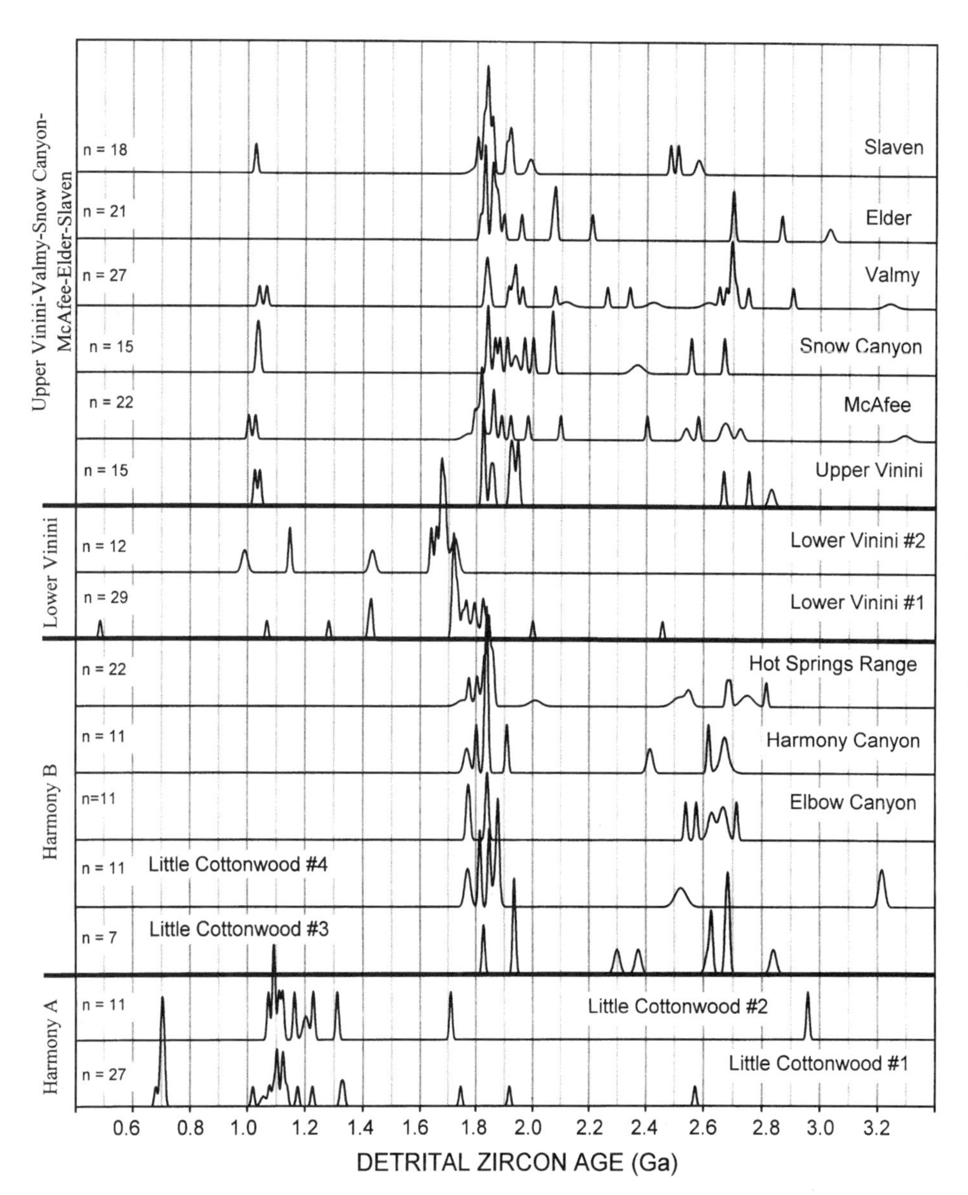


Figure 8. Relative age-probability plots of single-grain detrital zircon ages for strata of Roberts Mountains allochthon. In addition to samples from this study, samples reported by Smith and Gehrels (1994) from Harmony Formation (Little Cottonwood #1) and Valmy Formation (Valmy) are plotted. Ages for all grains plotted are either concordant or, if discordant, projected from 100 Ma. As discussed in text, samples are divided into four groups that have distinct age spectra.

**TABLE 2. COMPARISON OF AGE SPECTRA OF ALL SAMPLES AND SAMPLE GROUPS**

| | SLA | ELD | VAL | SC | MQ | VU | VL1 | VL2 | HOT | HAR | ELB | LC4 | LC3 | LC2 |
|---|---|---|---|---|---|---|---|---|---|---|---|---|---|---|
| ELD | 0.47 | | | | | | | | | | | | | |
| VAL | 0.40 | 0.42 | | | | | | | | | | | | |
| SC | 0.51 | 0.47 | 0.51 | | | | | | | | | | | |
| MQ | 0.62 | 0.45 | 0.41 | 0.42 | | | | | | | | | | |
| VU | 0.56 | 0.42 | 0.53 | 0.52 | 0.49 | | | | | | | | | |
| VL1 | 0.34 | 0.18 | .015 | 0.12 | 0.33 | 0.16 | | | | | | | | |
| VL2 | 0.00 | 0.00 | 0.00 | 0.00 | 0.06 | 0.00 | 0.33 | | | | | | | |
| HOT | 0.66 | 0.50 | 0.41 | 0.40 | 0.61 | 0.45 | 0.42 | 0.05 | | | | | | |
| HAR | 0.57 | 0.64 | 0.55 | 0.40 | 0.50 | 0.35 | 0.35 | 0.01 | 0.57 | | | | | |
| ELB | 0.36 | 0.20 | 0.46 | 0.28 | 0.45 | 0.22 | 0.23 | 0.00 | 0.54 | 0.67 | | | | |
| LC4 | 0.48 | 0.46 | 0.18 | 0.33 | 0.52 | 0.27 | 0.29 | 0.07 | 0.61 | 0.33 | 0.36 | | | |
| LC3 | 0.20 | 0.21 | 0.48 | 0.30 | 0.26 | 0.44 | 0.10 | 0.00 | 0.28 | 0.41 | 0.38 | 0.06 | | |
| LC2 | 0.00 | 0.00 | 0.04 | 0.00 | 0.00 | 0.00 | 0.16 | 0.11 | 0.02 | 0.00 | 0.00 | 0.00 | 0.00 | |
| LC1 | 0.15 | 0.00 | 0.11 | 0.11 | 0.13 | 0.14 | 0.10 | 0.07 | 0.05 | 0.06 | 0.06 | 0.03 | 0.02 | 0.48 |

Upper Vinini-Valmy-Snow Canyon-McAfee-Elder-Slaven = 0.48

Lower Vinini = 0.33

Harmony B = 0.42

Harmony A = 0.48

| VL | HB | HA | |
|---|---|---|---|
| 0.26 | 0.76 | 0.13 | VES |
| | 0.31 | 0.16 | VL |
| | | 0.06 | HB |

*Note:* Numerical values in lower left indicate the degree of similarity resulting from a comparison of the age spectra of two different samples. Gray shaded areas represent the sample groupings shown in Figures 8 and 9, with their average degree of similarity indicated. The average degree of similarity resulting from comparisons of samples belonging to two different groups is 0.21. Abbreviations for individual samples are as follows: LC1—Little Cottonwood #1; LC2—Little Cottonwood #2; LC3—Little Cottonwood #3; LC4—Little Cottonwood #4; ELB—Elbow Canyon; HAR—Harmony Canyon; HOT—Hot Springs Range; VL2—Lower Vinini #2; VL1—Lower Vinini #1; VU—Upper Vinini; MQ—McAfee Quartzite; SC—Snow Canyon Formation; VAL—Valmy Formation; ELD—Elder Sandstone; SLA—Slaven Chert. Numerical values in upper right are the degree of similarity values for comparisons of the age spectra of the four sample groups (Upper Vinini-Valmy-Snow Canyon-McAfee-Elder-Slaven—VES; Lower Vinini—VL; Harmony A—HA; Harmony B—HB).

Slaven comprises lower sandstone unit in northern Independence Range and upper sandstone units in the other Ordovician strata, and Silurian and Devonian sandstones, which yield a dominant age group of 1815–1860 Ma and subordinate but distinctive age groups of 1020–1045, 1905–1940, and ca. 2073 Ma.

Table 2 compares the age spectra of all of the samples in this study against each other to investigate the statistical validity of dividing the samples into these four groups. The statistical measure used for comparing each pair of samples is the degree of similarity, which, as described by Gehrels (this volume, Introduction), compares the proportions of overlapping ages of two relative age-probability curves. A value of 1.0 indicates a perfect match between the ages and relative abundances of ages in two samples, whereas a value of 0.0 indicates that there is no correlation of ages. The average degree of similarity when comparing samples from different groups is 0.21, a low value. In contrast, samples within the four groups yield average similarity values of 0.48, 0.33, 0.48, and 0.42. This generally supports the designation of four separate age groups, although there are high degrees of similarity for samples in the Harmony B and the Upper Vinini–Valmy–Snow Canyon–McAfee–Elder–Slaven groups.

The age-probability curves for samples in each of the four groups have been combined into four composite curves, one for each group (Fig. 9). These four composite curves are then compared with age-probability curves for strata of the Cordilleran miogeocline. Because Ordovician strata yield distinct detrital zircon ages (Gehrels et al., 1985), separate age spectra are shown for (1) the Eureka Quartzite (an Upper Ordovician quartzite in the miogeocline of central Nevada) and (2) the sum of three Middle-Upper Ordovician quartzites located between central Nevada and the Peace River arch region of Canada. The three include the Eureka Quartzite of Nevada, the Mount Wilson Quartzite of southeastern British Columbia, and the Monkman Quartzite of northeastern British Columbia, which accumulated along the flank of the Peace River arch (Gehrels and Dickinson, 1995; Gehrels and Ross, 1998). The ages of grains that accumulated in shelf facies strata of the Cordilleran margin during Cambrian and Devonian and later time are shown in the lower part of Figure 9, with a separate curve for each region of the miogeocline.

The interpretation that sandstones in the Roberts Mountains allochthon contain detritus derived from crystalline basement and/or overlying strata of the North American craton can be

TABLE 3. DOMINANT DETRITAL ZIRCON AGE GROUPS IN STRATA OF THE ROBERTS MOUNTAINS ALLOCHTHON

| Age range (Ma) | Peak ages (Ma) | Relative abundance |
|---|---|---|
| Upper Vinini-Valmy-Snow Canyon-McAfee-Elder-Slaven | | |
| 995-1070 *(mainly 1020-1045)* | 1004, 1028, 1065 | 0.07 |
| 1790-1895 *(mainly 1815-1860)* | 1828, 1858 | 0.36 |
| 1895-2010 *(mainly 1905-1940)* | 1922 | 0.18 |
| 2055-2110 | 2073 | 0.06 |
| 2645-2740 | 2671, 2697 | 0.12 |
| Lower Vinini | | |
| 1410-1445 | 1428 | 0.09 |
| 1630-1699 *(mainly 1665-1690)* | 1638, 1657, 1676 | 0.17 |
| 1700-1811 *(mainly 1705-1740)* | 1719, 1765, 1794 | 0.48 |
| 1812-1845 | 1825 | 0.08 |
| Harmony B | | |
| 1745-1790 | 1774 | 0.09 |
| 1790-1890 *(mainly 1820-1860)* | 1803, 1841 | 0.37 |
| 1900-1945 | 1910, 1936 | 0.05 |
| 2480-2585 | 2540, 2576 | 0.09 |
| 2595-2700 | 2624, 2681 | 0.22 |
| Harmony A | | |
| 675-720 *(mainly 690-715)* | 680, 703 | 0.23 |
| 1040-1145 *(mainly 1065-1130)* | 1075, 1093, 1103, 1123 | 0.39 |
| 1155-1240 | 1173, 1227 | 0.13 |
| 1305-1340 | 1313, 1330 | 0.08 |

*Note*: These ages and relative abundances refer to the relative age probability plots shown in Figure 9.

tested by comparing our detrital zircon data with detrital zircon ages from strata of the Cordilleran miogeocline. These comparisons can be made visually using the relative age-probability plots of Figure 9, and the comparisons can be done statistically, as described by Gehrels (this volume, Introduction). The statistical comparisons are shown in Figure 10, using the degree of overlap and the degree of similarity of age spectra from the Roberts Mountains allochthon and the miogeoclinal strata. The degree of overlap is a measure of the degree to which the ages from eugeoclinal sandstones match the range of ages in a suite of miogeoclinal samples from a particular region. This emphasizes whether similar ages of grains are present, but does not take into account the proportions of the different components in either sample suite. A value of 1.0 indicates that every age in a eugeoclinal suite is matched by one or more ages in a miogeoclinal suite. The degree of similarity is as described herein.

Some of the conclusions from qualitative and quantitative comparisons of miogeoclinal and eugeoclinal detrital zircon ages are given in the following.

1. The Harmony A signature is most similar to the miogeoclinal curve for Sonora (Fig. 9). The 1.0–1.35 Ga grains that dominate the Harmony A group are also abundant in the Sonoran strata. However, none of the miogeoclinal transects yields a match for the 10 ca. 0.7 Ga grains in the Harmony A samples (9 from Little Cottonwood #1 from Smith and Gehrels [1994] plus one grain from the Harmony Canyon sample of Wallin [1990]). It is interesting to note, however, that one ca. 0.7 Ga grain has been recovered from the Wood Canyon Formation, a miogeoclinal sandstone of Early Cambrian age from the Mojave Desert region of southern California. Hence, southwestern North America is a likely source for the detritus in some Harmony samples, as suggested by Wallin (1990).

2. The Harmony B signature is quite similar to the curves derived from miogeoclinal strata of southern (and to a lesser degree northern) British Columbia. Key similarities are the occurrences of abundant 1745–1790 and 1820–1860 Ma grains and subordinate 1.90–1.95 and 2.2–2.5 Ga, and >2.5 Ga grains in both groups. Direct links with southwestern North America are precluded by the lack of 1.0–1.4 Ga grains in the Harmony B samples. These conclusions are supported by the high degrees of overlap and similarity with miogeoclinal samples from British Columbia and the low values for Nevada and Sonora (Fig. 10).

3. The Lower Vinini curve is generally similar to the miogeoclinal curve from Nevada, but there is little similarity with the other miogeoclinal curves. Specific similarities with Nevada are the dominance of ca. 1.43 and 1.65–1.85 Ga grains and the low proportion in both suites of <1.4 and >2.0 Ga grains. Quantitative comparisons also show high degrees of overlap and similarity for Nevada, low values for the three northern transects, and moderate values for Sonora (Fig. 10).

4. The combined curve for samples of the Upper Vinini–Valmy–Snow Canyon–McAfee–Elder–Slaven group is quite similar to the curve for miogeoclinal strata in northern British Columbia, somewhat similar to the curves for Alaska and southern British Columbia, and distinct from the curves for miogeoclinal strata of Nevada and Sonora (Fig. 9). Specific matches with the northern British Columbia data include the dominance of ca. 1.82–1.86 and 1.89–1.93 Ga grains and the presence of subordinate ca. 1.03 and 2.06 Ga grains. The curve for the Upper Vinini–Valmy–Snow Canyon–McAfee–Elder–Slaven group is also quite similar to curves for both the Eureka Quartzite and the set of all three Ordovician miogeoclinal quartzites of Nevada and British Columbia (Fig. 9). These visual similarities are supported by high degrees of overlap (0.82) and similarity (0.85) for comparisons with the Ordovician miogeocline curve. Correlation of this provenance group with the miogeoclinal strata is reasonable given the evidence that detritus in Ordovician miogeoclinal strata all along the Cordilleran margin was derived from the Peace River arch region of northern British Columbia (Ketner, 1968; Gehrels et al., 1995; Gehrels, this volume, Introduction), and the direct stratigraphic ties between shelf, slope, and basinal strata (Finney and Perry, 1991).

## PROVENANCE IMPLICATIONS

The detrital zircon age spectra shown in Figures 8 and 9 define important variations in the provenance of detritus in Cambrian through Devonian eugeoclinal strata of the Roberts Mountains allochthon. These variations in detrital zircon provenance,

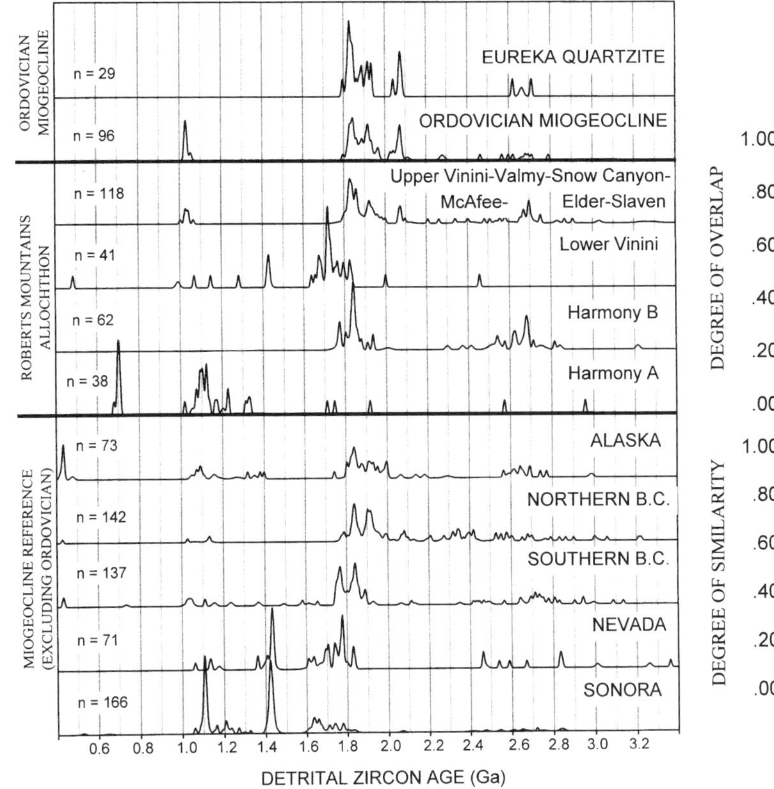


Figure 9. Relative age-probability plots comparing ages of detrital zircons in strata of Roberts Mountains and Cordilleran miogeocline. Miogeoclinal samples are shown separately for Ordovician strata in Nevada and British Columbia (B.C.) (top) and for Neoproterozoic, Cambrian, Devonian, upper Paleozoic, and Triassic strata in five different areas of Cordilleran margin (bottom). Each miogeoclinal curve is composite of detrital zircon ages from strata of various ages because there are no significant changes in age spectra through time (Gehrels et al., 1995; Gehrels, this volume). Curves for eugeoclinal strata are combinations of curves in four groups shown in Figure 8. All curves have been normalized for number of grains included.

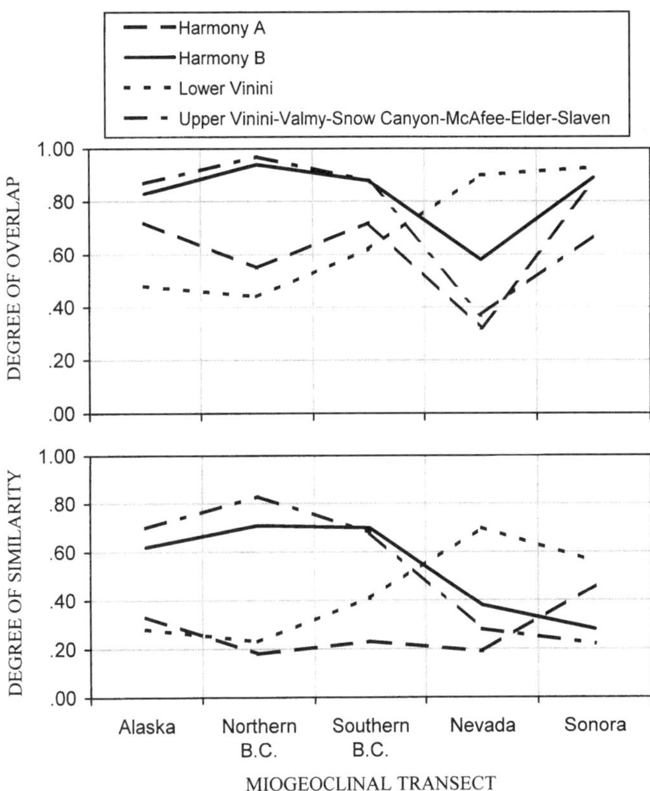


Figure 10. Results of statistical comparisons of detrital zircon ages from four groups of eugeoclinal samples with ages of grains in Neoproterozoic, Cambrian, and Devonian through Triassic miogeoclinal strata along Cordilleran margin. Degree of overlap is sum of age probability that is matched by age probability in miogeoclinal transect. For example, degree of overlap of 0.50 indicates that approximately half of ages in set of eugeoclinal samples overlap with ages of grains in set of miogeoclinal samples. Degree of similarity is comparison of proportions of different ages in sets of eugeoclinal and miogeoclinal samples. Value of 1.0 indicates that proportions of ages in two sets of samples are identical, whereas value of 0.0 reflects no similarity in proportions of ages. B.C.—British Columbia.

when compared with the provenance of detrital zircons in strata of the Cordilleran miogeocline, and combined with existing stratigraphic and paleontologic constraints, yield new insights into the tectonic history of the Cordilleran margin. Our conclusions are discussed chronologically and are keyed to a series of schematic paleodispersal maps shown in Figure 11.

Beginning by Late Cambrian(?) time, coarse immature detritus of the Harmony Formation was shed from basement rocks with dominant ages of 1.8–2.8 Ga and subordinate ages of 1.0–1.35 Ga and 690–715 Ma (Fig. 9). The strong correlation of >1.8 Ga zircon ages in most Harmony samples with detrital zircon ages in miogeoclinal strata from southern and northern British Columbia suggests that rocks of the western Canadian shield are a likely source area for most Harmony strata. The two samples with mainly <1.4 Ga ages were apparently derived mainly from the Grenville Province in southern or southwestern North America.

On the basis of these interpreted provenance links, we propose that detritus in the Harmony Formation was derived from basement rocks exposed both to the north and south of the present location of the Roberts Mountains allochthon. Harmony B detritus was most likely shed from basement rocks exposed along the northern part of the continental margin, and was transported southward by turbidity currents to a site of accumulation offshore from present-day western Nevada. This is similar to the model of Suczek (1977), Stewart and Suczek (1977), Rowell et al. (1979), and Schweickert and Snyder (1981), except that the distance of transport is considerably greater. Cambrian(?) feldspathic grits of the Lardeau and Covada Groups along the Washington–British Columbia border, which apparently have detrital zircon ages similar to those of the Harmony B samples (Smith and Gehrels, 1991, 1992), may be intermediate deposits of this broad dispersal system (Fig. 11A). This interpretation supports the Late Cambrian age for the Harmony Formation, because feldspathic grits

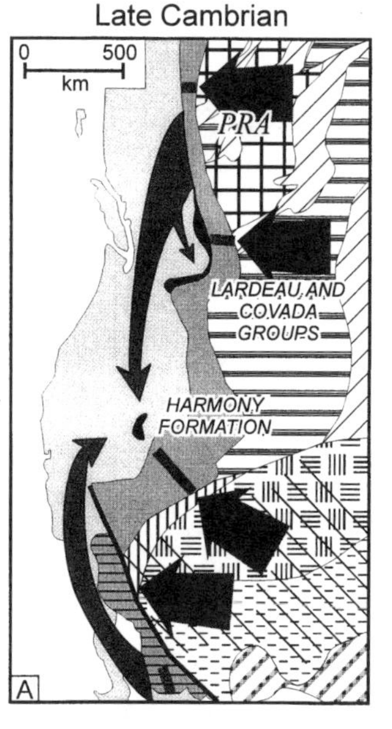


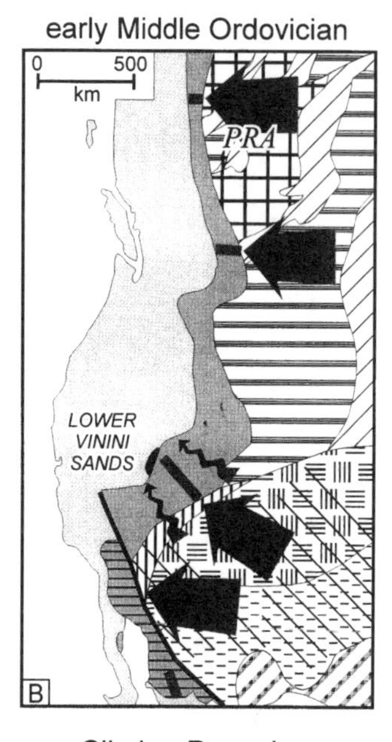


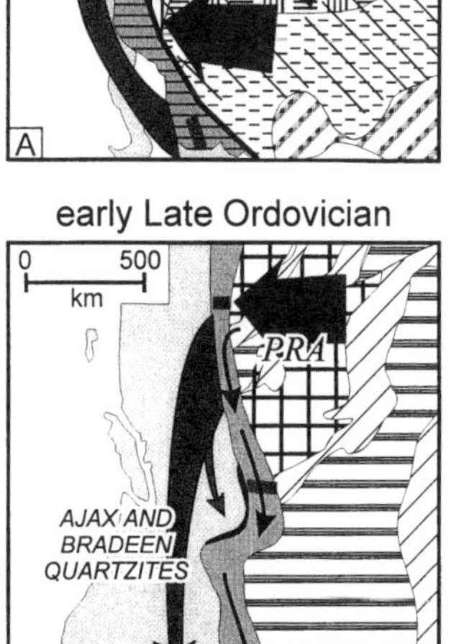


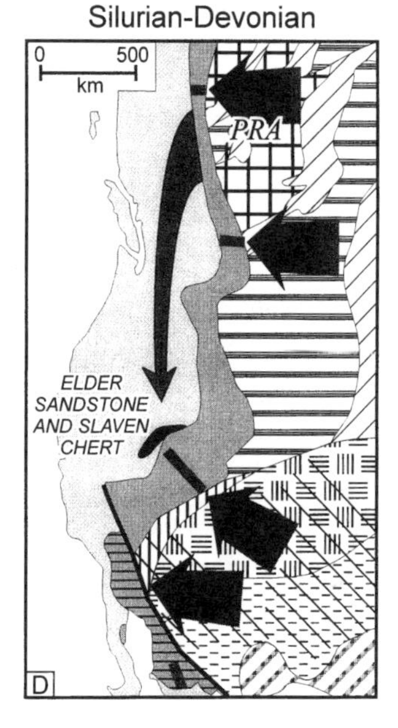


Figure 11. Schematic maps showing interpreted provenance and dispersal patterns of clastic detritus in eugeoclinal strata of Roberts Mountains allochthon and in miogeoclinal strata of western North America. Base map is adapted from Figure 1, which defines symbols. B.C.—British Columbia; PRA—Peace River arch.

in the Covada Group are depositionally overlain by well dated Lower Ordovician strata (Smith and Gehrels, 1992).

Harmony A detritus may have been derived from Grenville age and currently unrecognized ca. 0.7 Ga basement rocks exposed along the early Paleozoic continental margin in western Mexico (as shown in Fig. 11A), or it may have been transported from the Grenville belt across the southwestern part of the craton in large river systems (J.H. Stewart, 1998, written commun.). Depositional overlap of the two submarine fan systems, rather than tectonic interleaving, is recorded by the relationship of Harmony B sandstones depositionally overlying Harmony A sandstones in the Little Cottonwood Canyon area of Battle Mountain (Fig. 3B; Dickinson and Gehrels, this volume, Chapter 11).

Following deposition of the Harmony Formation, the next regionally extensive sandstone units to have been deposited are in the lower part of the Vinini Formation (Fig. 4). These Whiterockian (Middle Ordovician) sandstones yielded detrital zircons that were derived from basement rocks of the southwestern United States (Figs. 9 and 11B), a result consistent with the recent findings of Finney and Perry (1991) and Finney et al. (1997).

Beginning in Mohawkian (early Late Ordovician) time, mature quartz sand blanketed much of the shelf and accumulated to great thicknesses in widespread off-shelf basins of Nevada. Regional stratigraphic and detrital zircon age data demonstrate that the miogeoclinal detritus was shed from the Peace River arch region of northwestern Canada (Ketner, 1968; Gehrels et al., 1995), but the sources for the eugeoclinal detritus have been controversial.

Our age spectra for the Upper Ordovician eugeoclinal sandstones are very similar to the detrital zircon ages from miogeoclinal strata of the same age, but are significantly different from ages in Harmony A strata and the Middle Ordovician sandstones (Figs. 8 and 9). We suggest that the detritus in Upper Ordovician eugeoclinal units was also derived from the Peace River arch region, and that miogeoclinal and eugeoclinal quartzites of Nevada were therefore derived from the same source region. Although the miogeoclinal detritus likely traveled southward via longshore currents, we suggest that detritus in the eugeoclinal sandstones was transported southward largely in off-shelf troughs, basins, and perhaps trenches along and outboard of the continental margin (Fig. 11, C and D). Long-distance transport in deep water seems necessary to explain the presence of (1) significantly coarser quartz grains (Ketner, 1966) and zircon grains in the Ordovician eugeoclinal units, (2) interbedded quartzose and feldspathic sandstones in the Ordovician and Devonian units (Madrid, 1987), and (3) dominantly feldspathic units like the Silurian Elder Sandstone. Off-shelf transport of detritus in the Slaven Chert is also required by the fact that a Middle Devonian sandstone from the miogeocline in central Nevada contains almost exclusively <1.8 Ga grains (Gehrels and Dickinson, 1995), whereas the Slaven Chert has abundant 1.8–2.0 Ga grains.

The model outlined in Figure 11, which emphasizes large-scale, mainly southward sediment dispersal along the Cordilleran margin, requires a large volume of clastic sediment to have been eroded from basement and cover rocks of western Canada. During Cambrian time, this involved erosion of igneous rocks, yielding first-cycle detritus, whereas detritus in the Ordovician through Devonian strata is predominantly multicyclic (Ketner, 1968; Madrid, 1987; Dickinson and Gehrels, this volume, Chapter 11).

The most likely northern source for detritus in Ordovician-Devonian strata is the Peace River arch region (Fig. 11), which was emergent from Early Cambrian through Late Devonian time (McMechan, 1990; Norford, 1990; Moore, 1993). As shown in various compilations of platformal stratigraphy (Cook and Bally, 1973; Ricketts, 1989; Stott and Aitken, 1993), the Peace River arch extended from the continental margin in northern British Columbia eastward to connect with a large region of exposed basement rocks of the Canadian shield in the continental interior. The Peace River arch was gradually onlapped during early Paleozoic time, with several phases of regional transgression and regression. Basement rocks of the arch, and to a lesser degree the cratonal interior, were the source of widespread quartz-rich sands that accumulated in a platformal setting during Cambrian, Ordovician, and Devonian time. These widespread sandstones, many of which were exposed during cycles of uplift and erosion during early Paleozoic time, provide a likely source for the shelf facies sandstones in the miogeocline of western Canada, and for some of the eugeoclinal sandstones in the Roberts Mountains allochthon.

In contrast, the feldspathic detritus of the Harmony Formation was probably derived directly from granitoids exposed along the continental margin. Such granitoids could have been exposed along a western, offshore continuation of the Peace River arch, although there is no stratigraphic evidence for such a feature in shelf facies strata of the miogeocline. Perhaps the stratigraphic record of this feature was located sufficiently to the west that it has been incorporated into the Yukon-Tanana terrane of the northern Cordillera (Monger and Berg, 1987) prior to ~1000 km of Mesozoic right-lateral slip on the Tintina and related faults (Gabrielse, 1985).

Alternatively, extensional block faulting, such as that recorded by Middle and Upper Cambrian clastic strata in the Roosevelt graben, northwest of the Peace River arch (Fritz et al., 1991; Aitken, 1993), could have exposed crystalline rocks outboard of the shelf farther south along the margin. Bond and Kominz (1984) and Turner et al. (1989) reported stratigraphic evidence for extensional tectonism during early Paleozoic time, but there is no stratigraphic record of basement rocks being exposed along such faults. This lack of an obvious crystalline source located outboard of the Peace River arch poses a serious challenge to our interpretation of a northern source for Harmony B detritus.

## SUMMARY

U-Pb analysis of 205 individual detrital zircon grains from eugeoclinal strata of the Roberts Mountains allochthon yields important new constraints on the early Paleozoic sediment dispersal history of the western United States. The detrital zircon ages define several disparate signatures that reflect several modes of sediment dispersal along the Cordilleran margin during Cambrian through Devonian time.

The Upper Cambrian(?) Harmony Formation, long one of the more enigmatic parts of the allochthon due to its outboard position and immature composition, yields two detrital zircon age spectra. One group of sandstones yields mainly <1.4 Ga grains that were derived mainly from the Grenville Province in southern or southwestern North America. A second group, which is apparently the dominant component in the Harmony Formation, yields >1.75 Ga ages that were most likely derived from a westward continuation of the Peace River arch along the Canadian continental margin. This requires ~1200 km of southward transport, which, following the ideas of Suczek (1977), Stewart and Suczek (1977), Rowell et al. (1979), and Schweickert and Snyder (1981), is interpreted to have occurred by southward-flowing turbidity currents in a broad submarine fan. The depositional contact between the two sandstone types in the Harmony indicates that the Harmony Formation contains remnants of two overlapping submarine fan systems that were shed from two very different portions of the Cordilleran margin.

Sandstones in Ordovician strata of the allochthon were also derived from two fundamentally different source regions, as recognized by Finney and Perry (1991). The lower of two Ordovician sandstone units, which generally occurs in eastern parts of the allochthon, yields mainly <1.8 Ga zircon ages that are similar to the ages of basement rocks of the southwestern United States. During a phase of low sea level, sands were eroded from these rocks and/or overlying platformal strata, carried across the shelf in isolated channels and canyons, and dispersed into proximal parts of the eugeoclinal basin. This provenance link leads to the critical conclusion that at least parts of the Roberts Mountains allochthon formed near their present position, and are accordingly not exotic or far traveled (Finney and Perry, 1991).

The higher of the two Ordovician sandstone units, in contrast, contains mainly >1.8 Ga detrital zircons that are similar in age to basement rocks of the Peace River arch region of western Canada (Fig. 11). Stratigraphic relations (Ketner, 1968) and detrital zircon data (Gehrels et al., 1995) indicate that miogeoclinal quartzites of Nevada were also derived from the Peace River arch, with southward dispersal driven by longshore transport. The eugeoclinal detritus was apparently transported southward largely by turbidity currents in off-shelf basins or trenches, rather than by longshore currents, because the eugeoclinal sandstones are consistently coarser, less well sorted, and less mature compositionally than the miogeoclinal sands (Ketner, 1966; Madrid, 1987). It is also possible that the eugeoclinal sandstones contain a mixture of detritus transported southward largely in off-shelf environments and detritus transported southward within the miogeocline and then westward off of the shelf, as described by Miller and Larue (1983).

Detrital zircon grains in Silurian and Devonian strata of the allochthon yield mainly >1.8 Ga ages that are very similar to the ages of detrital zircon grains in the upper Ordovician quartzites. We accordingly interpret these grains to also have originated in the Peace River arch region. The detritus in these units could not have been transported southward on the shelf because feldspathic sandstones similar to the Silurian Elder Sandstone are not common in the miogeocline, and because Devonian strata of the miogeocline in Nevada yield mainly <1.8 Ga detrital zircons.

As described in more detail by Gehrels et al. (this volume, Chapter 10), the provenance links interpreted from our detrital zircon data have important implications for the tectonic evolution of the Cordilleran margin. The main conclusion resulting from this study is that most sandstones in the Roberts Mountains allochthon consist of detritus that originated in Precambrian basement rocks along the western margin of North America. There is no evidence that strata of the allochthon are far traveled, or that they contain detritus derived from rocks that have been removed from the Cordilleran orogen. Furthermore, comparison of the detrital zircon ages in strata of the allochthon with ages in strata of the Cordilleran miogeocline supports previous interpretations (e.g., Finney and Perry, 1991; Miller et al., 1992; Burchfiel et al., 1992; Finney et al., 1997) that the eugeoclinal strata have not been transported significantly along the Cordilleran margin. Unfortunately, the lack of arc-derived detritus within the eugeoclinal strata precludes interpretations concerning the facing direction(s) of early Paleozoic island arcs that may have existed outboard of the Roberts Mountains allochthon.

## ACKNOWLEDGMENTS

We thank Jack Stewart, Gerry Ross, Gary Girty, and Chris Suczek for sharing their views on sediment dispersal patterns along the Cordilleran margin. Angela Smith was of great assistance in collecting and processing some of our samples. This research was supported by National Science Foundation grants EAR-9116000 and EAR-9416933.

## REFERENCES CITED

Aitken, J.D., 1993, Cambrian and Lower Ordovician—Sauk sequence, *in* Stott, D.F., and Aitken, J.D., eds., Sedimentary cover of the craton in Canada: Geological Survey of Canada, Geology of Canada, no. 5, p. 96–124.

Blake, M.C., Jr., Bruhn, R.L., Miller, E.L., Moores, E.M., Smithson, S.B., and Speed, R.C., 1985, C-1 Menodocino triple junction to North American craton: Geological Society of America Centennial Continent-Ocean Transect 12, scale 1:500 000, 30 p.

Bond, G., and Kominz, M.A., 1984, Construction of tectonic subsidence curves for the early Paleozoic miogeocline, southern Rocky Mountains: Implications for subsidence mechanisms, age of breakup, and crustal thinning: Geological Society of America Bulletin, v. 95, p. 155–173.

Burchfiel, B.C., and Davis, G.A., 1972, Structural framework and evolution of the southern part of the Cordilleran orogen, western United States: American Journal of Science, v. 272, p. 97–118.

Burchfiel, B.C., and Davis, G.A., 1975, Nature and controls of Cordilleran orogenesis, western United States: Extensions of an earlier synthesis: American Journal of Science, v. 275–A, p. 363–396.

Burchfiel, B.C., and Royden, L.H., 1991, Antler orogeny: A Mediterranean-type orogeny: Geology, v. 19, p. 66–69.

Burchfiel, B.C., Cowan, D.S., and Davis, G.A., 1992, Tectonic overview of the Cordilleran orogen in the western United States, *in* Burchfiel, B.C., et al., eds., The Cordilleran orogen: Conterminous U.S.: Boulder, Colorado, Geological Society of America, Geology of North America, v. G-3, p. 407–480.

Churkin, M., Jr., 1974, Paleozoic marginal ocean basin-volcanic arc in the Cordilleran foldbelt, *in* Dott, R.H., and Shaver, R.H., eds., Modern and ancient geosynclinal sedimentation: Society of Economic Paleontologists and Mineralogists Special Publication 19, p. 174–192.

Churkin, M., Jr., and Kay, M., 1967, Graptolite-bearing Ordovician siliceous and volcanic rocks, northern Independence Range, Nevada: Geological Society of America Bulletin, v. 78, p. 651–668.

Cluer, J.K., Keith, S.B., Cellura, B.R., Finney, S.C., and Bellert, S., 1997, Stratigraphy and structure of the Bell Creek Nappe, Ren Property, northern Carlin Trend, Nevada, *in* Perry, A.J., and Abbott, W.E., eds., The Roberts Mountains thrust, Elko and Eureka Counties, Nevada: Reno, Nevada Petroleum Society 1997 Field Trip Guidebook, p. 41–53.

Cook, T.D., and Bally, A.W., (eds.), 1973, Stratigraphic atlas of north and central North America: Princeton, New Jersey, Princeton University Press, 272 p.

Dickinson, W.R., 1977, Paleozoic plate tectonics and the evolution of the Cordilleran continental margin, *in* Stewart, J.H., et al., eds., Paleozoic paleogeography of the western United States: Society of Economic Paleontologists and Mineralogists, Pacific Coast Paleogeography Symposium I, p. 137–155.

Dickinson, W.R., Harbaugh, D.W., Saller, A.H., Heller, P.L., and Snyder, W.S., 1983, Detrital modes of upper Paleozoic sandstones derived from Antler orogen in Nevada: Implications for the nature of the Antler orogeny: American Journal of Science, v. 282, p. 481–509.

Eisbacher, G.H., 1983, Devonian-Mississippian sinistral transcurrent faulting along the cratonic margin of western North America: Geology, v. 11, p. 7–10.

Evans, K.V., and Fischer, L.B., 1986, U-Pb geochronology of two augen gneiss terranes, Idaho—New data and tectonic implications: Canadian Journal of Earth Sciences, v. 23, p. 1919–1927.

Evans, K.V., and Zartman, R.E., 1988, Early Paleozoic alkalic plutonism in east-central Idaho: Geological Society of America Bulletin, v. 100, p. 1981–1987.

Finney, S.C., and Perry, B.D., 1991, Depositional setting and paleogeography of Ordovician Vinini Formation, central Nevada, *in* Cooper, J.D., and Stevens, C.H., eds., Paleozoic paleography of the western United States—II, Volume 2: Pacific Section, Society of Economic Paleontologists and Mineralogists book 67, p. 747–766.

Finney, S.C., Cooper, J.D., and Berry, W.B.N., 1997, Late Ordovician mass extinction: Sedimentologic, cyclostratigraphic, biostratigraphic, and chemostratigraphic records from platform and basin successions, central Nevada, *in* Link, P.K., and Kowallis, B.J., eds., Proterozoic to recent stratigraphy, tectonics, and volcanology, Utah, Nevada, southern Idaho, and central Mexico: Brigham Young University Geology Studies, v. 42, p. 79–104.

Fritz, W.H., Cecile, M.P., Norford, B.S., Morrow, D., and Geldsetzer, H.H.J., 1991, Cambrian to Middle Devonian assemblages, *in* Gabrielse, H., and Yorath, C.J., eds., Geology of the Cordilleran orogen in Canada: Geological Survey of Canada, Geology of Canada, no. 5, p. 153–218.

Gabrielse, H., 1985, Major dextral transcurrent displacements along the Northern Rocky Mountain Trench and related lineaments in north-central British Columbia: Geological Society of America Bulletin, v. 96, p. 1–14.

Gehrels, G.E., and Dickinson, W.R., 1995, Detrital zircon provenance of Cambrian to Triassic miogeoclinal and eugeoclinal strata in Nevada: American Journal of Science, v. 295, p. 18–48.

Gehrels, G.E., and Ross, G.M., 1998, Detrital zircon geochronology of Neoproterozoic to Triassic miogeoclinal strata in British Columbia and Alberta: Canadian Journal of Earth Sciences, v. 35, p. 1380–1401.

Gehrels, G.E., Dickinson, W.R., Ross, G.M., Stewart, J.H., and Howell, D.G., 1995, Detrital zircon reference for Cambrian to Triassic miogeoclinal strata of western North America: Geology, v. 23, p. 831–834.

Gilluly, J., and Gates, O., 1965, Tectonic and igneous geology of the northern Shoshone Range, Nevada: U.S. Geological Survey Professional Paper 465, 153 p.

Girty, G.H., Reiland, D.N., and Wardlaw, M.S., 1985, Provenance of Silurian Elder Sandstone, north-central Nevada: Geological Society of America Bulletin, v. 96, p. 925–930.

Hoffman, P.F., 1989, Precambrian geology and tectonic history of North America, *in* Bally, A.W., and Palmer, A.R., eds., The geology of North America—An overview: Boulder, Colorado, Geological Society of America, Geology of North America, v. A, p. 447–512.

Hotz, P.E., and Willden, R., 1964, Geology and mineral deposits of the Osgood Mountains quadrangle, Humboldt County, Nevada: U.S. Geological Survey Professional Paper 431, 128 p.

Jones, A.E., 1997, Geology of the Delvada Spring quadrangle, Humboldt County,

Nevada: Nevada Bureau of Mines and Geology Field Studies Map 13, scale 1: 24 000.
Kay, M., 1951, North American geosynclines: Geological Society of America Memoir 48, 143 p.
Ketner, K.B., 1966, Comparison of Ordovician eugeosynclinal and miogeosynclinal quartzites of the Cordilleran geosyncline: U.S. Geological Survey Professional Paper 550-C, p. C54–C60.
Ketner, K.B., 1968, Origin of Ordovician quartzite in the Cordilleran miogeosyncline: U.S. Geological Survey Professional Paper 600-B, p. 169–177.
Ketner, K.B., 1977, Deposition and deformation of lower Paleozoic western facies rocks, northern Nevada, *in* Stewart, J.H., et al., eds., Pacific Section, Society of Economic Paleontologists and Mineralogists, Paleozoic paleogeography of the western United States: Pacific Coast Paleogeography Symposium I, p. 251–258.
Ketner, K.B., 1984, Recent studies indicate that major structures in northeastern Nevada and the Golconda thrust in north-central Nevada are of Jurassic or Cretaceous age: Geology, v. 12, p. 483–486.
Ketner, K.B., 1993, Paleozoic and Mesozoic rocks of Mount Ichabod and Dorsey canyon, Elko County, Nevada—Evidence for post-Early Triassic emplacement of the Roberts Mountains and Golconda allochthons: U.S. Geological Survey Bulletin 1988-D, 12 p.
Kistler, R.W., 1991, Chemical and isotopic characteristics of plutons in the Great Basin, *in* Raines, G.L., et al., eds., Geology and ore deposits of the Great Basin: Reno, Geological Society of Nevada, p. 107–109.
Kistler, R.W., and Peterman, Z.E., 1973, Variations in Sr, Rb, K, Na, and initial $^{87}Sr/^{86}Sr$ in Mesozoic granitic rocks and intruded wall rocks in central California: Geological Society of America Bulletin, v. 84, p. 3489–3512.
Little, T.A., 1987, Stratigraphy and structure of metamorphosed upper Paleozoic rocks near Mountain City, Nevada: Geological Society of America Bulletin, v. 98, p. 1–17.
Ludwig, K.R., 1991a, A computer program for processing Pb-U-Th isotopic data: U.S. Geological Survey Open-File Report 88-542.
Ludwig, K.R., 1991b, A plotting and regression program for radiogenic-isotopic data: U.S. Geological Survey Open-File Report 91-445.
Madden-McGuire, D.J., Hutter, T.J., and Suczek, C.A., 1991, Late Cambrian–Early Ordovician microfossils from the Harmony Formation at its type locality, northern Sonoma Range, Humboldt County, Nevada: Geological Society of America Abstracts with Programs, v. 23, no. 2, p. 75.
Madrid, R.J., 1987, Stratigraphy of the Roberts Mountains allochthon in north-central Nevada [Ph.D. thesis]: Stanford, California, Stanford University, 336 p.
McMechan, M.E., 1990, Upper Proterozoic to Middle Cambrian history of the Peace River arch: Evidence from the Rocky Mountains: Bulletin of Canadian Petroleum Geology, v. 38A, p. 36–44.
Miller, E.L., and Larue, D.K, 1983, Ordovician quartzite in the Roberts Mountains allochthon, Nevada; Deep-sea fan deposits derived from cratonal North America, *in* Stevens, C.H., ed., Pre-Jurassic rocks in western North American suspect terranes: Pacific Section, Society of Paleontologists and Mineralogists, p. 91–102.
Miller, E.L., Holdsworth, B.K., Whiteford, W.B., and Rodgers, D., 1984, Stratigraphy and structure of the Schoonover sequence, northeastern Nevada: Implications for Paleozoic plate–margin tectonics: Geological Society of America Bulletin, v. 95, p. 1063–1076.
Miller, E.L., Miller, M.M., Stevens, C.H., Wright, J.E., and Madrid, R., 1992, Late Paleozoic paleogeographic and tectonic evolution of the western U.S. Cordillera, *in* Burchfiel, B.C., et al., eds., The Cordilleran orogen: Conterminous U.S.: Boulder, Colorado, Geological Society of America, Geology of North America, v. G-3, p. 57–106.
Monger, J.W.H., and Berg, H.C., 1987, Lithotectonic terrane map of western Canada and southeastern Alaska: U.S. Geological Survey Miscellaneous Field Studies Map MF-1874-B, scale 1: 2 500 000.
Moore, P.F., 1993, Devonian, *in* Stott, D.F., and Aitken, J.D., eds., Sedimentary cover of the craton in Canada: Geological Survey of Canada, Geology of Canada, no. 5, p. 150–201.
Nilsen, T.H., and Stewart, J.H., 1980, The Antler orogeny—Mid-Paleozoic tectonism in western North America (Penrose Conference Report): Geology, v. 8, p. 298–302.
Norford, B.S., 1990, Ordovician and Silurian stratigraphy, paleogeography, and depositional history in the Peace River arch area, Alberta and British Columbia: Bulletin of Canadian Petroleum Geology, v. 38A, p. 45–54.
Oldow, J.S., 1984, Evolution of a late Mesozoic back-arc fold and thrust belt, northwestern Great Basin, U.S.A.: Tectonophysics, v. 102, p. 245–274.
Palmer, A.R., 1971, Cambrian of the Great Basin and adjacent areas, western United States, in Holland, C.H., ed., Cambrian of the new world: New York, John Wiley Interscience, p. 1–78.
Poole, F.G., 1974, Flysch deposits of Antler foreland basin, western United States, *in* Dickinson, W.R., ed., Tectonics and sedimentation: Society of Economic Paleontologists and Mineralogists Special Publication 22, p. 58–82.
Poole, F.G., Stewart, J.H., Palmer, A.R., Sandberg, C.A., Madrid, R.J., Ross, R.J., Hintze, L.F., Miller, M.M., and Wrucke, C.T., 1992, Latest Precambrian to latest Devonian time; Development of a continental margin, *in* Burchfiel, B.C., et al., eds., The Cordilleran orogen: Conterminous U.S.: Boulder, Colorado, Geological Society of America, Geology of North America, v. G-3, p. 9–56.
Ricketts, B.D., 1989, Western Canada sedimentary basin; a case history: Calgary, Alberta, Canadian Society of Petroleum Geology, 320 p.
Roberts, R.J., 1964, Stratigraphy and structure of the Antler Peak quadrangle, Humboldt and Lander Counties, Nevada: U.S. Geological Survey Professional Paper 459-A, p. A1–A90.
Roberts, R.J., Hotz, P.E., Gilluly, J., and Ferguson, H.G., 1958, Paleozoic rocks of north-central Nevada: American Association of Petroleum Geologists Bulletin, v. 42, p. 2813–2857.
Ross, G.M., 1991, Precambrian basement in the Canadian Cordillera: An introduction: Canadian Journal of Earth Sciences, v. 28, p. 1133–1139.
Rowell, A.J., Rees, M.N., and Suczek, C.A., 1979, Margin of the North American continent in Nevada during Late Cambrian time: American Journal of Science, v. 279, p. 1–18.
Saleeby, J.B., 1986, C-2 Central California offshore to Colorado Plateau: Geological Society of America Centennial Continent-Ocean Transect 10, scale 1:500 000, 63 p.
Schuchert, C., 1923, Sites and nature of the North American geosynclines: Geological Society of America Bulletin, v. 34, p. 151–230.
Schweickert, R.A., and Snyder, W.S., 1981, Paleozoic plate tectonics of the Sierra Nevada and adjacent regions, *in* Ernst, W.G., ed., The geotectonic development of California: Englewood Cliffs, New Jersey, Prentice-Hall, p. 183–201.
Silberling, N.J., 1991, Allochthonous terranes of western Nevada: Current status, *in* Raines, G.L., et al., eds., Geology and ore deposits of the Great Basin: Reno, Geological Society of Nevada, p. 101–102.
Silberling, N.J., Jones, D.L., Blake, M.C., Jr., and Howell, D.G., 1987, Lithotectonic terrane map of the western conterminous United States: U.S. Geological Survey Miscellaneous Field Studies Map MF-1874-C, scale 1:2 500 000.
Smith, M.T., and Gehrels, G.E., 1991, Detrital zircon geochronology of Upper Proterozoic to lower Paleozoic continental margin strata of the Kootenay arc: Implications for the early Paleozoic tectonic development of the eastern Canadian Cordillera: Canadian Journal of Earth Sciences, v. 28, p. 1271–1284.
Smith, M.T., and Gehrels, G.E., 1992, Stratigraphy and tectonic significance of lower Paleozoic continental margin strata in northeastern Washington: Tectonics, v. 11, p. 607–620.
Smith, M.T., and Gehrels, G.E., 1994, Detrital zircon geochronology and the provenance of the Harmony and Valmy Formations, Roberts Mountains allochthon, Nevada: Geological Society of America Bulletin, v. 106, p. 968–979.
Speed, R.C., and Sleep, N.H., 1982, Antler orogeny and foreland basin: A model: Geological Society of America Bulletin, v. 93, p. 815–828.
Stewart, J.H., and Carlson, J.E., 1978, Geologic map of Nevada: U.S. Geological Survey, 1:500 000, 1 sheet.

Stewart, J.H., and Suczek, C.A., 1977, Cambrian and latest Precambrian paleogeography and tectonics in the western United States, *in* Stewart, J.H., et al., eds., Society of Economic Paleontologists and Mineralogists, Paleozoic paleogeography of the western United States: Pacific Coast Paleogeography Symposium I, p. 1–17.

Stewart, J.H., Poole, F.G., Ketner, K.B., Madrid, R.J., Roldan-Quintana, J., and Amaya-Martinez, R., 1990, Tectonics and stratigraphy of the Paleozoic and Triassic southern margin of North America, Sonora, Mexico, *in* Gehrels, G.E., and Spencer, J.E., eds., Geologic excursions through the Sonoran Desert region, Arizona and Sonora: Arizona Geological Survey Special Paper 7, p. 183–202.

Stott, D.F., and Aitken, J.D. (eds.), 1993, Sedimentary cover of the craton: Geological Survey of Canada, Geology of Canada, no. 5, 826 p.

Suczek, C.A., 1977, Tectonic relations of the Harmony Formation, northern Nevada [Ph.D. thesis]: Stanford, California, Stanford University, 96 p.

Turner, R.J.W., Madrid, R.J., and Miller, E.L., 1989, Roberts Mountains allochthon: Stratigraphic comparison with lower Paleozoic outer continental margin strata of the northern Canadian Cordillera: Geology, v. 17, p. 341–344.

Van Schmus, W.R., and 24 others, 1993, Transcontinental Proterozoic Provinces, *in* Reed, J.C., et al., eds., Precambrian: Conterminous U.S.: Boulder, Colorado, Geological Society of America, Geology of North America, v. C-2, p. 171–334.

Villeneuve, M.E., Ross, G.M., Theriault, R.J., Miles, W., Parrish, R.R., and Broome, J., 1993, Tectonic subdivision and U-Pb geochronology of the crystalline basement of the Alberta basin, western Canada: Geological Survey of Canada Bulletin 447, 86 p.

Wallin, E.T., 1990, Provenance of selected lower Paleozoic siliciclastic rocks in the Roberts Mountains allochthon, Nevada, *in* Harwood, D.S., and Miller, M.M., eds., Paleozoic and early Mesozoic paleogeographic relations; Sierra Nevada, Klamath Mountains, and related terranes: Geological Society of America Special Paper 255, p. 17–32.

Wallin, E.T., 1993, Sonomia revisited: Evidence for a western Canadian provenance of the eastern Klamath and northern Sierra Nevada terranes: Geological Society of America Abstracts with Programs, v. 25, no. 6, p. A-173.

Webby, B., 1998, Steps toward a global standard for Ordovician stratigraphy: Newsletters in Stratigraphy, v. 36, p. 1–33.

Manuscript Accepted by the Society January 24, 2000

Printed in U.S.A.

Geological Society of America
Special Paper 347
2000

# *Detrital zircon geochronology of the Shoo Fly Complex, northern Sierra terrane, northeastern California*

**James P. Harding, George E. Gehrels**
*Department of Geosciences, University of Arizona, Tucson, Arizona 85721, USA*
**David S. Harwood**
*U.S. Geological Survey, 345 Middlefield Road, Menlo Park, California 94025, USA*
**Gary H. Girty**
*Department of Geological Sciences, San Diego State University, San Diego, California 92182, USA*

## ABSTRACT

**U-Pb analyses have been conducted on 92 individual detrital zircon grains from 4 of the main thrust sheets of the Shoo Fly Complex. Samples from the Culbertson Lake allochthon, Duncan Peak allochthon, and Lang sequence yield mainly 1.80–2.10, 2.20–2.45, and 2.55–2.70 Ga ages, which suggests that sediments in these units originated in a cratonal region containing Paleoproterozoic and Archean igneous rocks. These ages match those of basement provinces from the northwestern Canadian shield, suggesting a provenance link with northwestern North America during early Paleozoic time. The Sierra City melange, however, has significantly different zircon ages of 551–635 and 1170–1319 Ma, with only a subordinate population of >1.8 Ga grains. These grains apparently were derived originally from an outboard Neoproterozoic-Cambrian(?) volcanic arc and from 1.0–1.7 Ga basement rocks of southwestern North America. The occurrence of all three sets of ages in a sandstone that accumulated outboard of the Lang, Culbertson Lake, and Duncan Peak thrust sheets indicates that most rocks of the Shoo Fly Complex formed inboard of a volcanic arc located in proximity to the southern portion of the Cordilleran margin.**

## INTRODUCTION

The Shoo Fly Complex, located in northeastern California (Fig. 1), is an intensely deformed assemblage of eugeoclinal rocks of Cambrian(?) to pre-Late Devonian age that is unconformably overlain by upper Paleozoic arc deposits of the Taylorsville sequence (Girty et al., 1990; Harwood, 1988, 1992). Collectively the Shoo Fly Complex and overlying strata are referred to as the northern Sierra terrane (Schweickert et al., 1984; Harwood, 1988).

The Shoo Fly Complex is composed of four major thrust sheets, arranged in a northeast-dipping structural stack (Fig. 2) (Girty and Schweickert, 1984; Schweickert et al., 1984; Harwood, 1988). From east to west, and top to bottom, the thrust sheets include: (1) the Sierra City melange, (2) the Culbertson Lake allochthon, (3) the Duncan Peak allochthon, and (4) the Lang (or Lang-Halsted) sequence (Schweickert et al., 1984; Harwood, 1988). All four units and the thrusts that separate them are truncated by a major pre-Late Devonian unconformity that has been interpreted as a shallow-marine planation surface (Schweickert and Girty, 1981). The original configuration of strata in the thrust sheets is uncertain, but Schweickert et al. (1984) envisioned an arrangement opposite to the present order, with Sierra City melange to the west and Lang sequence toward the east.

The complex consists mainly of deep marine siliciclastic rocks that include bedded quartz- and feldspar-rich sandstone, siltstone, mudstone, chert, argillite, limestone, and basalt (Girty et al., 1996). Blocks of limestone, sandstone, bedded chert, gabbro, basalt, and volcanogenic sedimentary rocks also have been

Harding, J.P., et al., 2000, Detrital zircon geochronology of the Shoo Fly Complex, northern Sierra terrane, northeastern California, *in* Soreghan, M.J., and Gehrels, G.E., eds., Paleozoic and Triassic paleogeography and tectonics of western Nevada and northern California: Boulder, Colorado, Geological Society of America Special Paper 347, p. 43–55.

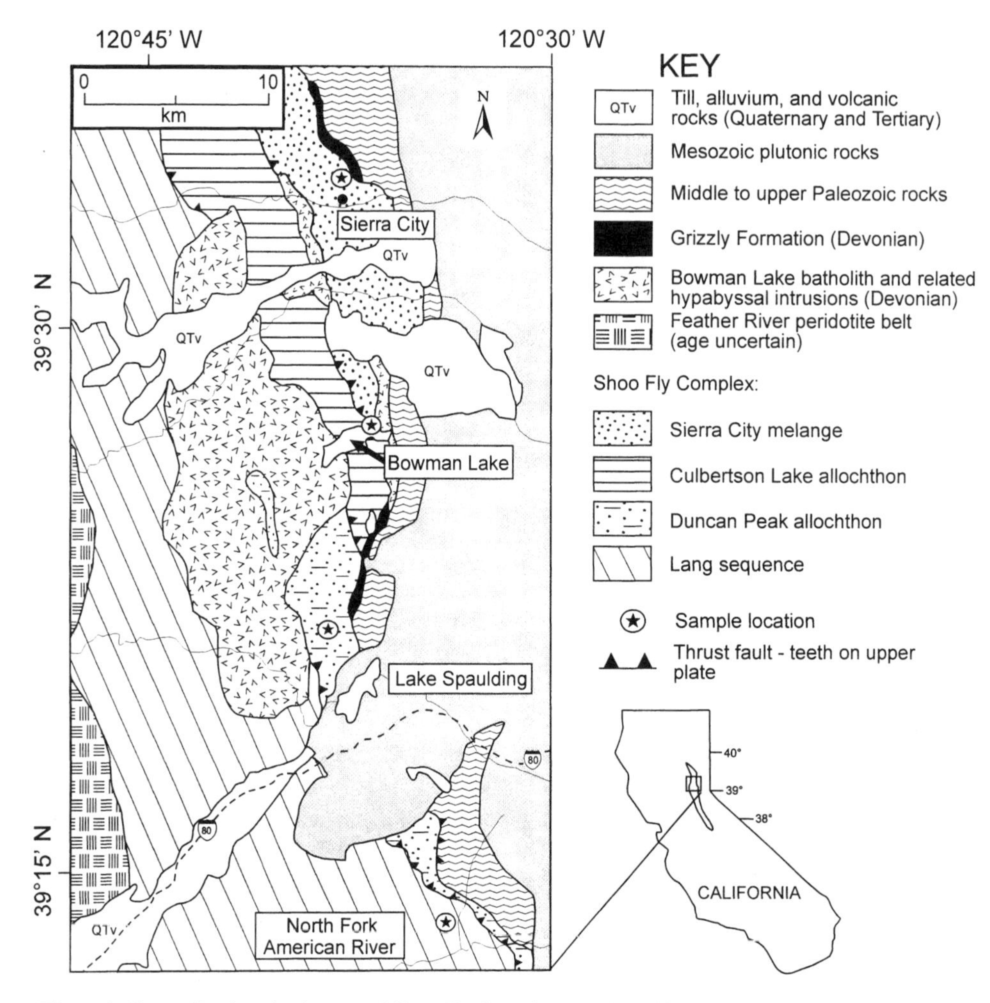


Figure 1. Generalized geologic map of Shoo Fly Complex, northern Sierra terrane. Modified from Harwood (1992) and Girty et al. (1996).

identified (Girty et al., 1996). Two major episodes of deformation—the first during early Paleozoic time, the second during the Mesozoic—have metamorphosed these rocks to a chlorite-grade greenschist facies, and workers have identified higher metamorphic grades in southern exposures of the complex (D'Allura et al., 1977; Bond and DeVay, 1980; Girty and Schweickert, 1979, 1984).

Although the Shoo Fly Complex is interpreted to have been structurally accreted to North America during the Permian-Triassic Sonoma orogeny (Speed, 1979; Schweickert and Snyder, 1981), its preaccretionary history is still debated. Some workers have argued that the complex developed in a continental slope-rise setting along the Cordilleran margin, deposited there by submarine fans (D'Allura et al., 1977; Schweickert and Snyder, 1981; Varga, 1982; Hannah and Moores, 1986). Others have argued in favor of deposition in a trench setting by submarine fans and trench-parallel axial channels (Girty et al., 1991, 1996). Both models agree that the Shoo Fly Complex was in close enough proximity to a continental provenance to receive sand-sized detritus. Moreover, petrological analyses conducted by Girty and Pardini (1987) and multigrain detrital zircon work by Girty and Wardlaw (1984) and Pardini (1986) strongly suggest that sediment was derived from both a continental block and a dissected magmatic arc of predominantly Cambrian age. This arc was originally hypothesized to be the Alexander terrane (Schweickert et al., 1984; Girty and Wardlaw, 1984), but detrital zircon work conducted by Gehrels et al. (1996) indicates that the Alexander terrane was probably not in the vicinity of the southwestern Cordilleran margin during Ordovician-Devonian time. It has also been suggested that the arc may reside somewhere along the Pacific rim after having been rifted away from its Ordovician-Devonian position off the coast of North America, or may remain buried under strata of western Nevada and/or northern California (Girty and Pardini, 1987; Girty et al., 1996).

In this chapter we present and summarize the results of recent U-Pb detrital zircon analyses that have been conducted on the four main units of the Shoo Fly Complex, and use the U-Pb data to attempt to identify potential original source areas from which the clastic strata may have been derived.

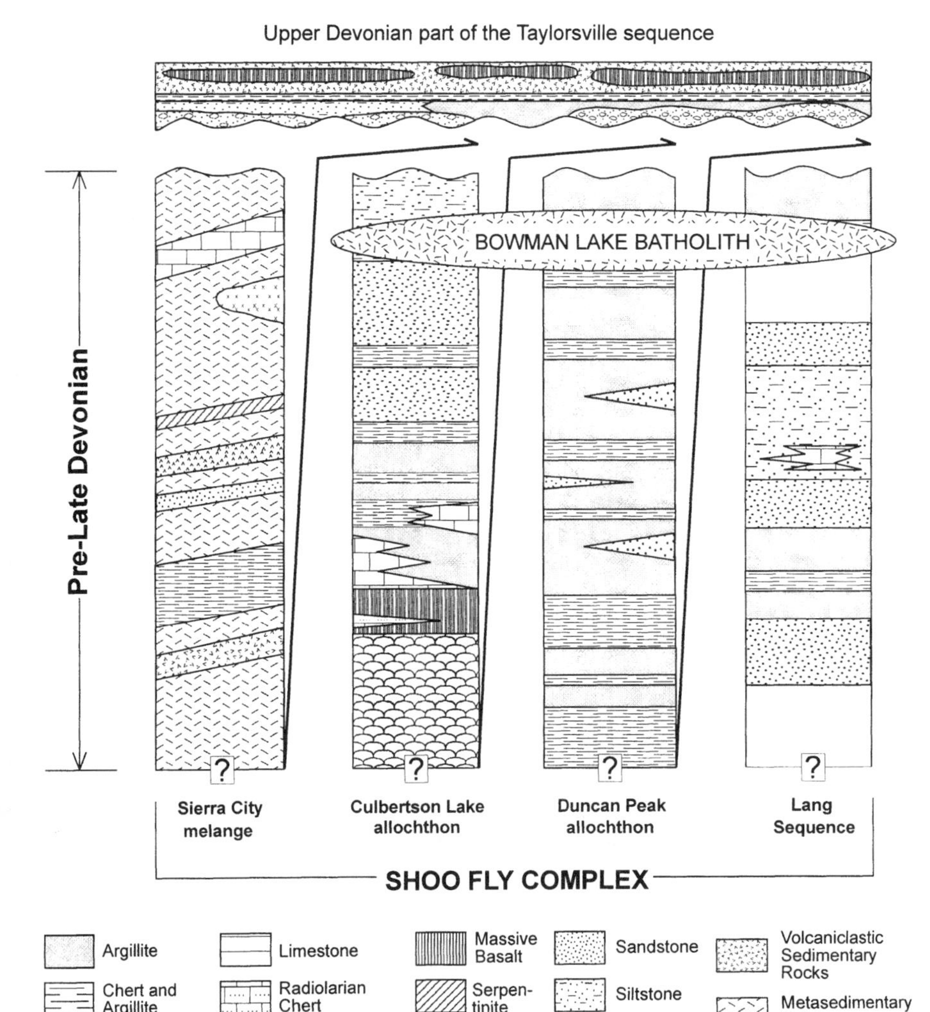


Figure 2. Schematic stratigraphic columns showing main thrust sheets of Shoo Fly Complex. Modified from Poole et al. (1992). Also shown are unconformably overlying Upper Devonian strata of Taylorsville sequence. Barbed lines between columns represent thrust faults that separate each of major sheets.

## SAMPLE PREPARATION AND SELECTION TECHNIQUES

We collected ~20 kg of rock from each of the 4 thrust sheets of the complex. Each sample yielded between 140 and 860 mg of detrital zircon grains that were isolated using standard mechanical, Wilfley table, heavy liquid, and magnetic techniques as outlined by Gehrels (this volume, Introduction). Zircons were washed in dilute $HNO_3$ for ~20 min, and then separated into several size fractions using disposable nylon sieve screens. Grains coarser than 125 μm in diameter were sorted into different groups based on their color and morphological characteristics; at least two representatives from each group were selected for analysis. Smaller size fractions were also investigated for grains with colors and/or morphologies not represented in the larger grain sizes. This was done for the purpose of maximizing the number of age groups recognized. Next, using standard air abrasion techniques, zircons were abraded for 3–6 hr at 3.0 psi, removing the outer ~30% of each grain. Zircons were then dissolved, U and Pb were isolated chemically, and the samples were analyzed by thermal ionization mass spectrometry (Gehrels, this volume, Introduction).

## RESULTS

### *Sierra City melange*

Schweickert et al. (1984) and Girty and Pardini (1987) described the Sierra City melange as consisting of a variably sheared mudstone or serpentinite matrix containing large slabs of sandstone, limestone, bedded chert, gabbro, and basalt. In the area north of Bowman Lake, Girty and Pardini (1987) identified (1) pods and lenses of chert, siltstone, and sandstone; (2) kilometer-size slices of chert; (3) a volcanogenic pebbly mudstone unit; and (4) a mass of conglomeratic mudstone.

Five lines of evidence indicate that the melange is mainly Ordovician to Devonian age. First, Late Ordovician megafossils and conodonts with North American affinity were identified within the Montgomery Limestone by Hannah and Moores (1986). Located in areas to the north of Sierra City, the Montgomery Limestone occurs either as exotic tectonic blocks (D'Allura et al., 1977) or as submarine slide blocks (Bond and DeVay, 1980). Second, Varga and Moores (1981) and Varga (1982) identified marine phosphate layers and nodules within black chert and phyllite that have yielded Ordovician-Silurian radiolarians. Third, U-Pb analyses on zircons obtained from a submarine tuff, also located to the north of Sierra City, yielded a Silurian age of 423 +5/–15 Ma (Saleeby et al., 1987). This tuff has been interpreted by other workers as a dike; if this is correct, the age of the melange is older than Late Silurian (see Girty et al., 1996). Fourth, the melange is unconformably overlain by Upper Devonian to Jurassic arc deposits that make up the rest of the northern Sierra terrane. Fifth, the Sierra City melange is intruded by the Late Devonian Wolf Creek stock (U-Pb zircon age of 378 +5/–10 Ma, as reported by Saleeby et al. [1987]).

Multigrain detrital zircon analyses from feldspar-rich and feldspar-poor sandstones within the melange yielded different U-Pb ages (Girty and Wardlaw, 1984; Pardini, 1986). The feldspathic sands yielded a Cambrian upper-intercept age (ca. 506 ± 22 Ma) but the quartz-rich sands yielded much older Precambrian ages. The feldspathic sandstones were interpreted to have originated from the plutonic roots of a dissected Cambrian magmatic arc, whereas the quartz-rich sandstones were interpreted to have an original provenance from a Precambrian craton (Girty and Pardini, 1987; Pardini, 1986; Girty et al., 1996).

Our sandstone sample was collected from an outcrop ~0.5 km north of Sierra City (Fig. 1), along the road to the Colombo Mine. The sample, from a bed about 1 m thick, is a medium-grained sandstone with vitreous quartz grains. Sandstone from this locality is feldspar poor, with moderate amounts of quartz grains and discernible volcanic, metavolcanic, plutonic, sedimentary, and metasedimentary lithic fragments (Dickinson and Gehrels, this volume). The sandstone sampled is moderately deformed, and beds are broken into meter-scale blocks enclosed in a more highly disrupted, finer grained matrix.

Zircons of several distinct morphologies occur within this sample. The majority of the grains are translucent, colorless to light pink or purple, and subrounded to very well rounded. A few euhedral grains are also present. The zircons analyzed for this study were all between 175 and 125 μm in diameter prior to abrasion.

Of the 30 zircon grains analyzed, 28 yielded ages that are concordant to slightly discordant and of moderate to high precision (Table 1; Fig. 3). The ages obtained are 551–635 Ma (n = 14), ca. 1086 Ma (n = 1), 1170–1281 Ma (n = 8), ca. 1318 Ma (n = 2), ca. 1666 Ma (n = 1), ca. 1760 Ma (n = 1), ca. 1835 Ma (n = 1), ca. 1971 Ma (n = 1), and ca. 2666 Ma (n = 1). In terms of age probability, there are four age groups: 540–585 Ma, 595-635 Ma, 1160–1195 Ma, and 1305–1330 Ma, with peaks at 561, 609, 1180, and 1318 Ma, respectively (Table 2).

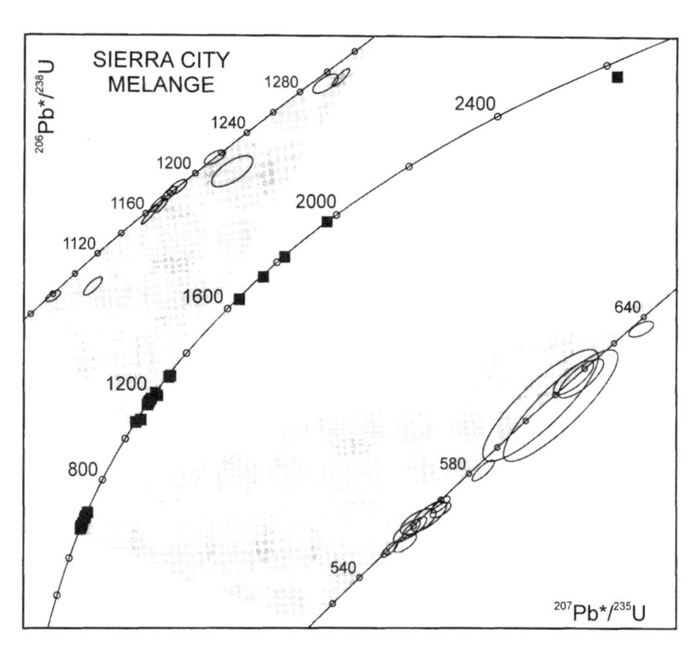


Figure 3. U-Pb concordia diagram of individual detrital zircon grains from Sierra City melange. All concordia diagrams were generated using programs of Ludwig (1991a, 1991b), and error ellipses show uncertainties at 95% confidence level.

### *Culbertson Lake allochthon*

Detailed studies by Schweickert (1981), Girty and Wardlaw (1985), and Girty et al. (1990, 1991) led to the recognition of six stratigraphic units within the Culbertson Lake allochthon. From structurally lowest to highest, these units are described as follows: (1) the ~1200-m-thick Bullpen Lake sequence, composed primarily of alkalic basalt interstratified with rhythmically bedded radiolarian chert (Texas Creek chert), both of which are, in turn, overlain by allodapic limestone; (2) a 75–150-m-thick section of chert and argillite, informally named the McMurray Lake chert; (3) the Poison Canyon and Red Hill units, which are dominated by massive to graded sandstone interlayered with thin beds of argillite; (4) the Toms Creek chert, a ~20–50-m-thick sequence of chert and argillite that separates the Poison Canyon and Red Hill units; and (5) a 2–15-cm-thick section of rhythmically bedded chert and argillite, referred to as the Quartz Hill chert.

The maximum age of the Culbertson Lake allochthon is not well constrained. Radiolarians and sponge spicules that are probably Cambrian to Late Devonian in age occur within bedded cherts (Girty and Wardlaw, 1985). The minimum age is constrained by the Middle to Late Devonian Bowman Lake batholith (U-Pb zircon age of 375 ± 10 Ma; Hanson et al., 1988), which intrudes the allochthon, and by the unconformably overlying deposits of the Upper Devonian Grizzly and Sierra Buttes Formations.

The quartz-rich framework of sandstones in the Culbertson Lake allochthon has led workers (e.g., Girty and Wardlaw, 1985) to conclude that detritus was derived from Precambrian crystalline rocks or recycled from older supracrustal rocks that originated from a Precambrian basement source. U-Pb analyses of multigrain detrital zircon fractions yielded an upper intercept age of 2087 ± 47 Ma, confirming a continental basement source (Girty and Wardlaw, 1985).

We collected 20 kg of rock along the northeastern shore of Bowman Lake (Fig. 1) from a continuous outcrop of well-

**TABLE 1. U-Pb ISOTOPIC DATA AND AGES**

| | | | | | | Apparent ages (Ma) | | | Interpreted age (Ma) |
|---|---|---|---|---|---|---|---|---|---|
| Grain type | Grain wt. (μg) | $Pb_c$ (pg) | U (ppm) | $^{206}Pb_m/^{204}Pb$ | $^{206}Pb_c/^{208}Pb$ | $^{206}Pb^*/^{238}U$ | $^{207}Pb^*/^{235}U$ | $^{207}Pb^*/^{206}Pb^*$ | |
| Sierra City melange (39°34'04"N, 120°38'58"W) | | | | | | | | | |
| **MsR** | **13** | **9** | **1086** | **8990** | **7.3** | **551 ± 3** | **551 ± 3** | **553 ± 8** | **551 ± 10** |
| DE | 13 | 15 | 1256 | 5920 | 6.3 | 553 ± 3 | 557 ± 4 | 572 ± 11 | 553 ± 20 |
| **MsR** | **9** | **14** | **627** | **2280** | **3.9** | **558 ± 4** | **558 ± 5** | **556 ± 13** | **558 ± 10** |
| **LsR** | **9** | **9** | **196** | **3230** | **5.3** | **559 ± 4** | **560 ± 5** | **565 ± 11** | **559 ± 10** |
| **LsR** | **3** | **6** | **589** | **3650** | **5.9** | **562 ± 4** | **562 ± 7** | **562 ± 21** | **561 ± 10** |
| **LsR** | **5** | **6** | **424** | **2200** | **5.9** | **564 ± 5** | **566 ± 7** | **573 ± 18** | **563 ± 10** |
| LsR | 12 | 8 | 618 | 5180 | 3.4 | 566 ± 3 | 569 ± 4 | 585 ± 11 | 566 ± 20 |
| **LsR** | **12** | **3** | **378** | **8850** | **3.5** | **568 ± 3** | **569 ± 4** | **576 ± 9** | **568 ± 10** |
| LsR | 10 | 4 | 505 | 7150 | 5.5 | 581 ± 3 | 585 ± 4 | 600 ± 9 | 581 ± 20 |
| CE | 7 | 16 | 36 | 460 | 5.0 | 606 ± 18 | 612 ± 21 | 633 ± 37 | 606 ± 20 |
| **CsR** | **13** | **7** | **38** | **498** | **4.3** | **606 ± 18** | **605 ± 22** | **600 ± 39** | **606 ± 10** |
| CsR | 2 | 4 | 403 | 1403 | 3.0 | 616 ± 6 | 619 ± 9 | 628 ± 18 | 616 ± 20 |
| CsR | 10 | 5 | 183 | 2380 | 3.6 | 617 ± 5 | 616 ± 7 | 615 ± 17 | **617 ± 10** |
| DE | 17 | 14 | 618 | 5020 | 7.6 | 635 ± 3 | 639 ± 5 | 652 ± 12 | 635 ± 20 |
| **DwR** | **8** | **19** | **736** | **3500** | **9.9** | **1078 ± 5** | **1080 ± 9** | **1085 ± 13** | **1086 ± 10** |
| **LwR** | **9** | **10** | **173** | **1990** | **4.4** | **1169 ± 7** | **1169 ± 8** | **1170 ± 8** | **1170 ± 10** |
| DsR | 9 | 15 | 1679 | 12200 | 20.3 | 1158 ± 9 | 1163 ± 9 | 1173 ± 5 | 1174 ± 10 |
| MwR | 10 | 13 | 124 | 1144 | 2.3 | 1088 ± 8 | 1116 ± 10 | 1171 ± 12 | 1177 ± 20 |
| **DsR** | **10** | **16** | **195** | **1540** | **4.8** | **1179 ± 6** | **1179 ± 9** | **1179 ± 9** | **1179 ± 10** |
| MwR | 10 | 5 | 533 | 12700 | 7.1 | 1168 ± 6 | 1173 ± 7 | 1183 ± 5 | 1184 ± 10 |
| **MsR** | **4** | **10** | **618** | **3100** | **14.6** | **1186 ± 7** | **1186 ± 11** | **1185 ± 12** | **1185 ± 10** |
| **MwR** | **10** | **29** | **199** | **900** | **3.2** | **1216 ± 6** | **1215 ± 11** | **1214 ± 7** | **1215 ± 10** |
| CwR | 8 | 18 | 87 | 520 | 3.7 | 1202 ± 14 | 1228 ± 22 | 1276 ± 26 | 1281 ± 20 |
| DwR | 6 | 9 | 385 | 3760 | 10.3 | 1288 ± 9 | 1299 ± 13 | 1317 ± 15 | 1318 ± 10 |
| CwR | 14 | 12 | 379 | 6200 | 3.0 | 1294 ± 8 | 1303 ± 9 | 1318 ± 5 | 1319 ± 10 |
| DsR | 12 | 34 | 1164 | 7330 | 13.1 | 1641 ± 10 | 1651 ± 12 | 1665 ± 7 | 1666 ± 10 |
| DwR | 16 | 34 | 422 | 3805 | 8.9 | 1738 ± 15 | 1748 ± 17 | 1759 ± 7 | 1760 ± 10 |
| CsR | 10 | 4 | 658 | 34000 | 12.1 | 1823 ± 7 | 1829 ± 8 | 1835 ± 4 | 1835 ± 10 |
| **LsR** | **15** | **9** | **108** | **3884** | **7.4** | **1974 ± 13** | **1974 ± 16** | **1971 ± 7** | **1971 ± 10** |
| LsR | 11 | 13 | 100 | 2380 | 5.0 | 2559 ± 14 | 2617 ± 15 | 2665 ± 4 | 2666 ± 20 |
| Culbertson Lake allochthon (39°27'32"N, 120°37'18"W) | | | | | | | | | |
| **MwR** | **26** | **29** | **497** | **8360** | **9.6** | **1790 ± 11** | **1790 ± 12** | **1789 ± 6** | **1790 ± 10** |
| **LsR** | **17** | **8** | **180** | **7600** | **4.9** | **1835 ± 8** | **1836 ± 10** | **1836 ± 6** | **1836 ± 10** |
| **LE** | **42** | **8** | **73** | **8200** | **6.5** | **1913 ± 17** | **1914 ± 18** | **1915 ± 6** | **1915 ± 10** |
| **LsR** | **17** | **37** | **183** | **1690** | **15.1** | **1918 ± 8** | **1917 ± 10** | **1916 ± 6** | **1916 ± 10** |
| **MsR** | **28** | **12** | **218** | **10100** | **7.9** | **1916 ± 10** | **1917 ± 12** | **1919 ± 6** | **1919 ± 10** |
| MwR | 17 | 13 | 376 | 10180 | 4.3 | 1918 ± 13 | 1923 ± 12 | 1929 ± 4 | 1929 ± 10 |
| MwR | 19 | 14 | 489 | 13800 | 5.0 | 1900 ± 13 | 1917 ± 13 | 1935 ± 4 | 1936 ± 10 |
| **LsR** | **15** | **8** | **110** | **4400** | **5.5** | **1998 ± 10** | **2000 ± 12** | **2003 ± 6** | **2003 ± 10** |
| LwR | 17 | 9 | 88 | 3550 | 2.2 | 2045 ± 13 | 2057 ± 14 | 2069 ± 7 | 2070 ± 10 |
| MwR | 17 | 13 | 405 | 12800 | 17.8 | 2196 ± 13 | 2242 ± 14 | 2283 ± 4 | 2285 ± 20 |
| **LE** | **44** | **9** | **81** | **9800** | **5.1** | **2305 ± 12** | **2308 ± 13** | **2310 ± 4** | **2310 ± 10** |
| MsR | 32 | 8 | 254 | 28000 | 3.9 | 2319 ± 16 | 2336 ± 18 | 2350 ± 5 | 2351 ± 10 |
| **MsR** | **26** | **8** | **264** | **23700** | **4.8** | **2352 ± 14** | **2353 ± 16** | **2355 ± 5** | **2355 ± 10** |
| MsR | 43 | 17 | 389 | 25600 | 4.5 | 2343 ± 20 | 2358 ± 21 | 2370 ± 5 | 2371 ± 10 |
| LwR | 15 | 11 | 215 | 7400 | 9.3 | 2378 ± 11 | 2397 ± 12 | 2413 ± 4 | 2414 ± 10 |
| **MwR** | **32** | **12** | **111** | **8500** | **3.3** | **2595 ± 12** | **2596 ± 13** | **2597 ± 4** | **2597 ± 10** |
| MsR | 37 | 7 | 173 | 3756 | 14.0 | 2589 ± 11 | 2596 ± 13 | 2603 ± 4 | 2603 ± 10 |
| **MsR** | **25** | **22** | **291** | **34000** | **9.8** | **2645 ± 18** | **2650 ± 20** | **2653 ± 5** | **2654 ± 10** |
| LsR | 19 | 32 | 139 | 2410 | 4.3 | 2632 ± 16 | 2661 ± 19 | 2682 ± 6 | 2683 ± 10 |
| CwR | 48 | 10 | 30 | 4160 | 16.7 | 2616 ± 20 | 2654 ± 21 | 2684 ± 5 | 2685 ± 10 |

exposed sandstone turbidites. The sample is poorly sorted and coarse grained, and contains a few quartz granules and almost no feldspar and/or micaceous material. The rocks from this locality are dominated by monocrystalline quartz grains (91%) and polycrystalline quartzose grains (6%), with only a minor amount (3%) of sedimentary and metasedimentary lithic fragments (Dickinson and Gehrels, this volume).

Translucent, colorless to medium pinkish-purple zircons that are mostly subrounded to well rounded were isolated. Grains with preserved crystal faces, labeled as euhedral in Table 1, are

**TABLE 1. U-Pb ISOTOPIC DATA AND AGES (continued)**

| Grain type | Grain wt. (μg) | $Pb_c$ (pg) | U (ppm) | $^{206}Pb_m/^{204}Pb$ | $^{206}Pb_c/^{208}Pb$ | Apparent ages (Ma) $^{206}Pb^*/^{238}U$ | $^{207}Pb^*/^{235}U$ | $^{207}Pb^*/^{206}Pb^*$ | Interpreted age (Ma) |
|---|---|---|---|---|---|---|---|---|---|
| Duncan Peak allochthon (39°21'28"N, 120°38'24"W) | | | | | | | | | |
| LsR | 23 | 12 | 114 | 4 | 2.0 | 1776 ± 10 | 1794 ± 12 | 1815 ± 7 | 1817 ± 10 |
| LsR | 30 | 14 | 96 | 4140 | 6.1 | 1860 ± 12 | 1872 ± 14 | 1886 ± 6 | 1887 ± 10 |
| LsR | 55 | 13 | 145 | 12160 | 3.8 | 1884 ± 9 | 1893 ± 10 | 1903 ± 4 | 1903 ± 10 |
| LsR | 27 | 33 | 124 | 2050 | 2.1 | 1895 ± 9 | 1900 ± 12 | 1904 ± 7 | 1906 ± 10 |
| CwR | 13 | 18 | 91 | 1288 | 8.3 | 1882 ± 12 | 1894 ± 14 | 1908 ± 7 | 1909 ± 10 |
| MwR | 77 | 8 | 212 | 39600 | 6.2 | 1851 ± 13 | 1879 ± 14 | 1910 ± 6 | 1912 ± 10 |
| LsR | 19 | 7 | 84 | 4690 | 3.5 | 1886 ± 10 | 1902 ± 14 | 1919 ± 9 | 1921 ± 10 |
| LsR | 36 | 27 | 273 | 7312 | 4.3 | 1877 ± 12 | 1897 ± 14 | 1919 ± 6 | 1921 ± 10 |
| MwR | 56 | 17 | 169 | 11640 | 1.7 | 1906 ± 15 | 1615 ± 15 | 1923 ± 4 | 1924 ± 10 |
| MwR | 27 | 85 | 318 | 2004 | 4.4 | 1886 ± 11 | 1907 ± 14 | 1930 ± 7 | 1931 ± 10 |
| MwR | 18 | 9 | 312 | 13820 | 3.9 | 1917 ± 12 | 1923 ± 13 | 1931 ± 4 | 1932 ± 10 |
| CwR | 25 | 14 | 119 | 3605 | 4.3 | 1696 ± 10 | 1817 ± 13 | 1959 ± 7 | 1968 ± 40 |
| MsR | 15 | 13 | 462 | 9590 | 9.9 | 1710 ± 8 | 1835 ± 10 | 1981 ± 6 | 1990 ± 40 |
| MwR | 39 | 16 | 135 | 7220 | 4.1 | 1998 ± 11 | 2034 ± 14 | 2070 ± 6 | 2071 ± 10 |
| LsR | 27 | 9 | 135 | 8690 | 4.8 | 2133 ± 11 | 2178 ± 14 | 2220 ± 7 | 2222 ± 10 |
| CwR | 27 | 9 | 130 | 9050 | 5.6 | 2137 ± 16 | 2196 ± 18 | 2251 ± 6 | 2254 ± 20 |
| LsR | 34 | 12 | 326 | 23520 | 11.9 | 2299 ± 16 | 2310 ± 18 | 2321 ± 5 | 2322 ± 10 |
| **LsR** | **37** | **12** | **160** | **14730** | **5.8** | **2563 ± 15** | **2567 ± 19** | **2570 ± 7** | **2571 ± 10** |
| LsR | 51 | 8 | 88 | 14940 | 3.1 | 2460 ± 12 | 2522 ± 15 | 2573 ± 5 | 2576 ± 20 |
| LsR | 17 | 19 | 304 | 7530 | 7.6 | 2610 ± 15 | 2619 ± 17 | 2627 ± 5 | 2628 ± 10 |
| *CwR* | *17* | *55* | *110* | *702* | *5.8* | *1950 ± 11* | *2322 ± 18* | *2668 ± 9* | *2686 ± 40* |
| CsR | 17 | 16 | 217 | 5125 | 6.2 | 2170 ± 10 | 2467 ± 14 | 2722 ± 6 | *2735 ± 40* |
| Lang sequence (39°13'43"N, 120°34'41"W) | | | | | | | | | |
| MwR | 11 | 12 | 139 | 2450 | 4.2 | 1763 ± 17 | 1793 ± 19 | 1829 ± 8 | 1831 ± 10 |
| **LwR** | **49** | **58** | **203** | **3140** | **3.4** | **1835 ± 15** | **1834 ± 18** | **1833 ± 7** | **1833 ± 10** |
| LwR | 130 | 19 | 59 | 7700 | 3.6 | 1825 ± 13 | 1830 ± 14 | 1836 ± 6 | 1836 ± 10 |
| MwR | 51 | 15 | 175 | 11500 | 2.4 | 1830 ± 10 | 1834 ± 11 | 1838 ± 4 | 1839 ± 10 |
| MwR | 8 | 195 | 251 | 208 | 3.9 | 1832 ± 18 | 1837 ± 33 | 1844 ± 26 | 1844 ± 20 |
| **MwR** | **56** | **13** | **127** | **11400** | **7.6** | **1870 ± 11** | **1871 ± 12** | **1872 ± 6** | **1872 ± 10** |
| LsR | 27 | 9 | 85 | 4470 | 4.7 | 1859 ± 15 | 1878 ± 15 | 1899 ± 7 | 1900 ± 10 |
| **LsR** | **47** | **9.7** | **80** | **7980** | **3.5** | **1909 ± 12** | **1912 ± 13** | **1916 ± 5** | **1917 ± 10** |
| MsR | 30 | 15 | 358 | 15300 | 7.5 | 1936 ± 1 | 1951 ± 13 | 1966 ± 6 | 1967 ± 10 |
| CsR | 32 | 9 | 30 | 2280 | 2.9 | 1953 ± 9 | 1960 ± 11 | 1967 ± 5 | 1967 ± 10 |
| **CwR** | **46** | **21** | **41** | **2000** | **8.8** | **2070 ± 12** | **2073 ± 15** | **2075 ± 6** | **2075 ± 10** |
| LwR | 102 | 36 | 42 | 2600 | 6.0 | 1973 ± 15 | 2023 ± 17 | 2075 ± 7 | 2078 ± 20 |
| MwR | 121 | 63 | 233 | 10700 | 3.3 | 2290 ± 25 | 2295 ± 26 | 2301 ± 4 | 2301 ± 10 |
| MwR | 9 | 7 | 285 | 10400 | 6.5 | 2301 ± 18 | 2343 ± 19 | 2379 ± 6 | 2382 ± 10 |
| MsR | 44 | 33 | 114 | 4170 | 6.9 | 2534 ± 10 | 2549 ± 12 | 2561 ± 5 | 2562 ± 10 |
| **CsR** | **26** | **30** | **29** | **660** | **3.4** | **2568 ± 24** | **2570 ± 28** | **2572 ± 9** | **2572 ± 10** |
| **MwR** | **32** | **11** | **292** | **27070** | **4.1** | **2671 ± 16** | **2669 ± 17** | **2668 ± 4** | **2668 ± 10** |
| LsR | 46 | 16 | 75 | 6400 | 2.6 | 2673 ± 16 | 2682 ± 19 | 2689 ± 5 | 2689 ± 10 |
| MsR | 75 | 14 | 131 | 20510 | 2.0 | 2672 ± 14 | 2683 ± 15 | 2691 ± 4 | 2692 ± 10 |
| MwR | 9 | 7 | 197 | 10700 | 7.7 | 3199 ± 15 | 3243 ± 18 | 3271 ± 5 | 3272 ± 20 |

*Note:*

Grain with $^{206}Pb^*/^{238}U$ age not within 15% of $^{207}Pb^*/^{206}Pb^*$ age in italics.
Grain with $^{206}Pb^*/^{238}U$ age within 15% of $^{207}Pb^*/^{206}Pb^*$ age.
Grain with concordant age and moderate to high precision analysis in bold.
* radiogenic Pb
Grain type: L—light pink; M—medium pink; D—dark pink; C—colorless; sR—slightly rounded; wR—well rounded; E—euhedral. Unless otherwise noted, all grains are clear and/or translucent and abraded by air abrasion.
$^{206}Pb/^{204}Pb$ is measured ratio, uncorrected for blank, spike, or fractionation.
$^{206}Pb/^{208}Pb$ is corrected for blank, spike, and fractionation.
Most concentrations have an uncertainty of 25% due to uncertainty in weight of grain.
Constants used: $^{238}U/^{235}U$ = 137.88. Decay constant for $^{235}U = 9.8485 \times 10^{-10}$. Decay constant for $^{238}U = 1.55125 \times 10^{-10}$.
All uncertainties are at the 95% confidence level.
Pb blank ranged from 2 to 10 pg. U blank was consistently <1 pg.
Interpreted ages for concordant grains are $^{206}Pb^*/^{238}U$ ages if <1.0 Ga and $^{207}Pb^*/^{206}Pb^*$ ages if >1.0 Ga.
Interpreted ages for discordant grains are projected from 100 Ma, which is a reasonable approximate age for Pb loss in the region (Gehrels, this volume, Introduction).
All analyses conducted using conventional isotope dilution and thermal ionization mass spectrometry, as described by Gehrels (this volume, Introduction).

rare. The analysis included only grains with diameters between 250 and 175 μm prior to abrasion.

We analyzed 20 zircons, all of which yielded concordant to slightly discordant ages (Table 1; Fig. 4). U-Pb data yield the following age distribution: ca. 1790 Ma (n = 1), ca. 1836 Ma (n = 1), 1915–1936 Ma (n = 5), ca. 2003 Ma (n = 1), ca. 2070 Ma (n = 1), 2285–2414 Ma (n = 6), and 2597–2685 Ma (n = 5). These ages define several groups of age probability: 1905–1945 Ma, 2265–2320 Ma, 2340–2380 Ma, 2585–2610 Ma, and 2670–2695 Ma, with a dominant peak at 1917 Ma and subordinate peaks at 1931, 2310, 2353, 2370, 2600, and 2689 Ma (Table 2).

### *Duncan Peak allochthon*

The Duncan Peak allochthon structurally overlies the Lang sequence and includes strata of the Zion Hill and Fuller Lake sequences (Taylor, 1986; Richards, 1990). In general, this allochthon consists predominantly of chert, chert breccia, and black siliceous argillite (Schweickert et al., 1984; Harwood, 1992). The Zion Hill sequence consists of ~90 m of quartzose sandstone and argillite, and the Fuller Lake sequence is composed of ~90 m of bedded chert and argillite with only minor sandstone (Taylor, 1986; Richards, 1990).

Like the Culbertson Lake allochthon, the maximum age of strata in this allochthon is not well constrained. Although radiolarian and sponge spicule-bearing chert occurs in the Duncan Peak allochthon, attempts to extract and date the fossils were unsuccessful (Schweickert et al., 1984; Girty et al., 1996). Minimum age constraints include the pre-Late Devonian unconformity that separates the Shoo Fly Complex from the overlying Taylorsville sequence, and the Middle to Late Devonian Bowman Lake batholith, which also intrudes this allochthon.

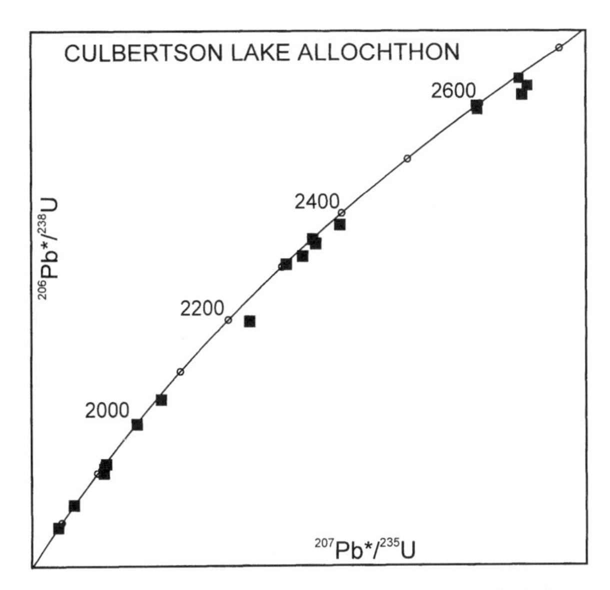


Figure 4. U-Pb concordia diagram of individual detrital zircon grains from Culbertson Lake allochthon.

Richards (1990) conducted multigrain U-Pb analyses on detrital zircons collected from quartz-rich sands of the Zion Hill and obtained predominantly Proterozoic $^{207}Pb^*/^{206}Pb^*$ ages. These results, combined with rare earth element data (Girty et al., 1996), suggest that sandstones of the Duncan Peak allochthon contain detritus shed primarily from a Precambrian continental block.

Our sample was collected from the Zion Hill sequence (Fig. 1) near Lake Spaulding, from a 1-m-thick, quartz-rich sandstone interlayered with pelite. A few chert layers interbedded with pelite and quartzite are also observed at this locality. The sandstone from this locality is coarse grained and poorly sorted.

The sample yielded translucent, colorless to pinkish-purple zircons that are subrounded to well rounded. Zircon grains ranging from 250 to 175 μm in diameter prior to abrasion were analyzed.

The results of our U-Pb analyses are presented in Table 1 and Figure 5. Of 22 grains analyzed, 20 are concordant to slightly discordant and two are highly discordant. The following ages were obtained: ca. 1817 Ma (n = 1), ca. 1887 Ma (n = 1), 1903–1932 Ma (n = 9), 1968–1990 Ma (n = 2), ca. 2071 Ma (n = 1), 2222–2322 Ma (n = 3), and 2571–2735 Ma (n = 5). There is one main age probability group of 1820-1960 Ma, with dominant peaks at 1908 and 1923 Ma, and subordinate groups of

**TABLE 2. DOMINANT DETRITAL ZIRCON AGE GROUPS IN STRATA OF THE SHOO FLY COMPLEX**

| Age range (Ma) | Peak ages (Ma) | Relative abundance |
|---|---|---|
| Sierra City melange | | |
| 540–585 | 561 | 0.29 |
| 595–635 | 609 | 0.15 |
| 1160–1195 | 1180 | 0.20 |
| 1305–1330 | 1318 | 0.07 |
| Culbertson Lake allochthon | | |
| 1905–1945 | 1917*, 1931 | 0.25 |
| 2265–2320 | 2310 | 0.10 |
| 2340–2380 | 2353*, 2370 | 0.15 |
| 2585–2610 | 2600 | 0.10 |
| 2670–2695 | 2689 | 0.10 |
| Duncan Peak allochthon | | |
| 1880–1940 | 1887, 1908, 1923* | 0.45 |
| 1955–2005 | 1979 | 0.07 |
| 2555–2595 | 2572 | 0.09 |
| 2650–2760 | 2705 | 0.09 |
| Lang sequence | | |
| 1820–1860 | 1835 | 0.25 |
| 1955–1975 | 1967 | 0.10 |
| 2060–2095 | 2075 | 0.10 |
| 2555–2580 | 2567 | 0.10 |
| 2660–2700 | 2690 | 0.10 |

*Note*: These ages and relative abundances refer to the age-probability plots shown in Figure 7.
* Indicates the peak with the highest abundance.

1955–2005 Ma, 2555–2595 Ma, and 2650–2760 Ma, with peaks at 1979, 2600, and 2689 Ma, respectively (Table 2).

### *Lang sequence*

The stratigraphy of part of the complexly deformed Lang sequence was described by Girty et al. (1991, 1996). It includes an ~300-m-thick section of massive and graded sandstone beds interbedded with thin (~2.5–12.0 cm) layers of mudstone and argillite.

The Lang sequence is unfossiliferous (Girty et al., 1991, 1996). However, units considered to be part of the Lang sequence ~150 km south of Bowman Lake are intruded by Late Devonian granitic gneisses (Sharp et al., 1982). Furthermore, this unit is unconformably overlain by the Upper Devonian Grizzly and Sierra Buttes Formations. The maximum age for this unit is uncertain.

Previous multigrain analyses of detrital zircons from the Lang sequence yielded $^{207}Pb^{*}/^{206}Pb^{*}$ages from 1.8 to 2.4 Ga (Girty et al., 1991). Point count and chemical analyses by Girty et al. (1991, 1996), Taylor (1986), and Richards (1990) confirm that the detritus was derived from a continental source.

We collected a sample of poorly sorted sandstone near the North Fork of the American River (Fig. 1). In this area the sandstones are amalgamated turbidites ~1–2 m thick. Our sample is very similar to sandstone of the Culbertson Lake allochthon, and is dominated by monocrystalline quartz (90%) and polycrystalline quartz (7%), with only a small amount (3%) of sedimentary and metasedimentary lithic fragments (Dickinson and Gehrels, this volume).

Translucent, colorless to pinkish-purple zircons were recovered from this sample, and most are subrounded to well rounded. Zircons analyzed ranged from 250 to 125 μm in diameter prior to abrasion. Zircons that contained large and/or numerous opaque inclusions were avoided.

Of the 20 grains analyzed, all yielded concordant to moderately discordant ages (Table 1; Fig. 6). The following zircon age distribution was obtained: 1831–1967 Ma (n = 10), ca. 2077 Ma (n = 2), ca. 2301 Ma (n = 1), ca. 2382 Ma (n = 1), 2562–2692 Ma (n = 5), and ca. 3272 Ma (n = 1). Age-probability groups include a single dominant group of 1820–1860 Ma, with a peak at 1835 Ma, and subordinate groups of 1955–1975 Ma, 2060–2095 Ma, 2555–2580 Ma, and 2660–2700 Ma with peaks at 1967, 2075, 2567, and 2690 Ma, respectively (Table 2).

## TECTONIC IMPLICATIONS

Sandstones in the Culbertson Lake allochthon, Duncan Peak allochthon, and Lang sequence share a similar suite of detrital zircon ages: 1.80–2.10 Ga, 2.20–2.45 Ga, and 2.55–2.70 Ga (Fig. 7; Tables 1 and 2). Detrital zircon ages for all three units accordingly are grouped together into one age-probability curve in Figure 8, which we interpret as representative of much of the Shoo Fly Complex. The presence of Paleoproterozoic and Archean zircons in these samples supports the conclusions of Girty and Wardlaw (1985), Pardini (1986), and Girty et al. (1991), that quartz-rich sandstones of the Shoo Fly Complex were derived from a cratonic source region.

To test for possible provenance links with the North American craton, the ages of detrital zircons in the Shoo Fly Complex are compared with the ages of grains from miogeoclinal strata (Figs. 8 and 9). A temporal and latitudinal reference for the ages

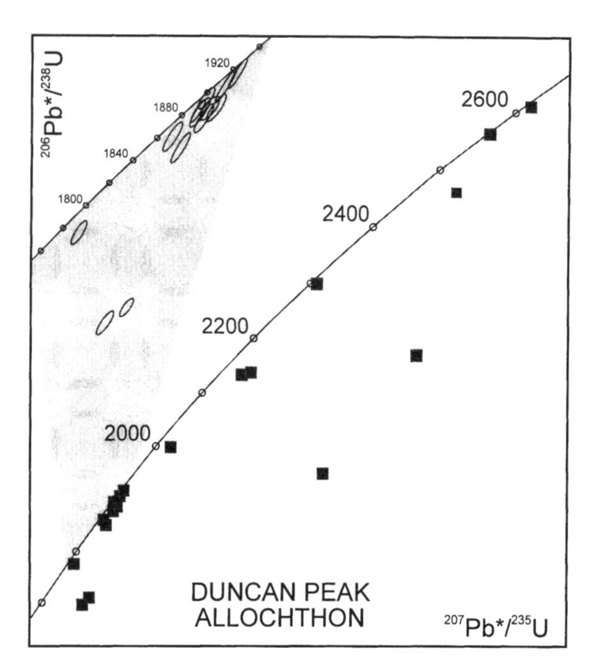


Figure 5. U-Pb concordia diagram of individual detrital zircon grains from Duncan Peak allochthon. Error ellipses are shown at 95% confidence level.

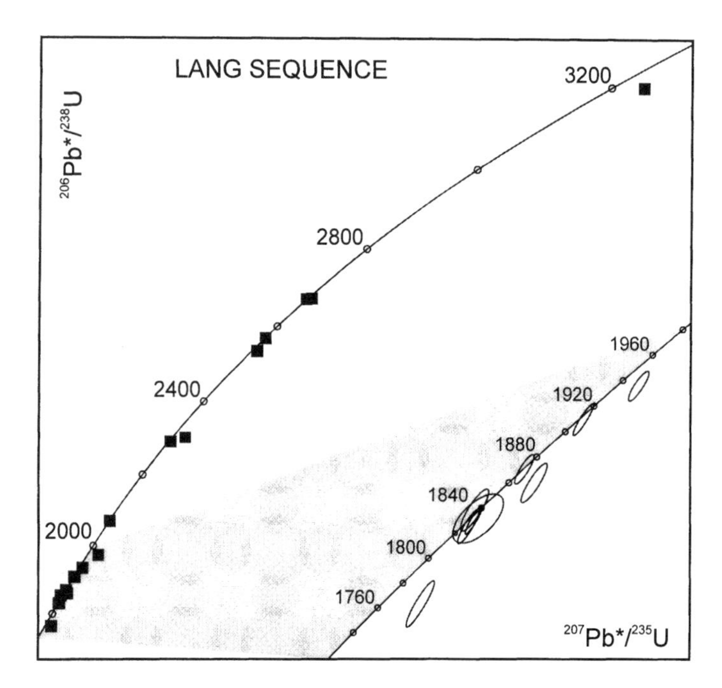


Figure 6. U-Pb concordia diagram of individual detrital zircon grains from Lang sequence.

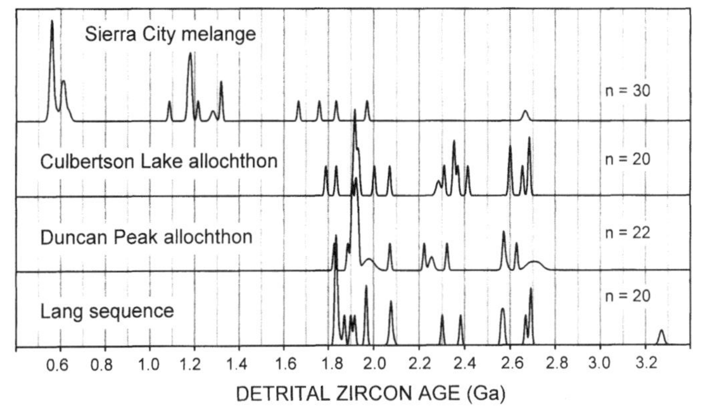


Figure 7. Relative age-probability diagrams showing detrital zircon ages from sandstones of Shoo Fly Complex. N is number of grains analyzed.

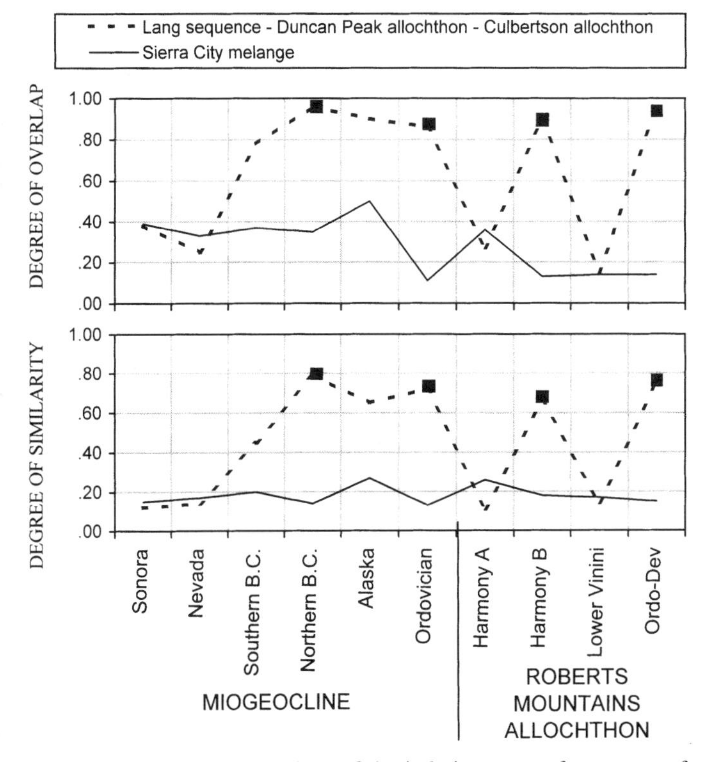


Figure 9. Statistical comparison of detrital zircon ages from strata of Shoo Fly Complex, Roberts Mountains allochthon, and Cordilleran miogeocline. Degree of overlap indicates whether ages in Shoo Fly samples are present in reference age spectra, whereas degree of similarity also factors in whether proportions of ages are similar. Filled squares emphasize high degrees of overlap and similarity. B.C.—British Columbia.

of zircon grains that accumulated within the Cordilleran miogeocline during Paleozoic to early Mesozoic time has been reported by Gehrels et al. (1995; Gehrels, this volume, Introduction). Neoproterozoic-Cambrian and Devonian through Triassic miogeoclinal strata generally yield zircon ages that are similar to the ages of nearby Precambrian basement rocks (Fig. 10). In contrast, Ordovician miogeoclinal strata along the Cordilleran margin, from northern British Columbia to northern Mexico, yield detrital

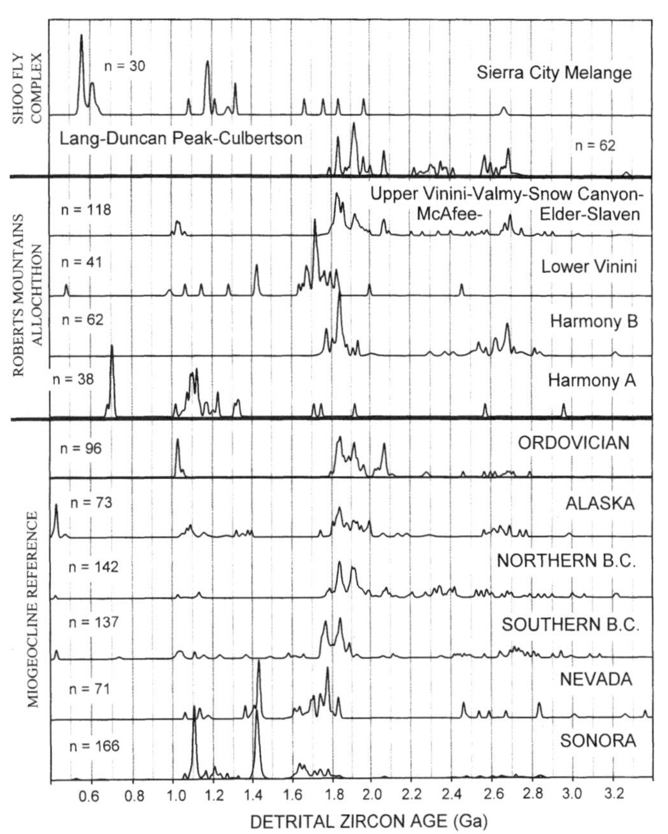


Figure 8. Relative age-probability diagrams for strata of Shoo Fly Complex, Roberts Mountains allochthon, and Cordilleran miogeocline. Samples from Lang sequence, Duncan Peak allochthon, and Culbertson Lake allochthon are sufficiently similar (Fig. 7) that they have been combined into one composite curve. Miogeoclinal reference consists of ages of 564 single detrital zircons analyzed from Neoproterozoic-Cambrian and Devonian through Triassic strata all along Cordilleran margin (Fig. 10), and 96 grains from Ordovician strata of Nevada and British Columbia (B.C.) (Gehrels et al., 1995). Ages for strata of Roberts Mountains allochthon are from Gehrels et al. (this volume, Chapter 1).

zircon ages that match the ages of basement rocks in northern British Columbia. Hence, the ages of grains in Ordovician strata are shown as a separate curve in Figure 8.

Qualitatively the composite Lang–Culbertson Lake–Duncan Peak curve appears quite different from curves for the southern Cordillera, but resembles curves for northern British Columbia (Fig. 8). The results of two statistical comparisons of these curves are shown in Figure 9. As described by Gehrels (this volume, Introduction), degree of overlap indicates whether similar ages are present in the Shoo Fly curve and a miogeoclinal curve, whereas degree of similarity factors in the proportions of different ages in the two curves. In both cases, a high value indicates that similar ages are present in both sets of samples. The high values for northern British Columbia suggest that basement rocks in northwestern Canada are a likely ultimate source for the detritus in sandstones of much of the Shoo Fly Complex. However, the maturity of Shoo Fly sandstones suggests that their detritus has been recycled through older platformal and/or shelf facies strata prior to accu-

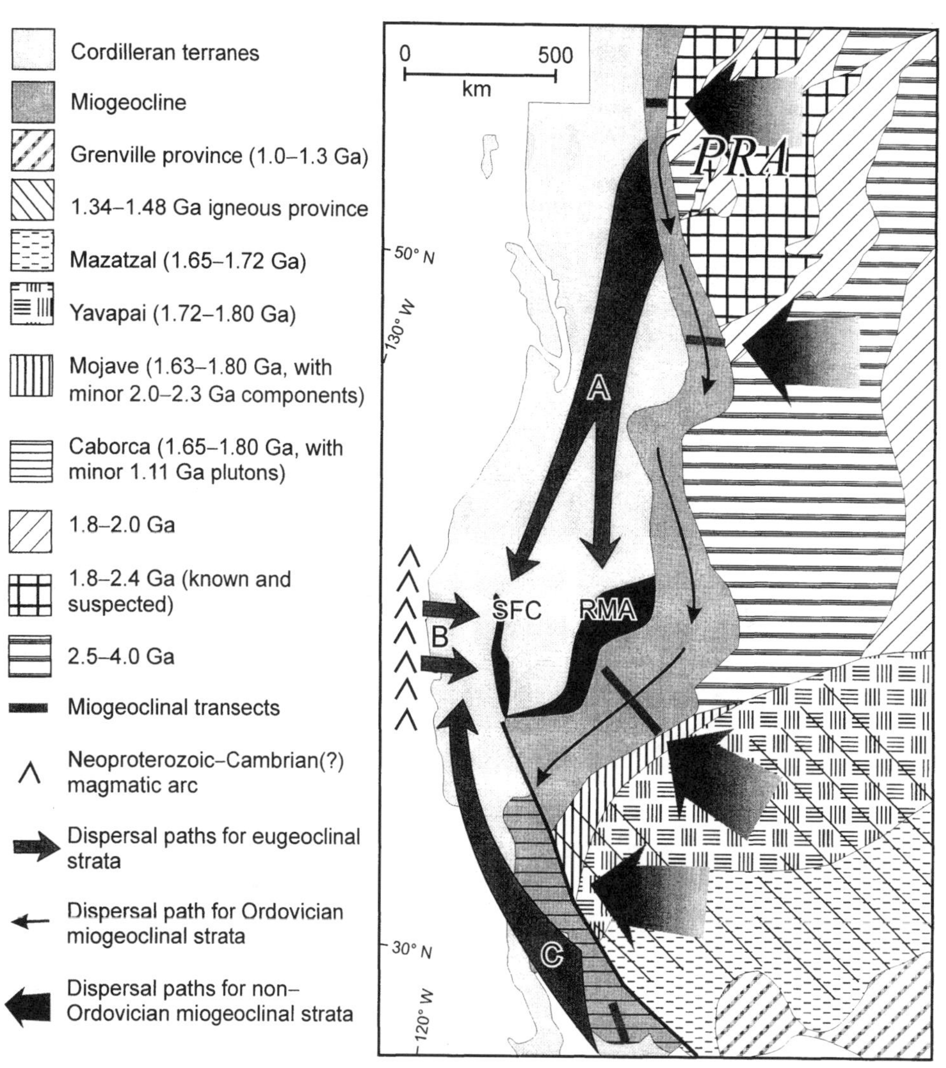


Figure 10. Schematic map showing present location of Shoo Fly Complex in relation to Cordilleran miogeocline, first-order basement provinces of western North America, and interpreted sedimentary dispersal patterns for lower Paleozoic miogeoclinal and eugeoclinal strata. Also shown are locations of four transects that yielded miogeoclinal detrital zircon reference (Gehrels et al., 1995). Basement provinces are simplified from Hoffman (1989) for cratonic interior, Ross (1991) and Villeneuve et al. (1993) for western Canadian shield, Van Schmus et al. (1993) for southwestern United States, and Stewart et al. (1990) for northwestern Mexico. PRA—Peace River arch, SFC—Shoo Fly Complex, RMA—Roberts Mountains allochthon, A—transport of detritus from Peace River arch region of western Canada, B—transport of detritus from outboard Neoproterozoic-Cambrian(?) magmatic arc into Sierra City melange, C—transport of detritus into Sierra City melange from 1.0–1.7 Ga basement rocks of southwestern North America.

mulation in deep-sea settings. This recycling probably occurred along the margins of the Peace River arch (Fig. 10), which was emergent during much of early Paleozoic time (Fritz et al., 1991).

The Lang–Culbertson Lake–Duncan Peak curve is also compared with age-probability curves for strata of the Roberts Mountains allochthon (Gehrels et al., this volume, Chapter 1) (Fig. 9). This comparison is appropriate because the Roberts Mountains allochthon includes abundant lower Paleozoic eugeoclinal strata in the Nevada-California region, and because Schweickert and Snyder (1981) and many subsequent workers proposed that strata of the Shoo Fly Complex accumulated adjacent to strata within the Roberts Mountains allochthon. High degrees of overlap and similarity between the Shoo Fly samples and two assemblages in the Roberts Mountains allochthon (Harmony B and Upper Vinini–Slaven; Fig. 9) strongly support scenarios in which the Shoo Fly Complex and Roberts Mountains allochthon were originally contiguous or at least in close proximity. The similarity in these detrital zircon age spectra cannot be used to resolve the depositional age of Shoo Fly strata, however, because units of the Roberts Mountains allochthon that yield zircons of these ages are of Cambrian(?), Ordovician, Silurian, and Devonian age.

The high degrees of similarity and overlap between detrital zircon ages from lower Paleozoic strata of the Shoo Fly Complex, Roberts Mountains allochthon, and miogeoclinal strata of the northern Cordillera lead to the interpretation that much of the detritus in the Shoo Fly Complex was shed from basement and cover rocks in northwestern Canada. One possible scenario is that the eugeoclinal strata were deposited along the northern Cordilleran margin and were subsequently transported southward (e.g., Eisbacher, 1983; Wallin, 1993) prior to mid-Paleozoic emplacement of the Roberts Mountains allochthon.

However, some Ordovician strata of the Roberts Mountains allochthon (Lower Vinini curve in Fig. 8) are clearly tied to basement rocks of the southwestern United States (Finney and Perry, 1991; Gehrels et al., this volume, Chapter 1). Therefore, a more likely scenario is that detritus in the eugeoclinal strata was derived from the northern part of the margin and transported southward by sedimentary processes (Fig. 10). Southward transport of sand-sized detritus from the Peace River arch region to Nevada-California, a distance of ~1800 km, is best explained by turbidity currents flowing within a marginal basin or trench that deepened southward.

An alternative source area may have included an outboard continental or microcontinental sliver (Madrid, 1987; Burchfiel et al., 1992) containing large rifted fragments of the >1.8 Ga Canadian shield provinces. No such basement rocks, however, are currently recognized outboard of the Cordilleran margin.

The Sierra City melange contains two younger populations of zircons (551–635 and 1170–1319 Ma), which the other Shoo Fly units lack (Fig. 7). The 506 ± 22 Ma multigrain age reported by Girty and Wardlaw (1984) probably represents a suite of 551–635 Ma grains that underwent high degrees and/or a complex history of Pb loss. These Neoproterozoic grains in the Shoo Fly Complex are similar in age to a ca. 600 Ma plagiogranite block in the Sierra City melange (Saleeby, 1990) and to a ca. 565 Ma tonalite body that intrudes the Trinity ultramafic sheet of the eastern Klamath Mountains (Wallin, 1988). The eastern Klamath Mountains include units that probably are northern continuations of units in the northern Sierra terrane (Davis, 1969; and most subsequent syntheses). The presence of these young ages within the Sierra City melange is consistent with petrographic and geochemical data that suggest derivation largely from a volcanic arc (Girty and Pardini, 1987; Pardini, 1986; Girty et al., 1996; Dickinson and Gehrels, this volume, Chapter 11).

The source of 1.17–1.32 Ga grains in the Sierra City melange is uncertain. Grains of similar age occur in two samples from the Roberts Mountains allochthon (Harmony Formation), but the source of these grains is also uncertain (Wallin, 1990; Gehrels et al., this volume, Chapter 1). One possibility, suggested by Wallin (1990), is that grains of these ages were derived from Grenville-age basement rocks of the southwestern United States and northwestern Mexico (Fig. 10). The ca. 1666 and ca. 1760 Ma grains in our sample could also have come from this region.

In contrast to the <1.8 Ga detrital zircon grains, the three >1.8 Ga grains from the Sierra City melange are similar to the ages of grains from the other thrust sheets (Fig. 7). The Sierra City melange therefore contains a mixture of grains that appear to have originated in several different regions: (1) the Peace River arch area of northwestern Canada; (2) an offshore magmatic arc of Neoproterozoic-Cambrian(?) age; and (3) 1.0–1.8 Ga basement rocks of the southwestern United States–northwestern Mexico. If correct, the mixture of these grains places the Sierra City melange and a Neoproterozoic-Cambrian(?) magmatic arc in proximity to western North America during early Paleozoic time.

## CONCLUSIONS

We analyzed 92 single detrital zircons from sandstones of the Shoo Fly Complex. Our analyses have allowed us to draw three main conclusions.

1. On the basis of similarities of detrital zircon ages (1.80–2.10, 2.20–2.45, and 2.55–2.70 Ga), the Culbertson Lake allochthon, Duncan Peak allochthon, and Lang sequence all contain detritus that originated in the same (or a very similar) cratonic source area.

2. The ages of grains in the Culbertson Lake allochthon, Duncan Peak allochthon, and Lang sequence are very similar to the ages of grains in parts of the Roberts Mountains allochthon, and in miogeoclinal strata of northwestern Canada. A likely scenario to explain these similarities involves large-scale, margin-parallel sedimentary transport of detritus from the Peace River arch region of northwestern Canada to an extensive submarine fan system located at the latitude of Nevada and California.

3. Sandstones in the Sierra City melange contain a few grains derived from cratonic sources in northern Canada, but most grains have ages of 551–635 Ma, suggesting that they were derived from an offshore magmatic arc. Grains with 1170–1319 Ma ages may have originated in Grenville-age basement rocks of southwestern North America. The presence of grains from all three sources in one sample, together with the interpretation that the Sierra City melange formed outboard of other assemblages in the Shoo Fly Complex and the Roberts Mountains allochthon (Schweickert et al., 1984; Schweickert and Snyder, 1981), suggests that the Shoo Fly Complex, Roberts Mountains allochthon, and a Neoproterozoic-Cambrian(?) magmatic arc all formed in proximity to the southern part of the Cordilleran margin, perhaps near their present positions.

## ACKNOWLEDGMENTS

We thank Bill Dickinson for his guidance during all phases of this study, and John Cooper, Rich Schweickert, and Mike Soreghan for careful and constructive manuscript reviews. This study was supported from funds contributed by the National Science Foundation grant EAR-9416933.

## REFERENCES CITED

Bond, G.C., and DeVay, J.C., 1980, Pre-Upper Devonian quartzose sandstones in the Shoo Fly Formation of northern California—Petrology, provenance, and implications for regional tectonics: Journal of Geology, v. 88, p. 285–308.

Burchfiel, B.C., Cowan, D.S., and Davis, G.A., 1992, Tectonic overview of the Cordilleran orogen in the western United States, *in* Burchfiel, B.C., et al., eds., The Cordilleran orogen: Conterminous U.S.: Boulder, Colorado, Geological Society of America, Geology of North America, v. G-3, p. 407–479.

D'Allura, J.A., Moores, E.M., and Robinson, L., 1977, Paleozoic rocks of the northern Sierra Nevada: their structural and paleogeographic implications, *in* Stewart, J.H., et al., eds., Paleozoic paleogeography of the western United States: Pacific Section, Society of Economic Paleontologists and Mineralogists, Pacific Coast Paleogeography Symposium 1 p. 395–408.

Davis, G.A., 1969, Tectonic correlations, Klamath Mountains and western Sierra Nevada, California: Geological Society of America Bulletin, v. 80, p. 1095–1108.

Eisbacher, G.H., 1983, Devonian-Mississippian sinistral transcurrent faulting along the cratonic margin of western North America: Geology, v. 11, p. 7–10.

Finney, S.C., and Perry, B.D., 1991, Depositional setting and paleogeography of Ordovician Vinini Formation, central Nevada, *in* Cooper, J.D., and Stevens, C.H., eds., Paleozoic paleography of the Western United States–II, Volume 2: Pacific Section, Society of Economic Paleontologists and Mineralogists book 67, p. 747–766.

Fritz, W.H., Cecile, M.P., Norford, B.S., Morrow, D., and Geldsetzer, H.H.J., 1991, Cambrian to Middle Devonian assemblages, *in* Gabrielse, H., and Yorath, C.J., eds., Geology of the Cordilleran orogen in Canada: Geological Survey of Canada, Geology of Canada, no. 5, p. 153–218.

Gehrels, G.E., Dickinson, W.R., Ross, G.M., Stewart, J.H., and Howell, D.G., 1995, Detrital zircon reference for Cambrian to Triassic miogeoclinal strata of western North America: Geology, v. 23, p. 831–834.

Gehrels, G.E., Butler, R.F., and Bazard, D.R., 1996, Detrital zircon geochronology of the Alexander terrane, southeastern Alaska: Geological Society of America Bulletin, v. 108, p. 722–734.

Girty, G.H., and Pardini, C.H., 1987, Provenance of sandstone inclusions in the Paleozoic Sierra City melange, Sierra Nevada, California: Geological Society of America Bulletin, v. 108, p. 176–181.

Girty, G.H., and Schweickert, R.A., 1979, Preliminary results of a detailed study of the "lower" Shoo Fly Complex, Bowman Lake, northern Sierra Nevada, California: Geological Society of America Abstracts with Programs, v. 11, p. 79.

Girty, G.H., and Schweickert, R.A., 1984, The Culbertson Lake allochthon, a newly identified structure within the Shoo Fly Complex, California: Evidence for four phases of deformation and extension of the Antler orogeny to the northern Sierra Nevada: Modern Geology, v. 8, p. 181–198.

Girty, G.H., and Wardlaw, M.S., 1984, Was the Alexander terrane a source of feldspathic sandstones in the Shoo Fly Complex, northern Sierra Nevada, California?: Geology, v. 12, p. 339–342.

Girty, G.H., and Wardlaw, M.S., 1985, Petrology and provenance of pre-Late Devonian sandstones, Shoo Fly Complex, northern Sierra Nevada, California: Geological Society of America Bulletin, v. 96, p. 516–521.

Girty, G.H., Gester, K.C., and Turner, J.B., 1990, Pre-Late Devonian geochemical, stratigraphic, sedimentologic, and structural patterns, Shoo Fly Complex, northern Sierra Nevada, California, *in* Harwood, D.S., and Miller, M.M., eds., Paleozoic and early Mesozoic paleogeographic relations; Sierra Nevada, Klamath Mountains, and related terranes: Geological Society of America Special Paper 255, p. 43–56.

Girty, G.H., Gurrola, L.D., Taylor, G.W., Richards, M.J., and Wardlaw, M.S., 1991, The pre-Upper Devonian Lang and Black Oak Spring sequences, Shoo Fly Complex, northern Sierra Nevada, California: Trench deposits composed of continental detritus, *in* Cooper, J.D., and Stevens, C.H., eds., Paleozoic paleogeography of the western United States—II, Volume 2: Pacific Section, Society of Economic Paleontologists and Mineralogists, book 67, p. 703–716.

Girty, G.H., Lawrence, J., Burke, T., Fortin, A., Gallarano, C.S., Wirths, T.A., Lewis, J.G., Peterson, M.M., Ridge, D.L., Knaack, and C., Johnson, D., 1996, The Shoo Fly Complex: Its origin and tectonic significance, *in* Girty, G.H., et al., eds., The northern Sierra terrane and associated Mesozoic magmatic units: Implications for the tectonic history of the western Cordillera: Pacific Section, Society of Economic Paleontologists and Mineralogists book 81, p. 1–24.

Hannah, J.L., and Moores, E.M., 1986, Age relationships and depositional environments of Paleozoic strata, northern Sierra Nevada, California: Geological Society of America Bulletin, v. 97, p. 787–797.

Hanson R.E., Saleeby, J.B., and Schweickert, R.A., 1988, Composite Devonian island-arc batholith in the northern Sierra Nevada, California: Geological Society of America Bulletin, v. 100, p. 446–457.

Harwood, D.S., 1988, Tectonism and metamorphism in the northern Sierra terrane, northern California, *in* Ernst, W.D., ed., Metamorphism and crustal evolution of the western United States (Rubey Volume VII): Englewood Cliffs, New Jersey, Prentice-Hall, p. 765–788.

Harwood, D.S., 1992, Stratigraphy of Paleozoic and lower Mesozoic rocks in the northern Sierra terrane, California: U.S. Geological Survey Bulletin 1957, p. 1–19.

Hoffman, P.F., 1989, Precambrian geology and tectonic history of North America, *in* Bally, A.W., and Palmer, A.R., eds., The geology of North America—An overview: Boulder, Colorado Geological Society of America, Geology of North America, v. A, p. 447–512.

Ludwig, K.R., 1991a, A computer program for processing Pb-U-Th isotopic data: U.S. Geological Survey Open-File Report 88-542.

Ludwig, K.R., 1991b, A plotting and regression program for radiogenic-isotopic data: U.S. Geological Survey Open-File Report 91-445.

Madrid, R.J., 1987, Stratigraphy of the Roberts Mountains allochthon in north-central Nevada [Ph.D. thesis]: Stanford, California, Stanford University, 336 p.

Pardini, C.H., 1986, Petrological and structural analysis of the Sierra City melange, northern Sierra Nevada, California [M.S. thesis]: San Diego, California, San Diego State University, 87 p.

Poole, F.G., Stewart, J.H., Palmer, A.R., Sandberg, C.A., Madrid, R.J., Ross, R.J., Jr., Hintze, L.F., Miller, M.M., and Wrucke, C.T., 1992, Latest Precambrian to latest Devonian time; development of a continental margin, *in* Burchfiel, B.C., et al., eds., The Cordilleran orogen: Conterminous U.S.: Boulder Colorado, Geological Society of America, Geology of North America, v. G-3, p. 9-56.

Richards, M.J., 1990, Shoo Fly Complex, Lake Spaulding area, northern Sierra Nevada, California: An early Paleozoic accretionary complex? [M.S. thesis]: San Diego, California, San Diego State University, 78 p.

Ross, G.M., 1991, Precambrian basement in the Canadian Cordillera: An introduction: Canadian Journal of Earth Sciences, v. 28, p. 1133–1139.

Saleeby, J.B., 1990, Geochronologic and tectonostratigraphic framework of Sierran-Klamath ophiolitic assemblages, *in* Harwood, D.S., and Miller, M.M., eds., Paleozoic and early Mesozoic paleogeographic relations; Sierra Nevada, Klamath Mountains, and related terranes: Geological Society of America Special Paper 255, p. 93–114.

Saleeby, J., Hannah, J.L., and Varga, R.J., 1987, Isotopic age constraints on middle Paleozoic deformation in the northern Sierra Nevada, California: Geology, v. 15, p. 757–760.

Sharp, W.D., Saleeby, J.B., Schweickert, R.A., Merquerian, C., Kistler, R.W., Tobisch, O.T., and Wright, W.H., 1982, Age and tectonic significance of Paleozoic orthogneisses of the Sierra Nevada foothills metamorphic belt, California: Geological Society of America Abstracts with Programs, v. 14, p. 233.

Schweikert, R.A., 1981, Tectonic evolution of the Sierra Nevada, *in* Ernst, W.D., ed., The geotectonic development of California (Rubey Volume I): Englewood Cliffs, New Jersey, Prentice-Hall, p. 88–131.

Schweickert, R.A., and Girty, G.H., 1981, Significance of the unconformity between Shoo Fly Complex and Paleozoic island-arc sequences, north Sierra Nevada, California: Geological Society of America Abstracts with Programs, v. 13, p. 105.

Schweickert, R.A., and Snyder, W.S., 1981, Paleozoic plate tectonics of the Sierra Nevada and adjacent regions, *in* Ernst, W.D., ed., The geotectonic development of California (Rubey Volume I): Englewood Cliffs, New Jersey, Prentice-Hall, p. 182–202.

Schweickert, R.A., Harwood, D.S., Girty, G.H., and Hanson, R.E., 1984, Tectonic development of the northern Sierra terrane: An accreted late Paleozoic island arc and its basement, *in* Lintz, J., Jr., ed., Western geological excursions, Geological Society of America 1984 Annual Meeting Field Trip Guide, v. 4: Reno, University of Nevada Mackey School of Mines, p. 1–65.

Speed, R.C., 1979, Collided Paleozoic microplate in the western United States: Journal of Geology, v. 87, p. 279–292.

Stewart, J.H., Poole, F.G., Ketner, K.B., Madrid, R.J., Roldan-Quintana, J., and Amaya-Martinez, R., 1990, Tectonics and stratigraphy of the Paleozoic and Triassic southern margin of North America, Sonora, Mexico, *in* Gehrels, G.E., and Spencer, J.E., eds., Geologic excursions through the sonoran Desert region, Arizona and Sonora: Arizona Geological Survey Special Paper 7, p. 183–202.

Taylor, G.W., 1986, Structural, sedimentological, and petrological setting of the Lang-Halsted sequence and Duncan Peak chert, lower Shoo Fly Complex, northern Sierra Nevada, California [M.S. thesis]: San Diego, California, San Diego State University, 110 p.

Van Schmus, W.R., and 24 others, 1993, Transcontinental Proterozoic provinces, *in* Reed, J.C., et al., eds., Precambrian: Conterminous U.S.: Boulder, Colorado, Geological Society of America Geology of North America, v. C-2, p. 171–334.

Varga, R.J., 1982, Implications of Paleozoic phosphorites in the northern Sierra Nevada range: Nature, v. 297, p. 217–220.

Varga, R.J., and Moores, E.M., 1981, Age, origin, and significance of an unconformity that predates island-arc volcanism in the northern Sierra Nevada: Geology, v. 9, p. 512–518.

Villeneuve, M.E., Ross, G.M., Theriault, R.J., Miles, W., Parrish, R.R., and Broome, J., 1993, Tectonic subdivision and U-Pb geochronology of the crystalline basement of the Alberta basin, western Canada: Geological Survey of Canada Bulletin 447, 86 p.

Wallin, E.T., 1988, Early Paleozoic magmatic events in the eastern Klamath Mountains, northern California: Geology, v. 16, p. 144–148.

Wallin, E.T., 1990, Provenance of selected lower Paleozoic siliciclastic rocks in the Roberts Mountains allochthon, Nevada, *in* Harwood, D.S., and Miller, M.M., eds., Paleozoic and early Mesozoic paleogeographic relations; Sierra Nevada, Klamath Mountains, and related terranes: Geological Society of America Special Paper 255, p. 17–32.

Wallin, E.T., 1993, Sonomia revisited: Evidence for a western Canadian provenance of the eastern Klamath and northern Sierra Nevada terranes: Geological Society of America Abstracts with Programs, v. 25, no. 6, p. A-173.

MANUSCRIPT ACCEPTED BY THE SOCIETY JANUARY 24, 2000

Printed in U.S.A.

Geological Society of America
Special Paper 347
2000

# *Detrital zircon geochronology of the Antler overlap and foreland basin assemblages, Nevada*

**George E. Gehrels and William R. Dickinson**
*Department of Geosciences, University of Arizona, Tucson, Arizona 85721, USA*

## ABSTRACT

**U-Pb geochronologic analyses were conducted on 44 detrital zircon grains from two samples from Mississippian sandstones of the Inskip and Tonka formations of central Nevada. The Inskip Formation represents the westernmost exposure of clastic strata that accumulated on top of the Roberts Mountains allochthon following mid-Paleozoic emplacement, whereas the Tonka accumulated in a foreland basin inboard of the Antler orogen. This study complements a previous detrital zircon analysis of the Pennsylvanian Battle Formation, a unit that unconformably overlies central portions of the Roberts Mountains allochthon.**

**The main groups of ages in the Inskip and Tonka samples are ca. 1750–2070 Ma. The Tonka Formation also yielded a few grains that are <1.6 and >2.2 Ga. The detrital zircon ages from these two units and from the Battle Formation are similar to the ages of detrital zircons in the main units of the Roberts Mountains allochthon. This result is consistent with stratigraphic evidence suggesting that detritus in all three upper Paleozoic units was derived primarily from the Antler orogen. The age spectra for the three samples accordingly provide a detrital zircon reference for the ages of zircon grains accumulating along the Cordilleran margin in proximity to the Antler orogen during Carboniferous time.**

## INTRODUCTION

The mid-Paleozoic Antler orogeny records the earliest phase of accretion of Paleozoic eugeoclinal strata along the Cordilleran margin (Roberts et al., 1958; Burchfiel and Davis, 1972; Miller et al., 1992; Poole et al., 1992). During the Antler orogeny, lower Paleozoic eugeoclinal strata of the Roberts Mountains allochthon (Figs. 1 and 2) were thrust 75–200 km eastward onto Devonian and older shelf facies strata (Roberts et al., 1958; Miller et al., 1992; Poole et al., 1992). Emplacement of the eugeoclinal rocks produced a broad subaerial region, the Antler highland, from which detritus was shed eastward onto the shelf and craton, and westward into off-shelf basins (Roberts et al., 1958; Poole, 1974; Dickinson et al., 1983).

Upper Paleozoic strata to the east of the Antler highland accumulated primarily in marine settings within a foreland basin (Fig. 3) (Poole, 1974). Conglomerate and sandstone, to several kilometers in thickness, have been referred to as the Tonka Formation (Dott, 1955) and Diamond Peak Formation (Roberts et al., 1958; Brew, 1971). Poole (1974) and Harbaugh and Dickinson (1981) reported that the strata were deposited in submarine fan and deltaic settings. Most or all of the detritus in these units was derived from uplifted eugeoclinal rocks several kilometers to several tens of kilometers west of the basin.

Within the Antler highland, clastic strata of the Antler overlap assemblage accumulated in fluvial, deltaic, and marginal marine settings directly upon deformed eugeoclinal strata of the Roberts Mountains allochthon (Roberts et al., 1958; Saller and Dickinson, 1982; Dickinson et al., 1983; Madrid, 1987). Locally, overlap strata are as old as Late Mississippian (Little, 1987). The principal clastic unit of the Antler overlap sequence in central Nevada is the Pennsylvanian Battle Formation. In general, paleo-

Gehrels, G.E., and Dickinson, W.R., 2000, Detrital zircon geochronology of the Antler overlap and foreland basin assemblages, Nevada, *in* Soreghan, M.J., and Gehrels, G.E., eds., Paleozoic and Triassic paleogeography and tectonics of western Nevada and northern California: Boulder, Colorado, Geological Society of America Special Paper 347, p. 57–63.

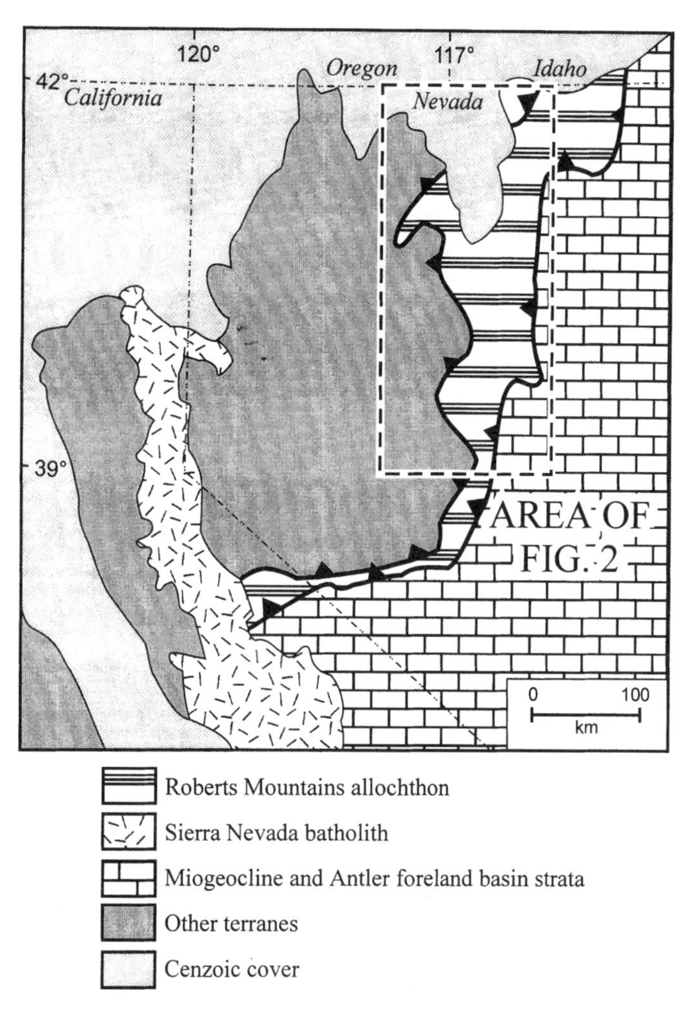


Figure 1. Schematic map of study area in western Nevada (adapted primarily from Silberling et al., 1987).

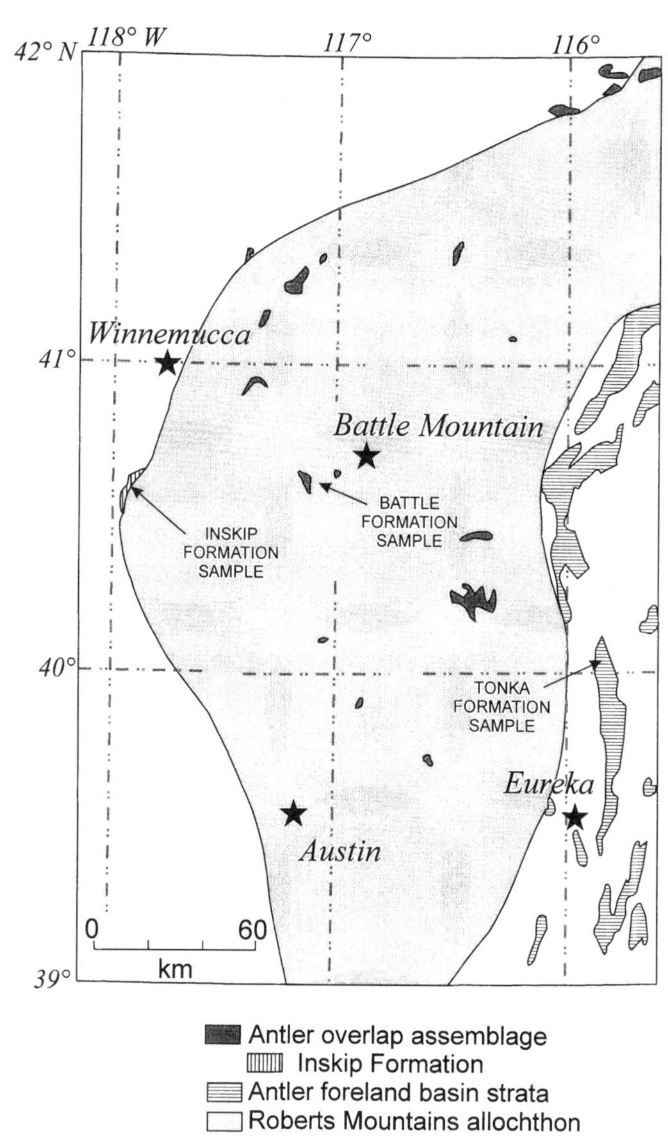


Figure 2. Map of central Nevada showing general outcrop pattern of Antler overlap and foreland basin assemblages and the Roberts Mountains allochthon (from Stewart and Carlson, 1978 and Stewart, 1980). Also shown are approximate locations of our detrital zircon samples and of related sample of Battle Formation from Gehrels and Dickinson (1995).

currents in the Battle Formation record transport toward the west-southwest (Saller and Dickinson, 1982).

The westernmost clastic unit of the Antler overlap assemblage in central Nevada is the Mississippian Inskip Formation, which consists mainly of phyllite, quartzofeldspathic turbidites, volcanic rocks, and limestone (Whitebread, 1994). These units unconformably overlie the Ordovician Valmy Formation of the Roberts Mountains allochthon.

We have analyzed 44 detrital zircon grains from the Tonka and Inskip Formations (Figs. 2 and 3) in an effort to constrain the ages of zircon grains that were accumulating along the Cordilleran margin, near or adjacent to the Roberts Mountains allochthon, during Mississippian time. Our data complement previous analyses of detrital zircons in the Battle Formation, which were presented by Gehrels and Dickinson (1995). Together, the ages of detrital zircon grains in these three units of the Antler overlap sequence and foreland basin provide a critical reference for comparison with ages of zircons in allochthonous(?) upper Paleozoic and younger sandstones to the west, as described by Riley et al. (this volume, Chapter 4), Darby et al. (this volume, Chapter 5), Spurlin et al. (this volume, Chapter 6), and Gehrels and Miller (this volume, Chapter 7).

## RESULTS

### *Inskip Formation*

The Inskip Formation occurs only in the East Range, where it unconformably overlies Ordovician strata of the Valmy Formation (Whitebread, 1994). As described by Whitebread (1994), the Inskip Formation can be divided into a lower and an upper unit. The lower unit consists of phyllite, wacke, quartzite, conglomerate, and minor limestone, and averages 900 m in thickness. Sand-

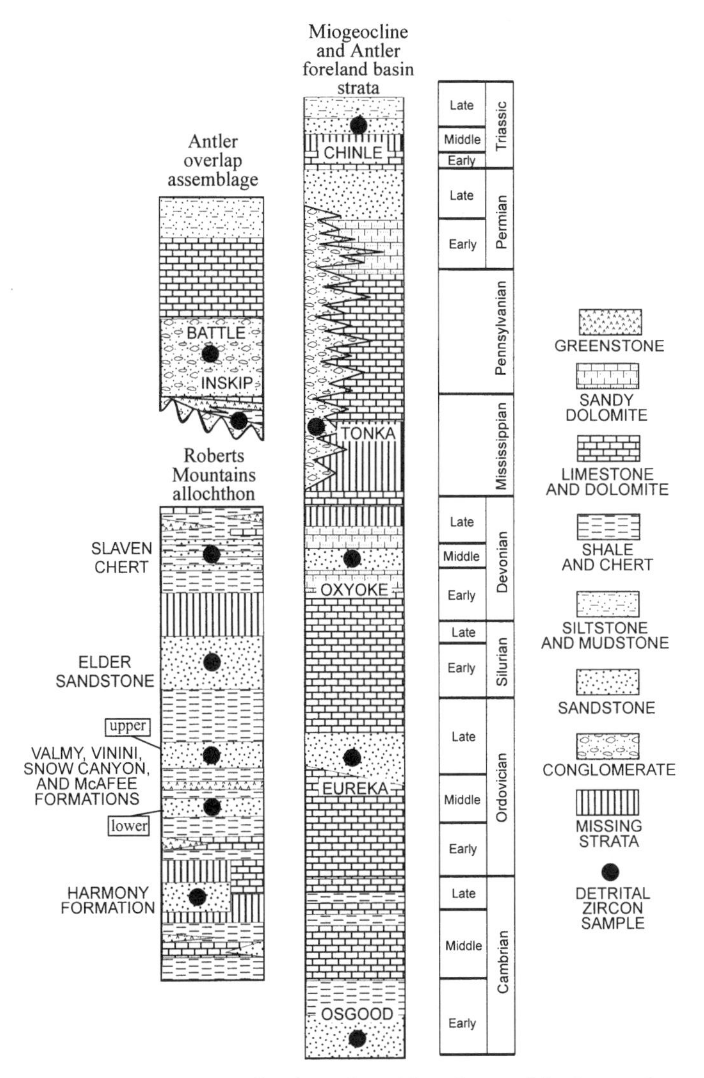


Figure 3. Highly generalized stratigraphic column of Antler overlap and foreland basin assemblages in relation to Roberts Mountains allochthon and miogeocline (compiled mainly from Madrid, 1987; Miller et al., 1992).

stone beds containing coarse grains and granules of quartz and quartzite characterize the unit. A <2.5-km-thick upper unit comprises phyllite, quartzite, and wacke, and minor volcanic rocks and limestone. Our sample was collected from an outcrop of coarse-grained sandstone exposed along the north side of Inskip Canyon. Although foliation in the unit is penetrative, bedding is well preserved and generally homoclinal. The bed is part of the lower unit, and is ~40 m stratigraphically above the base of the unit. In this area, the Inskip Formation unconformably overlies overturned quartzite beds of the Valmy Formation.

The sample of sandstone contains large grains and granules of quartz and feldspar set in a phyllitic matrix (Dickinson and Gehrels, this volume, Chapter 11). We processed 25 kg of sandstone, and 1.12 gm of zircon was recovered. A few grains range from 175 to 250 μm in size, and abundant grains are between 145 and 175 μm in size. In both size fractions, the grains have

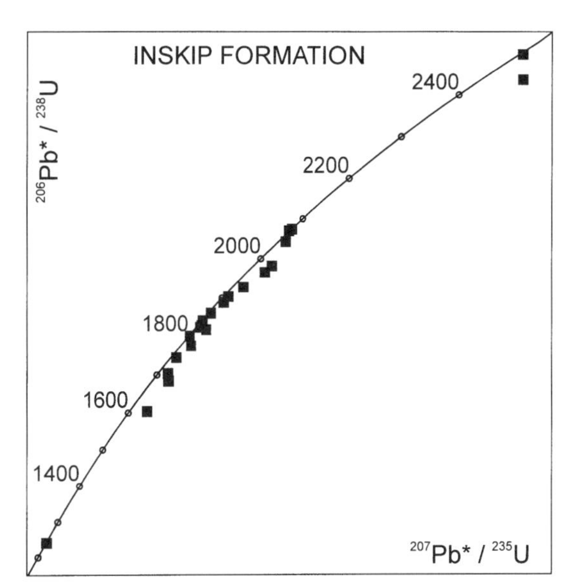


Figure 4. U-Pb concordia diagrams showing single-grain detrital zircon analyses from Mississippian strata of Inskip Formation. All data reduction and concordia plots are from programs of Ludwig (1991a, 1991b).

highly variable morphology and color. The most abundant grains are colorless to light pink, and moderately to well-rounded. These grains strongly resemble the well-rounded, moderately to highly spherical, and highly polished grains that are common in the Valmy and related Middle-Upper Ordovician quartzites of the Roberts Mountains allochthon (Gehrels et al., this volume, Chapter 1). Slightly rounded to euhedral grains, which range from colorless to medium pink, are less abundant.

We analyzed 22 detrital zircon grains from this sample utilizing the analytical techniques described by Gehrels (this volume, Introduction). The grains selected represent all color and morphology groups observed among grains greater than 63 μm in sieve size. The U-Pb analyses are all of medium to high precision and range from concordant to moderately discordant (Fig. 4; Table 1). The interpreted ages reported in Table 1 are $^{207}Pb^{*}/^{206}Pb^{*}$ ages for concordant grains and concordia intercepts projected from 100 Ma for discordant grains. As described by Gehrels (this volume, Introdction), these upper intercept ages are a better approximation of the true crystallization age than $^{207}Pb^{*}/^{206}Pb^{*}$ ages because the main phase(s) of metamorphism in the region are at least 100 Ma, whereas there is little evidence for recent isotopic disturbance (which is implicit in the use of $^{207}Pb^{*}/^{206}Pb^{*}$ ages). For the purposes of this study, however, the difference between $^{207}Pb^{*}/^{206}Pb^{*}$ and upper intercept ages is insignificant, because the two ages differ only by an average of 2.6 m.y. for this sample.

The dominant ages in this sample range between 1830 and 1940 Ma. A main peak in age probability occurs at 1849 Ma and a lesser peak occurs at 1928 Ma (Fig. 5; Table 2). Subordinate age groups range from 1740 to 1820 Ma and 2055 to 2085 Ma, and peaks occur at 1782, 1808, and 2076 Ma.

## TABLE 1. U-Pb ISOTOPIC DATA AND AGES

| Grain type | Grain wt. (µg) | $Pb_c$ (pg) | U (ppm) | $^{206}Pb_m/^{204}Pb$ | $^{206}Pb_c/^{208}Pb$ | Apparent ages (Ma) $^{206}Pb^*/^{238}U$ | $^{207}Pb^*/^{235}U$ | $^{207}Pb^*/^{206}Pb^*$ | Interpreted age (Ma) |
|---|---|---|---|---|---|---|---|---|---|
| Inskip Formation (40°33'23"N, 117°56'41"W) | | | | | | | | | |
| **CE** | **9** | **8** | **222** | **3430** | **6.5** | **1241 ± 8** | **1245 ± 11** | **1251 ± 12** | **1252 ± 20** |
| LE | 13 | 11 | 424 | 8690 | 18.3 | 1604 ± 6 | 1666 ± 8 | 1746 ± 5 | 1752 ± 20 |
| LsE | 16 | 9 | 234 | 7670 | 9.3 | 1704 ± 7 | 1736 ± 10 | 1775 ± 8 | 1778 ± 10 |
| LsE | 11 | 5 | 148 | 5810 | 8.1 | 1745 ± 18 | 1763 ± 19 | 1785 ± 9 | 1787 ± 10 |
| **CE** | **8** | **7** | **207** | **7500** | **7.2** | **1802 ± 9** | **1804 ± 11** | **1808 ± 6** | **1808 ± 10** |
| MwR | 36 | 9 | 131 | 8900 | 5.9 | 1683 ± 8 | 1738 ± 9 | 1804 ± 5 | 1809 ± 20 |
| **LsE** | **10** | **7** | **223** | **6520** | **10.5** | **1833 ± 9** | **1835 ± 12** | **1838 ± 8** | **1838 ± 10** |
| **CE** | **10** | **10** | **127** | **2520** | **8.0** | **1841 ± 18** | **1844 ± 20** | **1847 ± 8** | **1847 ± 10** |
| CwR | 15 | 8 | 75 | 2762 | 10.9 | 1776 ± 10 | 1807 ± 12 | 1847 ± 6 | 1850 ± 10 |
| CsR | 10 | 6 | 103 | 3800 | 6.1 | 1824 ± 12 | 1836 ± 15 | 1849 ± 8 | 1851 ± 10 |
| LsR | 9 | 6 | 146 | 4200 | 21.5 | 1860 ± 14 | 1868 ± 16 | 1877 ± 8 | 1878 ± 10 |
| CsR | 20 | 11 | 366 | 13520 | 3.2 | 1888 ± 7 | 1905 ± 9 | 1923 ± 4 | 1924 ± 10 |
| CwR | 16 | 8 | 87 | 3720 | 2.8 | 1904 ± 9 | 1917 ± 11 | 1931 ± 5 | 1933 ± 10 |
| LsE | 15 | 9 | 187 | 6750 | 7.7 | 1927 ± 11 | 1957 ± 13 | 1988 ± 5 | 1989 ± 10 |
| CwR | 13 | 4 | 114 | 7370 | 3.2 | 1818 ± 8 | 1854 ± 9 | 1895 ± 5 | 1898 ± 10 |
| CsR | 22 | 14 | 331 | 11430 | 6.9 | 1966 ± 8 | 2010 ± 9 | 2057 ± 4 | 2060 ± 10 |
| **LwR** | **24** | **16** | **280** | **965** | **3.1** | **2071 ± 15** | **2069 ± 20** | **2068 ± 11** | **2068 ± 10** |
| **CsR** | **7** | **18** | **249** | **2210** | **5.8** | **2073 ± 11** | **2074 ± 14** | **2074 ± 7** | **2074 ± 10** |
| **MwR** | **19** | **6** | **167** | **12200** | **7.5** | **1981 ± 10** | **2027 ± 12** | **2074 ± 4** | **2077 ± 10** |
| LsE | 8 | 10 | 205 | 3820 | 7.3 | 2043 ± 13 | 2060 ± 18 | 2077 ± 10 | 2079 ± 10 |
| **CsR** | **11** | **8** | **90** | **3640** | **7.2** | **2496 ± 24** | **2502 ± 26** | **2506 ± 8** | **2506 ± 10** |
| **CsE** | **12** | **6** | **64** | **3500** | **7.0** | **2436 ± 15** | **2502 ± 19** | **2556 ± 6** | **2559 ± 10** |
| Tonka Formation (40°43'30"N, 116°01'15"W) | | | | | | | | | |
| LsR | 24 | 6 | 53 | 1290 | 2.6 | 583 ± 8 | 590 ± 10 | 615 ± 23 | 583 ± 20 |
| **LwR** | **12** | **8** | **166** | **2690** | **5.8** | **1074 ± 8** | **1075 ± 10** | **1077 ± 10** | **1077 ± 10** |
| **LsR** | **28** | **8** | **59** | **3210** | **4.7** | **1522 ± 8** | **1521 ± 11** | **1521 ± 10** | **1521 ± 10** |
| LwR | 15 | 19 | 311 | 3920 | 7.6 | 1522 ± 8 | 1619 ± 12 | 1747 ± 8 | 1757 ± 20 |
| CwR | 17 | 6 | 70 | 3700 | 4.8 | 1796 ± 14 | 1806 ± 13 | 1817 ± 7 | 1818 ± 10 |
| **LwR** | **14** | **7** | **96** | **3910** | **4.9** | **1839 ± 9** | **1840 ± 13** | **1841 ± 8** | **1841 ± 10** |
| **LsR** | **26** | **9** | **64** | **2190** | **2.8** | **1845 ± 11** | **1846 ± 14** | **1847 ± 8** | **1847 ± 10** |
| CwR | 25 | 12 | 178 | 6700 | 5.3 | 1674 ± 8 | 1758 ± 10 | 1862 ± 5 | 1868 ± 20 |
| LsR | 20 | 11 | 80 | 2690 | 4.4 | 1778 ± 9 | 1820 ± 14 | 1868 ± 10 | 1872 ± 10 |
| LwR | 18 | 8 | 134 | 5950 | 12.1 | 1749 ± 8 | 1810 ± 10 | 1882 ± 7 | 1887 ± 20 |
| **CwR** | **34** | **9** | **98** | **8040** | **3.2** | **1936 ± 12** | **1937 14** | **1938 ± 6** | **1938 ± 10** |
| *CsR* | *25* | *11* | *210* | *7750* | *3.5* | *1525 ± 9* | *1710 12* | *1946 ± 6* | *1962 ± 40* |
| *LwR* | *21* | *9* | *119* | *4660* | *5.1* | *1492 ± 7* | *1704 10* | *1977 ± 7* | *1996 ± 40* |
| CsR | 30 | 11 | 149 | 8220 | 7.9 | 1958 ± 12 | 1979 14 | 2000 ± 6 | 2002 ± 10 |
| CsR | 18 | 18 | 107 | 2250 | 5.8 | 1975 ± 9 | 1988 12 | 2002 ± 7 | 2004 ± 10 |
| CsR | 17 | 14 | 157 | 4730 | 5.9 | 2212 ± 9 | 2256 ± 13 | 2296 ± 6 | 2298 ± 10 |
| CwR | 28 | 7 | 103 | 10300 | 4.4 | 2312 ± 15 | 2324 16 | 2334 ± 4 | 2335 ± 10 |
| CwR | 27 | 10 | 115 | 9260 | 3.5 | 2620 ± 16 | 2645 18 | 2664 ± 4 | 2665 ± 10 |
| **LsR** | **16** | **7** | **56** | **1930** | **15.9** | **2659 ± 21** | **2667 23** | **2673 ± 6** | **2673 ± 10** |
| LwR | 24 | 9 | 139 | 10880 | 6.8 | 2420 ± 15 | 2648 17 | 2827 ± 4 | 2835 ± 40 |
| CsR | 18 | 7 | 92 | 4130 | 6.8 | 2856 ± 15 | 2872 17 | 2882 ± 4 | 2884 ± 10 |

*Note:*

Grain with $^{206}Pb^*/^{238}U$ age within 15% of $^{207}Pb^*/^{206}Pb^*$ age in plain type.

Grain with $^{206}Pb^*/^{238}U$ age not within 15% of $^{207}Pb^*/^{206}Pb^*$ age in italics.

Grain with concordant age and moderate to high precision analysis in bold.

* Radiogenic Pb.

Grain type: L—light pink, M—medium pink, C—colorless; sR—slightly rounded, wR—well rounded, E—euhedral, sE—subeuhedral.

Unless otherwise noted, all grains are clear and/or translucent.

$^{206}Pb/^{204}Pb$ is measured ratio, uncorrected for blank, spike, or fractionation.

$^{206}Pb/^{208}Pb$ is corrected for blank, spike, and fractionation.

Most concentrations have an uncertainty to 25% due to uncertainty in weight of grain.

Constants used: $^{238}U/^{235}U$ = 137.88. Decay constant for $^{235}U$ = $9.8485 \times 10^{-10}$. Decay constant for $^{238}U$ = $1.55125 \times 10^{-10}$.

All uncertainties are at the 95% confidence level.

Pb blank ranged from 2 to 10 pg. U blank was consistently <1 pg.

Interpreted ages for concordant grains are $^{206}Pb^*/^{238}U$ ages if <1.0 Ga and $^{207}Pb^*/^{206}Pb^*$ ages if >1.0 Ga.

Interpreted ages for discordant grains are projected from 100 Ma.

All analyses conducted using conventional isotope dilution and thermal ionization mass spectrometry, as described by Gehrels (this volume, Introduction).

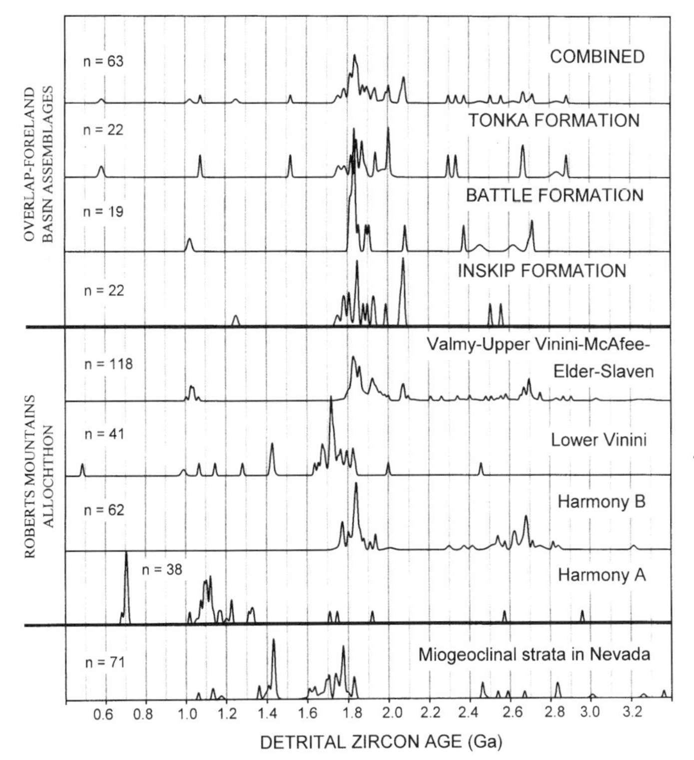


Figure 5. Cumulative probability plots of detrital zircon ages for strata of Antler overlap and foreland basin assemblages, Roberts Mountains allochthon, and miogeoclinal strata in Nevada. Combined curve at top represents age spectra of combination of Tonka, Battle, and Inskip Formations. Strata of Roberts Mountains allochthon are divided into four groups based on their age spectra (Gehrels et al., this volume, Chapter 1). Age spectra for miogeoclinal strata in Nevada, which includes detrital zircon ages from Cambrian, Devonian, and Triassic strata, provides reference for ages of detrital zircon grains that could have been shed from basement rocks of southwestern United States (Gehrels et al., 1995; Gehrels, this volume, Introduction). Ages for all grains plotted are either concordant or, if discordant, projected from 100 Ma.

TABLE 2. DOMINANT DETRITAL ZIRCON AGE GROUPS IN STRATA OF THE ANTLER OVERLAP AND FORELAND BASIN

| Age range (Ma) | Peak ages (Ma) | Relative abundance |
|---|---|---|
| Inskip Formation | | |
| 1740–1820 | 1782, 1808* | 0.22 |
| 1830–1940 | 1849*, 1928 | 0.36 |
| 2055–2085 | 2076 | 0.21 |
| Tonka Formation | | |
| 1740–1805 | 1785 | 0.09 |
| 1830–1900 | 1844*, 1872 | 0.22 |
| 1925–2020 | 2003 | 0.22 |

*Note:* Ages and relative abundances refer to the relative age–probability plots shown in Figure 5. Ages with greatest relative abundance are indicated with an asterisk.

### *Tonka Formation*

The Tonka Formation (Dott, 1955), or Diamond Peak Formation (Roberts et al., 1958; Brew, 1971), consists of several kilometers of interbedded conglomerate, sandstone, and mudstone that accumulated in basinal to deltaic settings (Poole, 1974; Harbaugh and Dickinson, 1981). Paleocurrent and provenance data from the Tonka Formation indicate that the detritus in the sandstones and conglomerates was derived from eugeoclinal strata exposed in the Antler highland to the west (Brew, 1971; Poole, 1974; Harbaugh and Dickinson, 1981).

We collected a sample of coarse-grained sandstone from an interval of interbedded sandstone and conglomerate in the upper part of the Tonka Formation from the north side of Interstate 80, ~300 m west of the tunnel east of Carlin, Nevada. Fossils in overlying strata indicate that the sandstone sampled is Late Mississippian age (Dott, 1955). The presence of boulder conglomerates in fluvial strata in the section suggests that the unit sampled accumulated within several kilometers of the thrust front of the Antler allochthon.

The sandstone is poorly sorted, feldspathic, and contains ~10% reworked argillite and siltstone grains (Dickinson and Gehrels, this volume, Chapter 11). The 25 kg of coarse-grained sandstone yielded 850 mg of zircon grains. The grains ranged to 175 μm in length; a few grains were 145–175 μm, and hundreds of grains range from 125 to 145 μm. In all size fractions, both slightly rounded and well-rounded grains were present. Colorless to light pink, well rounded grains are dominant, and slightly rounded grains ranging from colorless to light pink form subordinate populations.

Most grains yielded concordant to moderately discordant analyses (Fig. 6; Table 1). The dominant ages range between 1830–1900 and 1925–2020 Ma, and age-probability peaks occur at 1844 (greatest abundance), 1872, and 2003 Ma (Fig. 5; Table 2). A smaller group of ages ranges from 1740 to 1805 Ma, with a peak at 1785 Ma. Individual grains yielded interpreted ages of ca. 583, 1077, 1521, and 2298–2884 (n = 6) Ma.

## COMPARISON AND SIGNIFICANCE OF DETRITAL ZIRCON AGES

The detrital zircon ages reported here are compared with each other and with previous detrital zircon data from the Battle Formation (Gehrels and Dickinson, 1995) in Figure 5. Although there are differences in the relative proportions of ages in the three samples, all samples are dominated by grains ranging from 1750 to 2100 Ma, and all contain a significant number of 1800–1900 Ma grains. Because the age spectra from the three samples are similar, they are combined into one composite curve in Figure 5.

The detrital zircon ages from these three units are compared to detrital zircon ages from potential source regions in Figure 5. Possible provenance links with basement rocks of the southwestern United States can be evaluated by comparison of the composite age-probability curve for the samples in this study with the composite age spectra for miogeoclinal strata in Nevada, most of which contain clastic detritus derived from southwestern North

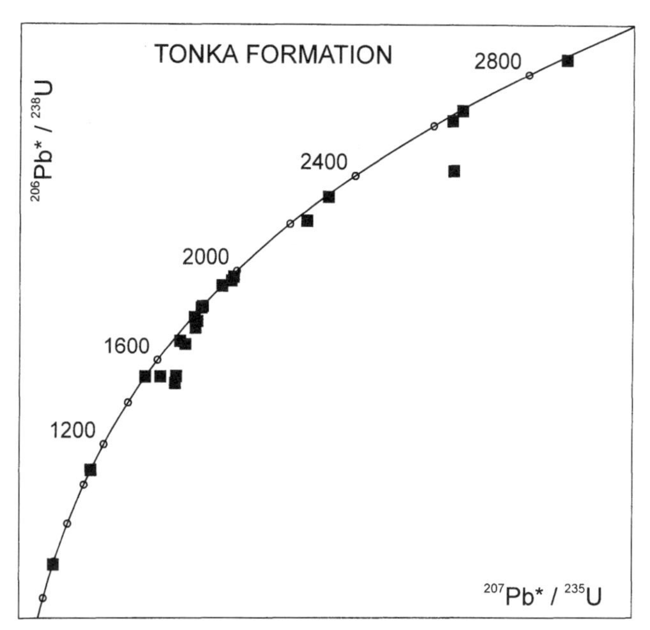


Figure 6. U-Pb concordia diagrams showing single-grain detrital zircon analyses from Upper Mississippian strata of Tonka Formation.

America. The units incorporated into this miogeoclinal reference include strata of Cambrian, Devonian, and Triassic age, which collectively provide a record of the ages of detrital grains that would have been derived from nearby basement provinces (Gehrels et al., 1995; Gehrels, this volume, Introduction). The detrital zircon ages from Ordovician miogeoclinal strata are excluded from this reference curve because detritus in these units was most likely derived from much farther north along the Cordilleran margin (Ketner, 1968; Gehrels et al., 1995).

As a complement to visual comparisons of the age spectra, Table 3 presents the results of two different statistical comparisons of the age spectra. As described by Gehrels (this volume, Introduction), the degree of overlap indicates whether the grains a sample overlap in age with the ages of grains in a particular reference. A value of 1.0 indicates that all grains in the sample could have been derived from the region characterized by the reference curve. The degree of similarity compares the shapes of two age-probability curves. A value of 1.0 indicates a perfect match between the relative abundances of ages in two samples, whereas a value of 0.0 indicates that there is no correlation of relative abundances of ages. The low values of both overlap and similarity (Table 3) suggest that little detritus in the Antler overlap sequence and foreland basin originated in basement rocks of the southwestern United States.

Previously interpreted provenance links with strata of the Roberts Mountains allochthon can be evaluated by comparing the composite age-probability curve for the samples in this study with the composite age-probability curves for the four main groups of strata from the Roberts Mountains allochthon (Gehrels et al., this volume, Chapter 1). The high degrees of overlap and similarity between our composite age spectra and the age-probability curves for at least two sets of units in the Roberts Mountains allochthon (Table 3) supports previous interpretations (Poole, 1974; Harbaugh and Dickinson, 1981; Saller and Dickinson, 1982; Dickinson et al., 1983) that most detritus in the Antler overlap sequence and foreland basin was derived from rocks of the Roberts Mountains allochthon. The low values of similarity and overlap for two sets of units in the allochthon (Table 3), suggest however, that strata of these units did not contribute a significant proportion of detritus to the overlap and foreland basin sequences.

Figure 7 shows that the best match between detrital zircon ages in the overlap and foreland basin sequences and in rocks of the Roberts Mountains allochthon is achieved by a combination of 66% Upper Vinini–Valmy–Snow Canyon–McAfee–Elder Slaven ages and 34% Harmony B ages. The introduction of either Harmony A or lower Vinini ages decreases the degree of similarity dramatically (Fig. 7), which indicates that sandstones within these units were not widespread during erosion of the Antler orogen.

## SUMMARY AND CONCLUSIONS

U-Pb ages have been determined for 22 detrital zircon grains from the Inskip Formation, part of the Antler overlap sequence, and for 22 grains from foreland basin strata of the Tonka Formation. Both of the sandstones analyzed are Mississippian age. These ages, together with data from a Pennsylvanian sandstone (Battle Formation; Gehrels and Dickinson, 1995), are dominated by 1750–2100 Ma grains, and abundant grains are in the 1800–1900 Ma range.

**TABLE 3. STATISTICAL COMPARISONS OF ANTLER OVERLAP AND FORELAND BASIN ASSEMBLAGES WITH ROBERTS MOUNTAINS ALLOCHTHON AND MIOGEOCLINAL STRATA IN NEVADA**

| | Nevada | Harmony A | Harmony B | Lower Vinini | Upper Vinini + |
|---|---|---|---|---|---|
| Degree of overlap | 0.42 | 0.12 | 0.82 | 0.43 | 0.92 |
| Degree of similarity | 0.28 | 0.12 | 0.75 | 0.34 | 0.82 |

*Note:* Numerical values are the results of comparing composite age-probability curve for the Antler overlap and foreland basin assemblages with curves for (1) miogeoclinal strata in Nevada, which record the ages of grains derived from basement rocks of the southwestern United States, and (2) four different stratigraphic components of the Roberts Mountains allochthon. All age-probability curves are shown in Figure 5. Degree of overlap indicates the degree to which ages in the two curves overlap, whereas degree of similarity is a measure of whether the curves contain similar proportions of ages.

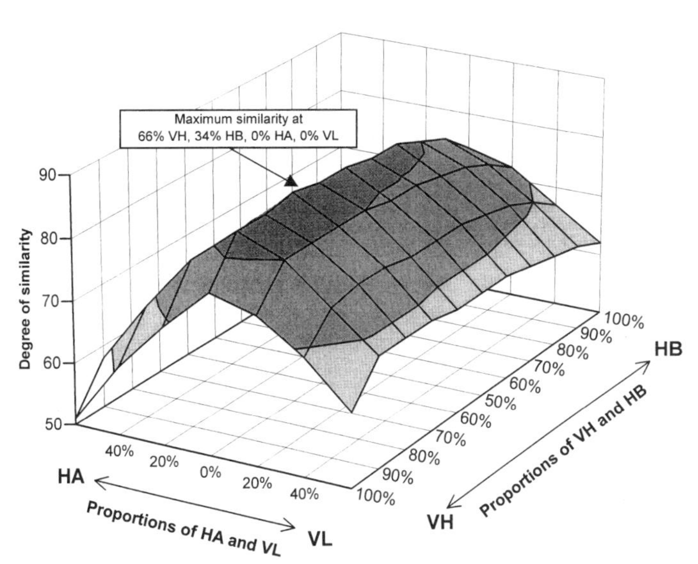


Figure 7. Plot showing degree of similarity between detrital zircon ages from Antler overlap and foreland basin assemblages (this paper) and from various proportions of four detrital zircon age components in Roberts Mountains allochthon. From front to back are variations in the proportions of two main components of allochthon, which include sandstones of Harmony Formation (HB) and upper Vinini, Valmy, Snow Canyon, McAfee, Elder, and Slaven formations (VH). Mixture of 66% VH and 34% HB has highest degree of similarity with our composite curve. From left to right are similarity values for mixtures of two other age components within allochthon, subset of Harmony Formation samples (HA) and lower sandstones of Vinini Formation (VL). Comparison suggests that little detritus in overlap and foreland basin assemblages was derived from HA and VL.

The ages in these samples are very similar to the ages of detrital zircon grains from sandstones of the Roberts Mountains allochthon (Gehrels et al., this volume, Chapter 1). Dominant sources apparently included upper Ordovician sandstones of the Valmy, Vinini, Snow Canyon, and McAfee formations, subarkosic sandstones of the Harmony Formation, and perhaps the Elder Sandstone and Slaven Chert. This result confirms previous stratigraphic studies that indicated that most detritus in the overlap and foreland basin assemblages was shed from allochthonous eugeoclinal strata in the Antler highland (Poole, 1974; Harbaugh and Dickinson, 1981; Saller and Dickinson, 1982; Dickinson et al., 1983).

## ACKNOWLEDGMENTS

We thank Walt Snyder for guidance in sampling the Tonka Formation. Brook Riley and Jim Harding did much of the mineral separation on our two samples. This research was supported by National Science Foundation grants EAR-9116000 and EAR-9416933. Reviewed by Katherine Giles, Richard Schweickert, and Arthur Saller.

## REFERENCES CITED

Brew, D.A., 1971, Mississippian stratigraphy of the Diamond Peak area, Eureka County, Nevada: U.S. Geological Survey Professional Paper 661, 84 p.

Burchfiel, B.C., and Davis, G.A., 1972, Structural framework and evolution of the southern part of the Cordilleran orogen, western United States: American Journal of Science, v. 272, p. 97–118.

Dickinson, W.R., Harbaugh, D.W., Saller, A.H., Heller, P.L., and Snyder, W.S., 1983, Detrital modes of upper Paleozoic sandstones derived from Antler orogen in Nevada: Implications for the nature of the Antler orogeny: American Journal of Science, v. 282, p. 481–509.

Dott, R.H., Jr., 1955, Pennsylvanian stratigraphy of Elko and northern Diamond ranges, northeastern Nevada: American Association of Petroleum Geologists Bulletin, v. 39, p. 2211–2305.

Gehrels, G.E., and Dickinson, W.R., 1995, Detrital zircon provenance of Cambrian to Triassic miogeoclinal and eugeoclinal strata in Nevada: American Journal of Science, v. 295, p. 18–48.

Gehrels, G.E., Dickinson, W.R., Ross, G.M., Stewart, J.H., and Howell, D.G., 1995, Detrital zircon reference for Cambrian to Triassic miogeoclinal strata of western North America: Geology, v. 23, p. 831–834.

Harbaugh, D.W., and Dickinson, W.R., 1981, Depositional facies of Mississippian clastics, Antler foreland basin, central Diamond Mountains, Nevada: Journal of Sedimentary Petrology, v. 51, p. 1223–1234.

Ketner, K.B., 1968, Origin of Ordovician Quartzite in the Cordilleran miogeosyncline: U.S. Geological Survey Professional Paper 600-B, p. 169–177.

Little, T.A., 1987, Stratigraphy and structure of metamorphosed upper Paleozoic rocks near Mountain City, Nevada: Geological Society of America Bulletin, v. 98, p. 1–17.

Ludwig, K.R., 1991a, A computer program for processing Pb-U-Th isotopic data: U.S. Geological Survey Open-File Report 88-542.

Ludwig, K.R., 1991b, A plotting and regression program for radiogenic-isotopic data: U.S. Geological Survey Open-File Report 91-445.

Madrid, R.J., 1987, Stratigraphy of the Roberts Mountains allochthon in north-central Nevada [Ph.D. thesis]: Stanford, California, Stanford University, 336 p.

Miller, E.L., Miller, M.M., Stevens, C.H., Wright, J.E., and Madrid, R., 1992, Late Paleozoic paleographic and tectonic evolution of the western U.S. Cordillera, *in* Burchfiel, B.C., et al., eds., The Cordilleran orogen: Conterminous U.S.: Boulder, Colorado, Geological Society of America, Geology of North America, v. G-3, p. 57–106.

Poole, F.G., 1974, Flysch deposits of Antler foreland basin, western United States: *in* Dickinson, W.R., Tectonics and Sedimentation: Society of Economic Paleontologists and Mineralogists Special Publication 22, p. 58–82.

Poole, F.G., Stewart, J.H., Palmer, A.R., Sandberg, C.A., Madrid, R.J., Ross, R.J., Hintze, L.F., Miller, M.M., and Wrucke, C.T., 1992, Latest Precambrian to latest Devonian time, Development of a continental margin, *in* Burchfiel, B.C., Lipman, P.W., and Zoback, M.L., eds., The Cordilleran orogen: Conterminous U.S.: Geological Society of America, The Geology of North America, v. G-3, p. 9–56.

Roberts, R.J., Hotz, P.E., Gilluly, J., and Ferguson, H.G., 1958, Paleozoic rocks of north-central Nevada: American Association of Petroleum Geologists Bulletin, v. 42, p. 2813–2857.

Saller, A.H., and Dickinson, W.R., 1982, Alluvial to marine facies transition in the Antler overlap sequence, Pennsylvanian and Permian of north-central Nevada: Journal of Sedimentary Petrology, v. 52, p. 925–940.

Silberling, N.J., Jones, D.L., Blake, M.C., Jr., and Howell, D.G., 1987, Lithotectonic terrane map of the western conterminous United States: U.S. Geological Survey Miscellaneous Field Studies Map MF-1874-C, scale 1:2,500,000.

Stewart, J.H., 1980, Geology of Nevada: Nevada Bureau of Mines and Geology Special Publication 4, 136 p.

Stewart, J.H., and Carlson, J.E., 1978, Geologic map of Nevada: U.S. Geological Survey, scale 1:500,000, 1 sheet.

Whitebread, D.H., 1994, Geologic map of the Dun Glen quadrangle, Pershing County, Nevada: U.S. Geological Survey Miscellaneous Investigations Map I-2409, scale 1:48,000.

MANUSCRIPT ACCEPTED BY THE SOCIETY JANUARY 24, 2000

Printed in U.S.A.

Geological Society of America
Special Paper 347
2000

# *U-Pb detrital zircon geochronology of the Golconda allochthon, Nevada*

**Brook C.D. Riley***
*Department of Geosciences, University of Arizona, Tucson, Arizona 85721, USA*
**Walter S. Snyder**
*Department of Geological Sciences, Boise State University, Boise, Idaho 83725, USA*
**George E. Gehrels**
*Department of Geosciences, University of Arizona, Tucson, Arizona 85721, USA*

## ABSTRACT

**The Golconda allochthon consists of deformed and imbricated deep-marine strata and volcanic rocks of late Paleozoic age that structurally overlie lower Paleozoic rocks of the Roberts Mountains allochthon and overlying upper Paleozoic strata. We analyzed 86 detrital zircon grains from 4 sandstone units of the Golconda allochthon in an effort to help reconstruct the paleogeographic and tectonic setting of the allochthon prior to Permian(?)-Triassic thrusting onto the continental margin. Of these, 81 grains yielded concordant to moderately discordant ages that define three main groups: 338–358 Ma (n = 7), 1770–1922 Ma (n = 38), and 2474–2729 Ma (n = 19).**

**Comparison of these ages with detrital zircon age spectra from adjacent terranes indicates that most of the detritus in Golconda sandstones was probably shed from rocks of both the Roberts Mountains allochthon to the east and the northern Sierra terrane to the west. Some grains may also have been shed from basement rocks of the southwestern United States. These relations support previously proposed tectonic models in which strata in the Golconda allochthon were deposited between the Sierra-Klamath magmatic arc to the west and the previously emplaced Roberts Mountains allochthon to the east.**

## INTRODUCTION

The Golconda allochthon (Figs. 1–3) includes complexly deformed marine sedimentary and volcanic rocks of latest Devonian through middle Permian age. These rocks were thrust onto the Roberts Mountains allochthon during the latest Permian–Early Triassic Sonoma orogeny (Silberling and Roberts, 1962; Silberling, 1973; Stewart et al., 1977; Miller et al., 1981, 1982, 1984, 1992; Schweickert and Snyder, 1981; Gabrielse et al., 1983; Snyder and Brueckner, 1983; Brueckner and Snyder, 1985; Babaie, 1987; Brueckner et al., 1987; Jones, 1991; Jones and Jones, 1991). The allochthon consists of two general lithostratigraphic assemblages: the Schoonover sequence (Fig. 4), deposited from latest Devonian through earliest Permian time; and the Havallah sequence, which ranges in age from Early Mississippian through Permian (Miller et al., 1981, 1982, 1984, 1992; Brueckner and Snyder, 1985; Murchey, 1990; Harwood and Murchey, 1990).

In the Independence Mountains in northern Nevada, the Schoonover sequence consists of chert-lithic and volcaniclastic sandstones interbedded with voluminous chert, argillite, and basaltic-andesitic greenstone, and subordinate pebbly mudstone (Figs. 3 and 4) (Fagan, 1962; Miller et al., 1984). The lower part of the section is mainly volcaniclastic and volcanic rocks, whereas the upper part is dominated by chert, argillite, and limestone turbidites (Miller et al., 1984). Samples analyzed from the Schoonover

*Current address: Department of Geological Sciences,University of Texas, Austin, Texas 78712, USA.

Riley, B.C.D., Snyder, W.S., and Gehrels, G.E., 2000, U-Pb zircon geochronology of the Golconda allochthon, Nevada, *in* Soreghan, M.J., and Gehrels, G.E., eds., Paleozoic and Triassic paleogeography and tectonics of western Nevada and northern California: Boulder, Colorado, Geological Society of America Special Paper 347, p. 65–75.

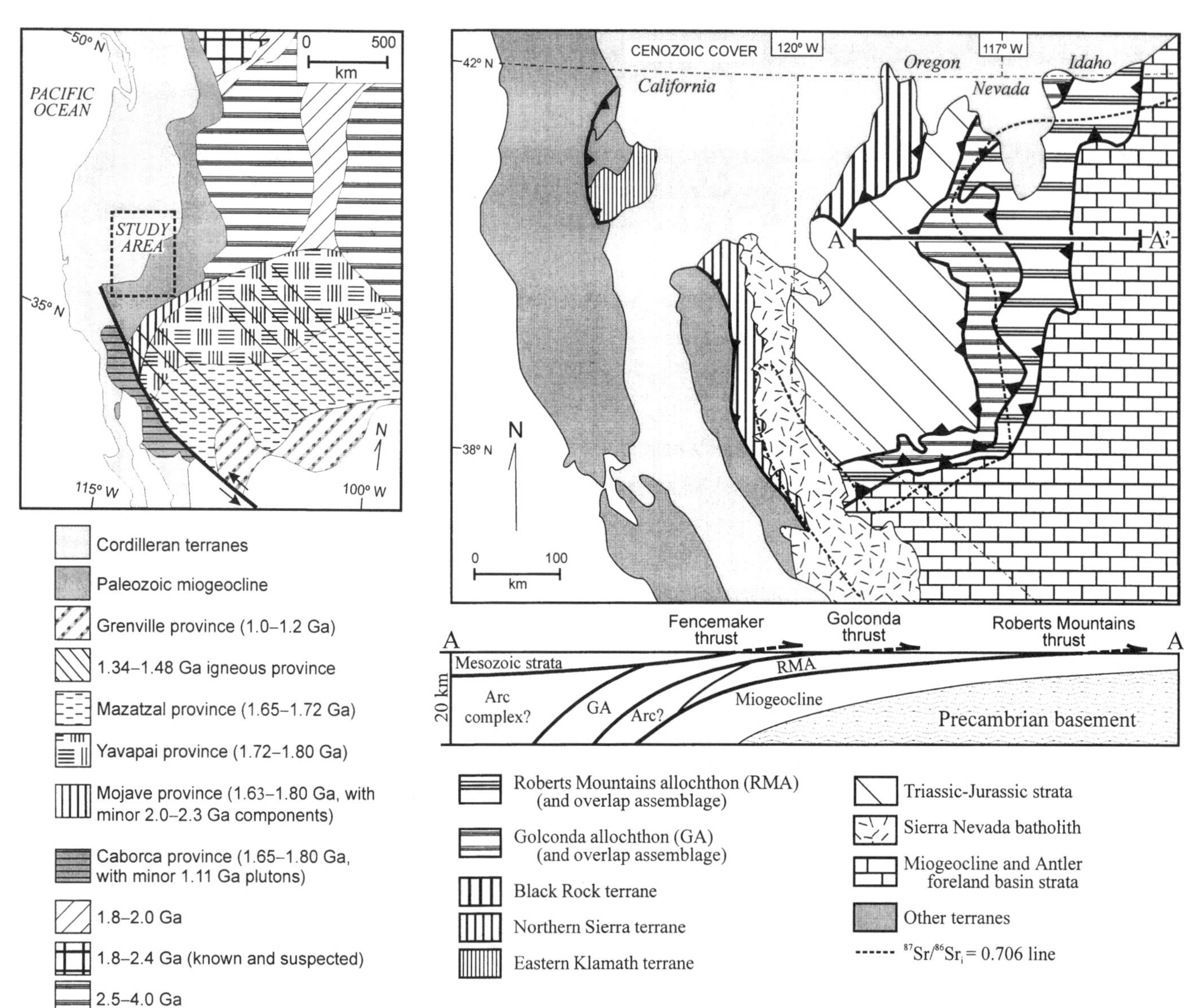


Figure 1. Schematic map showing location of study area in relation to Cordilleran miogeocline and first-order basement provinces of western North America. Basement provinces are simplified from Hoffman (1989) for cratonal interior, various sections of Van Schmus et al. (1993) for southwestern United States, and Stewart et al. (1990) for northwestern Mexico.

Figure 2. Schematic map and cross section of main terranes and/or assemblages and their bounding faults in western Nevada and northern California. Map configuration of terranes and assemblages is adapted primarily from Oldow (1984), Silberling (1991), and Silberling et al. (1987), and cross section is highly simplified from Blake et al. (1985) and Saleeby (1986). Location of $^{87}Sr/^{86}Sr_i$ = 0.706 line is from Kistler and Peterman (1973) and Kistler (1991). Note that widespread Cretaceous and younger rocks and structures are ignored in map and sections in effort to emphasize configuration of pre-Cretaceous features.

sequence during this study are from Lower Mississippian rocks within the volcanic-rich part of the sequence (Fig. 4).

The Havallah sequence in the Galena Range of central Nevada (Fig. 3) is dominated by chert, argillite, and limestone turbidites, with some interbedded shale, volcaniclastic and chert-lithic sandstone, and pebbly conglomerate. Basalt flows and fault-bounded slivers of greenstone appear both at the tectonic base of the sequence and at the base of certain thrust packets (Miller et al., 1982, 1992; Murchey, 1990). In the original description of the Golconda allochthon in the Antler Peak quadrangle (Roberts, 1964), the term Havallah Formation was used for the younger part of the Golconda allochthon (Jory and Trenton Members), whereas the name Pumpernickel Formation was applied to the structurally lowest unit of the allochthon. Terminology in this paper follows that of Silberling and Roberts (1962), Miller et al. (1982), and later authors, in which the Havallah sequence encompasses the Pumpernickel, Jory, and Trenton units.

The Pumpernickel unit is a heterogeneous sequence of chert and argillite, with interbedded siltstone toward the top of the unit. The Jory unit is characterized by chert-rich conglomerate with

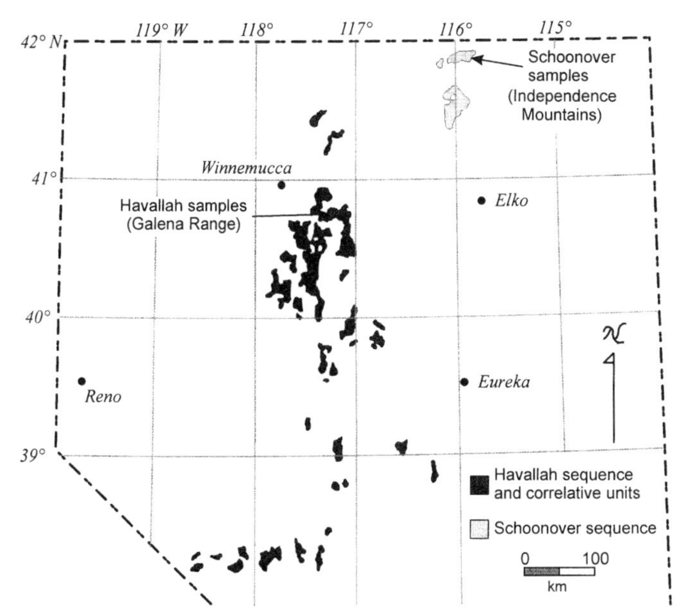


Figure 3. Map of central Nevada showing general outcrop patterns of main units of Golconda allochthon (from Stewart and Carlson, 1978).

sandstone interbeds. The Trenton unit is composed mainly of chert and shale. In other exposures of the Havallah sequence, the uppermost unit is the Mill Canyon Member, which includes interbedded turbidites, chert, and shale (Miller et al., 1982). Sandstones from the Middle Pennsylvanian–Lower Permian Jory unit and the Lower Permian Pumpernickel unit were analyzed during this study (Fig. 4).

In general, the monotonous lithology and structural complexity of the Havallah and Schoonover sequences preclude a full understanding of the structure of the Golconda allochthon. Both units are multiply deformed in most exposures, and structures include boudins, mullions, and open to tight folds. During progressive deformation of the Golconda allochthon, bedding was disrupted and thrust faults imbricated all rock types. According to a number of workers (e.g., Miller et al., 1982; Snyder and Brueckner, 1983; Babaie, 1987), a significant amount of the deformation occurred prior to emplacement of the allochthon onto the continental margin, as the subjacent autochthon is not as pervasively deformed as the Golconda allochthon.

A first-order problem in Cordilleran tectonics concerns the pre-emplacement position of the Golconda allochthon with respect to nearby terranes as well as North America. One model suggests that strata of the Golconda allochthon were deposited in a relatively wide basin and were subsequently incorporated into an accretionary prism built along the inboard margin of a magmatic arc that faced eastward, or toward the North American continent (Speed, 1979; Schweickert and Snyder, 1981; Speed and Sleep, 1982; Snyder and Brueckner, 1983; Dickinson et al., 1983; Brueckner and Snyder, 1985; Babaie, 1987). An opposing view asserts that the strata were deposited within a relatively narrow basin separating the North American margin from an outboard, west-facing magmatic arc (Burchfiel and Davis, 1972, 1975; Silberling, 1973; Miller et al., 1984, 1992; Harwood and Murchey, 1990; Burchfiel and Royden, 1991; Burchfiel et al., 1992). In addition, the possibility exists that strata within the allochthon accumulated a great distance from their present position and are far traveled (Coney et al., 1980).

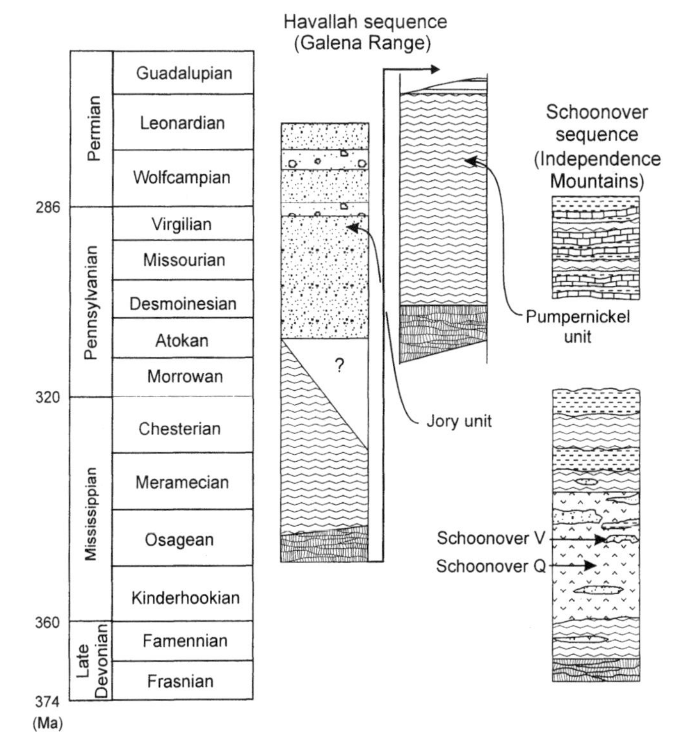


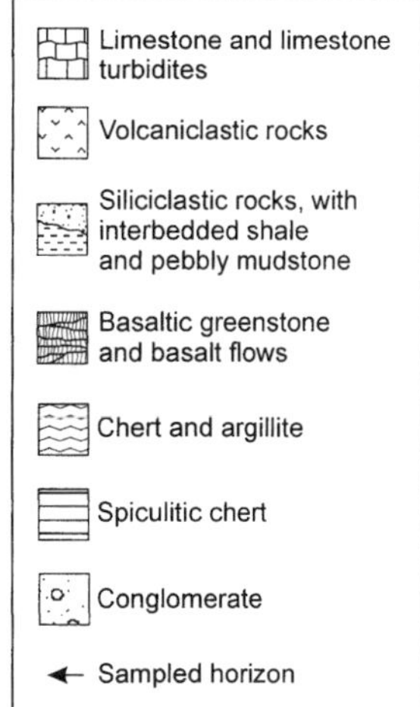


Figure 4. Generalized stratigraphic columns for Havallah and Schoonover sequences of Golconda allochthon. Schoonover sequence shown is middle thrust plate of Miller et al. (1984). Modified from Miller et al. (1984, 1992).

The first two models predict that different sets of source terranes contributed detritus to the Golconda basin. If the Golconda allochthon evolved in an east-facing accretionary prism, most strata in the allochthon should contain predominantly Precambrian detritus derived from continental regions to the east. Source regions for clastic detritus would have included 1.0–1.8 Ga rocks of the Mojave, Yavapai, Mazatzal, and Grenville basement provinces (Fig. 1), and/or uplifted rocks of the Roberts Mountains allochthon (Fig. 2). As described by Gehrels et al. (this volume, Chapter 1), the latter would have yielded detrital zircons primarily >1.8 Ga.

If the Golconda allochthon evolved in a relatively narrow backarc basin, one might expect to see Precambrian zircon populations derived from the east as well as from a volcanic arc terrane to the west. The work of Harwood and Murchey (1990) suggests specifically that siliciclastic detritus in the Golconda allochthon would have been derived from the uplifted Roberts Mountains allochthon to the east, whereas volcanic-rich detritus would have been shed from Devonian-Mississippian and Permian igneous rocks in the Sierra-Klamath arc to the west (Fig. 2). The third model raises the possibility that sandstones in the Golconda allochthon could contain zircons with ages not found in rocks in adjacent regions.

In an effort to identify the sources of clastic sedimentary rocks in the allochthon, we have analyzed representative populations of detrital zircons from four different units of Mississippian through Permian age. These zircon ages are then compared with North American reference populations and with those of the adjacent Roberts Mountains allochthon and northern Sierra terrane (Fig. 2) to resolve possible provenance linkages.

## ANALYTICAL METHODS

For each sample, ~20 kg of clean, essentially unweathered rock was removed from a narrow (~1 m) stratigraphic interval. Each sample was crushed and pulverized, and heavier minerals were removed with a Wilfley table. Zircon was further isolated with a Frantz LB-1 magnetic separator and heavy liquids. The zircons from each sample were separated into size fractions using disposable nylon sieve screens, and the >100 μm grains were separated on the basis of color, shape, and degree of rounding. Zircons from each of these populations were abraded in an air abrader for varying amounts of time depending upon their size. At least 18 grains were selected from the various color and morphology groups in each sample and analyzed by isotope dilution-thermal ionization mass spectrometry. The accompanying chapter by Gehrels (this volume, Introduction) describes the analytical techniques in detail, and also discusses potential biases resulting from our grain selection techniques.

Results from the four samples are presented in Table 1 and plotted on concordia diagrams (Fig. 5) and a relative age-probability plot (Fig. 6). Table 1 divides the analyses from each sample into four groups, depending on their degree of discordance. Of the 86 analyses presented, 26 are concordant and of moderate to

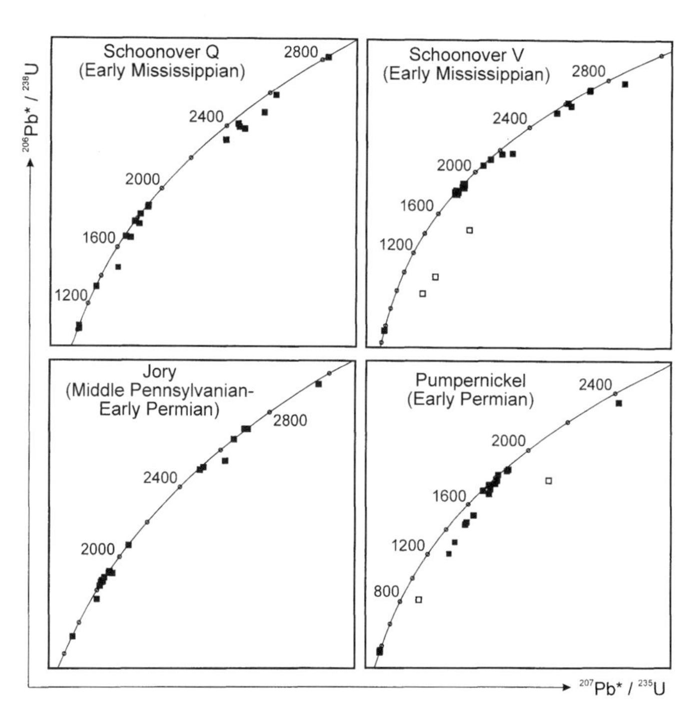


Figure 5. U-Pb concordia diagrams showing single-grain detrital zircon analyses from Havallah and Schoonover sequences. Open squares represent analyses that are sufficiently discordant that they are not assigned ages. All data reduction and concordia plots are from programs of Ludwig (1991a, 1991b).

high precision. The interpreted ages of concordant grains are $^{206}Pb^*/^{238}U$ ages for <1.0 Ga grains and $^{207}Pb^*/^{206}Pb^*$ ages for older grains. There are 49 grains that are slightly to moderately discordant ($^{206}Pb^*/^{238}U$ age within 15% of $^{207}Pb^*/^{206}Pb^*$ age), and an additional 6 grains are highly discordant but still informative. Five grains, shown with open squares in Figure 5, are sufficiently discordant that they are not considered further.

The interpreted ages for discordant grains are determined by (1) assuming that discordance results entirely from Pb loss, and (2) projecting from 100 Ma, because this is a more reasonable age of Pb loss than 0 Ma, which is implicit in the use of $^{207}Pb^*/^{206}Pb^*$ ages. However, the average difference between the $^{207}Pb^*/^{206}Pb^*$ ages and the projected ages is <3 m.y., which is not significant for the purposes of this study. To incorporate the uncertainty that results from discordance, highly discordant grains are assigned a much larger uncertainty (±40 m.y.) than concordant (±10 m.y.) or slightly discordant (generally ±10 or ±20 m.y.) grains. The interpreted ages and their uncertainties are incorporated into the age-probability plots shown in Figure 6.

## RESULTS

### *Schoonover Q (Lower Mississippian)*

The Schoonover Q sample was collected from Lower Mississippian strata that have been assigned to map unit Mss by Miller et al. (1981, 1984). It was collected from Cole Canyon in the northern Independence Mountains, where the sequence is dominated by black shale and siltstone. The sandstone bed sampled is 0.5 m thick and is graded from medium to fine sand. The material collected from this bed is a medium-grained chert-lithic sandstone (Dickinson and Gehrels, this volume, Chapter 11).

The sample yielded 400 mg of zircon grains, most of which are <175 μm in sieve size. The zircons represent six distinct color groups, the majority being colorless or yellowish, with minor populations of light to medium pink zircons. Most grains are highly spherical and well rounded, with crystal faces and tips preserved on only a small proportion of grains. All grains analyzed were >145 μm prior to abrasion, which removed the outer 20% of each grain.

We analyzed 18 detrital zircons, of which 17 are concordant to slightly discordant (Table 1; Fig. 5). The dominant age groups are 1770–1920 Ma (n = 7) and 2460–2675 Ma (n = 7), and there are scattered single ages of approximately 1033, 1071, 1345, 1713, 2080, and 2827 Ma (Tables 1 and 2). The dominant peaks in age probability are at 1828, 1912, 2496, and 2656 Ma (Fig. 6; Table 2).

### *Schoonover V (Lower Mississippian)*

This sample was collected from unit Msv of Miller et al. (1981, 1984), and is also of Early Mississippian age. It was collected from the north side of Cole Canyon, ~1 km west of Schoonover Q, but in the same lithotectonic sequence. The sandstone bed sampled is 1 m thick and is interbedded with more strongly deformed mudstone and siltstone. As described by Dickinson and Gehrels (this volume, Chapter 11), the sample analyzed is a medium- to coarse-grained, chert-rich, volcaniclastic sandstone.

The sample yielded 435 mg of zircons, most of which are <175 μm in sieve size. The majority of zircons from the unit are colorless to light yellowish, and there are subordinate populations of light to medium pink grains. The darker colored grains are somewhat cloudy, whereas all other grains are highly translucent and generally free of inclusions. Each of the color groups displays a wide range of morphologies, including well rounded and highly spherical grains, grains with slightly rounded edges and tips, and euhedral grains that show no sign of rounding. All grains analyzed were >145 μm prior to abrasion, which removed the outer 20% of each grain.

Ages of 22 detrital zircons from this sample are listed in Table 1 and shown on a concordia diagram in Figure 5. Most are concordant to slightly discordant, except for three pink, cloudy grains that are highly discordant and are not assigned ages (Fig. 5). Three grains yield young ages of ca. 341, 347, and 351 Ma. Similar age populations as those in Schoonover Q are also present in this unit: 1805–1935 Ma (n = 6), and 2575–2735 Ma (n = 5). Peaks in age probability are present at 347, 1848, 1907, and 2727 Ma (Fig. 6; Table 2).

### *Jory unit (Middle Pennsylvanian–Lower Permian)*

This sample is a medium- to coarse-grained chert-lithic sandstone (Dickinson and Gehrels, this volume, Chapter 11) of Middle Pennsylvanian–Early Permian age (Roberts, 1964; Murchey, 1990). It was collected from the Galena Range, a few kilmeters northwest of Antler Peak (Fig. 3). The sample was collected from the pebbly base of a 2-m-thick graded bed in a sequence of sandstone and conglomerate. Clasts in the conglomerate layers are predominantly quartzite, chert, and argillite.

The sample yielded 275 mg of zircon grains, most of which are <175 μm in sieve size. The grains are distributed evenly among light and medium pink populations, with a smaller population of yellowish to colorless zircons. Most grains are well rounded and highly spherical, with crystal faces and tips preserved in only a small proportion of grains. All grains analyzed were >145 μm prior to removal of the outer 20% of each grain by abrasion.

Ages of 22 zircons analyzed from this sample are listed in Table 1 and shown on a concordia diagram in Figure 5. All of the grains from this sample are concordant to slightly discordant and yield well-constrained ages. Most are 1825–1930 Ma (n = 12) and 2645–2730 Ma (n = 4) (Fig. 6). Peaks in age probability occur at 1838, 1860, 1922, and 2720 Ma (Table 2).

### *Pumpernickel unit (Lower Permian)*

The youngest part of the Havallah sequence is represented in this study by a sample of the Pumpernickel unit, which was collected from the west side of Willow Canyon in the southern part of the Galena Range. This is unit LU-1B of Murchey (1990) and is late Early Permian (Leonardian) age. The sample was collected from a 3-m-thick layer of conglomeratic sandstone within a sequence of mainly chert and argillite. The sandstone is generally rich in chert and volcanic lithic fragments (Dickinson and Gehrels, this volume, Chapter 11), and there are also centimeter-scale volcanic fragments contained within the adjacent chert and argillite layers.

This sample yielded 180 mg of zircon grains, most of which are <145 μm in sieve size. Grains from this unit display a wide range of morphologies, from well rounded to euhedral. Color populations are evenly distributed among colorless, light pink, and medium pinkish-purple grains.

Zircons from the Pumpernickel unit (n = 22) yield a range of concordant to highly discordant ages (Table 1; Fig. 5). Five grains are young, with interpreted ages of ca. 338, 340, 341, 350, and 358 Ma, and a peak in age probability at 340 Ma. Older grains are mainly 1725–1920 (n = 15), with one additional grain of ca. 2484 Ma. Peaks in age probability are present at 1779, 1818, 1852, and 1911 Ma (Fig. 6; Table 2).

## INTERPRETATION

Samples from all four sandstones within the Golconda allochthon are dominated by detrital zircons with ages of 1770–1922 and 2474–2729 Ma (Fig. 6). As shown in Figure 6, these ages contrast with the generally <1.8 Ga grains in miogeoclinal strata of Nevada, which were derived from basement rocks

**TABLE 1. U-Pb ISOTOPIC DATA AND AGES**

| Grain type | Grain wt. (µg) | $Pb_c$ (pg) | U (ppm) | $^{206}Pb_m$/$^{204}Pb$ | $^{206}Pb_c$/$^{208}Pb$ | Apparent ages (Ma) $^{206}Pb^*$/$^{238}U$ | $^{207}Pb^*$/$^{235}U$ | $^{207}Pb^*$/$^{206}Pb^*$ | Interpreted age (Ma) |
|---|---|---|---|---|---|---|---|---|---|
| Schoonover Q (41°34'32.1"N, 115°57'27.3"W) | | | | | | | | | |
| **YsR** | **16** | **4** | **231** | **8230** | **3.4** | **1034 ± 4** | **1034 ± 5** | **1033 ± 10** | **1033 ± 10** |
| LwR | 8 | 5 | 43 | 690 | 3.1 | 1013 ± 3 | 1031 ± 3 | 1071 ± 21 | 1071 ± 20 |
| CwR | 18 | 8 | 266 | 8060 | 5.8 | 1324 ± 6 | 1332 ± 7 | 1344 ± 6 | 1345 ± 10 |
| CwR | 16 | 6 | 188 | 9380 | 5.7 | 1673 ± 7 | 1690 ± 8 | 1712 ± 6 | 1713 ± 10 |
| **YsR** | **15** | **7** | **167** | **7210** | **7.1** | **1779 ± 7** | **1778 ± 9** | **1777 ± 7** | **1777 ± 10** |
| *MsR* | *9* | *8* | *893* | *16320* | *6.0* | *1459 ± 6* | *1607 ± 7* | *1807 ± 8* | *1822 ± 40* |
| YsR | 12 | 6 | 383 | 14540 | 10.1 | 1667 ± 7 | 1736 ± 8 | 1821 ± 6 | 1826 ± 20 |
| **LwR** | **13** | **6** | **488** | **23440** | **8.1** | **1829 ± 7** | **1828 ± 8** | **1828 ± 6** | **1828 ± 10** |
| LwR | 7 | 5 | 734 | 19790 | 10.0 | 1762 ± 7 | 1819 ± 8 | 1884 ± 6 | 1889 ± 20 |
| CwR | 19 | 5 | 226 | 17560 | 4.4 | 1887 ± 7 | 1898 ± 9 | 1911 ± 7 | 1912 ± 10 |
| YwR | 14 | 7 | 79 | 3330 | 3.9 | 1877 ± 10 | 1894 ± 11 | 1912 ± 7 | 1913 ± 10 |
| **CwR** | **18** | **7** | **87** | **5500** | **3.3** | **2078 ± 13** | **2079 ± 16** | **2080 ± 8** | **2080 ± 10** |
| CwR | 27 | 10 | 41 | 2870 | 3.7 | 2313 ± 12 | 2398 ± 13 | 2471 ± 6 | 2474 ± 20 |
| YsR | 23 | 10 | 44 | 2730 | 3.0 | 2413 ± 12 | 2458 ± 14 | 2495 ± 8 | 2496 ± 10 |
| CsR | 9 | 6 | 139 | 5760 | 8.2 | 2394 ± 11 | 2464 ± 13 | 2522 ± 4 | 2524 ± 20 |
| YsR | 11 | 4 | 103 | 7390 | 5.4 | 2382 ± 10 | 2490 ± 12 | 2579 ± 4 | 2583 ± 20 |
| **YwR** | **13** | **9** | **107** | **5020** | **3.1** | **2645 ± 17** | **2650 ± 19** | **2654 ± 6** | **2655 ± 10** |
| YwR | 17 | 9 | 151 | 840 | 10.5 | 2479 ± 32 | 2575 ± 36 | 2653 ± 8 | 2655 ± 20 |
| CwR | 9 | 6 | 145 | 6150 | 4.4 | 2588 ± 12 | 2626 ± 13 | 2656 ± 7 | 2657 ± 10 |
| YwR | 14 | 10 | 159 | 7280 | 8.1 | 2814 ± 11 | 2821 ± 13 | 2826 ± 4 | 2827 ± 10 |
| Schoonover V (41°34'37.0"N, 115°58'06.3"W) | | | | | | | | | |
| **LsE** | **13** | **15** | **1706** | **5190** | **5.2** | **341 ± 2** | **343 ± 2** | **358 ± 8** | **341 ± 10** |
| **YsR** | **24** | **13** | **558** | **3660** | **4.6** | **347 ± 4** | **347 ± 4** | **349 ± 10** | **347 ± 10** |
| **YsR** | **31** | **12** | **148** | **1380** | **5.4** | **351 ± 3** | **352 ± 4** | **359 ± 17** | **351 ± 10** |
| YsE | 20 | 79 | 265 | 1240 | 7.2 | 1804 ± 11 | 1808 ± 14 | 1813 ± 8 | 1813 ± 10 |
| **CwR** | **30** | **14** | **209** | **8300** | **11.1** | **1818 ± 11** | **1821 ± 12** | **1825 ± 6** | **1825 ± 10** |
| CwR | 36 | 11 | 135 | 8550 | 12.2 | 1792 ± 8 | 1817 ± 9 | 1845 ± 4 | 1847 ± 10 |
| CwR | 29 | 16 | 116 | 4020 | 7.4 | 1813 ± 8 | 1829 ± 10 | 1847 ± 4 | 1849 ± 10 |
| YsE | 16 | 8.4 | 60 | 2370 | 2.5 | 1883 ± 11 | 1894 ± 13 | 1906 ± 6 | 1907 ± 10 |
| LwR | 29 | 23 | 113 | 2700 | 3.3 | 1854 ± 9 | 1888 ± 10 | 1925 ± 5 | 1927 ± 10 |
| LwR | 13 | 125 | 26 | 74 | 1.4 | 2058 ± 41 | 2055 ± 92 | 2053 ± 65 | 2053 ± 40 |
| LwR | 22 | 14 | 45 | 1610 | 4.9 | 2113 ± 23 | 2130 ± 26 | 2146 ± 5 | 2148 ± 10 |
| YsE | 27 | 41 | 224 | 3320 | 4.6 | 2161 ± 8 | 2216 ± 10 | 2266 ± 5 | 2269 ± 20 |
| LsE | 23 | 31 | 254 | 4030 | 4.3 | 2169 ± 13 | 2291 ± 15 | 2401 ± 4 | 2407 ± 40 |
| CwR | 40 | 135 | 86 | 680 | 2.6 | 2522 ± 17 | 2557 ± 20 | 2584 ± 6 | 2585 ± 10 |
| MOE | 39 | 275 | 665 | 1270 | 12.0 | 1436 ± 17 | 1950 ± 25 | 2551 ± 5 | N.A. |
| MOE | 28 | 103 | 569 | 1050 | 3.8 | 766 ± 3 | 1366 ± 8 | 2489 ± 6 | N.A. |
| MOE | 30 | 86 | 781 | 2320 | 6.3 | 949 ± 4 | 1561 ± 8 | 2526 ± 5 | N.A. |
| *LsE* | *18* | *13* | *101* | *400* | *7.7* | *2607 ± 50* | *2614 ± 52* | *2620 ± 9* | *2621 ± 10* |
| LwR | 23 | 28 | 216 | 4700 | 4.7 | 2580 ± 15 | 2629 ± 17 | 2668 ± 4 | 2669 ± 20 |
| **CwR** | **28** | **13.4** | **51** | **3150** | **4.6** | **2717 ± 33** | **2721 ± 33** | **2724 ± 4** | **2725 ± 10** |
| LwR | 20 | 22 | 192 | 4960 | 3.7 | 2706 ± 16 | 2719 ± 17 | 2728 ± 4 | 2729 ± 10 |
| YsE | 27 | 17 | 109 | 5100 | 8.6 | 2773 ± 11 | 2867 ± 13 | 2934 ± 4 | 2936 ± 20 |
| Jory Formation (40°36'21"N, 117°06'43"W) | | | | | | | | | |
| LwR | 15 | 11 | 551 | 11410 | 6.6 | 1512 ± 7 | 1517 ± 8 | 1523 ± 5 | 1524 ± 10 |
| **LwR** | **12** | **9** | **203** | **5360** | **5.2** | **1827 ± 9** | **1830 ± 10** | **1833 ± 7** | **1833 ± 10** |
| LsR | 21 | 17 | 71 | 4970 | 3.3 | 1821 ± 10 | 1828 ± 11 | 1835 ± 5 | 1836 ± 10 |
| LwR | 16 | 16 | 109 | 5800 | 4.7 | 1815 ± 9 | 1826 ± 10 | 1839 ± 4 | 1840 ± 10 |
| MsR | 25 | 8 | 185 | 11110 | 7.7 | 1829 ± 8 | 1835 ± 9 | 1843 ± 4 | 1844 ± 10 |
| **MwR** | **12** | **9** | **95** | **2290** | **3.4** | **1851 ± 11** | **1852 ± 12** | **1852 ± 5** | **1852 ± 10** |
| **MsR** | **16** | **13** | **143** | **3430** | **3.9** | **1855 ± 9** | **1856 ± 11** | **1858 ± 5** | **1858 ± 10** |
| **LsR** | **21** | **11** | **156** | **6110** | **20.4** | **1856 ± 7** | **1858 ± 9** | **1859 ± 7** | **1859 ± 10** |
| MsR | 40 | 5 | 61 | 17460 | 12.0 | 1797 14 | 1827 ± 15 | 1861 ± 6 | 1863 ± 10 |
| **MsR** | **28** | **12** | **240** | **11050** | **12.9** | **1858 ± 11** | **1861 ± 12** | **1863 ± 5** | **1864 ± 10** |
| LsR | 15 | 15 | 204 | 3700 | 7.0 | 1745 ± 8 | 1799 ± 10 | 1862 ± 6 | 1866 ± 20 |
| YwR | 19 | 7 | 67 | 4100 | 1.9 | 1905 ± 11 | 1912 ± 13 | 1920 ± 5 | 1921 ± 20 |
| **MsR** | **28** | **20** | **425** | **12000** | **2.9** | **1914 ± 11** | **1918 ± 12** | **1922 ± 6** | **1922 ± 10** |
| LsR | 22 | 10 | 119 | 5180 | 5.7 | 1903 ± 9 | 1941 ± 10 | 1983 ± 5 | 1984 ± 10 |
| **LwR** | **20** | **8** | **118** | **7050** | **1.2** | **2068 ± 9** | **2070 ± 10** | **2073 ± 6** | **2073 ± 10** |
| YwR | 10 | 26 | 196 | 2090 | 5.8 | 2492 ± 12 | 2502 ± 14 | 2511 ± 4 | 2512 ± 10 |

**TABLE 1. U-Pb ISOTOPIC DATA AND AGES (continued)**

| Grain type | Grain wt. (μg) | $Pb_c$ (pg) | U (ppm) | $\frac{^{206}Pb_m}{^{204}Pb}$ | $\frac{^{206}Pb_c}{^{208}Pb}$ | Apparent ages (Ma) $\frac{^{206}Pb^*}{^{238}U}$ | $\frac{^{207}Pb^*}{^{235}U}$ | $\frac{^{207}Pb^*}{^{206}Pb^*}$ | Interpreted age (Ma) |
|---|---|---|---|---|---|---|---|---|---|
| YsR | 18 | 12 | 119 | 4940 | 6.7 | 2507 ± 12 | 2520 ± 12 | 2532 ± 6 | 2532 ± 10 |
| **MsR** | **36** | **12** | **185** | **17160** | **6.7** | **2657 ± 16** | **2657 ± 17** | **2657 ± 5** | **2657 ± 10** |
| LwR | 12 | 8 | 328 | 14000 | 4.3 | 2542 ± 11 | 2620 ± 13 | 2680 ± 5 | 2682 ± 20 |
| **VsR** | **12** | **8** | **48** | **2270** | **6.3** | **2711 ± 23** | **2716 ± 24** | **2719 ± 4** | **2719 ± 10** |
| **YwR** | **11** | **8** | **118** | **5310** | **3.9** | **2712 ± 15** | **2717 ± 16** | **2721 ± 4** | **2722 ± 10** |
| LwR | 41 | 12 | 131 | 15000 | 5.4 | 2944 ± 15 | 2967 ± 18 | 2983 ± 5 | 2984 ± 10 |
| Pumpernickel Formation (40°36'13"N, 117°09'41"W) | | | | | | | | | |
| **CsR** | **22** | **12** | **244** | **1590** | **4.5** | **338 ± 2** | **338 ± 3** | **337 ± 15** | **338 ± 10** |
| **CsR** | **87** | **16** | **813** | **14800** | **3.3** | **340 ± 2** | **340 ± 2** | **343 ± 5** | **340 ± 10** |
| **CsR** | **111** | **24** | **1203** | **19350** | **2.2** | **341 ± 2** | **342 ± 2** | **344 ± 5** | **341 ± 10** |
| **CwR** | **18** | **21** | **297** | **905** | **7.1** | **350 ± 2** | **351 ± 5** | **354 ± 28** | **350 ± 10** |
| **CE** | **25** | **24** | **307** | **1185** | **7.4** | **358 ± 2** | **357 ± 4** | **349 ± 20** | **358 ± 10** |
| CsE | 20 | 19 | 31 | 612 | 3.5 | 1701 ± 16 | 1714 ± 21 | 1730 ± 14 | 1731 ± 10 |
| ME | 22 | 58 | 422 | 1270 | 12.0 | 816 ± 5 | 1090 ± 9 | 1687 ± 9 | N.A. |
| CsR | 20 | 25 | 59 | 870 | 6.0 | 1744 ± 21 | 1756 ± 23 | 1769 ± 11 | 1770 ± 10 |
| *MsR* | *12* | *24* | *249* | *1840* | *7.4* | *1440 ± 6* | *1573 ± 10* | *1757 ± 8* | *1771 ± 20* |
| *ME* | *21* | *15* | *67* | *1420* | *4.7* | *1456 ± 8* | *1585 ± 12* | *1761 ± 8* | *1774 ± 40* |
| *ME* | *24* | *32* | *203* | *1970* | *7.4* | *1298 ± 6* | *1484 ± 9* | *1760 ± 6* | *1783 ± 40* |
| *ME* | *25* | *34* | *252* | *2230* | *7.4* | *1203 ± 6* | *1429 ± 9* | *1785 ± 7* | *1816 ± 40* |
| **CsR** | **18** | **15** | **23** | **546** | **3.4** | **1820 ± 22** | **1819 ± 24** | **1818 ± 11** | **1818 ± 10** |
| *MsE* | *12* | *19* | *232* | *2230* | *14.6* | *1512 ± 7* | *1642 ± 10* | *1813 ± 8* | *1825 ± 40* |
| MwR | 21 | 20 | 96 | 1800 | 7.7 | 1714 ± 8 | 1766 ± 10 | 1828 ± 6 | 1832 ± 20 |
| LwR | 21 | 17 | 63 | 1450 | 17.6 | 1779 ± 20 | 1812 ± 22 | 1850 ± 7 | 1852 ± 10 |
| CsR | 19 | 20 | 88 | 1545 | 7.2 | 1754 ± 8 | 1799 ± 11 | 1851 ± 6 | 1854 ± 20 |
| MsE | 16 | 37 | 429 | 3180 | 7.0 | 1679 ± 8 | 1756 ± 9 | 1850 ± 4 | 1856 ± 20 |
| MsE | 24 | 22 | 56 | 1200 | 2.9 | 1848 ± 10 | 1875 ± 13 | 1905 ± 7 | 1907 ± 10 |
| MsR | 16 | 15 | 87 | 1795 | 4.0 | 1859 ± 10 | 1885 ± 11 | 1913 ± 5 | 1915 ± 10 |
| ME | 23 | 60 | 260 | 1760 | 6.3 | 1780 ± 8 | 2109 ± 11 | 2448 ± 5 | N.A. |
| LsR | 17 | 34 | 84 | 1045 | 5.5 | 2334 ± 11 | 2413 ± 15 | 2481 ± 6 | 2484 ± 20 |

*Note:*

Grain with $^{206}Pb^*/^{238}U$ age within 15% of $^{207}Pb^*/^{206}Pb^*$ age in plain type.

Grain with $^{206}Pb^*/^{238}U$ age not within 15% of $^{207}Pb^*/^{206}Pb^*$ age in italics.

Grain with concordant age and moderate to high precision analysis in bold.

* Radiogenic Pb.

Grain type: L—light pink, M—medium pink, C—colorless, Y—yellow, O—cloudy, sR—slightly rounded, wR—well rounded, E—euhedral, sE—subeuhedral.

Unless otherwise noted, all grains are clear and/or translucent.

$^{206}Pb/^{204}Pb$ is measured ratio, uncorrected for blank, spike, or fractionation.

$^{206}Pb/^{208}Pb$ is corrected for blank, spike, and fractionation.

Most concentrations have an uncertainty of 25% due to uncertainty in weight of grain.

Constants used: $^{238}U/^{235}U$ = 137.88. Decay constant for $^{235}U = 9.8485 \times 10^{-10}$. Decay constant for $^{238}U = 1.55125 \times 10^{-10}$.

All uncertainties are at the 95% confidence level.

Pb blank ranged from 2 to 10 pg. U blank was consistently <1 pg.

Interpreted ages for concordant grains are $^{206}Pb^*/^{238}U$ ages if <1.0 Ga and $^{207}Pb^*/^{206}Pb^*$ ages if >1.0 Ga.

Interpreted ages for discordant grains are projected from 100 Ma.

All analyses conducted using conventional isotope dilution and thermal ionization mass spectrometry, as described by Gehrels (this volume, Introduction).

of the southwestern United States (Gehrels, this volume, Introduction). This is not surprising, however, because the Roberts Mountains allochthon formed a highland along the continental margin (the Antler Highland) during late Paleozoic time, apparently shielding the oceanic realm from craton-derived detritus (Roberts et al., 1958; Nilsen and Stewart, 1980; Dickinson et al., 1983).

Derivation of strata in the Golconda allochthon from rocks in the Antler Highland was proposed by Burchfiel and Davis (1972, 1975), Schweickert and Snyder (1981), Dickinson et al. (1983), Miller et al. (1984), Harwood and Murchey (1990), and most subsequent workers. This provenance link can be evaluated by comparison of our Golconda ages with detrital zircon ages in strata of the Roberts Mountains allochthon and in upper Paleozoic strata that overlie the allochthon (Antler overlap sequence). As shown in Figure 6, there is a strong similarity with both sequences, which supports the conclusion that strata of the Golconda allochthon accumulated in proximity to the Nevada segment of the continental margin. The presence of a ca. 1731 Ma grain in the Pumper-

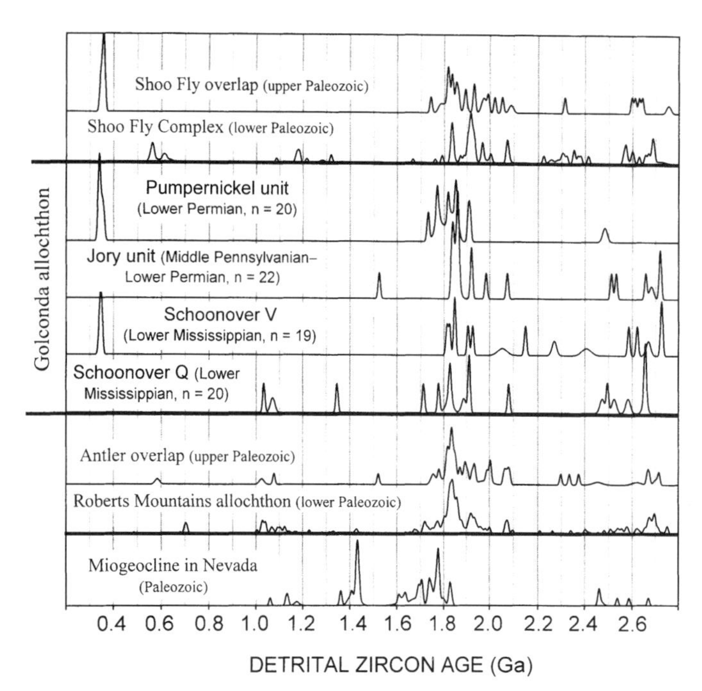


Figure 6. Relative age-probability plots that compare detrital zircon ages from Golconda allochthon, Roberts Mountains allochthon and its overlap sequence, Shoo Fly Complex and overlying upper Paleozoic strata, and Paleozoic miogeoclinal strata in Nevada. Each curve shows sum of age probability of all ages in sample or set of samples, where each sample was collected from ~1-m-thick stratigraphic interval at single locality. Sum of age probability for each sample has been divided by number of grains analyzed, so that curves are normalized to contain same area. Curve for Roberts Mountains allochthon represents 259 analyses from 15 different samples, and is composite of four main groups of detrital zircon ages in following proportions: 10% Harmony A, 26% Harmony B, 10% Lower Vinini, and 54% Upper Vinini–Valmy–McAfee–Elder–Slaven (Gehrels et al., this volume, Chapter 1; Gehrels and Dickinson, this volume, Chapter 3). Antler overlap curve summarizes 63 analyses from three samples (Gehrels and Dickinson, this volume, Chapter 3). Shoo Fly curve is composite of 92 grains analyzed from four different stratigraphic units (Harding et al., this volume, Chapter 2). Curve for strata overlying Shoo Fly Complex includes 36 analyses from 2 different samples (Spurlin et al., this volume, Chapter 6), and miogeocline in Nevada curve includes 71 analyses from three samples (Gehrels and Dickinson, 1995; Gehrels, this volume, Introduction).

**TABLE 2. DOMINANT DETRITAL ZIRCON AGE GROUPS IN STRATA OF THE GOLCONDA ALLOCHTHON**

| Age range (Ma) | Peak ages (Ma) | Relative abundance |
|---|---|---|
| Schoonover Q | | |
| 1025-1080 | 1033, 1071 | 0.09 |
| 1770-1920 | 1828, 1912 | 0.34 |
| 2460-2675 | 2496, 2656* | 0.34 |
| Schoonover V | | |
| 330-360 | 347* | 0.16 |
| 1805-1935 | 1814, 1824, 1848*, 1907, 1927 | 0.31 |
| 2575-2735 | 2727 | 0.26 |
| Jory | | |
| 1825-1930 | 1838, 1860*, 1922 | 0.54 |
| 2645-2730 | 2657, 2720 | 0.18 |
| Pumpernickel | | |
| 330-365 | 340 | 0.24 |
| 1725-1920 | 1770, 1818, 1852*, 1911 | 0.69 |

*Note:* These ages and relative abundances refer to the relative age-probability plots shown in Figure 6.
* Indicates the peak with the highest probability.

nickel and a ca. 1713 Ma grain in Schoonover Q, which are younger than most grains in the Roberts Mountains allochthon, suggests that some detritus may also have originated in basement rocks of the southwestern United States. If this is the case, however, the lack of ca. 1.43 Ga grains in our samples is puzzling.

It is also possible that the Precambrian grains in our samples were derived from outboard rocks of the Shoo Fly Complex in the northern Sierra terrane (Fig. 2). As shown in Figure 6, pre-Late Devonian rocks of the Shoo Fly Complex and overlying Mississippian through Permian strata yield detrital zircon ages that are similar to the ages from strata of the Schoonover and Havallah sequences. Because of this possibility, the >1.77 Ga ages in our samples do not provide a definitive provenance link with the craton margin to the east.

In an attempt to discriminate between Antler Highland and northern Sierra sources, we use two statistical measures to compare the age spectra of strata from the Golconda allochthon and these two regions. The comparisons were made using relative age-probability curves of: (1) all ages from our four Golconda samples; (2) a curve that combines the Roberts Mountains allochthon (50%) and Antler overlap sequence (50%); and (3) a curve that combines the Shoo Fly Complex (50%) and its overlap sequence (50%). In terms of degree of overlap, a measure of whether Golconda ages overlap with the ages of grains in the two references (Gehrels, this volume, Introduction), the potential sources achieve equally high scores of 0.88 (Roberts Mountains allochthon) and 0.81 (northern Sierra terrane), where a perfect match equals 1.00 (Fig. 7). For reference, a comparison of ages from the Golconda allochthon and the Nevada miogeocline yields a degree of overlap value of 0.45. In terms of degree of similarity, a measure of whether the proportions of ages in two curves are similar (Gehrels, this volume, Introduction), high scores of 0.77 and 0.71, respectively, are achieved. A comparison with miogeoclinal strata in Nevada yields a much lower value of 0.37.

For both statistical measures, values are slightly higher for the Roberts Mountains allochthon than the northern Sierra terrane (Fig. 7). This is due primarily to the presence of ca. 2.5 Ga grains in samples from the Golconda and Roberts Mountains allochthons, but not in samples from the northern Sierra terrane (Fig. 6). The statistical values suggest that an eastern source is slightly more likely or dominant than a western source, but that our data are consistent with either source or a combination of the two sources.

The presence of 338–358 Ma grains in the Schoonover V and Pumpernickel samples indicates that sandstones in the Golconda allochthon also contain detritus from igneous rocks of latest Devo-

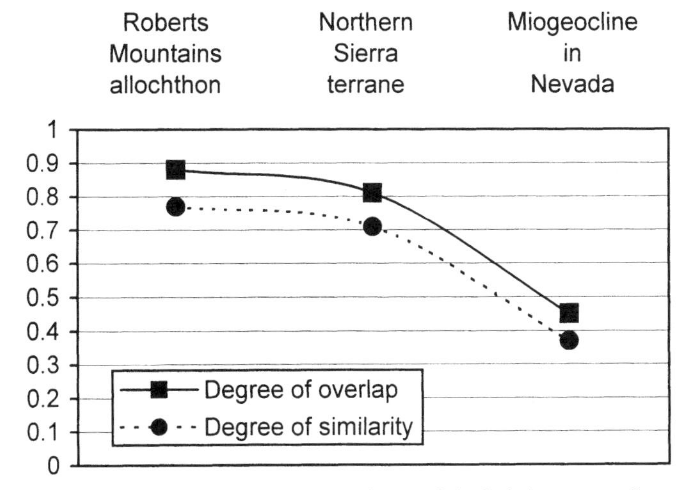


Figure 7. Results of statistical comparisons of detrital zircon ages from Golconda allochthon with age spectra from Roberts Mountains allochthon and overlying upper Paleozoic strata, northern Sierra terrane, and Paleozoic miogeocline in Nevada (Fig. 6). Degree of overlap is measure of degree to which two age spectra contain overlapping ages, and degree of similarity evaluates degree to which overlapping ages have similar proportions (Gehrels, this volume, Introduction). Values of 1.00 reflect perfect overlap and proportions, respectively, of two age spectra.

nian–Early Mississippian age. The most likely source for this detritus is the northern Sierra terrane, which underwent widespread magmatism during this time interval (Burchfiel and Davis, 1972, 1975; Hanson and Schweikert, 1986; Hanson et al., 1988; Harwood and Murchey, 1990; Miller and Harwood, 1990; Harwood, 1992; Spurlin et al., this volume). It is possible, however, that the young grains were shed from Devonian-Mississippian igneous rocks within the allochthon (Fig. 4), which were described by Miller et al. (1984), or in a hypothetical magmatic arc that may have separated the Antler orogen from the Golconda allochthon prior to the Sonoma orogeny (Speed and Sleep, 1982; Dickinson et al., 1983). Unfortunately, our data are not sufficient to discriminate between these possibilities, largely because latest Devonian–Early Mississippian magmatism is known or suspected to have occurred in all three potential source regions. Our data also allow for other potential sources, given the widespread extent of mid-Paleozoic igneous rocks in western North America (Rubin et al., 1990).

## CONCLUSIONS

Our data suggest that detrital zircons in sandstones of the Golconda allochthon were probably derived from both the Roberts Mountains allochthon to the east and the northern Sierra terrane to the west. The older (>1.77 Ga) grains were most likely recycled from lower Paleozoic rocks of the Roberts Mountains allochthon following their mid-Paleozoic emplacement onto the Cordilleran margin. This provenance tie would require the Golconda allochthon to have formed and evolved in close proximity to its present position outboard of the Roberts Mountains allochthon. However, it is also possible that the older grains were recycled from rocks of the Shoo Fly Complex, which is currently located outboard (west) of the Golconda allochthon (Fig. 2). These two potential source regions cannot be distinguished with confidence using our detrital zircon data because they contain very similar suites of zircon ages, as shown in Figure 6.

Zircon grains of 338–358 Ma that are present in Mississippian and Permian sandstones could also have come from several recognized sources, including volcanic rocks present within lower parts of the allochthon, and volcanic and plutonic rocks in the northern Sierra–eastern Klamath terranes. In addition, it is possible that these grains were derived from a magmatic arc that was originally along the outer margin of the Roberts Mountains allochthon, but was subsequently buried and overridden by the Golconda allochthon (Speed and Sleep, 1982; Dickinson et al., 1983; Burchfiel and Royden, 1991).

The only grains that may have a unique source are ca. 1713 and ca. 1731 Ma (Fig. 6). These grains were most likely derived from basement rocks of the southwestern United States (Fig. 1).

Because of the uncertainties outlined here, at least three alternatives exist for the paleogeography and provenance of the Golconda allochthon.

1. Deposition was in a basinal setting outboard of the Antler highland, with little or no input from arc terranes to the west. In this case the detrital zircons would have been shed mainly from the Roberts Mountains allochthon, with minor contributions from southwestern U.S. basement provinces and from either local, extension-related(?) volcanism or an extinct arc separating the Golconda and Roberts Mountains allochthons. Emplacement of the allochthon would have been related to subduction along the inboard margin of a possibly far-traveled, east-facing magmatic arc, now preserved as the northern Sierra and eastern Klamath terranes (Speed, 1979; Speed and Sleep, 1982; Dickinson et al., 1983).

2. Deposition was in a backarc basin that opened as the northern Sierra terrane pulled away from previously contiguous rocks of the Roberts Mountains allochthon. Most detritus would have come from the Shoo Fly Complex, exposed to the west, and/or the Roberts Mountains allochthon, exposed in the Antler highland to the east. Some detritus would also have been shed from the southwestern U.S. craton and from the magmatic arc to the west. Final emplacement of the Golconda allochthon would have occurred in response to the collapse, subduction, or backarc thrusting of this basin (Burchfiel and Davis, 1972, 1975; Schweickert and Snyder, 1981; Snyder and Brueckner, 1983; Miller et al., 1984, 1992; Brueckner and Snyder, 1985; Burchfiel and Royden, 1991; Burchfiel et al., 1992).

3. Deposition of strata was in proximity to the northern Sierra terrane, potentially far from the Nevada continental margin. Most detritus would have been recycled from the Shoo Fly Complex and latest Devonian–Mississippian igneous rocks, but the source for the two 1713–1731 Ma grains would be uncertain.

While all of these scenarios remain viable, model 2 is preferred because it offers the simplest explanation for the similarity in detrital zircon ages from the Roberts Mountains allochthon and northern Sierra terrane (Fig. 6), and is also consistent with a variety of stratigraphic links between the Golconda allochthon and both the northern Sierra terrane and Roberts Mountains allochthon (Miller et al., 1984, 1992; Harwood and Murchey, 1990). In both models 1 and 3, rocks of the northern Sierra terrane would not have been in proximity to the Roberts Mountains allochthon until their accretion during the Sonoma orogeny.

## ACKNOWLEDGMENTS

We thank Bill Dickinson for his guidance in the field and in interpreting our results. This research was supported by National Science Foundation grant EAR-9416933. Reviewed by Rich Schweickert, Hassan Babaie, and James Trexler.

## REFERENCES CITED

Babaie, H.A., 1987, Paleogeographic and tectonic implications of the Golconda allochthon, southern Toiyabe Range, Nevada: Geological Society of America Bulletin, v. 99, p. 231–243.

Blake, M.C., Jr., Bruhn, R.L., Miller, E.L., Moores, E.M., Smithson, S.B., and Speed, R.C., 1985, C-1 Menodocino triple junction to North American craton: Geological Society of America Centennial Continent-Ocean Transect #12.

Brueckner, H.K., and Snyder, W.S., 1985, Structure of the Havallah sequence, Golconda allochthon, Nevada: Evidence for prolonged evolution in an accretionary prism: Geological Society of America Bulletin, v. 96, p. 1113–1130.

Brueckner, H.K., Snyder, W.S., and Boudreau, M., 1987, Diagenetic controls on the structural evolution of siliceous sediments in the Golconda allochthon, Nevada, U.S.A.: Journal of Structural Geology, v. 9, p. 403–417.

Burchfiel, B.C., and Davis, G.A., 1972, Structural framework and evolution of the southern part of the Cordilleran orogen, western United States: American Journal of Science, v. 272, p. 97–118.

Burchfiel, B.C., and Davis, G.S., 1975, Nature and controls of Cordilleran orogenesis, western United States: Extensions of an earlier synthesis: American Journal of Science, v. 275A, p. 363–396.

Burchfiel, B.C., and Royden, L.H., 1991, Antler orogeny: A Mediterranean-type orogeny: Geology, v. 19, p. 66–69.

Burchfiel, B.C., Cowan, D.S., and Davis, G.A., 1992, Tectonic overview of the Cordilleran orogen in the western United States, *in* Burchfiel, B.C., et al., eds., The Cordilleran orogen: Conterminous U.S.: Boulder Colorado, Geological Society of America, Geology of North America, v. G-3, p. 407–480.

Coney, P.J., Jones, D.L., and Monger, J.W.H., 1980, Cordilleran suspect terranes: Nature, v. 288, p. 329–333.

Dickinson, W.R., Harbaugh, D., Saller, A.H., Heller, P.L., and Snyder, W.S., 1983, Detrital modes of upper Paleozoic sandstones derived from the Antler orogen in Nevada: Implications for nature of Antler orogeny: American Journal of Science, v. 283, p. 481–509.

Fagan, J.J., 1962, Carboniferous cherts, turbidites, and volcanic rocks in northern Independence Range, Nevada: Geological Society of America Bulletin, v. 73, p. 595–612.

Gabrielse, H., Snyder, W.S., and Stewart, J.H., 1983, Sonoma orogeny and Permian to Triassic tectonism in western North America (Penrose Conference Report): Geology, v. 11, p. 484–486.

Gehrels, G.E., and Dickinson, W.R., 1995, Detrital zircon provenance of Cambrian to Triassic miogeoclinal and eugeoclinal strata in Nevada: American Journal of Science, v. 295, p. 18–48.

Hanson, R.E., and Schweickert, R.A., 1986, Stratigraphy of mid-Paleozoic island-arc rocks in part of the northern Sierra Nevada, Sierra and Nevada counties, California: Geological Society of America Bulletin, v. 97, p. 986–998.

Hanson R.E., Saleeby, J.B., and Schweickert, R.A., 1988, Composite Devonian island-arc batholith in the northern Sierra Nevada, California: Geological Society of America Bulletin, v. 100, p. 446–457.

Harwood, D.S., 1992, Stratigraphy of Paleozoic and lower Mesozoic rocks in the northern Sierra terrane, California: U.S. Geological Survey Bulletin 1957, 78 p.

Harwood, D.S., and Murchey, B.L., 1990, Biostratigraphic, tectonic, and paleogeographic ties between upper Paleozoic volcanic and basinal rocks in the northern Sierra terrane, California, and the Havallah Sequence, Nevada, *in* Harwood, D.S., and Miller, M.M., eds., Paleozoic and early Mesozoic paleogeographic relations; Sierra Nevada, Klamath Mountains, and related terranes: Geological Society of America Special Paper 255, p. 157–173.

Hoffman, P.F., 1989, Precambrian geology and tectonic history of North America, *in* Bally, A.W., and Palmer, A.R., eds., The geology of North America—An overview: Boulder, Colorado, Geological Society of America, Geology of North America, v. A, p. 447–512.

Jones, A.E., 1991, Sedimentary rocks of the Golconda terrane: Provenance and paleogeographic implications, *in* Cooper, J.D., and Stevens, C.H., eds., Paleozoic paleogeography of the western United States—II: Pacific Section, Society of Economic Paleontologists and Mineralogists book 67, p. 783–800.

Jones, A.E., and Jones, D.L., 1991, Paleogeographic significance of subterranes of the Golconda allochthon, northern Nevada, *in* Raines, G.L., et al., eds., Geology and ore deposits of the Great Basin: Reno, Geological Society of Nevada, p. 21–23.

Kistler, R.W., 1991, Chemical and isotopic characteristics of plutons in the Great Basin, *in* Raines, G.L., et al., eds., Geology and ore deposits of the Great Basin: Reno, Geological Society of Nevada, p. 107–109.

Kistler, R.W., and Peterman, Z.E., 1973, Variations in Sr, Rb, K, Na, and initial $^{87}Sr/^{86}Sr$ in Mesozoic granitic rocks and intruded wall rocks in central California: Geological Society of America Bulletin, v. 84, p. 3489–3512.

Ludwig, K.R., 1991a, A computer program for processing Pb-U-Th isotopic data: U.S. Geological Survey Open-File Report 88-542.

Ludwig, K.R., 1991b, A plotting and regression program for radiogenic-isotopic data: U.S. Geological Survey Open-File Report 91-445.

Miller, E.L., Bateson, J., Dinter, D., Dyer, J.R., Harbaugh, D., and Jones, D.L., 1981, Thrust emplacement of the Schoonover sequence, northern Independence Mountains, Nevada: Geological Society of America Bulletin, v. 92, p. 730–737.

Miller, E.L., Kanter, L.R., Larue, D.K., Turner, R.J., Murchey, B., and Jones, D.L., 1982, Structural fabric of the Paleozoic Golconda allochthon, Antler Peak quadrangle, Nevada: Progressive deformation of an oceanic sedimentary assemblage: Journal of Geophysical Research, v. 87, p. 3795–3804.

Miller, E.L., Holdsworth, B.K., Whiteford, W.B., and Rodgers, D., 1984, Stratigraphy and structure of the Schoonover sequence, northeastern Nevada: Implications for Paleozoic plate margin tectonics: Geological Society of America Bulletin, v. 95, p. 1063–1076.

Miller, E.L., Miller, M.M., Stevens, C.H., Wright, J.E., and Madrid, R., 1992, Late Paleozoic paleogeographic and tectonic evolution of the western U.S. Cordillera, *in* Burchfiel, B.C., et al., eds., The Cordilleran orogen: Conterminous U.S.: Boulder, Colorado, Geological Society of America, Geology of North America, v. G-3, p. 57–106.

Miller, M.M., and Harwood, D.S., 1990, Paleogeographic setting of upper Paleozoic rocks in the northern Sierra and eastern Klamath terranes, northern California, *in* Harwood, D.S., and Miller, M.M., eds., Paleozoic and early Mesozoic paleogeographic relations; Sierra Nevada, Klamath Mountains, and related terranes: Geological Society of America Special Paper 255, p. 175–192.

Murchey, B.L., 1990, Age and depositional setting of siliceous sediments in the upper Paleozoic Havallah sequence near Battle Mountain, Nevada; implications for the paleogeography and structural evolution of the western margin of North America, *in* Harwood, D.S., and Miller. M.M., eds., Paleozoic and early Mesozoic paleogeographic relations; Sierra Nevada,

Klamath Mountain, and related terranes: Geological Society of America Special Paper 255, p. 137–155.
Nilsen, T.H., and Stewart, J.H. 1980, The Antler orogeny—Mid-Paleozoic tectonism in western North America (Penrose Conference Report): Geology, v. 8, p. 298–302.
Oldow, J.S., 1984, Evolution of a late Mesozoic back-arc fold and thrust belt, northwestern Great Basin, U.S.A.: Tectonophysics, v. 102, p. 245–274.
Roberts, R.J., 1964, Stratigraphy and structure of the Antler Peak quadrangle, Humboldt and Lander counties, Nevada: U.S. Geological Survey Professional Paper 495A, 93 p.
Roberts, R.J., Hotz, P.E., Gilluly, J., and Ferguson, H.G., 1958, Paleozoic rocks of north-central Nevada: American Association of Petroleum Geologists Bulletin, v. 42, p. 2813–2857.
Rubin, C.M., Miller, M.M., and Smith, G.M., 1990, Tectonic development of Cordilleran mid-Paleozoic volcanoplutonic complexes: Evidence for convergent margin tectonism, *in* Harwood, D.S., and Miller, M.M., eds., Paleozoic and early Mesozoic paleogeographic relations; Sierra Nevada, Klamath Mountains, and related rocks: Geological Society of America Special Paper 255, p. 1–16.
Saleeby, J.B., 1986, C-2 Central California offshore to Colorado Plateau: Geological Society of America Centennial Continent-Ocean Transect #10.
Schweickert, R.A., and Snyder, W.S., 1981, Paleozoic plate tectonics of the Sierra Nevada and adjacent regions, *in* Ernst, W.G., ed., The geotectonic evolution of California: (Rubey Volume 1): Englewood Cliffs, New Jersey, Prentice-Hall, Inc., p. 609-627.
Silberling, N.J., 1973, Geologic events during Permian-Triassic time along the Pacific margin of the United States, *in* Logan, A., and Hills, L.V., eds., The Permian and Triassic Systems and their mutual boundary: Calgary, Alberta Society of Petroleum Geologists, p. 345–362.
Silberling, N.J., 1991, Allochthonous terranes of western Nevada: Current status, *in* Raines, G.L., et al., eds., Geology and ore deposits of the Great Basin: Reno, Geological Society of Nevada, p. 101–102.
Silberling, N.J., and Roberts, R.T., 1962, Pre-Tertiary stratigraphy and structure of northwestern Nevada: Geological Society of America Special Paper 163, 28 p.
Silberling, N.J., Jones, D.L., Blake, M.C., Jr., and Howell, D.G., 1987, Lithotectonic terrane map of the western conterminous United States: U.S. Geological Survey Miscellaneous Field Studies Map MF-1874-C, scale 1:2,500,000.
Snyder, W.S., and Brueckner, H.K., 1983, Tectonic evolution of the Golconda allochthon, Nevada: Problems and perspectives, *in* Stevens, C.H., ed., Paleozoic and early Mesozoic rocks in microplates of western North America: Pacific Section, Society of Economic Paleontologists and Mineralogists, p. 103–123.
Speed, R.C., 1979, Collided Paleozoic microplate in the western U.S.: Journal of Geology, v. 87, p. 279-292.
Speed, R.C., and Sleep, N.H., 1982, Antler orogeny and foreland basin: A model: Geological Society of America Bulletin, v. 93, p. 815–828.
Stewart, J.H., and Carlson, J.E., 1978, Geologic map of Nevada: U.S. Geological Survey, scale 1:500,000, 1 sheet.
Stewart, J.H., MacMillan, J.R., Nichols, K.M., and Stevens, C.H., 1977, Deep-water upper Paleozoic rocks in north-central Nevada—A study of the type area of the Havallah formation, *in* Stewart, J.H., et al., eds., Paleozoic paleogeography of the western United States—I: Pacific Section, Society of Economic Paleontologists and Mineralogists p. 337–347.
Stewart, J.H., Poole, F.G., Ketner, K.B., Madrid, R.J., Roldan-Quintana, J., and Amaya-Martinez, R., 1990, Tectonics and stratigraphy of the Paleozoic and Triassic southern margin of North America, Sonora, Mexico, *in* Gehrels, G.E., and Spencer, J.E., eds., Geologic excursions through the Sonoran Desert region, Arizona and Sonora: Arizona Geological Survey Special Paper 7, p. 183–202.
Van Schmus, W.R., and 24 others, 1993, Transcontinental Proterozoic Provinces, *in* Reed, J.C., et al., eds., Precambrian: Conterminous U.S.: Boulder, Colorado, Geological Society of America, The Geology of North America, v. C-2, p. 171–334.

Manuscript Accepted by the Society January 24, 2000

Geological Society of America
Special Paper 347
2000

# *Provenance and paleogeography of the Black Rock terrane, northwestern Nevada: Implications of U-Pb detrital zircon geochronology*

**Brian J. Darby***
*Department of Geosciences, University of Arizona, Tucson, Arizona, 85721, USA*
**Sandra J. Wyld**
*Department of Geology, University of Georgia, Athens, Georgia 30602, USA*
**George E. Gehrels**
*Department of Geosciences, University of Arizona, Tucson, Arizona, 85721, USA*

## ABSTRACT

**U-Pb ages have been determined for 50 detrital zircon grains from Mississippian and Triassic strata of the Black Rock terrane, northwestern Nevada. The Devonian(?) to Mississippian Pass Creek unit has three broad age groups: 976–1132 Ma (n = 8), 1595–1927 Ma (n = 10), and 2504–2660 Ma (n = 3). The Triassic Bishop Canyon formation contains a dominant group of grains between 268 and 441 Ma (n = 11), a cluster of ages between 1868 and 1925 Ma (n = 5), and scattered ages between 1184 and 1813 Ma (n = 10) and between 2183 and 3183 Ma (n = 3).**

**Most of the ages in these samples match well with the ages of grains present in basement provinces and off-shelf assemblages in the western United States. Grains in the Pass Creek unit were most likely recycled from lower Paleozoic strata of the Roberts Mountains allochthon and from strata exposed in the Salmon River arch region of Idaho, western Montana, and eastern Washington. These provenance links suggest that the Black Rock terrane was located along the northern Nevada–Idaho segment of the Cordilleran margin, near its current location, during late Paleozoic time. Because the upper Paleozoic stratigraphy of the Black Rock terrane is similar to that found in more outboard arc assemblages, including those of the Klamath Mountains and Sierra Nevada, this relation provides an indirect but important link between the U.S. continental margin and the more outboard arc assemblages. Zircon grains in the Upper Triassic Bishop Canyon formation were derived from a source containing both upper Paleozoic igneous rocks and clastic strata bearing 1.1–3.2 Ga detrital zircons. The most likely source for this combination of grains is Paleozoic basement rocks of Mesozoic arc assemblages in the eastern Klamath Mountains and Black Rock terrane. These provenance links provide evidence of uplift and erosion of arc basement in the western Cordillera during early Mesozoic time, and support interpretations that lower Mesozoic arc assemblages in this region of the Cordillera were isolated from the continental margin by a Triassic backarc basin.**

*Present address: Department of Earth Sciences, University of Southern California, Los Angeles, California 90089, USA

Darby, B.J., Wyld, S.J., and Gehrels, G.E., 2000, Provenance and paleogeography of the Black Rock terrane, northwestern Nevada: Implications of U-Pb detrital zircon geochronology, *in* Soreghan, M.J., and Gehrels, G.E., eds., Paleozoic and Triassic paleogeography and tectonics of western Nevada and northern California: Boulder, Colorado, Geological Society of America Special Paper 347, p. 77–87.

## INTRODUCTION

The Black Rock terrane of northwest Nevada consists of Paleozoic and Mesozoic sedimentary and volcanic rocks of island-arc affinity (Silberling et al., 1987). Rocks of the Black Rock terrane have been studied in the most detail near the Oregon border in the Pine Forest Range (Wyld, 1990), Jackson Mountains (Russell, 1984; Quinn et al., 1997), and Bilk Creek Mountains (Ketner and Wardlaw, 1981; Jones, 1990). In this chapter we refer to relations in these three ranges, particularly in the Pine Forest Range, as representative of the Black Rock terrane.

Various stratigraphic studies indicate that the middle and upper Paleozoic stratigraphy of the Black Rock terrane is similar to that of the eastern Klamath and northern Sierra terranes (Fig. 1) (Speed, 1978; Ketner and Wardlaw, 1981; Jones, 1990; Wyld, 1990, 1992). The Black Rock terrane is separated from more outboard arc assemblages, however, by large expanses of Cenozoic cover (Fig. 1), and the exact nature of paleogeographic or paleotectonic links between the terrane and other arc sequences is therefore uncertain.

Recent studies indicating that the Jurassic deformational history of the Black Rock terrane is very different from, and incompatible with, that of the Klamath Mountains or Sierra Nevada (Wyld and Wright, 1997), have led to the conclusion that the current geographic location of outboard arc assemblages relative to the Black Rock terrane likely does not reflect their pre-Cretaceous paleogeographic location. In particular, the Klamath Mountains and northern Sierra are interpreted to have been displaced northward relative to the Black Rock terrane by late Mesozoic dextral strike-slip faulting (Lahren et al., 1990; Schweickert and Lahren, 1990; Wyld and Wright, 1997).

The relation of the Black Rock terrane to more inboard continental margin assemblages is less controversial. East of the Black Rock terrane is a succession of lower Mesozoic basinal strata (Lovelock assemblage; Fig. 1) that is generally interpreted to have been deposited in a marine basin behind an active volcanic arc (e.g., Speed, 1978). All prior studies have inferred that lower Mesozoic igneous rocks in the Black Rock terrane represent part of the arc system that was adjacent to this basin (Speed, 1978; Oldow, 1984; Russell, 1984; Wyld, 1990; Saleeby and Busby-Spera, 1992). Because strata of the basinal assemblage can be linked stratigraphically and sedimentologically to continental sources farther east (Lupe and Silberling, 1985), these relations imply that the Black Rock terrane evolved in proximity to the continental margin of north-central Nevada, at least during Mesozoic time (see also Wyld, this volume, Chapter 13). Where the Black Rock terrane resided relative to the continental margin during Paleozoic time, however, is not well constrained.

This study is an attempt to place tighter constraints on the paleogeographic and paleotectonic setting of the Black Rock terrane by analyzing detrital zircons from middle Paleozoic and Triassic clastic strata in the Pine Forest Range. The methodology used in this study is a simple comparison of ages in our samples with those in potential source areas, such as the northern Sierra and eastern Klamath terranes, the Golconda and Roberts Mountains allochthons, and various portions of the western North American continent. The occurrence of similar ages of grains is used as evidence for proximity at the time of deposition, whereas the occurrence of disparate ages requires either geographic separation or complex dispersal systems. Topics of particular interest are: (1) whether the Black Rock terrane can be linked to outboard arc assemblages of the Klamath Mountains and/or Sierra Nevada in the Paleozoic; (2) whether the terrane can be linked to the continent or continental margin in the Paleozoic, and if so, what part of the margin; (3) what sediment sources provided material to the Black Rock terrane in the early Mesozoic; and (4) what these sources imply about the paleogeography and paleotectonics of the arc and backarc system.

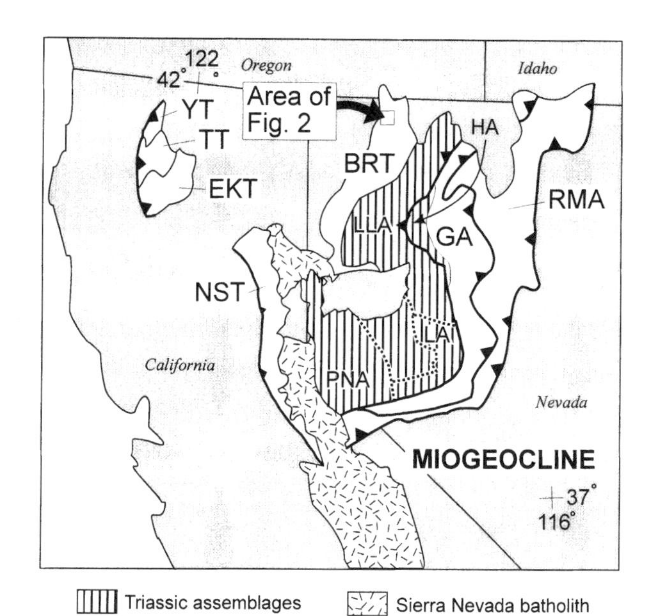


Figure 1. Location map of selected terranes and miogeoclinal strata in western North America (modified from Oldow, 1984; Silberling, 1991; Silberling et al., 1987). RMA—Roberts Mountains allochthon, GA—Golconda allochthon, PNA—Pine Nut assemblage, LA—Luning assemblage, LLA—Lovelock assemblage, HA—Humboldt assemblage, BRT—Black Rock terrane, NST—Northern Sierra terrane, EKT—Eastern Klamath terrane, TT—Trinity terrane, YT—Yreka terrane.

## GEOLOGIC FRAMEWORK OF THE BLACK ROCK TERRANE

Most detailed work on the Black Rock terrane has focused on Paleozoic and Mesozoic rocks in three areas in the northern part of the terrane; the Jackson Mountains (Russell, 1981, 1984; Maher, 1989; Quinn, 1996; Quinn et al., 1997; Wyld, this volume, Chapter 13), Pine Forest Range (Wyld, 1990, 1996), and Bilk Creek Mountains (Ketner and Wardlaw, 1981; Jones, 1990). These studies have yielded four important conclusions.

1. Clear correlations exist between Paleozoic and Mesozoic strata in all three mountain ranges; thus, although the ranges are now separated by Cenozoic basins, strata contained within them represent a coherent geologic province.

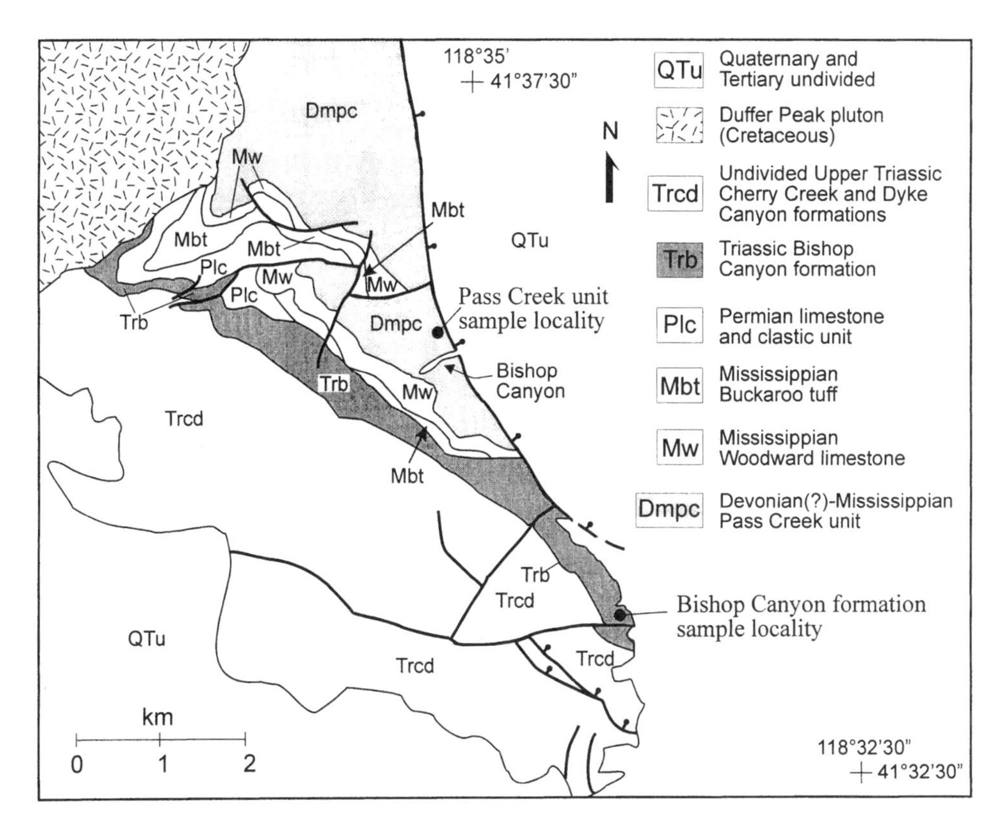


Figure 2. Geologic map of southeastern Pine Forest Range (after Wyld, 1990, 1992).

2. Paleozoic strata in this province range in age from Devonian(?) or Early Mississippian to Late Permian, consist mostly of siliciclastic sedimentary rocks and carbonate strata, with less abundant volcanogenic rocks, and reflect sedimentation and intermittent volcanism (or volcanogenic input) within a marine environment. Although generally similar to upper Paleozoic rocks in the eastern Klamath Mountains and northern Sierra terrane (Miller and Harwood, 1990), strata of the Black Rock terrane contain fewer volcanic rocks and may therefore have formed farther from centers of active volcanism.

3. Mesozoic strata in the terrane are mostly volcanogenic and generally reflect proximal arc volcanism; Mesozoic arc plutons are likewise abundant. These volcanogenic strata are mostly late Late Triassic (Norian) age, but ages extend into the Early Jurassic in some areas. Lower to lower Upper Triassic and Lower Cretaceous units, in contrast, are mostly sedimentary strata.

4. Several unconformities occur, at least locally, in the Paleozoic to Mesozoic successions of the Black Rock terrane and these generally span Late Carboniferous, latest Permian to late Middle Triassic, and Early Jurassic to Early Cretaceous time.

We have analyzed samples from the eastern Pine Forest Range (Fig. 2), where a structurally intact succession of Paleozoic and Triassic rocks has been studied in detail (Wyld, 1990, 1992). A stratigraphic column showing units in the eastern Pine Forest Range is shown in Figure 3; unit names are modified from those in Wyld (1990) and are based on a revised nomenclature defined in Wyld (1992, 1996). Paleozoic and Mesozoic strata in the eastern Pine Forest Range have been metamorphosed and deformed in such a way that strain and metamorphic grade increase with increasing stratigraphic (structural) depth (Wyld, 1996). Because of this, the protoliths of the oldest rocks are less certain than those of the younger rocks, and the oldest rocks are described in terms of their metamorphic character (Fig. 3).

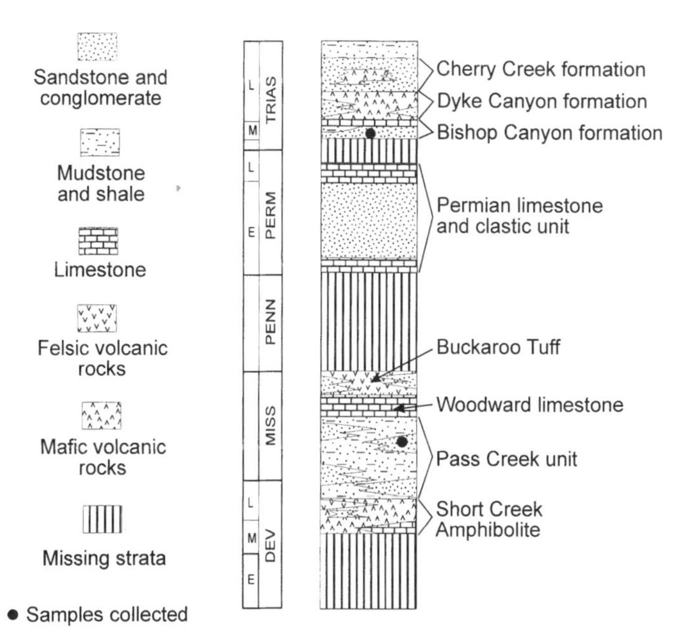


Figure 3. Generalized stratigraphic column of Black Rock terrane (after Wyld, 1990, 1992, 1996).

The oldest stratigraphic unit exposed in the eastern Pine Forest Range is the Short Creek amphibolite, which consists mostly of mafic schist interpreted to have been derived from mafic lava and volcaniclastic protoliths. A Devonian(?) age is assigned to this unit because it is stratigraphically beneath the Devonian(?) to Mississippian Pass Creek unit (Fig. 3). The Pass Creek unit, which is one of the units analyzed for detrital zircons, consists of metasedimentary rocks derived mostly from shale, siltstone, and sandstone, and minor conglomerate and rare limestone. Fossils from this unit and its correlatives in the Jackson and Bilk Creek Mountains indicate Late Devonian(?) to Late Mississippian ages (Russell, 1981; Jones, 1990; Wyld, 1990; Quinn, 1996). Graded beds and partial Bouma sequences in the Pass Creek unit, coupled with facies distribution, suggest deposition in a submarine fan environment. Clasts in this unit, and in its somewhat less recrystallized correlative in the Jackson Mountains (McGill Canyon Formation), consist mostly of chert and sedimentary rocks (argillite >> quartzose clastics), with less common monocrystalline quartz and polycrystalline quartz, and rare feldspar and volcanic rock fragments (Wyld, 1990; Quinn, 1996; Dickinson and Gehrels, this volume, Chapter 11).

Two additional units were deposited during Late Mississippian time. The Woodward limestone, which consists of shallow-marine carbonate and minor interbedded clastic rocks, conformably overlies the Pass Creek unit. This unit is in turn overlain conformably by dacitic tuff and volcaniclastic strata of the Buckaroo tuff (Fig. 3). A U-Pb zircon age of ca. 327 Ma has been determined for the Buckaroo tuff (Wyld, 1992).

A thin sequence of Permian and Permian(?) strata overlies the older Paleozoic rocks along an unconformity that appears to span most or all of Pennsylvanian time. This sequence consists mostly of shallow-marine, fossiliferous limestone and clastic rocks containing sedimentary and volcanic detritus. Collectively, these Permian units and the intervening unconformity appear to reflect shoaling, uplift, erosion, and then subsidence in the region during the late Paleozoic (Wyld, 1990, 1991).

Paleozoic rocks in the Pine Forest Range are overlain by Middle(?) to Upper Triassic strata across a slightly angular unconformity that is interpreted to reflect uplift and erosion related to the Permian-Triassic Sonoma orogeny (Wyld, 1991; Figs. 2 and 3). This unconformity cuts down to the level of Upper Mississippian strata (Wyld, 1990, 1992). All Triassic units were deposited in a relatively deep marine environment formed by subsidence of the region following the Sonoma orogeny. The oldest Triassic unit is the Bishop Canyon formation (Figs. 2 and 3), which ranges in age from Ladinian or Carnian (late Middle or early Late Triassic) at the base to early Norian (late Late Triassic) at the top. This unit contains three members: a lower and an upper member consisting mostly of limestone, and a middle member consisting mostly of coarse sandstone and conglomerate with minor siltstone and limestone lenses (Fig. 3). Clastic rocks in the middle member contain abundant sedimentary lithic material (argillite and siliceous argillite > quartzose clastic material > carbonate), volcanic lithic material (intermediate or silicic composition) and plagioclase, and minor chert and quartz (Wyld, 1992; Dickinson and Gehrels, this volume, Chapter 11). A sample from this member was analyzed for detrital zircons. Younger Triassic rocks have been divided into the Dyke Canyon and Cherry Creek formations (Fig. 3), both of which consist mostly of basaltic to andesitic volcanic flows and volcaniclastic sedimentary rocks. These units reflect active arc volcanism in the region during Norian time.

## ANALYTICAL METHODS

Our samples consisted of ~20 kg each of coarse sandstone. The samples were crushed, pulverized, and separated on a Wilfley table, and the heavy mineral fractions were further processed with a Frantz LB-1 magnetic separator and heavy liquids. This yielded fairly pure zircon separates, which were then washed with warm $HNO_3$ and sieved into size fractions with disposable nylon screens. Grains greater than 100–125 μm in sieve size were sorted into groups based on color and grain morphology, the grains were abraded to ~75% of their original size using air abrasion, and then zircons representative of each group were selected for analysis. The individual zircons were analyzed using conventional isotope dilution and thermal ionization mass spectrometry as described by Gehrels (this volume, Introduction).

As reported in Table 1 and shown in Figure 4, our samples yield two groups of analyses: <500 Ma grains that are apparently concordant, and >900 Ma grains that in most cases are slightly to moderately discordant. For the younger grains, the interpreted ages shown in Table 1 are $^{206}Pb^{*}/^{238}U$ ages. As described by Gehrels (this volume, Introduction), $^{206}Pb^{*}/^{238}U$ ages are generally more precise than $^{207}Pb^{*}/^{206}Pb^{*}$ ages for samples that are younger than about 1.0 Ga. For the older grains, interpreted ages for concordant analyses are based on the $^{207}Pb^{*}/^{206}Pb^{*}$ ages. Most of the older grains are discordant, however, presumably due to Pb loss. Ages for these grains are assigned on the basis of a projection from 100 Ma, which is a reasonable minimum age for metamorphism and isotopic disturbance in the region (Wyld, 1996). The projected ages are interpreted to be more reliable than the $^{207}Pb^{*}/^{206}Pb^{*}$ ages because there is no evidence for recent isotopic disturbance in the region, which is implicit in the use of $^{207}Pb^{*}/^{206}Pb^{*}$ ages. Because projected ages become less reliable with increasing degree of discordance, the ages assigned in Table 1 and used in constructing Figures 5 and 6 have uncertainties that increase with the degree of discordance (Gehrels, this volume, Introduction).

## RESULTS

### *Devonian(?) to Mississippian Pass Creek unit*

Zircons were extracted from a coarse-grained sandstone with chert-pebble conglomerate horizons exposed north of Bishop Canyon on the eastern flank of the Pine Forest Range (Fig. 2).

**TABLE 1. U-Pb ISOTOPIC DATA AND AGES**

| Grain type | Grain wt. (µg) | $Pb_c$ (pg) | U (ppm) | $^{206}Pb_m$/$^{204}Pb$ | $^{206}Pb_c$/$^{208}Pb$ | Apparent ages (Ma) $^{206}Pb^*$/$^{238}U$ | $^{207}Pb^*$/$^{235}U$ | $^{207}Pb^*$/$^{206}Pb^*$ | Interpreted age (Ma) |
|---|---|---|---|---|---|---|---|---|---|
| Pass Creek Formation (41º35'18.1"N, 118º35'45.2"W) | | | | | | | | | |
| ME | 7 | 10 | 423 | 2660 | 5.1 | 857 ± 4 | 890 ± 6 | 973 ± 9 | 976 ± 20 |
| CR | 6 | 6 | 68 | 725 | 6.3 | 929 ± 22 | 960 ± 24 | 1035 ± 17 | 1038 ± 20 |
| LR | 22 | 9 | 54 | 1430 | 6.8 | 1035 ± 9 | 1045 ± 11 | 1064 ± 11 | 1066 ± 10 |
| LR | 20 | 7 | 60 | 1850 | 9.4 | 1032 ± 9 | 1043 ± 10 | 1066 ± 9 | 1068 ± 10 |
| CE | 4 | 10 | 227 | 980 | 8.4 | 1058 ± 12 | 1072 ± 14 | 1100 ± 15 | 1103 ± 20 |
| LR | 6 | 4 | 112 | 1820 | 5.7 | 1109 ± 8 | 1111 ± 9 | 1114 ± 8 | 1114 ± 10 |
| **MR** | **14** | **18** | **817** | **7650** | **32.8** | **1113 ± 4** | **1114 ± 6** | **1117 ± 7** | **1117 ± 10** |
| CE | 15 | 17 | 93 | 960 | 11.9 | 1114 ± 8 | 1120 ± 12 | 1132 ± 14 | 1132 ± 20 |
| LR | 18 | 9 | 171 | 5400 | 3.9 | 1542 ± 6 | 1564 ± 8 | 1593 ± 6 | 1595 ± 10 |
| LE | 5 | 3 | 151 | 3770 | 7.2 | 1546 ± 9 | 1574 ± 10 | 1610 ± 5 | 1612 ± 10 |
| MR | 15 | 5 | 276 | 15700 | 17.2 | 1565 ± 7 | 1620 ± 8 | 1692 ± 4 | 1695 ± 20 |
| CR | 7 | 6 | 230 | 4850 | 10.1 | 1673 ± 9 | 1700 ± 11 | 1733 ± 6 | 1735 ± 10 |
| LE | 16 | 17 | 130 | 2250 | 11.9 | 1769 ± 9 | 1774 ± 11 | 1780 ± 7 | 1780 ± 10 |
| CE | 15 | 13 | 50 | 1140 | 2.3 | 1789 ± 14 | 1794 ± 17 | 1800 ± 9 | 1801 ± 10 |
| MR | 12 | 2 | 114 | 11040 | 7.5 | 1809 ± 7 | 1832 ± 9 | 1859 ± 6 | 1861 ± 10 |
| LE | 4 | 5 | 102 | 1535 | 5.1 | 1640 ± 20 | 1749 ± 22 | 1882 ± 5 | 1862 ± 40 |
| LR | 10 | 4 | 200 | 9100 | 8.2 | 1579 ± 17 | 1718 ± 19 | 1892 ± 4 | 1899 ± 40 |
| LE | 9 | 6 | 83 | 2790 | 7.9 | 1836 ± 14 | 1878 ± 15 | 1924 ± 5 | 1927 ± 20 |
| ME | 11 | 8 | 197 | 7325 | 5.6 | 2461 ± 11 | 2483 ± 13 | 2501 ± 4 | 2504 ± 10 |
| CR | 9 | 8 | 158 | 4300 | 3.7 | 2221 ± 11 | 2400 ± 14 | 2556 ± 4 | 2562 ± 40 |
| LR | 6 | 8 | 74.4 | 1765 | 2.5 | 2612 ± 22 | 2638 ± 23 | 2658 ± 4 | 2660 ± 10 |
| Bishop Canyon Formation (41°33'32.0"N, 118°33'58.3"W) | | | | | | | | | |
| CS | 6 | 10 | 217 | 395 | 4.7 | 268 ± 7 | 272 ± 8 | 308 ± 32 | 268 ± 10 |
| CE | 10 | 45 | 290 | 230 | 4.4 | 319 ± 3 | 324 ± 10 | 355 ± 59 | 319 ± 10 |
| CE | 6 | 23 | 199 | 195 | 4.0 | 325 ± 7 | 333 ± 13 | 389 ± 65 | 325 ± 10 |
| LE | 12 | 11 | 264 | 900 | 7.0 | 327 ± 3 | 330 ± 5 | 353 ± 25 | 327 ± 10 |
| MS | 7 | 13 | 297 | 550 | 6.3 | 329 ± 5 | 324 ± 6 | 291 ± 24 | 329 ± 10 |
| CE | 6 | 13 | 153 | 260 | 5.1 | 339 ± 10 | 342 ± 12 | 369 ± 27 | 339 ± 10 |
| CE | 6 | 13 | 171 | 270 | 4.4 | 341 ± 10 | 345 ± 12 | 374 ± 49 | 341 ± 10 |
| CE | 5 | 11 | 174 | 260 | 4.0 | 345 ± 12 | 351 ± 14 | 389 ± 46 | 345 ± 10 |
| **LE** | **8** | **5** | **322** | **1880** | **7.6** | **361 ± 2** | **361 ± 4** | **356 ± 18** | **361 ± 10** |
| **LE** | **7** | **7** | **617** | **2590** | **6.4** | **371 ± 3** | **372 ± 4** | **378 ± 13** | **371 ± 10** |
| MS | 9 | 10 | 236 | 990 | 5.7 | 441 ± 4 | 444 ± 6 | 460 ± 22 | 441 ± 10 |
| LR | 7 | 12 | 151 | 1085 | 4.5 | 1137 ± 9 | 1153 ± 12 | 1182 ± 12 | 1184 ± 10 |
| CR | 7 | 7 | 11 | 165 | 3.9 | 1119 ± 42 | 1148 ± 48 | 1203 ± 22 | 1208 ± 40 |
| LR | 9 | 17 | 177 | 1160 | 9.4 | 1210 ± 9 | 1220 ± 12 | 1238 ± 12 | 1239 ± 10 |
| **LR** | **7** | **30** | **429** | **1415** | **8.0** | **1300 ± 6** | **1299 ± 9** | **1298 ± 9** | **1298 ± 10** |
| MR | 7 | 15 | 267 | 1700 | 10.9 | 1337 ± 8 | 1346 ± 11 | 1358 ± 11 | 1360 ± 10 |
| **LR** | **9** | **8** | **146** | **2720** | **8.6** | **1424 ± 9** | **1426 ± 11** | **1430 ± 8** | **1430 ± 10** |
| CS | 6 | 11 | 58 | 460 | 3.7 | 1485 ± 27 | 1530 ± 31 | 1593 ± 15 | 1597 ± 20 |
| **LR** | **10** | **11** | **193** | **3060** | **4.6** | **1633 ± 9** | **1634 ± 11** | **1636 ± 7** | **1637 ± 10** |
| LS | 9 | 9 | 68 | 1210 | 4.8 | 1676 ± 17 | 1706 ± 19 | 1743 ± 8 | 1747 ± 20 |
| CR | 8 | 14 | 49 | 520 | 12.9 | 1795 ± 25 | 1803 ± 28 | 1811 ± 11 | 1813 ± 10 |
| LR | 7 | 15 | 294 | 2245 | 13.0 | 1703 ± 10 | 1777 ± 12 | 1865 ± 6 | 1868 ± 20 |
| MR | 7 | 13 | 170 | 1810 | 3.4 | 1840 ± 11 | 1864 ± 11 | 1890 ± 3 | 1892 ± 10 |
| CR | 8 | 18 | 56 | 510 | 3.3 | 1885 ± 21 | 1892 ± 27 | 1900 ± 15 | 1901 ± 10 |
| LR | 7 | 13 | 324 | 3525 | 16.0 | 1806 ± 7 | 1857 ± 10 | 1915 ± 5 | 1919 ± 20 |
| CR | 6 | 17 | 24 | 210 | 3.9 | 1839 ± 52 | 1878 ± 67 | 1922 ± 27 | 1925 ± 20 |
| CR | 9 | 9 | 22 | 500 | 7.1 | 2130 ± 46 | 2157 ± 50 | 2181 ± 12 | 2183 ± 10 |
| LR | 9 | 9 | 287 | 5975 | 22.2 | 1948 ± 9 | 2150 ± 10 | 2349 ± 2 | 2354 ± 40 |
| CR | 9 | 15 | 62 | 1270 | 8.9 | 2954 ± 22 | 3090 ± 26 | 3180 ± 6 | 3183 ± 40 |

*Note:*

Grain with $^{206}Pb^*/^{238}U$ age within 15% of $^{207}Pb^*/^{206}Pb^*$ age in plain type.

Grain with concordant age and moderate to high precision analysis in bold.

* Radiogenic Pb.

Grain type: L—light pink, M—medium pink, C—colorless; sR—slightly rounded, wR—well rounded, E—euhedral, sE—subeuhedral.

Unless otherwise noted, all grains are clear and/or translucent.

$^{206}Pb/^{204}Pb$ is measured ratio, uncorrected for blank, spike, or fractionation.

$^{206}Pb/^{208}Pb$ is corrected for blank, spike, and fractionation.

Most concentrations have an uncertainty of 25% due to uncertainty in weight of grain.

Constants used: $^{238}U/^{235}U$ = 137.88. Decay constant for $^{235}U = 9.8485 \times 10^{-10}$. Decay constant for $^{238}U = 1.55125 \times 10^{-10}$.

All uncertainties are at the 95% confidence level.

Pb blank ranged from 2 to 10 pg. U blank was consistently <1 pg.

Interpreted ages for concordant grains are $^{206}Pb^*/^{238}U$ ages if <1.0 Ga and $^{207}Pb^*/^{206}Pb^*$ ages if >1.0 Ga.

Interpreted ages for discordant grains are projected from 100 Ma.

All analyses conducted using conventional isotope dilution and thermal ionization mass spectrometry, as described by Gehrels (this volume, Introduction).

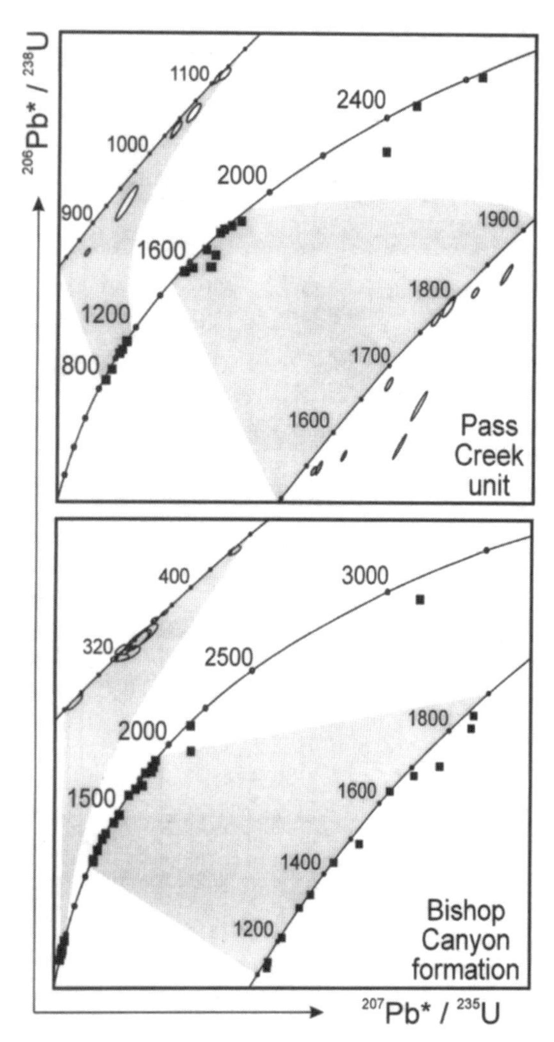


Figure 4. U-Pb concordia diagrams of detrital zircons in Pass Creek unit (Devonian[?] to Mississippian) and Bishop Canyon formation (Late Triassic). All analyses are of single zircon grains, and error ellipses are shown at 95% confidence level. Boxes are shown for analyses with error ellipses that are too small to be legible.

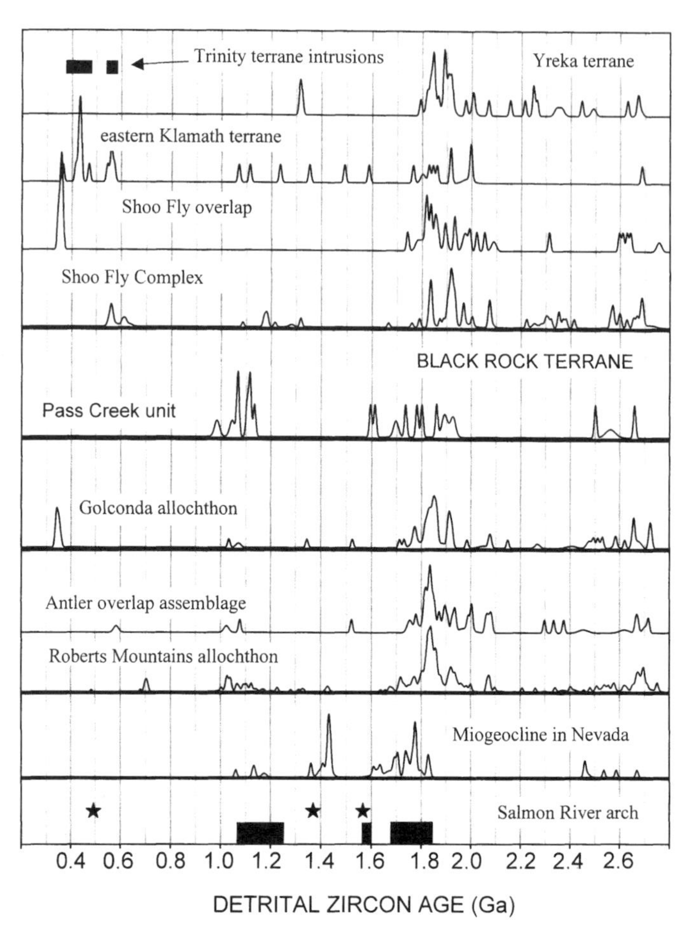


Figure 5. Relative age-probability plots comparing Pass Creek unit with age spectra from miogeoclinal strata in Nevada (Gehrels and Dickinson, 1995; Gehrels, this volume), Roberts Mountains allochthon (Gehrels et al., this volume, Chapter 1) and its overlap assemblage (Gehrels and Dickinson, this volume), Shoo Fly Complex (Harding et al., this volume) and overlying strata (Spurlin et al., this volume), eastern Klamath terrane (Gehrels and Miller, this volume), Yreka terrane (Wallin et al., this volume) and intrusions of Trinity terrane (Wallin and Metcalf, 1998), and Golconda allochthon (Riley et al., this volume). Curve for Roberts Mountains allochthon is composite of four main groups of detrital zircon ages in following proportions: 10% Harmony A, 26% Harmony B, 10% Lower Vinini, and 54% Upper Vinini–Valmy–McAfee–Elder–Slaven (Gehrels et al., this volume, Chapter 1). All other curves are simple composites of strata analyzed, except that grains younger than Mississippian have been excluded. Also shown, at bottom of diagram, are ages of igneous rocks (stars) and detrital zircons (filled boxes) from Salmon River arch region (Ross et al., 1992; Evans and Fischer, 1986; Evans and Zartman, 1988, 1990; Doughty and Chamberlain, 1996).

These rocks are close to the top of the unit and are therefore most likely Mississippian in age (Figs. 2 and 3).

The sample yielded 335 mg of zircon grains, most of which are <175 μm. The >125 μm grains were sorted into groups based on color (colorless, light pink, medium pink) and textural maturity (euhedral versus moderately to highly rounded grains). Most of the zircons are moderately rounded and light pink. Following abrasion, 21 grains representative of the original color and morphological groups were analyzed.

Figure 4 and Table 1 show that most of the grains yield slightly discordant ages, ranging from ca. 976 to 2660 Ma. The ages can be divided into two main groups of 1025–1140 Ma (n = 7) and 1850–1950 Ma (n = 4), and several groups of two or three grains each at 1595–1612, 1695–1735, 1780–1801, and 2504–2660 Ma (Fig. 5; Table 2). There is also a single grain with an interpreted age of ca. 976 Ma.

### *Triassic Bishop Canyon formation*

Our sample was collected from ~4 km south of Bishop Canyon along the eastern flank of the Pine Forest Range (Fig. 2). The sample analyzed is from the middle member of the formation, and is a medium-grained sandstone. This sample is more quartz rich and volcanic poor than most sandstones in the middle member of the Bishop Canyon formation (Wyld, 1992).

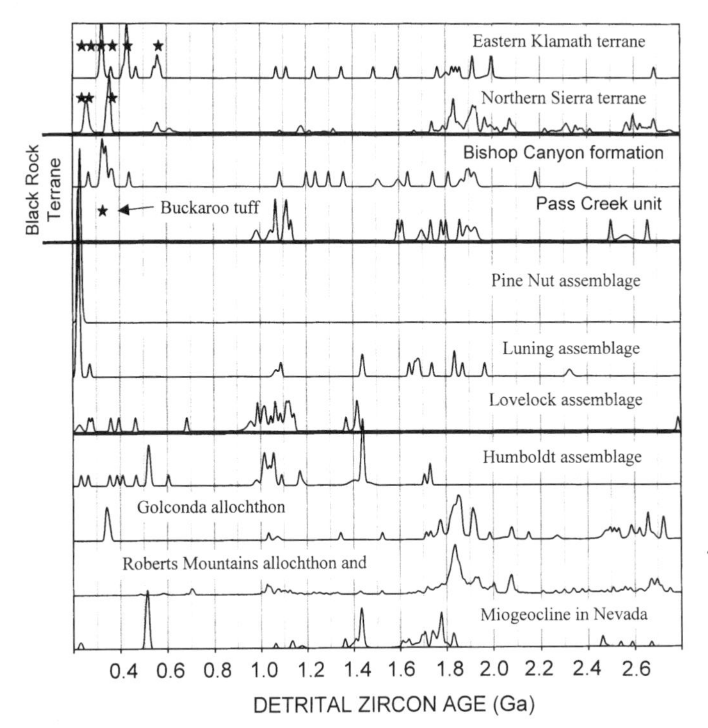


Figure 6. Relative age-probability plots comparing ages from Bishop Canyon formation with detrital zircon ages from strata in northern Sierra terrane (Harding et al., this volume; Spurlin et al., this volume), eastern Klamath terrane (Gehrels and Miller, this volume), Golconda allochthon (Riley et al., this volume) and overlying Triassic strata of Humboldt assemblage (Gehrels and Dickinson, 1995), Roberts Mountains allochthon (Gehrels et al., this volume, Chapter 1) and overlying strata (Gehrels and Dickinson, this volume), Triassic assemblages of western Nevada (Manuszak et al., this volume), and Cordilleran miogeocline (Gehrels and Dickinson, 1995). Ages of igneous rocks in Sierra Nevada and eastern Klamath terranes, indicated with stars, are from Hanson et al. (1988), Miller and Harwood (1990), and Saleeby and Busby-Spera (1992). Ca. 327 Ma age of Buckaroo tuff, shown with Pass Creek unit, is from Wyld (1992).

The sample yielded 280 mg of zircon grains, most of which are <100 μm in sieve size. All of the >100 μm grains in the sample were sorted into six groups defined by color (various shades of pink or colorless) and by degree of rounding (euhedral, subrounded, and well rounded). Following abrasion, 29 grains representative of the original color and/or morphology groups were selected for analysis.

Of the 29 grains analyzed, 11 are <450 Ma and the rest range from ca. 1.1 to 3.1 Ga. The younger grains are mainly concordant to slightly discordant, but of low to moderate precision because of the small amount of lead in each grain (Table 1; Fig. 4). The age interpretation of these grains is problematic due to their generally low $^{206}Pb/^{204}Pb$ and poor precision. One group of grains has $^{206}Pb^{*}/^{238}U$ ages between 268 and 371 Ma, but only two of these grains are analytically concordant and of moderate to high precision. The concordant grains (361 and 371 Ma) provide strong evidence for a ca. 360–375 Ma source, and the grains with younger $^{206}Pb^{*}/^{238}U$ ages may be of the same age but affected by lead loss. Alternatively, the younger ages could record igneous activity as young as ca. 320 Ma and perhaps as young as ca. 268 Ma. Assuming that the $^{206}Pb^{*}/^{238}U$ ages are reliable indicators of crystallization age, a dominant peak in age probability occurs at 326 Ma and subordinate peaks occur at 341 and 366 Ma (Fig. 6; Table 2).

Three of the older grains are apparently concordant and the rest are slightly to moderately discordant (Table 1). The interpreted ages range from ca. 1184 to ca. 3183 Ma, with only one apparent cluster between 1850 and 1940 Ma (n = 5). The other ages are scattered between 1184 and 1813 Ma, with an apparent gap between 1430 and 1597 Ma, and there are isolated older ages of 2183, 2354, and 3183 Ma (Table 1; Fig. 6).

**TABLE 2. DOMINANT DETRITAL ZIRCON AGE GROUPS IN STRATA OF THE BLACK ROCK TERRANE**

| Age range (Ma) | Peak ages (Ma) | Relative abundance |
|---|---|---|
| Pass Creek unit | | |
| 1025–1140 | 1067*, 1115* | 0.33 |
| 1585–1620 | | 0.09 |
| 1685–1745 | | 0.09 |
| 1775–1810 | | 0.09 |
| 1850–1950 | | 0.19 |
| 2495–2670 | | 0.14 |
| Bishop Canyon Formation | | |
| 310–380 | 326*, 341, 366 | 0.31 |
| 1850–1940 | 1900 | 0.17 |

*Note:* These ages and relative abundances refer to the relative age-probability plots shown in Figures 5 and 6.
* Indicates the peak with the highest probability.

## PROVENANCE OF DETRITAL ZIRCONS

### *Pass Creek unit*

Figure 5 is a comparison of detrital zircon ages from the Pass Creek unit, middle Paleozoic strata in the Antler overlap assemblage, Golconda allochthon, eastern Klamath terrane, and northern Sierra terrane, and potential source terranes (Nevada miogeocline, Roberts Mountains allochthon, and Yreka and Trinity terranes of the eastern Klamath Mountains). In terms of potential source terranes, zircon ages from the Pass Creek unit overlap in part with the ages of detrital grains in the Roberts Mountains allochthon (Fig. 5). Derivation of detritus from this allochthon is reasonable because rocks of the allochthon were exposed in a broad regional highland along the continental margin during the latest Devonian–Mississippian Antler orogeny (Roberts et al., 1958; Nilsen and Stewart, 1980; Burchfiel et al., 1992). This highland formed a significant source of sediment that was shed both to the east (Antler overlap and foreland basin assemblages) and to the west (Golconda allochthon and northern Sierra terrane) (Poole, 1974; Schweickert and Snyder, 1981; Dickinson et al., 1983; Harwood and Murchey, 1990).

Although Pass Creek zircons partly overlap with ages in the Roberts Mountains allochthon and in the overlying upper Paleozoic units, there are significant differences. First, the Pass Creek unit shows a greater abundance of 1.0–1.2 Ga grains and proportionately fewer 1.8–1.9 Ga grains than does the Roberts Mountains allochthon. Second, it contains several grains ca. 1595–1612 Ma and ca. 1695–1735 Ma, neither of which is a common age range for detrital zircons from the Roberts Mountains allochthon. Two conclusions are possible: the sediment source for the Pass Creek unit either (1) was not the Roberts Mountains allochthon, or (2) was some combination of the Roberts Mountains allochthon plus an additional source terrane. In either case, it is notable that Paleozoic rocks of the Black Rock terrane do not show the striking match in detrital zircon signature with the Roberts Mountains allochthon that is shown by the northern Sierra terrane and Golconda allochthon (Fig. 5). This points to contributions from additional source terranes.

Alternative potential sources include the Yreka and Trinity terranes, which are interpreted to have been an important source of Mississippian sediment in the eastern Klamath Mountains (Gehrels and Miller, this volume, Chapter 7). However, this source does not match particularly well with the 1.0–1.2 Ga zircon ages of the Pass Creek unit, and cannot explain the anomalies of the Pass Creek unit noted here (Fig. 5). The same problems arise when considering the Shoo Fly Complex in the northern Sierra terrane as a potential source for sediments in the Pass Creek unit (Fig. 5).

A more likely additional source terrane is the craton of the western United States, in particular the Salmon River arch region of Idaho, western Montana, and eastern Washington (Fig. 7). This region is a potential source of the two ca. 1.6 Ga grains (Evans and Fisher, 1986; Ross et al., 1992) and could also have contributed at least some of the 1.0–1.2 Ga and 1.7–1.8 Ga grains (Fig. 5). The latter sets of grains probably originated in basement rocks of the southwestern United States, but may then have been recycled through miogeoclinal or platformal strata in Nevada (Gehrels et al., 1995; Gehrels, this volume, Introduction) and/or strata in the Salmon River arch region (Ross et al., 1992). Derivation of sediment from the Salmon River arch region alone, however, cannot account for the complete detrital zircon signature of the Pass Creek unit. Instead, the Pass Creek unit zircon signature appears to be best accounted for by a combined source region, including the Salmon River arch and the Roberts Mountains allochthon (Fig. 5). The occurrence of detritus from the Salmon River arch in the Pass Creek unit, but not in the Golconda allochthon (Fig. 5), suggests that either the Black Rock terrane formed north of the Golconda allochthon or that upper Paleozoic paleogeography in this region of the Cordillera caused sediment from the Salmon River arch to bypass the Golconda basin and be shed directly into basins located farther west (Black Rock terrane).

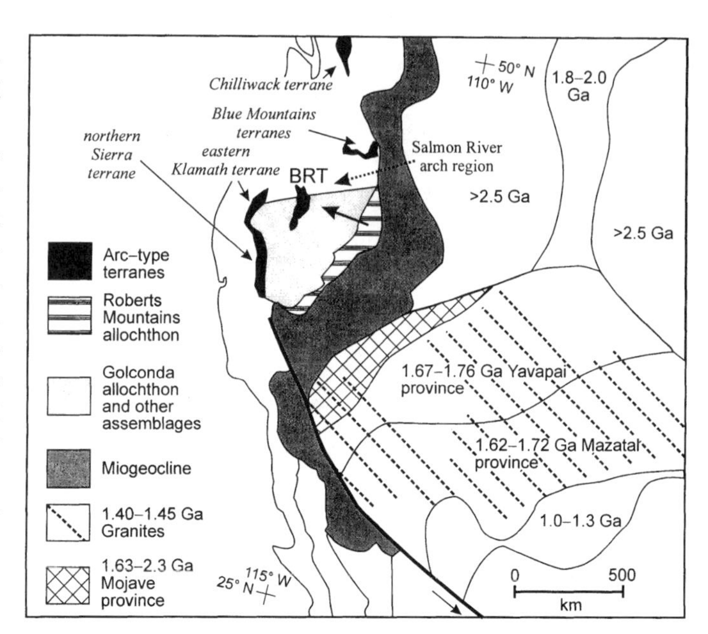


Figure 7. Schematic map showing main basement provinces of western North America, and interpreted provenance of detrital zircons in Devonian(?) to Mississippian Pass Creek unit. Note that map is not palinspastic. Distribution and age of basement provinces are adapted from Hoffman (1989) for cratonal interior, various sections of Van Schmus et al. (1993) for southwestern United States, and Stewart et al. (1990) for northwestern Mexico. Salmon River arch location is from Armstrong (1975). BRT is Black Rock terrane.

Collectively, these age comparisons suggest that detritus in our sample of the Pass Creek unit was probably recycled from older strata that were exposed in an uplifted region along the continental margin in the northwestern United States (Fig. 7). The main assemblages exposed in the source region would have included lower Paleozoic strata of the Roberts Mountains allochthon, which contributed the dominant chert and argillite clasts in the Pass Creek unit, and platformal or miogeoclinal strata and igneous rocks of the Salmon River arch region, which contributed mostly quartzose detritus.

These provenance links have several important implications, when considered in conjunction with other geologic relations.

1. They argue that Mississippian strata of the Black Rock terrane were deposited in proximity to the continental margin, near their current location, in late Paleozoic time. This is because the combined sediment sources identified by this study—Salmon River arch region plus Roberts Mountains allochthon—require that the depositional site was located in a position near both sources, consistent with the current location of the Black Rock terrane (Fig. 7).

2. They provide an indirect argument that outboard Paleozoic arc assemblages of the Cordillera also formed relatively close to the continental margin in late Paleozoic time. This is because the Black Rock terrane can be linked to more outboard arc assemblages of the eastern Klamath Mountains, Blue Mountains province, and Chilliwack terrane (Fig. 7) on the basis of similarities in upper Paleozoic stratigraphy (Wyld, 1990). These

mutual links support a model in which a number of upper Paleozoic arc assemblages in the western Cordillera evolved relatively near the continental margin.

3. Dissimilarities in the detrital zircon signatures of Mississippian strata between the Black Rock terrane and Klamath-Sierra terranes (Fig. 5) could be explained by formation of the Klamath-Sierran parts of the arc farther south along the margin than the Black Rock portion of the arc. According to this scenario, Mississippian sediment supplied from the continental margin to outboard terranes located farther south (Klamath-Sierran terranes) was derived primarily from the Roberts Mountains allochthon, whereas Mississippian sediment supplied from the continental margin to outboard terranes located farther north (Black Rock terrane) was derived both from the Roberts Mountains allochthon and from the more northern Salmon River arch source. This paleogeographic interpretation is consistent both with the detrital zircon data (Fig. 5; Riley et al., this volume, Chapter 4; Spurlin et al., this volume, Chapter 6), and with regional geologic evidence that the Klamath-Sierran terranes were displaced northward relative to the Black Rock terrane in Mesozoic time by strike-slip faulting (Lahren et al., 1990; Schweickert and Lahren, 1990; Wyld and Wright, 1997).

### *Bishop Canyon formation*

Detrital zircons in Triassic strata of the Bishop Canyon formation originated in Middle Proterozoic and older rocks of a continental region, and in middle (to late?) Paleozoic igneous rocks (Fig. 6). The most likely sources for the Paleozoic grains are middle to late Paleozoic igneous rocks that formed in arc-type terranes outboard of the continental margin; likely possibilities include the eastern Klamath, northern Sierra, and/or Black Rock terranes, all of which contain Paleozoic igneous rocks of the appropriate ages (Fig. 6; Miller and Harwood, 1990; Wyld, 1990).

Whereas Precambrian grains in the Bishop Canyon formation have ages compatible with basement provinces of the North American craton, it is unlikely that they were derived directly from the craton. First, the Precambrian and Paleozoic grains are found in the same sample. There is also no evidence within the Bishop Canyon formation for interlayering of sediments derived from different sources, because all sandstones and conglomerates in the formation are similar in composition (Wyld, 1992). Thus, the Precambrian grains most likely came from the same arc-type, outboard source as the Paleozoic grains. Second, an eastern source for the Bishop Canyon sandstone is unlikely, given the presence of significantly different detrital zircon ages in Triassic strata of the more inboard Lovelock and Humboldt assemblages, both of which were derived from the continent to the east (Fig. 6; Lupe and Silberling, 1985; Manuszak et al., this volume, Chapter 8).

Precambrian grains in the Bishop Canyon formation are therefore interpreted to have been derived from siliciclastic Paleozoic basement rocks of an outboard arc terrane. Detrital zircon ages from Paleozoic rocks of the eastern Klamath terrane are quite similar to those of the Bishop Canyon formation, suggesting that these Paleozoic rocks are a likely source terrane (Fig. 6). There is also significant overlap between zircon ages of the Pass Creek unit and those in the Bishop Canyon formation (Fig. 6), suggesting that Paleozoic basement of the Black Rock terrane could also have provided detritus to the Bishop Canyon formation. This possibility is supported by the coarse-grained, immature nature of Bishop Canyon sandstones, and by the presence of an unconformity at the base of the Triassic section in the Pine Forest Range that cuts down to the level of Mississippian strata (Fig. 2; Wyld, 1990).

There is considerably less overlap between the ages of Bishop Canyon zircons and the ages of detrital zircons from Paleozoic rocks of the northern Sierra terrane, however, suggesting that the latter are a less likely potential sediment source (Fig. 6).

Collectively, these provenance ties indicate that the Bishop Canyon formation consists of detritus derived from Paleozoic basement rocks of outboard arc terranes, probably the Black Rock terrane and the eastern Klamath terrane. These data provide new evidence that the Black Rock terrane was situated relatively near the eastern Klamath terrane in the early Mesozoic, and that outboard Paleozoic arc terranes of the Cordillera were at least locally uplifted and eroded following the Sonoma orogeny. This supports the previous conclusions of Silberling (1973), Schweickert (1981), Schweickert et al. (1984), Miller (1987), Harwood (1988, 1992), Jones (1990), Miller and Harwood (1990), and Wyld (1990, 1991).

The data also support a Triassic paleogeography in which sedimentary input to the Black Rock terrane was cannibalized from nearby arc basement rocks rather than having been derived from the continent to the east. These paleographic interpretations are consistent with evidence, discussed in more detail by Wyld (this volume, Chapter 13), that an extensional, deep-marine backarc basin opened in the Late Triassic between the Black Rock terrane and the continental margin. Extensional faulting along the basin margins is inferred to have resulted in exposure and erosion of arc basement rocks in the Black Rock terrane, while separation of the Black Rock terrane from the continental margin by an expanding backarc rift basin prevented continentally derived sediments from reaching depositional sites in the Black Rock terrane (Wyld, this volume, Chapter 13). Instead, deposition of continentally derived Late Triassic sediments was confined to the backarc basin, now represented by the Lovelock assemblage (Fig. 1; Manuszak et al., this volume, Chapter 8).

## CONCLUSIONS

U-Pb ages have been determined for 50 detrital zircon grains from Mississippian and Triassic sandstones of the Black Rock terrane, northwestern Nevada. Comparison of these ages with age spectra from strata in other regions yields several important conclusions about the late Paleozoic–early Mesozoic paleogeography and paleoposition of the Black Rock terrane.

During Mississippian time, the Black Rock terrane received detritus from the Roberts Mountains allochthon and from basement rocks or overlying strata in the Salmon River arch region of

Idaho, western Montana, and eastern Washington (Fig. 7). This suggests that the terrane was located along or near the northwest U.S. Cordilleran margin, at approximately its current latitude, during late Paleozoic time. It can also be linked on the basis of upper Paleozoic stratigraphic similarities with some of the more outboard arc assemblages of the Cordillera, such as the eastern Klamath region, Blue Mountains province, and Chilliwack terrane (Fig. 7). However, the Black Rock terrane apparently evolved north of the Klamath-Sierran portion of the Paleozoic arc because detrital zircons in Mississippian rocks of Klamath-Sierran terranes record somewhat different sediment source regions.

During Late Triassic time, the Black Rock terrane received detritus from uplifted Paleozoic basement sources in outboard Cordilleran terranes that most likely included the eastern Klamath Mountains and the Black Rock terrane. No sediment input from the continent is indicated during this time frame. These relations are consistent with independent evidence that lower Mesozoic arc assemblages in this part of the Cordillera were separated from the continental margin during the Late Triassic by a deep-marine backarc basin (e.g., Speed, 1978; Saleeby and Busby-Spera, 1992; Wyld, this volume, Chapter 13).

Collectively, these provenance links provide strong support for a tectonic scenario in which the Black Rock terrane formed and evolved near or north of its present position along the Cordilleran margin (Burchfiel and Davis, 1972, 1975; Miller et al., 1984, 1992; Wyld, 1990; Saleeby and Busby-Spera, 1992; Burchfiel et al., 1992). Our data are less consistent with suggestions that the Black Rock terrane is part of an arc system that formed much farther north along the continental margin (Wallin, 1993), or is far traveled (Speed, 1979; Coney et al., 1980).

## ACKNOWLEDGMENTS

We thank Jim Wright and Bill Dickinson for guidance and assistance in the field and in interpreting our results. Our studies have been supported by the National Science Foundation (EAR-9416933). Reviewed by Richard Schweickert and Michael Quinn.

## REFERENCES CITED

Armstrong, R.L., 1975, Precambrian (1,500 million year old) rocks of central Idaho—The Salmon River arch and its role in Cordilleran sedimentation and tectonics: American Journal of Science, v. 275-A, p. 437–467.

Burchfiel, B.C., and Davis, G.A., 1972, Structural framework and evolution of the southern part of the Cordilleran orogen, western United States: American Journal of Science, v. 272, p. 97–118.

Burchfiel, B.C., and Davis, G.A., 1975, Nature and controls of Cordilleran orogenesis, western United States: Extensions of an earlier synthesis: American Journal of Science, v. 275-A, p. 363–396.

Burchfiel, B.C., Cowan, D.S., and Davis, G.A., 1992, Tectonic overview of the Cordilleran orogen in the western United States, *in* Burchfiel, B.C., et al., eds., The Cordilleran orogen: Conterminous U.S.: Boulder, Colorado, Geological Society of America, The Geology of North America, v. G-3, p. 407–480.

Coney, P.J., Jones, D.L., and Monger, J.W.H., 1980, Cordilleran suspect terranes: Nature, v. 288, p. 329–333.

Dickinson, W.R., Harbaugh, D.W., Saller, A.H., Heller, P.L., and Snyder, W.S., 1983, Detrital modes of upper Paleozoic sandstones derived from Antler orogen in Nevada: Implications for the nature of the Antler orogeny: American Journal of Science, v. 282, p. 481–509.

Doughty, P.T., and Chamberlin, K.R., 1996, Salmon River Arch revisited: New evidence for 1370 Ma rifting near the end of deposition in the Middle Proterozoic Belt basin: Canadian Journal of Earth Sciences, v. 33, p. 1037–1052.

Evans, K.V., and Fischer, L.B., 1986, U-Pb geochronology of two augen gneiss terranes, Idaho—New data and tectonic implications: Canadian Journal of Earth Sciences, v. 23, p. 1919–1927.

Evans, K.V., and Zartman, R.E., 1988, Early Paleozoic alkalic plutonism in east-central Idaho: Geological Society of America Bulletin, v. 100, p. 1981–1987.

Evans, K.V., and Zartman, R.E., 1990, U-Th-Pb and Rb-Sr geochronology of Middle Proterozoic granite and augen gneiss, Salmon River Mountains, east-central Idaho: Geological Society of America Bulletin, v. 102, p. 63–73.

Gehrels, G.E., and Dickinson, W.R., 1995, Detrital zircon provenance of Cambrian to Triassic miogeoclinal and eugeoclinal strata in Nevada: American Journal of Science, v. 295, p. 18–48.

Gehrels, G.E., Dickinson, W.R., Ross, G.M., Stewart, J.H., and Howell, D.G., 1995, Detrital zircon reference for Cambrian to Triassic miogeoclinal strata of western North America: Geology, v. 23, p. 831–834.

Hanson, R.E., Saleeby, J.B., and Schweickert, R.A., 1988, Composite Devonian island-arc batholith in the northern Sierra Nevada, California: Geological Society of America Bulletin, v. 100, p. 446–457.

Harwood, D.S., 1988, Tectonism and metamorphism in the northern Sierra terrane, northern California, *in* Ernst, W.G., ed., Metamorphism and crustal evolution of the western United States (Rubey Volume 7): Englewood Cliffs, New Jersey, Prentice-Hall, p. 764–788.

Harwood, D.S., 1992, Stratigraphy of Paleozoic and lower Mesozoic rocks in the northern Sierra terrane, California: U.S. Geological Survey Bulletin 1957, 78 p.

Harwood, D.S., and Murchey, B.L., 1990, Biostratigraphic, tectonic, and paleogeographic ties between upper Paleozoic volcanic and basinal rocks in the northern Sierra terrane, California, and the Havallah sequence, Nevada, *in* Harwood, D.S., and Miller, M.M., eds., Paleozoic and early Mesozoic paleogeographic relations; Sierra Nevada, Klamath Mountains, and related terranes: Geological Society of America Special Paper 255, p. 157–174.

Hoffman, P.F., 1989, Precambrian geology and tectonic history of North America, *in* Bally, A.W., and Palmer, A.R., eds., The geology of North America—An overview: Boulder, Colorado, Geological Society of America, Geology of North America, v. A, p. 447–512.

Jones, A.E., 1990, Geology and tectonic significance of terranes near Quinn River Crossing, Humbolt County, Nevada, *in* Harwood, D.S., and Miller, M.M., eds., Paleozoic and early Mesozoic paleogeographic relations; Sierra Nevada, Klamath Mountains, and related terranes: Geological Society of America Special Paper 255, p. 239–253.

Ketner, K.B., and Wardlaw, B.R., 1981, Permian and Triassic rocks near Quinn River Crossing, Humbolt County, Nevada: Geology, v. 9, p. 123–126.

Lahren, M.M., Schweickert, R.A., Mattinson, J.E., and Walker, J.D., 1990, Evidence of uppermost Proterozoic to Lower Cambrian miogeoclinal rocks and the Mojave–Snow Lake fault: Snow Lake pendant, central Sierra Nevada: Tectonics, v. 9, p. 1585–1608.

Lupe, R., and Silberling, N.J., 1985, Genetic relationship between lower Mesozoic continental strata of the Colorado plateau and marine strata of the western Great Basin: Significance for accretionary history of Cordilleran lithotectonic terranes, *in* Howell, D.G., ed., Tectonostratigraphic terranes of the Circum-Pacific region: Houston, Texas, Circum-Pacific Council for Energy and Mineral Resources, p. 263–271.

Maher, K.A., 1989, Geology of the Jackson Mountains, northwest Nevada [Ph.D. thesis.]: Pasadena, California Institute of Technology, 491 p.

Miller, E.L., Holdsworth, B.K., Whiteford, W.B., and Rodgers, D., 1984, Stratigraphy and structure of the Schoonover sequence, northeastern Nevada: Implications for Paleozoic plate-margin tectonics: Geological Society of America Bulletin, v. 95, p. 1063–1076.

Miller, E.L., Miller, M.M., Stevens, C.H., Wright, J.E., and Madrid, R., 1992, Late Paleozoic paleographic and tectonic evolution of the western U.S. Cordillera, *in* Burchfiel, B.C., et al., eds., The Cordilleran orogen: Conterminous U.S.: Geological Society of America, Geology of North America, v. G-3, p. 57–106.

Miller, M.M., 1987, Dispersed remnants of a northeast Pacific fringing arc; upper Paleozoic terranes of Permian McCloud faunal affinity, western U.S.: Tectonics, v. 6, p. 807–830.

Miller, M.M., and Harwood, D.S., 1990, Paleographic setting of upper Paleozoic rocks in the northern Sierra and eastern Klamath terranes, northern California, *in* Harwood, D.S., and Miller, M.M., eds., Paleozoic and early Mesozoic paleogeographic relations; Sierra Nevada, Klamath Mountains, and related terranes: Geological Society of America Special Paper 255, p. 175–192.

Nilsen, T.H., and Stewart, J.H., 1980, The Antler orogeny—Mid-Paleozoic tectonism in western North America (Penrose Conference Report): Geology, v. 8, p. 298–302.

Oldow, J.S., 1984, Evolution of a late Mesozoic back-arc fold and thrust belt, northwestern Great Basin, U.S.A.: Tectonophysics, v. 102, p. 245–274.

Poole, F.G., 1974, Flysch deposits of Antler foreland basin, western United States, *in* Dickinson, W.R., ed., Tectonics and sedimentation: Society of Economic Paleontologists and Mineralogists Special Publication 22, p. 58–82.

Quinn, M.J., 1996, Pre-Tertiary stratigraphy, magmatism, and structural history of the central Jackson Mountains, Humboldt County, Nevada [Ph.D. thesis]: Houston, Texas, Rice University, 243 p.

Quinn, M.J., Wright, J.E., and Wyld, S.J., 1997, Happy Creek igneous complex and tectonic evolution of the early Mesozoic arc in the Jackson Mountains, northwest Nevada: Geological Society of America Bulletin, v. 109, p. 461–482.

Roberts, R.J., Hotz, P.E., Gilluly, J., and Ferguson, H.G., 1958, Paleozoic rocks of north-central Nevada: American Association of Petroleum Geologists Bulletin, v. 42, p. 2813–2857.

Ross, G.M., Parrish, R.R., and Winston, D., 1992, Provenance and U-Pb geochronology of the Mesoproterozoic Belt Supergroup (northwestern United States): Implications for age of deposition and pre-Panthalassa plate reconstructions: Earth and Planetary Science Letters, v. 113, p. 57–76.

Russell, B.J., 1981, Pre-Tertiary paleogeography and tectonic history of the Jackson Mountains, northwestern Nevada [Ph.D. thesis]: Evanston, Illinois, Northwestern University, 205 p.

Russell, B.J., 1984, Mesozoic geology of the Jackson Mountains, northwestern Nevada: Geological Society of America Bulletin, v. 95, p. 313–323.

Saleeby, J.B., and Busby-Spera, C., 1992, Early Mesozoic tectonic evolution of the western U.S. Cordillera, *in* Burchfiel, B.C., et al., eds., The Cordilleran orogen: Conterminous U.S.: Boulder, Colorado, Geological Society of America, Geology of North America, v. G-3, p. 107–168.

Schweikert, R.A., 1981, Tectonic evolution of the Sierra Nevada, *in* Ernst, W.D., ed., The geotectonic development of California (Rubey Volume I): Englewood Cliffs, New Jersey, Prentice-Hall, p. 88–131.

Schweickert, R.A., and Lahren, M.M., 1990, Speculative reconstruction of the Mojave–Snow Lake fault: Implications for Paleozoic and Mesozoic orogenesis in the western United States: Tectonics, v. 9, p. 1609–1629.

Schweickert, R.A., and Snyder, W.S., 1981, Paleozoic plate tectonics of the Sierra Nevada and adjacent regions, *in* Ernst, W.G., ed., The geotectonic development of California: Englewood Cliffs, New Jersey, Prentice-Hall, p. 183–201.

Schweikert, R.A., Harwood, D.S., Girty, G.H., and Hanson, R.E., 1984, Tectonic development of the northern Sierra terrane: An accreted late Paleozoic island arc and its basement, *in* Lintz, J., Jr., ed., Western geological excursions: Reno, Nevada, Mackay School of Mines, p. 1–65.

Silberling, N.J., 1973, Geologic events during Permian–Triassic time along the Pacific margin of the United States, *in* Logan, A., and Hills, L.V., eds., The Permian and Triassic Systems and their mutual boundary: Calgary, Alberta Society of Petroleum Geologists, p. 345–362.

Silberling, N.J., 1991, Allochthonous terranes of western Nevada, current status, *in* Raines, G.L., et al., eds., Geology and ore deposits of the Great Basin: Reno, Geological Society of Nevada, p. 101–102.

Silberling, N.J., Jones, D.L., Blake, M.C., Jr., and Howell, D.G., 1987, Lithotectonic terrane map of the western conterminous United States: U.S. Geological Survey Miscellaneous Field Studies Map MF-1874-C, scale: 1:2,500,000.

Speed, R.C., 1978, Paleogeographic and plate tectonic evolution of the early Mesozoic marine province of the western Great Basin, *in* Howell, D.G., and McDougall, K.A., eds., Mesozoic paleogeography of the western United States: Society of Economic Paleontologists and Mineralogists, Pacific Coast Paleogeography Symposium 2, p. 253–270.

Speed, R.C., 1979, Collided Paleozoic microplate in the western United States: Journal of Geology, v. 87, p. 279–292.

Stacey, J.S., and Kramers, J.D., 1975, Approximation of terrestrial lead isotope evolution by a two-stage model: Earth and Planetary Science Letters, v. 26, p. 207–221.

Stewart, J.H., Poole, F.G., Ketner, K.B., Madrid, R.J., Roldan-Quintana, J., and Amaya-Martinez, R., 1990, Tectonics and stratigraphy of the Paleozoic and Triassic southern margin of North America, Sonora, Mexico, *in* Gehrels, G.E., and Spencer, J.E., eds., Geologic excursions through the Sonoran Desert region, Arizona and Sonora: Arizona Geological Survey Special Paper 7, p. 183–202.

Van Schmus, W.R., and 24 others, 1993, Transcontinental Proterozoic provinces, *in* Reed, J.C., et al., eds., Precambrian: Conterminous U.S: Boulder, Colorado, Geological Society of America, Geology of North America, v. C-2, p. 171–334.

Wallin, E.T., 1993, Sonomia revisited: Evidence for a western Canadian provenance of the eastern Klamath and northern Sierra Nevada terranes: Geological Society of America Abstracts with Programs, v. 25, no. 6, p. A-173.

Wallin, E.T., and Metcalf, R.V., 1998, Supra-subduction zone ophiolite formed in an extensional forearc: Trinity terrane, Klamath Mountains, California: Journal of Geology, v. 106, p. 591–608.

Wyld, S.J., 1990, Paleozoic and Mesozoic rocks of the Pine Forest Range, northwest Nevada, and their relation volcanic arc assemblages of the western U.S. Cordillera, *in* Harwood, D.S., and Miller, M.M., eds., Paleozoic and early Mesozoic paleogeographic relations; Sierra Nevada, Klamath Mountains, and related terranes: Geological Society of America Special Paper 255, p. 219–238.

Wyld, S.J., 1991, Permo-Triassic tectonism in volcanic arc sequences of the western U.S. Cordillera and implications for the Sonoma orogeny: Tectonics, v. 10, p. 1007–1017.

Wyld, S.J., 1992, Geology and geochronology of the Pine Forest Range, northwest Nevada: Stratigraphic, structural, and magmatic history, and regional implications [Ph.D. thesis]: Stanford, California, Stanford University, 429 p.

Wyld, S.J., 1996, Early Jurassic deformation in the Pine Forest Range, northwest Nevada, and implications for Cordilleran tectonics: Tectonics, v. 15, p. 566–583.

Wyld, S.J., and Wright, J.E., 1997, Triassic-Jurassic tectonism and magmatism in the Mesozoic continental arc of Nevada: Classic relations and new developments, *in* Link, P.K., and Kowallis, B.J., eds., Proterozoic to recent stratigraphy, tectonics, and volcanology, Utah, Nevada, southern Idaho and central Mexico: Brigham Young University Geology Studies, v. 42, p. 197–224.

MANUSCRIPT ACCEPTED BY THE SOCIETY JANUARY 24, 2000

Printed in U.S.A.

Geological Society of America
Special Paper 347
2000

# *Detrital zircon geochronology of upper Paleozoic and lower Mesozoic strata of the northern Sierra terrane, northeastern California*

**Matthew S. Spurlin, George E. Gehrels**
*Department of Geosciences, University of Arizona, Tucson, Arizona 85721, USA*
**David S. Harwood**
*345 Middlefield Road, U.S. Geological Survey, Menlo Park, California 94025, USA*

## ABSTRACT

**U-Pb analyses of 56 individual detrital zircon grains from mid-Paleozoic through Lower Jurassic clastic strata of the northern Sierra terrane yield two distinct sets of ages: 1.7–2.8 Ga grains mainly in Upper Devonian–Mississippian strata of the Picayune Valley Formation, and ca. 370–185 Ma grains in strata of Permian through Jurassic age. The older ages are most similar to the ages of grains in the underlying Shoo Fly Complex and in the Roberts Mountains allochthon in Nevada. This age similarity, combined with stratigraphic relations that record a provenance link with the Roberts Mountains allochthon, are consistent with models in which the northern Sierra terrane was located in proximity to the Nevada continental margin during Late Devonian–Mississippian time.**

**The ca. 370–185 Ma ages of detrital grains in Permian, Triassic, and Jurassic strata are an excellent match for the ages of volcanic rocks within the northern Sierra terrane. The clastic detritus in these units was presumably derived from these intraterrane volcanic rocks.**

## INTRODUCTION

The Sierra Nevada of northern California consists of several terranes that evolved in volcanic arc and ocean-floor settings during Paleozoic and early Mesozoic time, and that were subsequently accreted to the North American continent (Fig. 1) (Burchfiel and Davis, 1972; Schweickert and Snyder, 1981). One of the larger terranes in the Sierra Nevada, the northern Sierra terrane, records three periods of island-arc volcanism from Late Devonian through Middle Jurassic time. Although all workers agree that the arc formed outboard of the Cordilleran miogeocline, whether it was a fringing arc located several hundred kilometers offshore or a far-traveled, exotic arc is still debated (Burchfiel and Davis, 1972, 1975; Speed, 1979; Schweickert and Snyder, 1981; Speed and Sleep, 1982; Miller et al., 1984, 1992; Miller, 1987; Burchfiel et al., 1992). Currently, the northern Sierra terrane is separated from miogeoclinal strata by the Roberts Mountains and Golconda allochthons, and relations with these units are obscured by assemblages of off-shelf Triassic strata that underlie much of western Nevada (Fig. 2).

## PREVIOUS WORK

Paleozoic volcanic-arc rocks in the northern Sierra Nevada were deposited during two distinct episodes of volcanism (D'Allura et al., 1977; Schweickert, 1981; Harwood, 1983, 1988, 1992). The first of these volcanic episodes, during Late Devonian–Early Mississippian time, predated or overlapped with the emplacement of lower Paleozoic eugeoclinal rocks in the Roberts Mountains allochthon onto miogeoclinal strata in Nevada during the Late Devonian–Early Mississippian Antler orogeny (Nilsen and Stewart, 1980; Speed and Sleep, 1982; Schweickert et al., 1984; Miller et al., 1984, 1992). This thrusting produced the Antler orogenic highlands, which shed chert- and quartz-rich debris both eastward and westward (Dickinson et al., 1983; Miller et al., 1984; Harwood and Murchey, 1990; Whiteford, 1990). Following a second

Spurlin, M.S., Gehrels, G.E., and Harwood, D.S., 2000, Detrital zircon geochronology of upper Paleozoic and lower Mesozoic strata of the northern Sierra terrane, northeastern California, *in* Soreghan, M.J., and Gehrels, G.E., eds., Paleozoic and Triassic paleogeography and tectonics of western Nevada and northern California: Boulder, Colorado, Geological Society of America Special Paper 347, p. 89–98.

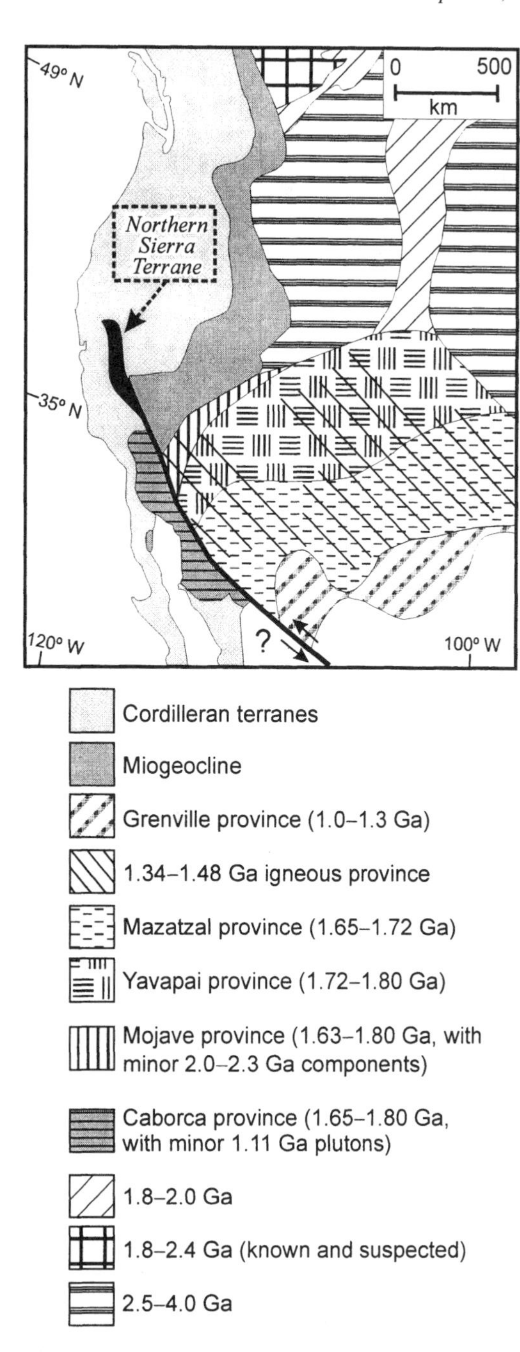


Figure 1. Schematic map showing location of northern Sierra terrane in relation to Cordilleran miogeocline and first-order basement provinces of western North America. Basement provinces are simplified from Hoffman (1989) for cratonal interior, various sections of Van Schmus et al. (1993) for southwestern United States, and Stewart et al. (1990) for northwestern Mexico.

volcanic episode, in Early Permian time, upper Paleozoic eugeoclinal rocks of the Golconda allochthon were thrust eastward over the Roberts Mountains allochthon and its overlap assemblage during the Late Permian to Early Triassic Sonoma orogeny (Gabrielse et al., 1983; Burchfiel et al., 1992).

Similarities in timing between Paleozoic arc volcanism in the northern Sierra Nevada and periods of thrusting in central Nevada have prompted numerous models relating emplacement of the Roberts Mountains and Golconda allochthons to various forms of arc-continent convergence. One model views Sierran volcanic rocks of mid-Paleozoic to early Mesozoic age as part of a long-lived, west-facing, continent-fringing arc system that formed, migrated, and reformed west of the continental margin of North America (Burchfiel and Davis, 1972, 1975; Miller et al., 1984; Miller, 1987). In this model, the Antler and Sonoma orogenies are generally viewed as backarc thrusting episodes, implying that stratigraphic ties should exist between Sierran volcanic-arc rocks and strata of the Roberts Mountains and Golconda allochthons.

A second set of models views the Paleozoic volcanic-arc rocks in the northern Sierra Nevada as part of a composite, potentially far traveled microplate that collided with the passive, western margin of North America during the Sonoma orogeny, emplacing the Golconda allochthon over the Roberts Mountains allochthon as an accretionary wedge (Dickinson, 1977; Speed, 1979; Speed and Sleep, 1982). If the northern Sierra terrane is far traveled, this model precludes any stratigraphic and paleogeographic ties between pre-Late Permian rocks in the Sierran arc terrane and allochthonous rocks in central Nevada.

A third set of models, proposed in several variations by Schweickert and Snyder (1981), Burchfiel and Royden (1981), and Burchfiel et al. (1992), involves arc-polarity reversals and repeated collisions of a fringing arc where the Roberts Mountains and Golconda allochthons were emplaced as accretionary wedges above west-dipping subduction zones. These models, like the first, predict close paleogeographic ties between the northern Sierra terrane and the Roberts Mountains allochthon.

The main objective of this study is to constrain the paleogeography and displacement history of the northern Sierra terrane using U-Pb ages of single detrital zircons obtained from clastic strata. In particular, we attempt to test the various models outlined here by comparing the ages of detrital zircons in the northern Sierra terrane with the ages of potential source rocks in nearby regions.

## STRATIGRAPHY OF THE NORTHERN SIERRA TERRANE

The northern Sierra terrane consists of arc-type rocks in the northeastern part of the Sierra Nevada that predate Late Jurassic and Cretaceous plutons of the Sierra Nevada batholith (Harwood, 1988). The stratigraphy of the northern Sierra terrane, as described in detail by Harwood (1992) and summarized in Figures 3 and 4, is as follows.

The Shoo Fly Complex, which contains the oldest rocks in the northern Sierra terrane, forms the basement to middle and upper Paleozoic volcanic and basinal rocks (Schweickert, 1981; Schweickert et al., 1984; Harwood, 1992; Girty et al., 1990, 1996). In most regions, the Shoo Fly Complex is unconformably overlain by Upper Devonian–Lower Mississippian volcanic and volcaniclastic rocks that form the lowest units of a submarine volcanic-arc sequence (D'Allura et al., 1977; Durrell and D'Allura, 1977; Hanson and Schweickert, 1986; Harwood, 1992; Hanson

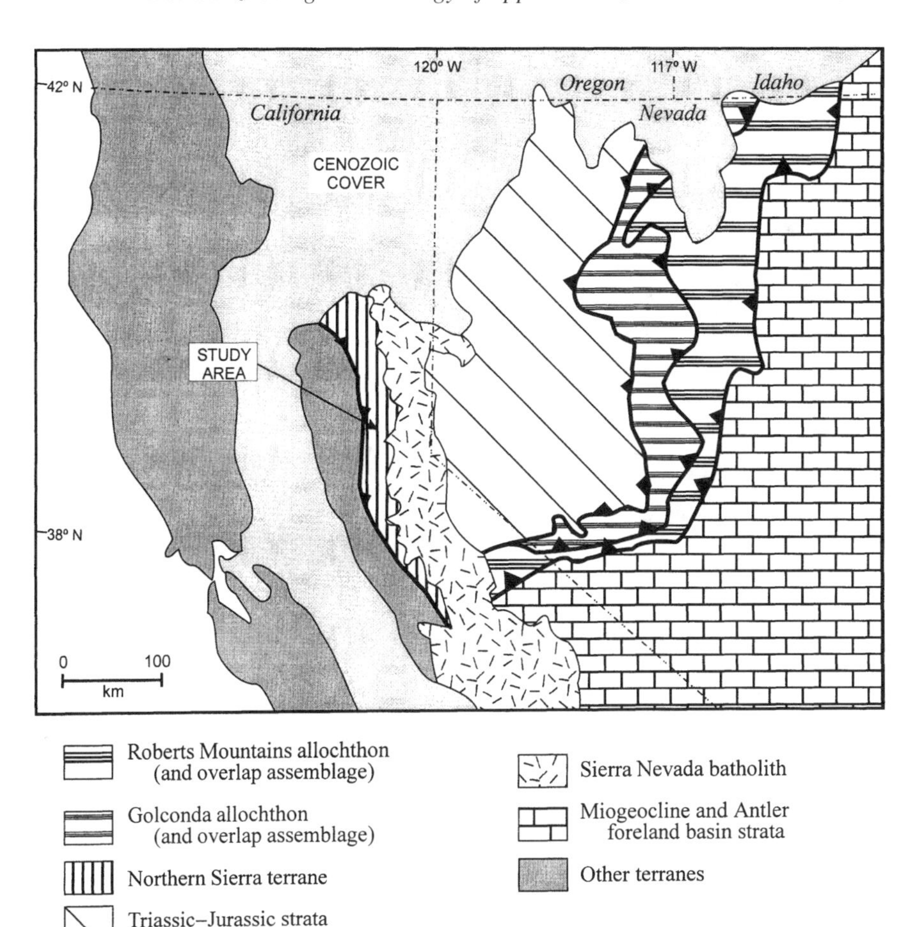


Figure 2. Schematic map of the main terranes and/or assemblages and their bounding faults in western Nevada and northern California. Map configuration of terranes and assemblages is adapted primarily from Oldow (1984), Silberling (1991), Silberling et al. (1987). Note that widespread Cretaceous and younger rocks and structures are ignored in the map in effort to emphasize configuration of pre-Cretaceous features.

et al., 1988, 1996). These rocks are gradationally overlain by mid-Mississippian to Middle Pennsylvanian cherts that mark a significant hiatus in arc volcanism (Schweickert et al., 1984; Miller and Harwood, 1990).

A regional unconformity separates the Devonian to Pennsylvanian rocks from an overlying Permian volcanic sequence consisting of a western, central, and eastern facies. In some regions, Upper Triassic conglomerate and sandstone unconformably overlie the Permian volcanic rocks and are overlain by Lower Jurassic volcaniclastic rocks. This Mesozoic cover represents a transgressive sequence that unconformably onlaps and progressively truncates older units toward the south, eventually resting directly on the Shoo Fly Complex (Schweickert, 1981; Schweickert et al., 1984; Harwood, 1983, 1992; Harwood and Murchey, 1990).

## DETRITAL ZIRCON GEOCHRONOLOGY

Detrital zircons have been analyzed from four different stratigraphic sequences, including the Upper Devonian to Mississippian Picayune Valley Formation, the Permian Arlington Formation, and two units from the lower Mesozoic onlap sequence. Zircons were extracted by traditional mineral separation procedures, as described by Gehrels (this volume, Introduction). Individual zircons from each sample were then separated into different populations on the basis of size, color, and shape. Representatives from each population were selected and abraded to two-thirds of their original diameter. The individual zircons were then analyzed utilizing isotope dilution and thermal ionization mass spectrometry (analytical techniques described by Gehrels, this volume, Introduction).

## RESULTS

### *Upper Devonian to Mississippian Picayune Valley Formation*

The Picayune Valley Formation is the oldest unit in the southeastern part of the northern Sierra terrane. It consists mainly of interbedded quartzite and black pelite that contain numerous lenses of chert-rich conglomerate. The western and stratigraphically lowest part of the formation contains fine-grained felsic tuff, volcaniclastic rocks, and andesitic tuff breccia that are interbedded with quartzite, pelite, and conglomerate. These felsic and andesitic rocks are lithologically identical to volcanic rocks (the Sierra Buttes and Taylor Formations) that elsewhere form the lowest parts of the sequence overlying the Shoo Fly Complex (Harwood, 1992). The upper part of the Picayune Valley Formation lacks volcanic rocks and contains numerous lenses of eastward-coarsening chert-rich conglomerate in a matrix of black slate. Grain size, sedimentary structures, bedding style, and the vertical

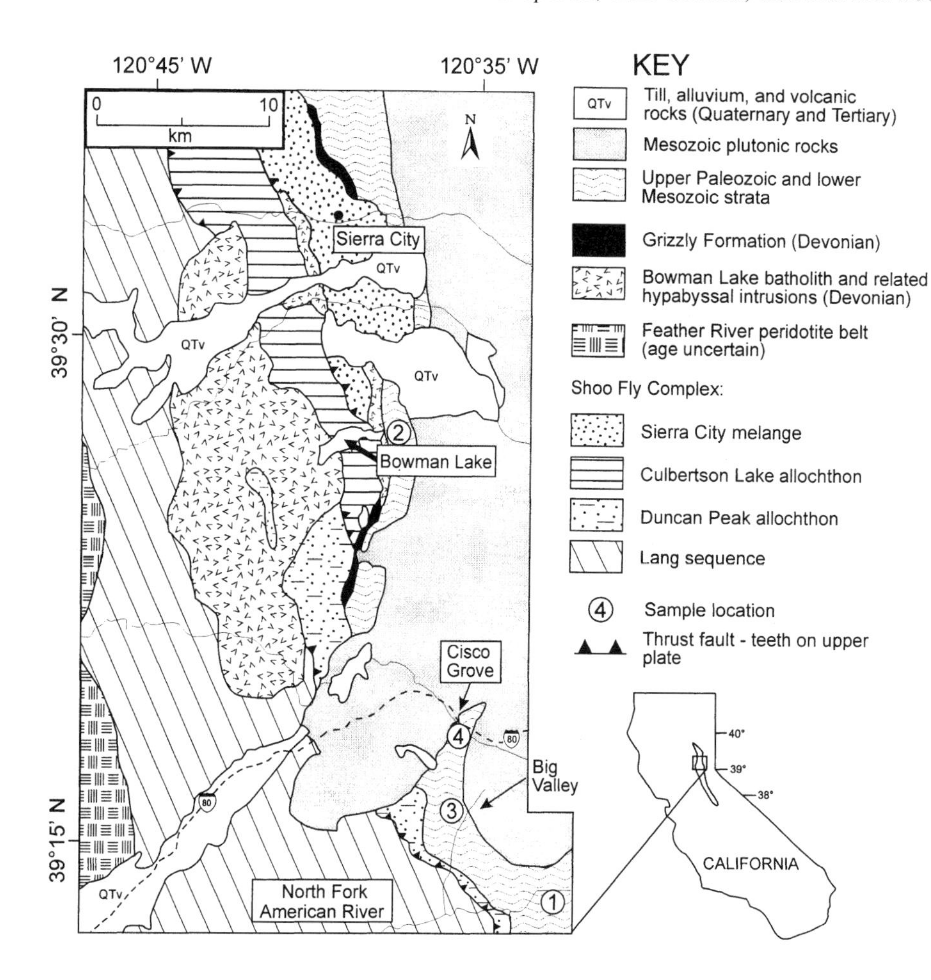


Figure 3. Geologic sketch map of central part of northern Sierra terrane, showing locations of detrital zircon samples, (adapted from Harwood, 1988). Sample 1 = Picayune Valley Formation, sample 2 = Arlington Formation, sample 3 = unnamed Upper Triassic strata, sample 4 = Sailor Canyon Formation.

arrangement of lithofacies in the Picayune Valley Formation show many of the features associated with channelized deep-sea fan environments (Harwood, 1992). The Picayune Valley Formation is conformably overlain by metachert (Serena Creek Formation) that is part of a widespread Mississippian-Pennsylvanian chert unit (Peale Formation) (Harwood and Murchey, 1990).

Our sample was collected from a glaciated outcrop near the unpaved road that crosses the North Fork of the American River (Fig. 3; lat 39°14′37.7″N, long 120°22′ 51.5″W). The sandstone is coarse grained and is interlayered with pebble to cobble conglomerate and finer grained sandstone. It contains large detrital quartz grains as well as abundant argillite rip-up clasts (Dickinson and Gehrels, this volume, Chapter 11). The age of the unit is constrained by correlation of the interbedded volcanic rocks with the Upper Devonian Sierra Buttes Formation.

Detrital zircon grains from the Picayune Valley Formation that were larger than 175 μm were divided into three color groups, consisting of dark pink, light pink, and colorless grains. Each of these three groups was then subdivided into two subgroups of well-rounded and subeuhedral grains. We abraded 21 grains from these 6 groups to 75% of their original size and analyzed them. Four additional grains were selected from the 100–125 μm range to determine whether grains of other ages were present in the smaller size fractions. This selection contained euhedral grains that were colorless, medium pink, and light pink.

Detrital zircons from the Picayune Valley Formation yield mainly concordant to moderately discordant results (Fig. 5). Apparent ages for these grains are Proterozoic and Late Archean, with apparent clusters at 1782–2052 (n = 20) and 2611–2642 Ma (n = 3), and two single grains with ages of ca. 2314 and ca. 2756 Ga (Table 1). The larger and rounded and smaller and euhedral grains in the sample yield similar ages.

### *Lower Permian Arlington Formation*

The Arlington Formation is one of three major facies of the Permian volcanic sequence deposited in the northwestern part of the northern Sierra terrane. These rocks were interpreted by Harwood (1992) to represent a western, predominantly volcaniclastic facies that is the lateral equivalent of the Goodhue, Robinson, and Reeve Formations. It consists of volcaniclastic sandstone, slate, conglomerate, and breccia. The lower part of the formation consists of chert-rich pebbly mudstone and conglomerate interbedded with volcaniclastic sandstone and slate. Blocks of chert, volcanic rocks, and quartzite, some as large as 50 m, occur within the conglomerate-rich lower part of the formation (Harwood, 1992). The conglomerate grades upward into volcaniclastic sand-

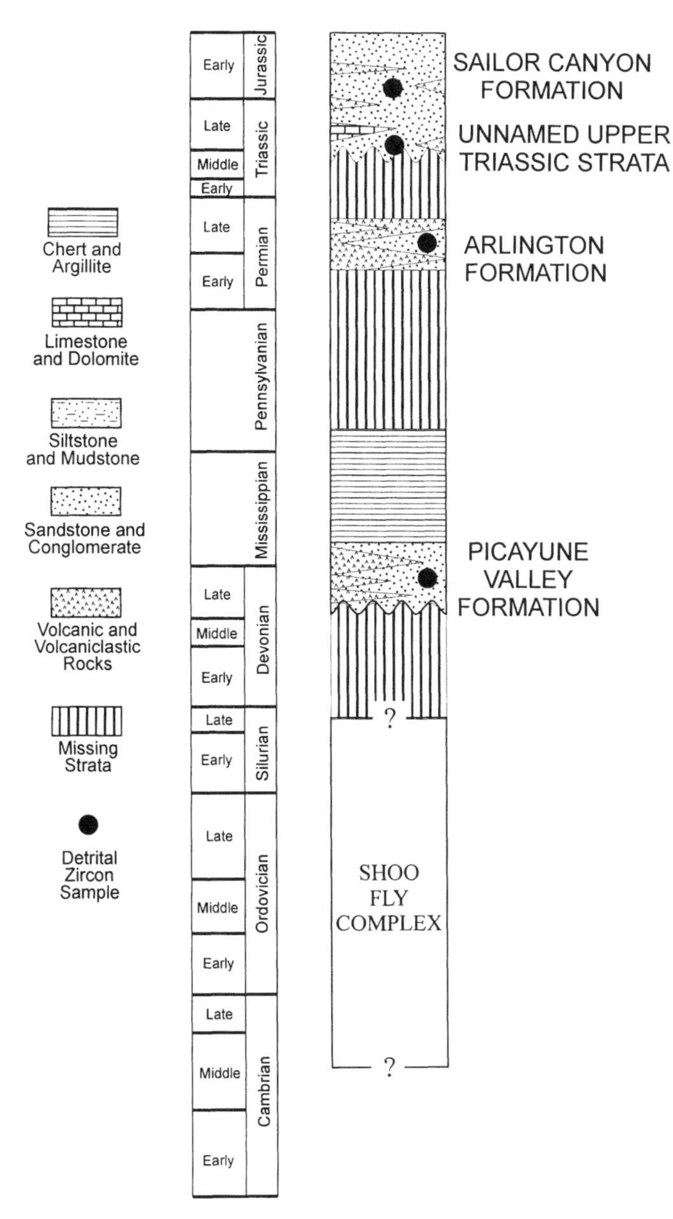


Figure 4. Schematic stratigraphic column of southeast part of northern Sierra terrane (compiled mainly from Harwood, 1992; Miller and Harwood, 1990).

stone and slate that contain only scattered lenses of chert-granule conglomerate. All of the Permian rocks are separated from the older strata by a pronounced Late Pennsylvanian–Early Permian regional unconformity. This break represents a period of uplift probably associated with crustal heating and extensional faulting prior to the initiation of late Early Permian volcanism (Schweickert, 1981; Schweickert et al., 1984; Harwood, 1983, 1992).

Detrital zircons were extracted from a coarse-grained sandstone consisting primarily of chert-argillite detritus with some volcanic rock fragments and feldspar grains (Dickinson and Gehrels, this volume, Chapter 11). Our sample was collected from Fir Hill (lat 39°28′16.6″N, long 120°36′2.7″W) near Bowman Lake

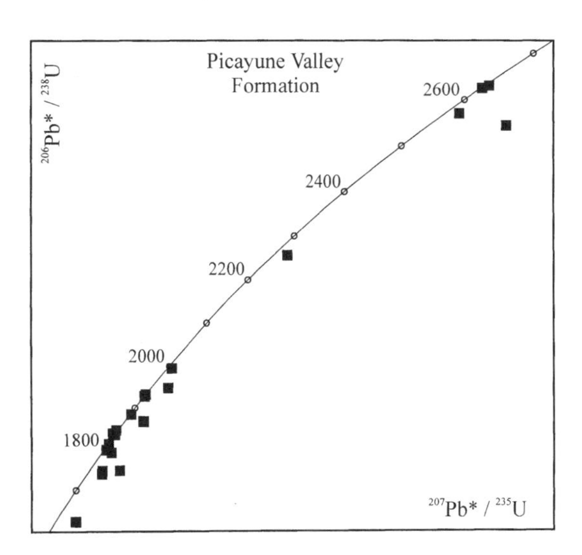


Figure 5. U-Pb concordia diagram showing single-grain detrital zircon analyses from the Picayune Valley Formation. All data reduction and concordia plots are from programs of Ludwig (1991a, 1991b).

(Fig. 3), from a section that was previously mapped as the Reeve Formation (Harwood, 1992), but which is herein interpreted to be more closely related to the Arlington Formation based on more detailed stratigraphic studies. This part of the Arlington unconformably overlies the Lower Mississippian to Pennsylvanian Peale Formation near the sample locality. Megafossils and conodonts date the lower part of the Arlington Formation as late Early Permian (late Wolfcampian) (Harwood, 1992).

Zircons from the sample range to 250 µm in sieve diameter, but most are in the 125–175 µm size range. The grains were divided into two groups according to color. These groups include a dominant population of colorless grains and a small population of dark pink grains. Most grains are translucent and contain few visible inclusions. Grains of the colorless group are primarily angular, but some are euhedral bipyramidal prisms, and there is a subordinate number of subrounded grains. Grains of the dark pink group are both rounded and euhedral.

As shown in Figure 6, all of the grains yielded concordant to slightly discordant ages. An apparent cluster of ages ranges from 344 to 362 Ma (n = 8). The dark pink grains and one subrounded colorless grain, however, are significantly older and yield apparent ages of ca. 1742, 2088, and 2597 Ma.

### *Upper Triassic–Middle Jurassic Strata*

A major unconformity separates the Permian arc sequence from overlying Middle Triassic through Middle Jurassic strata (Schweickert, 1981; Schweickert et al., 1984; Harwood, 1992; Hanson et al., 1996). The contact is a pronounced angular unconformity that truncates progressively older units toward the south, presumably reflecting a southward increase in intensity of deformation, uplift, and erosion related to the Sonoma orogeny (Harwood, 1983, 1992; Davis, 1989). The overlying Upper Triassic and Jurassic strata form a southward-transgressive onlap

**TABLE 1. U-Pb ISOTOPIC DATA AND AGES**

| Grain type | Grain wt. (μg) | $Pb_c$ (pg) | U (ppm) | $^{206}Pb_m/^{204}Pb$ | $^{206}Pb_c/^{208}Pb$ | Apparent ages (Ma) $^{206}Pb^*/^{238}U$ | $^{207}Pb^*/^{235}U$ | $^{207}Pb^*/^{206}Pb^*$ | Interpreted age (Ma) |
|---|---|---|---|---|---|---|---|---|---|
| Picayune Valley Formation (Upper Devonian-Mississippian) | | | | | | | | | |
| DEu | 8 | 27 | 263 | 1350 | 12.9 | 1646 ± 9 | 1704 ± 11 | 1777 ± 5 | 1782 ± 20 |
| DS | 29 | 12 | 490 | 19500 | 4.2 | 1644 ± 6 | 1712 ± 7 | 1798 ± 2 | 1803 ± 20 |
| LS | 23 | 8 | 318 | 17700 | 19.0 | 1809 ± 7 | 1811 ± 8 | 1815 ± 4 | 1815 ± 10 |
| **LS** | **26** | **9** | **189** | **10900** | **4.6** | **1813 ± 9** | **1815 ± 10** | **1817 ± 4** | **1818 ± 10** |
| LS | 48 | 9 | 171 | 17500 | 7.3 | 1812 ± 9 | 1816 ± 10 | 1820 ± 4 | 1821 ± 10 |
| *LEu* | *7* | *19* | *110* | *592* | *2.7* | *1385 ± 13* | *1564 ± 18* | *1814 ± 11* | *1834 ± 20* |
| DS | 55 | 19 | 733 | 40400 | 3.3 | 1810 ± 12 | 1821 ± 12 | 1834 ± 2 | 1835 ± 10 |
| **LS** | **34** | **13** | **380** | **19800** | **5.9** | **1836 ± 8** | **1836 ± 9** | **1836 ± 4** | **1836 ± 10** |
| LS | 26 | 15 | 334 | 10500 | 4.6 | 1747 ± 9 | 1793 ± 10 | 1847 ± 4 | 1851 ± 10 |
| LS | 24 | 10 | 336 | 14700 | 4.9 | 1774 ± 8 | 1811 ± 9 | 1853 ± 4 | 1856 ± 10 |
| LS | 21 | 145 | 363 | 1000 | 5.4 | 1801 ± 8 | 1829 ± 12 | 1862 ± 7 | 1864 ± 10 |
| CEu | 8 | 31 | 49 | 250 | 2.4 | 1809 ± 59 | 1848 ± 67 | 1891 ± 27 | 1894 ± 20 |
| CS | 24 | 32 | 63 | 956 | 3.5 | 1884 ± 9 | 1889 ± 12 | 1894 ± 7 | 1894 ± 10 |
| LS | 60 | 6 | 131 | 30700 | 2.5 | 1928 ± 10 | 1929 ± 11 | 1930 ± 4 | **1930 ± 10** |
| LS | 26 | 5 | 130 | 13950 | 4.3 | 1932 ± 8 | 1933 ± 9 | 1934 ± 4 | **1934 ± 10** |
| LEu | 6 | 32 | 133 | 475 | 5.1 | 1760 ± 54 | 1855 ± 58 | 1963 ± 10 | 1970 ± 20 |
| DS | 30 | 14 | 579 | 24500 | 8.4 | 1791 ± 8 | 1876 ± 9 | 1971 ± 4 | 1977 ± 20 |
| DS | 28 | 10 | 578 | 32000 | 8.9 | 1899 ± 8 | 1943 ± 8 | 1990 ± 2 | 1992 ± 10 |
| CS | 25 | 7 | 111 | 9500 | 6.2 | 1994 ± 8 | 2007 ± 9 | 2020 ± 4 | 2020 ± 10 |
| DS | 32 | 10 | 450 | 30800 | 11.5 | 1948 ± 9 | 1997 ± 9 | 2049 ± 2 | 2052 ± 10 |
| CS | 27 | 14 | 122 | 5700 | 5.5 | 2256 ± 8 | 2286 ± 9 | 2313 ± 2 | 2314 ± 10 |
| LS | 19 | 84 | 311 | 1910 | 3.6 | 2583 ± 12 | 2598 ± 12 | 2610 ± 3 | 2611 ± 10 |
| CS | 13 | 7 | 66 | 3950 | 7.7 | 2625 ± 17 | 2627 ± 18 | 2628 ± 4 | **2628 ± 10** |
| LS | 33 | 10 | 242 | 23950 | 7.6 | 2630 ± 11 | 2637 ± 11 | 2642 ± 2 | 2642 ± 10 |
| CS | 40 | 12 | 82 | 7800 | 3.9 | 2544 ± 16 | 2662 ± 22 | 2752 ± 8 | 2756 ± 20 |
| Arlington Formation (Lower Permian) | | | | | | | | | |
| CS | 23 | 19 | 356 | 1560 | 4.2 | 344 ± 4 | 340 ± 9 | 310 ± 15 | 344 ± 10 |
| CS | 3 | 7 | 442 | 710 | 5.0 | 346 ± 9 | 351 ± 13 | 381 ± 26 | 346 ± 10 |
| CE | 12 | 8 | 271 | 2390 | 5.8 | 353 ± 3 | 352 ± 4 | 346 ± 19 | 353 ± 10 |
| CE | 12 | 17 | 300 | 795 | 5.2 | 356 ± 3 | 357 ± 6 | 369 ± 29 | 356 ± 10 |
| CE | 10 | 6 | 938 | 5580 | 5.7 | 357 ± 3 | 356 ± 3 | 350 ± 13 | **357 ± 10** |
| CE | 19 | 10 | 287 | 2090 | 6.0 | 357 ± 4 | 356 ± 6 | 351 ± 20 | **357 ± 10** |
| CS | 12 | 9 | 736 | 3580 | 4.0 | 360 ± 3 | 362 ± 5 | 375 ± 20 | 360 ± 10 |
| CE | 11 | 21 | 200 | 409 | 3.8 | 362 ± 6 | 364 ± 10 | 379 ± 44 | 362 ± 10 |
| DS | 21 | 7 | 683 | 40030 | 6.3 | 1724 ± 12 | 1732 ± 14 | 1741 ± 6 | 1742 ± 10 |
| CS | 4 | 10 | 61 | 615 | 6.2 | 2043 ± 37 | 2065 ± 44 | 2087 ± 20 | 2088 ± 20 |
| DS | 13 | 9 | 477 | 20700 | 9.8 | 2508 ± 15 | 2557 ± 18 | 2596 ± 6 | 2597 ± 10 |

sequence of sandstone, conglomerate, and limestone that eventually directly overlies the Shoo Fly Complex in the southern part of the terrane (Harwood, 1983, 1992).

***Unnamed Upper Triassic strata.*** Detrital zircons were extracted from an Upper Triassic sandstone exposed along the north side of Big Valley near the North Fork of the American River (Fig. 3; lat 39°16′13″N, long. 120°33′13″W). The sandstone collected is coarse grained and rich in volcanic detritus (Dickinson and Gehrels, this volume, Chapter 11). Low-grade metamorphism is indicated by the presence of neomorphic mica in the sample. Clasts in Triassic chert-pebble conglomerate that directly overlies the sandstone unit generally were derived from underlying Paleozoic rocks, consisting of the Permian Reeve Formation at the sample site (Harwood, 1992). The conglomerate grades laterally into a limestone unit that has yielded Late Triassic conodonts (Harwood, 1992).

Detrital zircons analyzed from the sample were all >100 μm. The grains were divided into colorless versus honey colored grains. Each group was then subdivided into rounded and subhedral grains. All of the grains are translucent and contain very few inclusions. The nine grains analyzed yielded slightly to highly discordant ages (Fig. 7). Analyses of these grains are of fairly low precision due to the low $^{206}Pb/^{204}Pb$ measured in each grain. All of the interpreted ages are within a cluster from 254 to 270 Ma (n = 9) (Table 1).

***Sailor Canyon Formation (Lower and Middle Jurassic).*** Detrital zircons were extracted from a medium to coarse-grained, chert-rich sandstone that was collected from near Cisco Grove along Highway 80 west of Reno (Fig. 3; lat 39° 18′51.6″N, long 120°32′41.3″W). Zircons from the sample were divided into colorless versus light pink grains. Colorless grains were further subdivided into euhedral and rounded populations. All grains of the light pink population were rod shaped and euhedral. All of the

**TABLE 1. U-Pb ISOTOPIC DATA AND AGES (continued)**

| Grain type | Grain wt. (µg) | $Pb_c$ (pg) | U (ppm) | $^{206}Pb_m/^{204}Pb$ | $^{206}Pb_c/^{208}Pb$ | Apparent ages (Ma) $^{206}Pb^*/^{238}U$ | $^{207}Pb^*/^{235}U$ | $^{207}Pb^*/^{206}Pb^*$ | Interpreted age (Ma) |
|---|---|---|---|---|---|---|---|---|---|
| Unnamed Upper Triassic strata | | | | | | | | | |
| HS | 17 | 65 | 507 | 355 | 1.6 | 254 ± 3 | 256 ± 7 | 271 ± 54 | 254 ± 10 |
| CE | 8 | 14 | 183 | 296 | 3.8 | 255 ± 7 | 241 ± 14 | 103 ± 120 | *255 ± 20* |
| CS | 18 | 16 | 129 | 392 | 3.2 | 255 ± 5 | 250 ± 9 | 204 ± 62 | 255 ± 10 |
| CE | 9 | 29 | 340 | 295 | 3.0 | 257 ± 6 | 246 ± 14 | 143 ± 120 | *257 ± 20* |
| CS | 7 | 20 | 249 | 243 | 2.4 | 257 ± 6 | 239 ± 16 | 66 ± 140 | *257 ± 20* |
| CE | 5 | 15 | 137 | 143 | 2.6 | 262 ± 13 | 258 ± 27 | 224 ± 210 | 262 ± 10 |
| CS | 5 | 29 | 376 | 193 | 1.5 | 265 ± 6 | 268 ± 20 | 290 ± 160 | 265 ± 10 |
| CE | 6 | 22 | 169 | 145 | 2.4 | 268 ± 10 | 299 ± 27 | 553 ± 180 | *268 ± 20* |
| CE | 8 | 14 | 310 | 505 | 1.6 | 270 ± 5 | 279 ± 11 | 358 ± 79 | *270 ± 20* |
| Sailor Canyon Formation (Lower Jurassic) | | | | | | | | | |
| CE | 8 | 2 | 91 | 570 | 7.7 | 185 ± 4 | 186 ± 5 | 197 ± 46 | 185 ± 10 |
| CE | 9 | 2 | 89 | 1100 | 8.4 | 198 ± 3 | 199 ± 4 | 205 ± 27 | **198 ± 10** |
| CE | 12 | 2 | 76 | 1180 | 8.6 | 199 ± 3 | 198 ± 4 | 185 ± 33 | **199 ± 10** |
| CS | 8 | 3 | 148 | 790 | 7.9 | 199 ± 4 | 198 ± 5 | 184 ± 35 | **199 ± 10** |
| CE | 8 | 1 | 122 | 1850 | 7.4 | 200 ± 2 | 200 ± 2 | 207 ± 14 | **200 ± 10** |
| CS | 10 | 3 | 315 | 2450 | 9.1 | 203 ± 2 | 202 ± 3 | 196 ± 17 | **203 ± 10** |
| CS | 10 | 5 | 182 | 725 | 8.5 | 203 ± 6 | 203 ± 7 | 196 ± 39 | **203 ± 10** |
| CE | 11 | 2 | 221 | 2050 | 8.1 | 203 ± 2 | 203 ± 3 | 192 ± 18 | **203 ± 10** |
| LEr | 7 | 2 | 491 | 3800 | 8.7 | 204 ± 2 | 205 ± 2 | 208 ± 20 | **204 ± 10** |
| CS | 10 | 2 | 165 | 1930 | 7.5 | 205 ± 2 | 203 ± 3 | 187 ± 21 | 205 ± 10 |
| CS | 11 | 4 | 111 | 620 | 6.2 | 205 ± 4 | 205 ± 6 | 198 ± 41 | **205 ± 10** |

*Note:*

Grain with $^{206}Pb^*/^{238}U$ age within 15% of $^{207}Pb^*/^{206}Pb^*$ age in plain type.

Grain with $^{206}Pb^*/^{238}U$ age not within 15% of $^{207}Pb^*/^{206}Pb^*$ age in italics.

Grain with concordant age and moderate to high precision analysis in bold.

* Radiogenic Pb.

Grain type: L—light pink, D—dark pink, C—colorless, H—honey-colored; S—moderate to high sphericity, E—euhedral, u—unabraded, r—rod-shaped.

Unless otherwise noted, all grains are clear and/or translucent, and abraded with an air abrasion device.

$^{206}Pb/^{204}Pb$ is measured ratio, uncorrected for blank, spike, fractionation, or initial Pb.

$^{206}Pb/^{208}Pb$ is corrected for blank, spike, fractionation, and initial Pb.

Most concentrations have an uncertainty of 25% due to uncertainty in weight of grain.

Constants used: $^{238}U/^{235}U = 137.88$. Decay constant for $^{235}U = 9.8485 \times 10^{-10}$. Decay constant for $^{238}U = 1.55125 \times 10^{-10}$.

All uncertainties are at the 95% confidence level.

Pb blank ranged from 2 to 10 pg. U blank was consistently <1 pg.

Interpreted ages for concordant grains are $^{206}Pb^*/^{238}U$ ages if <1.0 Ga and $^{207}Pb^*/^{206}Pb^*$ ages if >1.0 Ga.

Interpreted ages for discordant grains are projected from 100 Ma.

All analyses conducted using conventional isotope dilution and thermal ionization mass spectrometry, as described by Gehrels (this volume, Introduction).

grains are translucent and appear to be free of inclusions and fractures. The 11 grains analyzed yielded concordant to slightly discordant ages (Fig. 8). There is a cluster of interpreted ages at 198–205 Ma (n = 10) and a single grain ca. 185 Ma (Table 1).

## PROVENANCE OF DETRITAL ZIRCONS

Strata from the northern Sierra terrane yield a combination of 2.8–1.7 Ga grains that originated in a region containing Precambrian continental crust, and ca. 370–180 Ma grains that were probably derived from volcanic arc rocks within or near the northern Sierra terrane. Although the older grains are continental in origin, basement rocks of the southwestern United States are not a likely source because rocks in these regions are generally <1.8 Ga (Fig. 1). The difference in age signature of these two regions is clearly shown by a comparison of the detrital zircon ages in the Picayune Valley Formation and in Paleozoic strata of the Cordilleran miogeocline in Nevada (Fig. 9). This difference can be quantified with statistical measures of the degree to which the two curves overlap and the degree to which the two curves have similar proportions of ages (Gehrels, this volume, Chapter 1). As shown in Table 2, the low values for these two comparisons indicate that the southwestern United States is not a likely source for the detrital zircon grains in the Picayune Valley Formation.

More likely sources for the >1.8 Ga grains in the Picayune Valley and Arlington Formations are underlying rocks of the Shoo Fly Complex and lower Paleozoic strata in the Roberts Mountains allochthon (Fig. 2). Both of these assemblages contain detrital zircons of the appropriate ages (Fig. 9), and both are known to have been uplifted and exposed to erosion during Late Devonian–Mississippian time (e.g., Miller et al., 1992). A statistical comparison between age spectra for the Picayune Valley and

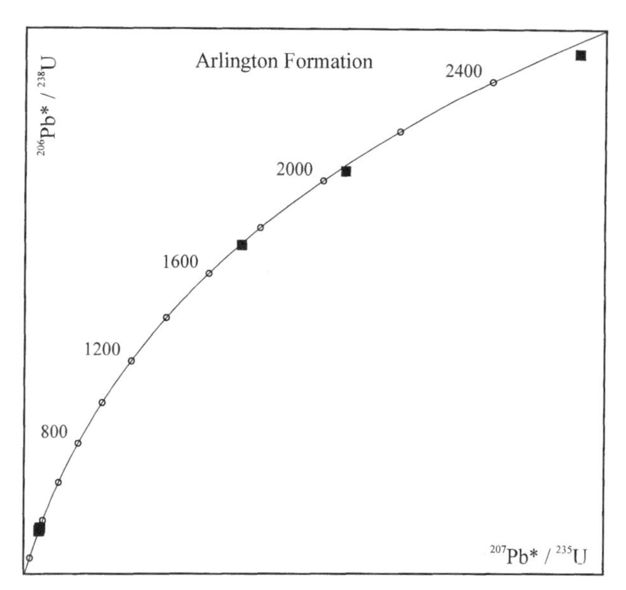


Figure 6. U-Pb concordia diagram showing single-grain detrital zircon analyses from Upper Devonian–Mississippian Arlington Formation.

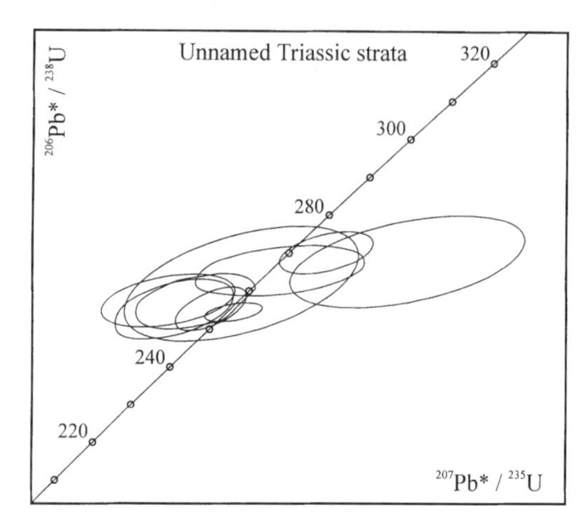


Figure 7. U-Pb concordia diagram showing single-grain detrital zircon analyses from unnamed Upper Triassic strata near Big Valley along North Fork of American River. Error ellipses are at 95% confidence level.

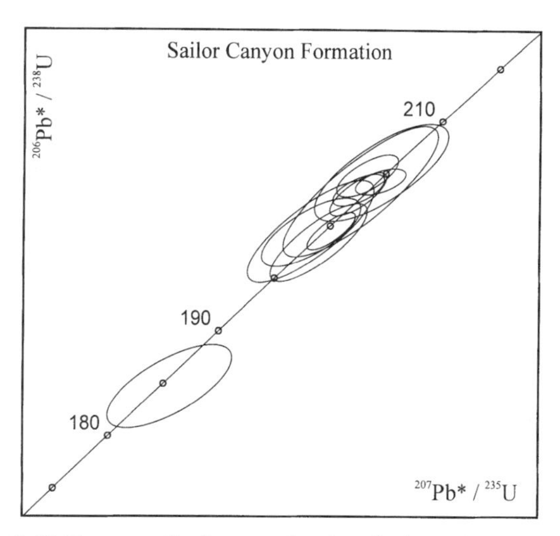


Figure 8. U-Pb concordia diagram showing single-grain detrital zircon analyses from Lower Jurassic strata of Sailor Canyon Formation near Cisco Grove. Error ellipses are at 95% confidence level.

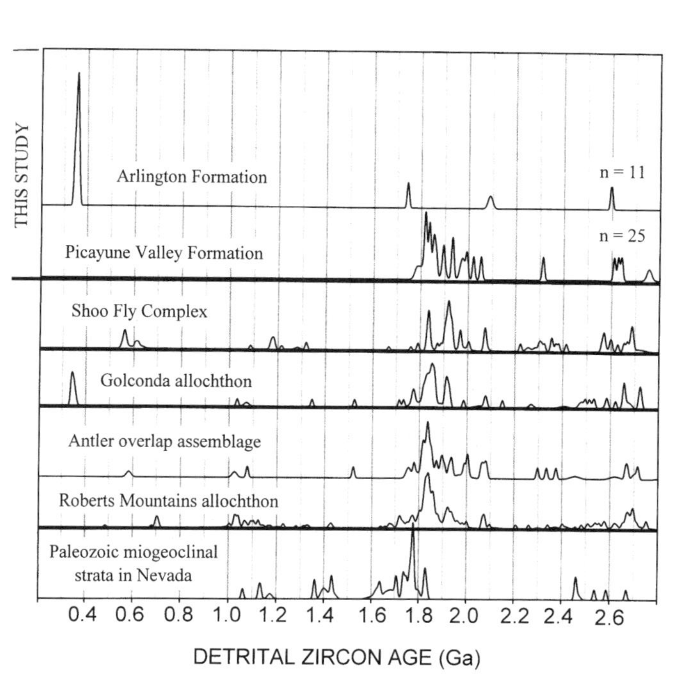


Figure 9. Relative age-probability plots of single-grain detrital zircon ages for strata of northern Sierra terrane. Ages for all grains plotted are either concordant or, if discordant, projected from 100 Ma. Also shown for comparison purposes are composite age spectra for Paleozoic miogeoclinal strata in Nevada (Gehrels et al., 1995; Gehrels, this volume), lower Paleozoic strata of Roberts Mountains allochthon (Gehrels et al. this volume, Chapter 1) and overlying upper Paleozoic strata (Gehrels and Dickinson, this volume), upper Paleozoic strata of Golconda allochton (Riley et al., this volume), and lower Paleozoic rocks of Shoo Fly Complex in northern Sierra terrane (Harding et al., this volume).

both the Shoo Fly Complex and Roberts Mountains allochthon reveals high degrees of overlap and similarity for both units, with slightly higher values for the Roberts Mountains allochthon (Table 2). The high values for both units support Schweickert and Snyder's (1981) conclusion that the Shoo Fly Complex formed in proximity to the Roberts Mountains allochthon, and the slightly higher values for the Roberts Mountains allochthon are consistent with stratigraphic relations reported by Harwood and Murchey (1990) and Harwood (1992) that suggest that Picayune Valley detritus was shed from an eastern source, most likely the Roberts Mountains allochthon.

These potential provenance links with the Shoo Fly Complex and the Roberts Mountains allochthon are important paleogeographically because the Roberts Mountains allochthon is known to have been emplaced onto the continental margin during latest Devonian–Early Mississippian time. Provenance ties between the northern Sierra terrane and the Roberts Mountains allochthon are therefore supportive of models that depict the northern Sierra terrane as part of a fringing arc along the Cordilleran margin, but are

**TABLE 2. STATISTICAL COMPARISONS OF DETRITAL ZIRCON AGE SPECTRA**

| | Overlap | Similarity |
|---|---|---|
| Roberts Mountains allochthon | 0.99 | 0.64 |
| Shoo Fly Complex | 0.97 | 0.55 |
| Nevada miogeocline | 0.35 | 0.24 |

*Note:* Degree of overlap is a measure of the degree to which ages in the Picayune Valley sample overlap with ages in the three references. Degree of similarity is a measure of the proportions of different ages in the Picayune Valley and the three reference spectra.

not consistent with arrival of an exotic Sierran arc during the Permian-Triassic Sonoma orogeny.

The zircon grains in our samples that are <370 Ma become progressively younger with depositional age of the host strata. As shown in Figure 10, the three younger units in our study contain zircons that are only slightly older than the depositional age of each unit. In addition, the ages of grains in these samples are an excellent reflection of the three main ages of volcanic rocks in the northern Sierra terrane. Hence, we conclude that the <370 Ma grains in the Arlington Formation, unnamed Triassic strata, and Sailor Canyon Formation were most likely shed from igneous rocks exposed within or near the northern Sierra terrane.

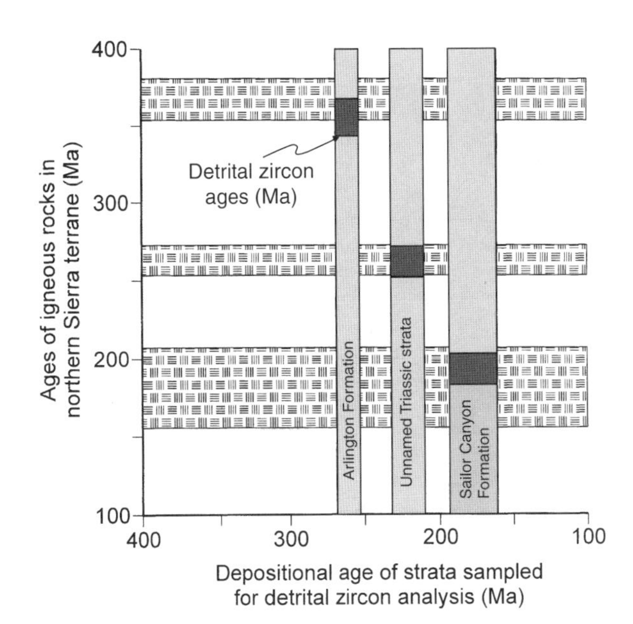


Figure 10. Comparison of ages of detrital zircons in Arlington Formation, unnamed Upper Triassic strata, and Sailor Canyon Formation with depositional ages of host strata and ages of volcanic rocks in northern Sierra terrane. Excellent match between detrital zircon ages and ages of volcanism in northern Sierra terrane indicates that grains in these units were derived locally.

## CONCLUSIONS

We have analyzed 56 detrital zircon grains from 4 different stratigraphic units within the northern Sierra terrane. The Upper Devonian–Mississippian Picayune Valley Formation yields ages of about 1782–2052, 2314, 2611–2642, and 2756 Ma. On the basis of a variety of stratigraphic and biostratigraphic relations (Harwood and Murchey, 1990), and the similarity in detrital zircon ages reported here (Fig. 9), we infer that these grains were shed from the Roberts Mountains allochthon and/or the Shoo Fly Complex during or soon after emplacement onto the Cordilleran margin. The similarity in detrital zircon ages of the Picayune Valley Formation, Roberts Mountains allochthon, and Shoo Fly Complex supports previously proposed ties between the northern Sierra terrane and the Nevada continental margin during Late Devonian–Carboniferous time.

The Lower Permian Arlington Formation yields a few grains ca. 1742, 2088, and 2597 Ma, but most grains have ages of 344–362 Ma. The younger ages match well with ages of mid-Paleozoic volcanism in the northern Sierra terrane, suggesting that the grains were locally derived.

An unnamed Upper Triassic sandstone from Big Valley yields zircons with ages clustering from 254 to 270 Ma. These grains were probably derived from the underlying Permian volcanic rocks. Uplift and erosion of the Permian source rocks may have been related to the Sonoma orogeny, as suggested by many previous workers.

A Lower Jurassic sandstone from the Sailor Canyon Formation, collected near Cisco Grove, contains detrital zircons with ages from 198 to 205 Ma, and perhaps as young as 185 Ma. This age range establishes proximity to an active Early Jurassic arc. Because volcanic rocks of these ages are common in the northern Sierra terrane, this set of grains was probably also derived locally.

## ACKNOWLEDGMENTS

We thank Gary Girty and Rich Schweickert for assistance in the field, and Rich Schweickert and George Dunne for their thorough and informative reviews. This research was supported by National Science Foundation grant EAR-9416933.

## REFERENCES CITED

Burchfiel, B.C., and Davis, G.A., 1972, Structural framework and evolution of the southern part of the Cordilleran orogen, western United States: American Journal of Science, v. 272, p. 97–118.

Burchfiel, B.C., and Davis, G.A., 1975, Nature and controls of Cordilleran orogenesis, western United States: Extensions of an earlier synthesis: American Journal of Science v. 275A, p. 363–396.

Burchfiel, B.C., and Royden, L.H., 1991, Antler orogeny: A Mediterranean-type orogeny: Geology, v. 19, p. 66–69.

Burchfiel, B.C., Cowan, D.S., and Davis, G.A., 1992, Tectonic overview of the Cordilleran orogen in the western United States, *in* Burchfiel, B.C., et al., eds., The Cordilleran orogen: Conterminous U.S.: Boulder, Colorado, Geological Society of America, Geology of North America, v. G-3, p. 407 480.

D'Allura, J.A., Moores, E.M., and Robinson, L., 1977, Paleozoic rocks of the northern Sierra Nevada: Their structural and paleogeographic implications, *in* Stewart, J.H., et al., eds., Paleozoic paleogeography of the western United States: Pacific Section, Society of Economic Paleontologists and Mineralogists, Pacific Coast Paleogeography Symposium 1, p. 395–408.

Davis, D.A., 1989, The Paleozoic-Mesozoic unconformity of the northern Sierra Nevada (NSN), California, and its significance: Geological Society of

America Abstracts with Programs, v. 21, no. 5, p. 71–72.

Dickinson, W.R., 1977, Paleozoic plate tectonics and the evolution of the Cordilleran continental margin, *in* Stewart, J.H., et al., eds., Paleozoic paleogeography of the western United States: Pacific Section, Society of Economic Paleontologists and Mineralogists, Pacific Coast Paleogeography Symposium 1, p. 137–155.

Dickinson, W.R., Harbaugh, D.W., Saller, A.H., Heller, P.L., and Snyder, W.S., 1983, Detrital modes of upper Paleozoic sandstones derived from Antler orogen in Nevada: Implications for the nature of the Antler orogeny: American Journal of Science, v. 282, p. 481–509.

Durrell, C., and D'Allura, J., 1977, Upper Paleozoic section in eastern Plumas and Sierra counties, northern Sierra Nevada, California: Geological Society of America Bulletin, v. 88, p. 844–852.

Gabrielse, H., Snyder, W.S., and Stewart, J.H., 1983, Sonoma orogeny and Permian to Triassic tectonism in western North America (Penrose Conference Report): Geology, v. 11, p. 484–486.

Gehrels, G.E., Dickinson, W.R., Ross, G.M., Stewart, J.H., and Howell, D.G., 1995, Detrital zircon reference for Cambrian to Triassic miogeoclinal strata of western North America: Geology, v. 23, p. 831–834.

Girty, G.H., Gester, K.C., and Turner, J.B., 1990, Pre-Late Devonian geochemical, stratigraphic, sedimentologic, and structural patterns, Shoo Fly Complex, northern Sierra Nevada, California, *in* Harwood, D.S., and Miller, M.M., eds., Paleozoic and early Mesozoic paleogeographic relations; Sierra Nevada, Klamath Mountains, and related terranes: Geological Society of America Special Paper 255, p. 43–56.

Girty, G.H., Lawrence, J., Burke, T., Fortin, A., Gallarano, C.S., Wirths, T.A., Lewis, J.G., Peterson, M.M., Ridge, D.L., Knaack, C., and Johnson, D., 1996, The Shoo Fly Complex: Its origin and tectonic significance, *in* Girty, G.H., et al., eds., The northern Sierra terrane and associated magmatic units: Implications for the tectonic history of the western Cordillera: Pacific Section, Society of Economic Paleontologists and Mineralogists book 81, p. 1–24.

Hanson, R.E., and Schweickert, R.A., 1986, Stratigraphy of mid-Paleozoic island-arc rocks in part of the northern Sierra Nevada, Sierra and Nevada counties, California: Geological Society of America Bulletin, v. 97, p. 986–998.

Hanson, R.E., Saleeby, J.B., and Schweickert, R.A., 1988, Composite Devonian island-arc batholith in the northern Sierra Nevada, California: Geological Society of America Bulletin, v. 100, p. 446–457.

Hanson, R.E., Girty, G.H., Girty, M.S., Hargrove, U.S., Harwood, D.S., Kulow, M.J., Mielke, K.L., Phillipson, S.E., Schweickert, R.A., and Templeton, J.H., 1996, Paleozoic and Mesozoic arc rocks in the northern Sierra terrane, *in* Girty, G. et al., eds., The northern Sierra terrane and associated magmatic units; implications for the tectonic history of the western Cordillera: Pacific Section, Society of Economic Paleontologists and Mineralogists book 81, p. 25–26.

Harwood, D.S., 1983, Stratigraphy of upper Paleozoic volcanic rocks and regional unconformities in part of the northern Sierra terrane, California: Geological Society of America Bulletin, v. 94, p. 413–422.

Harwood, D.S., 1988, Tectonism and metamorphism in the northern Sierra terrane, northern California, *in* Ernst W.G., ed., Metamorphism and crustal evolution of the western United States (Rubey Volume 7): Englewood Cliffs, New Jersey, Prentice-Hall, p. 764–788.

Harwood, D.S., 1992, Stratigraphy of Paleozoic and lower Mesozoic rocks in the northern Sierra terrane, California: U.S. Geological Survey Bulletin 1957, 78 p.

Harwood, D.S., and Murchey, B.L., 1990, Biostratigraphic, tectonic, and paleogeographic ties between upper Paleozoic volcanic and basinal rocks in the northern Sierra terrane, California, and the Havallah Sequence, Nevada, *in* Harwood, D.S., and Miller, M.M., eds., Paleozoic and early Mesozoic paleogeographic relations; Sierra Nevada, Klamath Mountains, and related terranes: Geological Society of America Special Paper 255, p. 157–173.

Hoffman, P.F., 1989, Precambrian geology and tectonic history of North America, *in* Bally, A.W., and Palmer, A.R., eds., The geology of North America—An overview: Boulder, Colorado, Geological Society of America, Geology of North America, v. A, p. 447–512.

Ludwig, K.R., 1991a, A computer program for processing Pb-U-Th isotopic data: U.S. Geological Survey Open-File Report 88-542.

Ludwig, K.R., 1991b, A plotting and regression program for radiogenic-isotopic data: U.S. Geological Survey Open-File Report 91-445.

Miller, E.L., Holdsworth, B.K., Whiteford, W.B., and Rogers, D., 1984, Stratigraphy and structure of the Schoonover Sequence, northeastern Nevada: Implications for Paleozoic plate margin tectonics: Geological Society of America Bulletin, v. 95, p. 1063–1076.

Miller, E.L., Miller, M.M., Stevens, C.H., Wright, J.E., and Madrid, R., 1992, Late Paleozoic paleographic and tectonic evolution of the western U.S. Cordillera, *in* Burchfiel, B.C., et al., eds., The Cordilleran orogen: Conterminous U.S.: Boulder, Colorado, Geological Society of America, Geology of North America, v. G-3, p. 57–106.

Miller, M.M., 1987, Dispersed remnants of a northeast Pacific fringing arc: Upper Paleozoic terranes of Permian McCloud faunal affinity, western U.S.: Tectonics, v. 6, p. 807–830.

Miller, M.M., and Harwood, D.S., 1990, Paleogeographic setting of upper Paleozoic rocks in the northern Sierra and eastern Klamath terranes, northern California, *in* Harwood, D.S., and Miller, M.M., eds., Paleozoic and early Mesozoic paleogeographic relations; Sierra Nevada, Klamath Mountains, and related terranes: Geological Society of America Special Paper 255, p. 175–192.

Nilsen, T.H., and Stewart, J.H., 1980, The Antler orogeny—Mid-Paleozoic tectonism in western North America: Geology, v. 8, p. 298–302.

Oldow, J.S., 1984, Evolution of a late Mesozoic back-arc fold and thrust belt, northwestern Great Basin, U.S.A.: Tectonophysics, v. 102, p. 245–274.

Schweickert, R.A., 1981, Tectonic evolution of the Sierra Nevada, *in* Ernst, W.G., ed., The geotectonic development of California (Rubey Volume I): Englewood Cliffs, New Jersey, Prentice-Hall, p. 88–131.

Schweickert, R.A., and Snyder, W.S., 1981, Paleozoic plate tectonics of the Sierra Nevada and adjacent regions, *in* Ernst, W.G., ed., The geotectonic development of California (Rubey Volume I): Englewood Cliffs, New Jersey, Prentice-Hall, p. 183–201.

Schweickert, R.A., Harwood, D.S., Girty, G.H., and Hanson, R.E., 1984, Tectonic development of the northern Sierra terrane: An accreted late Paleozoic island arc and its basement, *in* Lintz, J., Jr., ed., Western geological excursions: Reno, Nevada, Mackay School of Mines, p. 1–65.

Silberling, N.J., 1991, Allochthonous terranes of western Nevada: Current status, *in* Raines, G.L., et al., eds., Geology and ore deposits of the Great Basin: Reno, Geological Society of Nevada, p. 101–102.

Silberling, N.J., Jones, D.L., Blake, M.C., Jr., and Howell, D.G., 1987, Lithotectonic terrane map of the western conterminous United States: U.S. Geological Survey Miscellaneous Field Studies Map MF-1874-C, scale 1:2,500,000.

Speed, R.C., 1979, Collided Paleozoic microplate in the western U.S.: Journal of Geology, v. 87, p. 279–292.

Speed, R.C., and Sleep, N.H., 1982, Antler orogeny and foreland basin: A model: Geological Society of America Bulletin, v. 93, p. 815–828.

Stacey, J.S., and Kramers, J.D., 1975, Approximation of terrestrial lead isotope evolution by a two-stage model: Earth and Planetary Science Letters, v. 26, p. 207–221.

Stewart, J.H., Poole, F.G., Ketner, K.B., Madrid, R.J., Roldan-Quintana, J., and Amaya-Martinez, R., 1990, Tectonics and stratigraphy of the Paleozoic and Triassic southern margin of North America, Sonora, Mexico, *in* Gehrels, G.E., and Spencer, J.E., eds., Geologic excursions through the Sonoran Desert region, Arizona and Sonora: Arizona Geological Survey Special Paper 7, p. 183–202.

Van Schmus, W.R., and 24 others, 1993, Transcontinental Proterozoic provinces, *in* Reed, J.C., et al., eds., Precambrian: Conterminous U.S.: Boulder, Colorado, Geological Society of America, Geology of North America, v. C-2, p. 171–334.

Whiteford, W.B., 1990, Paleogeographic setting of the Schoonover sequence, Nevada, and implications for the late Paleozoic margin of western North America, *in* Harwood, D.S., and Miller, M.M., eds., Paleozoic and early Mesozoic paleogeographic relations; Sierra Nevada, Klamath Mountains, and related terranes: Geological Society of America Special Paper 255, p. 115–136.

Manuscript Accepted by the Society January 24, 2000

Printed in U.S.A.

Geological Society of America
Special Paper 347
2000

# Detrital zircon geochronologic study of upper Paleozoic strata in the eastern Klamath terrane, northern California

**George E. Gehrels**
*Department of Geosciences, University of Arizona, Tucson, Arizona 85721, USA*
**M. Meghan Miller**
*Department of Geology, Central Washington University, Ellensburg, Washington 98926, USA*

## ABSTRACT

**U-Pb analyses have been conducted on individual detrital zircon grains from Upper Devonian(?)–Carboniferous strata of the Bragdon Formation and Upper Carboniferous–Lower Permian strata of the Baird Formation. These are the two most important clastic units in the Redding section of the eastern Klamath terrane of northern California. There are 31 grains from the Bragdon Formation that yield mainly concordant to slightly discordant ages ranging from ca. 363 Ma to ca. 3.12 Ga. Grains with ages of ca. 363–572 Ma in this sample were apparently shed primarily from the nearby Trinity complex, which contains igneous rocks of these ages. Older grains, ranging from 1.07 to 3.12 Ga, but mostly between 1.76 and 1.99 Ga, were likely recycled from lower Paleozoic strata of the Yreka terrane or the Roberts Mountains allochthon. Six detrital zircons from the Baird Formation are all apparently concordant, with interpreted ages between 320 and 326 Ma. These grains are locally derived and probably record igneous activity represented by the widespread volcanic and volcaniclastic rocks within the Baird Formation.**

## INTRODUCTION

The Redding section of the eastern Klamath terrane consists of Devonian through Jurassic marine volcanic, carbonate, and clastic rocks that underlie much of the eastern Klamath Mountains of northern California (Figs. 1 and 2) (Irwin, 1977). Most workers interpret these rocks as having formed in a volcanic arc environment (e.g., Burchfiel and Davis, 1972). The original position of this arc system relative to the Cordilleran margin has been controversial because of a lack of firm ties with continental margin rocks to the east, the broad expanse of coeval ocean-floor strata of the Golconda allochthon in central and western Nevada (Fig. 2), and the lack of definitive evidence on facing direction(s) of the arc.

Uncertainties about the Paleozoic and Mesozoic paleogeography of the eastern Klamath terrane have led to two different hypotheses concerning the pre-Jurassic position and tectonic setting of the arc system. Burchfiel and Davis (1972) proposed a tectonically conservative scenario in which the arc system formed and evolved along a west-facing subduction zone near its present position along the Cordilleran margin. In this scenario, the arc is interpreted to have been separated from the continental margin by a backarc basin, represented by strata of the Golconda allochthon (Fig. 2). A Permian or Triassic accretionary complex that developed along the western margin of this arc is preserved in the North Fork and Hayfork terranes of the southwestern Klamath Mountains (Wright, 1982). Collapse of the backarc basin and final accretion of the arc system is interpreted to have occurred during the Permian-Triassic Sonoma orogeny. This scenario has been adopted by many subsequent workers, and was broadened by Miller (1987) and Rubin et al. (1990) to include several other Paleozoic arc-type terranes in the Cordillera.

A second scenario holds that the arc system is far traveled. One of the diagnostic characteristics of the eastern Klamath arc system is the presence in Permian limestones of a distinctive

Gehrels, G.E., and Miller, M.M., 2000, Detrital zircon geochronologic study of upper Paleozoic strata in the eastern Klamath terrane, northern California, *in* Soreghan, M.J., and Gehrels, G.E., eds., Paleozoic and Triassic paleogeography and tectonics of western Nevada and northern California: Boulder, Colorado, Geological Society of America Special Paper 347, p. 99–107.

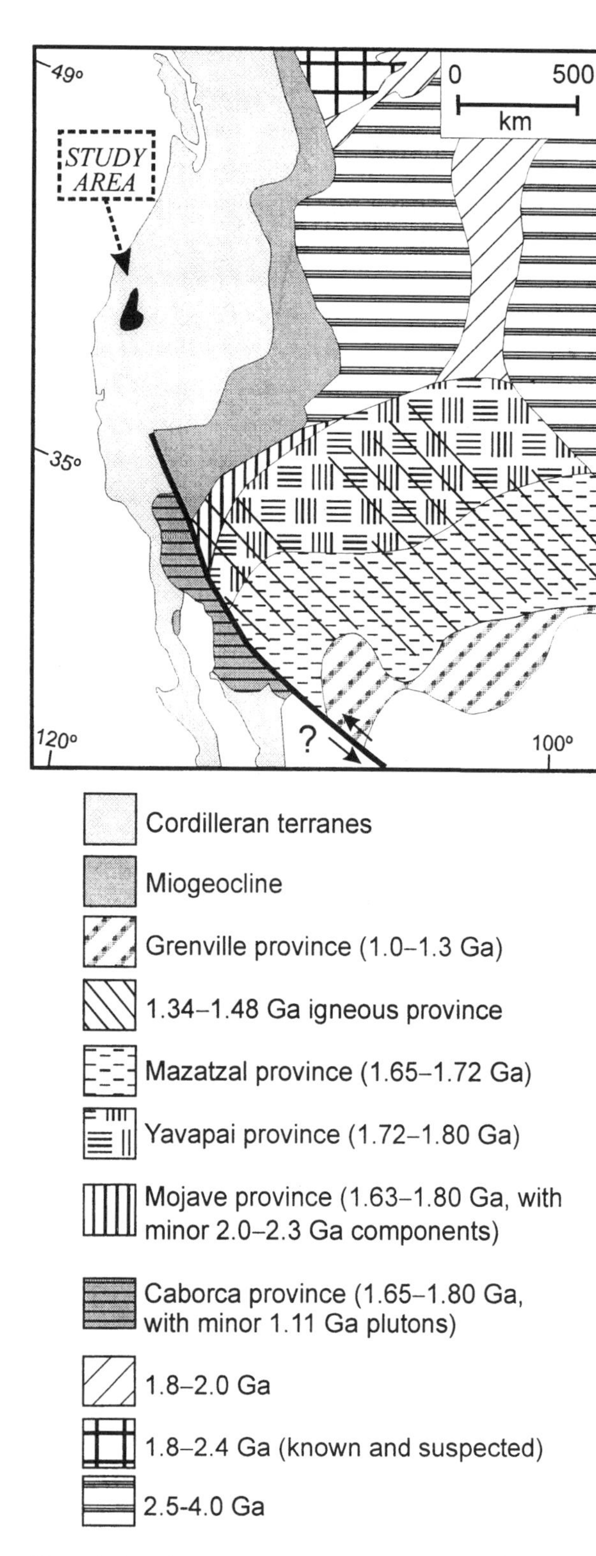


Figure 1. Location map of eastern Klamath terrane in relation to other terranes in Cordillera, miogeoclinal strata of western North America, and cratonal basement provinces. Adapted from Silberling et al. (1987), Hoffman (1989), Stewart et al. (1990), Ross (1991), and Van Schmus et al. (1993).

suite of fusulinids, termed the McCloud fauna (Irwin, 1977; Miller, 1987). Differences between the McCloud fauna and coeval faunas in miogeoclinal strata have led to the conclusion that the arc formed >3000 km (Belasky and Runnegar, 1994) or ~5000 km (Stevens et al., 1990) west of the North American margin during Permian time. This amount of geographic separation is more consistent with scenarios proposed by Dickinson (1977), Speed (1979), and Schweickert and Snyder (1981), in which late Paleozoic–early Mesozoic magmatism in the eastern Klamath terrane occurred in response to west-dipping subduction along the eastern margin of the terrane. In these models, the Permian-Triassic Sonoma orogeny is viewed as resulting from arc-continent collision.

This report uses detrital zircon geochronology to place tighter constraints on the late Paleozoic provenance and paleoposition of the eastern Klamath terrane. Our study builds on the work of Miller and Saleeby (1991), who analyzed many different size and morphological fractions of zircon from the Upper Devonian(?)–Mississippian Bragdon Formation. These workers demonstrated the existence of continent-derived zircons in the Bragdon Formation, but were unable to resolve specific source areas for the grains because their analyses were conducted on multigrain fractions of zircon that yielded a broad range of average ages. Due to improvements in analytical techniques, we have been able to determine ages for individual grains from the Bragdon Formation and thereby have identified specific age components in the source region. This chapter presents single-grain analyses of zircon grains from the Bragdon Formation, and compares these ages to detrital zircon age spectra from other regions of the Cordillera.

Samples from the Pennsylvanian–Lower Permian(?) Baird Formation and Middle-Upper Triassic Pit Formation were also collected in an effort to determine changes in provenance through late Paleozoic and early Mesozoic time. Analyses are presented herein for the Baird Formation, but our sample from the Pit Formation did not yield zircons of sufficient size for single grain analysis.

## REGIONAL GEOLOGY

The eastern part of the Klamath Mountains consists of lower Paleozoic igneous rocks of the Trinity terrane (or complex), lower Paleozoic strata and intrusions of the Yreka terrane, and Devonian through Jurassic strata of the Redding section (Irwin, 1966, 1977) (Fig. 2). All three of these terranes structurally overlie Devonian(?) metamorphic rocks of the central metamorphic belt, which are interpreted to have formed in a west-facing (Burchfiel and Davis, 1972; Peacock and Norris, 1989; Hacker and Goodge, 1990) or possibly southeast-facing (Dickinson, this volume, Chapter 14) subduction complex.

The Trinity complex consists of a variety of ultramafic rocks and mafic intrusions that range in age from ca. 400 to ca. 570 Ma (Wallin et al., 1988). Intrusive rocks have age groups of 398–431, 440–475, and 565–570 Ma (Wallin et al., 1988, 1995; Metcalf

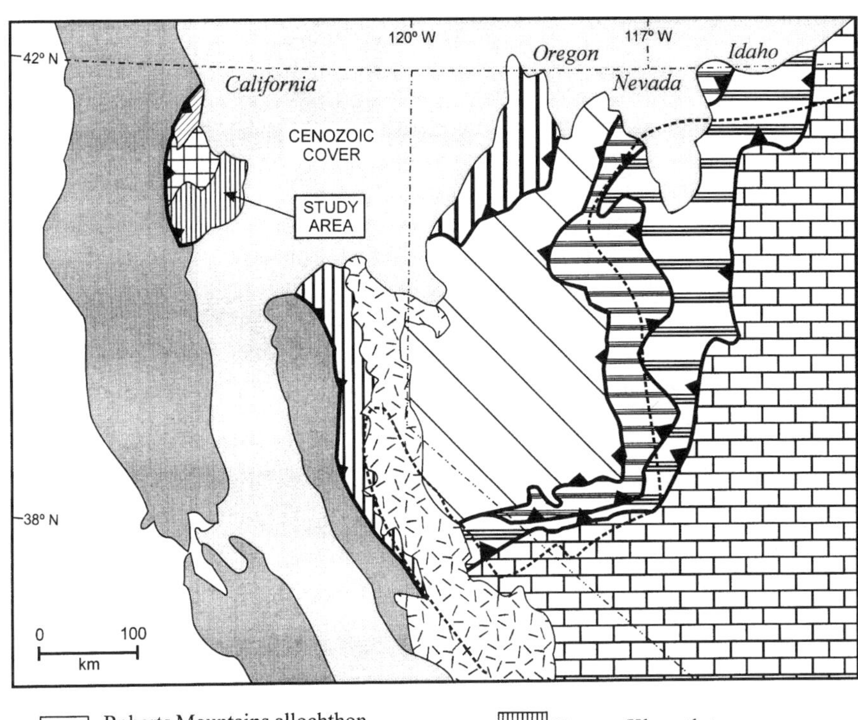


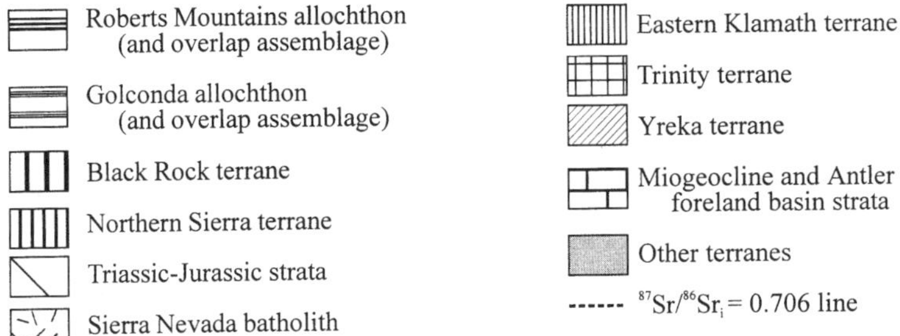


Figure 2. Schematic map of main terranes and/or assemblages and their bounding faults in western Nevada and northern California. Map is adapted primarily from Kistler and Peterman (1973), Oldow (1984), Silberling et al. (1987), Kistler (1991), and Silberling (1991). Note that widespread Cretaceous and younger rocks and structures are ignored in effort to emphasize configuration of pre-Cretaceous features.

et al., 1998). The Trinity complex is considered to be part of an ophiolite complex by Brouxel and Lapierre (1988) and Boudier et al. (1989), but other workers have interpreted the complex to have formed in an arc-type setting (Quick, 1981; Jacobsen et al., 1984; Saleeby, 1992; Miller and Harwood, 1990; Wallin et al., 1995; Metcalf et al., 1998).

The Yreka terrane structurally overlies both the central metamorphic belt and the Trinity complex. The main units in the Yreka terrane include Neoproterozoic–Cambrian plagiogranite bodies and Ordovician through Devonian volcanic rocks, limestone, and clastic strata (Fig. 3) (Irwin, 1966, 1977; Potter et al., 1977, 1990). One of the noteworthy units in the Yreka terrane is the Ordovician and/or Silurian Antelope Mountain Quartzite, which yields detrital zircons of Precambrian age (Wallin, 1989; Wallin et al., this volume, Chapter 9).

The Redding section of the eastern Klamath terrane consists of Devonian through Jurassic volcanic rocks with minor clastic strata and limestone. The oldest rocks recognized are Middle and possibly Lower Devonian volcanic rocks of the Copley Greenstone and Balaklala Rhyolite (Fig. 3). These rocks have geochemical signatures consistent with formation in an arc-type setting (Brouxel et al., 1988). Tuffaceous rocks and limestone of the overlying Middle Devonian Kennett Formation record the cessation of mid-Paleozoic arc magmatism. The next younger unit in the terrane is the Bragdon Formation, which is reported to unconformably overlie the older volcanic rocks (Diller, 1905; Kinkel et al., 1956) and ultramafic rocks of the Trinity complex (Brouxel et al., 1988; Charvet et al., 1989, 1990). The Bragdon Formation consists mainly of sandstone turbidites with subordinate channel-fill conglomerates and debris-flow deposits (Miller and Cui, 1989; Miller and Harwood, 1990). Sandstones are divided by these workers into three types: (1) quartz-, chert-, and sedimentary lithic–rich sandstone, (2) volcanic lithic– and feldspar-rich sandstone, and (3) crystal-rich and tuffaceous sandstone. These sandstones are interpreted to record derivation from uplifted quartz-rich basement and arc-type igneous rocks, as well as active volcanic centers. Limestone clasts in the Bragdon Formation yield Middle Devonian and Carboniferous fossils, and the minimum age of the unit is constrained by overlying middle-Upper Carboniferous strata of the Baird Formation.

Detrital zircons were previously analyzed from two samples of sedimentary lithic–rich sandstone within the Bragdon Forma-

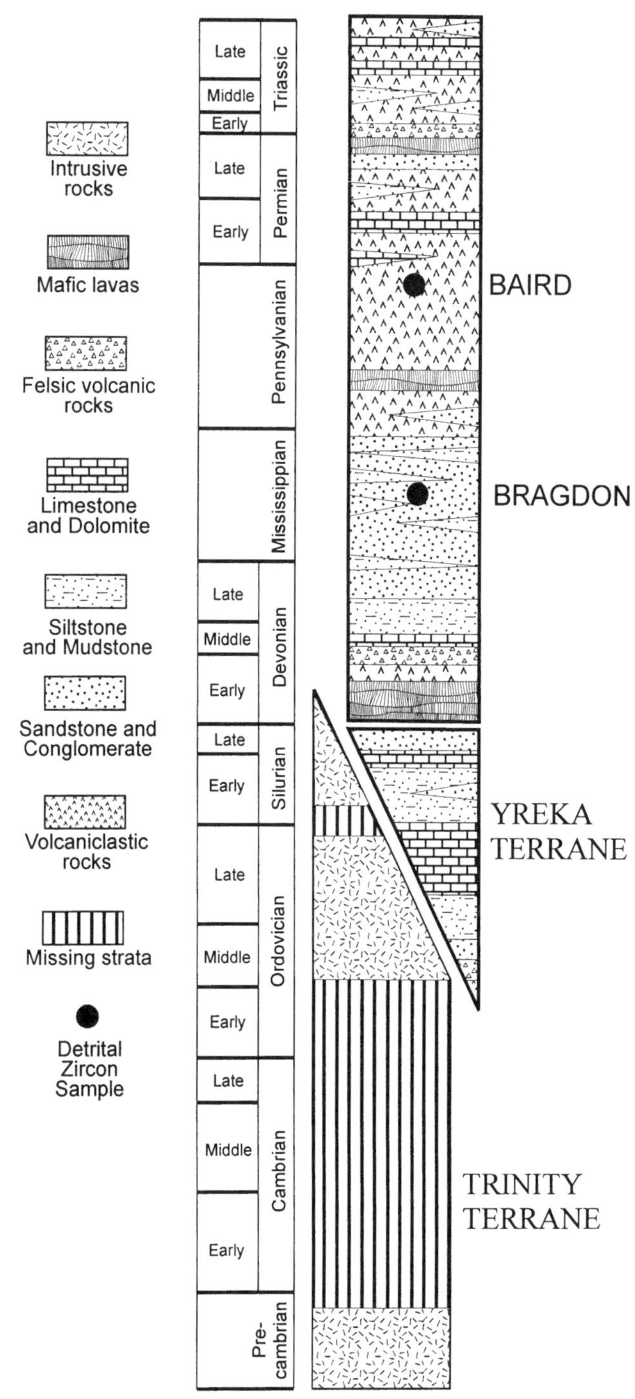


Figure 3. Generalized stratigraphic column of Redding section (eastern Klamath terrane), Yreka terrane, and Trinity terrane (after Miller and Harwood, 1990; Potter et al., 1977, 1990; Wallin et al., 1988, 1995; Metcalf et al., 1998).

tion (Miller and Saleeby, 1991). Multigrain fractions that were separated on the basis of morphology and color yielded highly discordant analyses with Middle and Early Proterozoic (1475–1941 Ma) $^{207}Pb^{*}/^{206}Pb^{*}$ ages. Miller and Saleeby (1991) interpreted these ages as evidence of derivation from Precambrian continental crust, thereby requiring the eastern Klamath terrane to have resided in proximity to continental or continental-margin rocks during part of late Paleozoic time.

The Baird Formation consists of a lower sequence of argillite and an upper sequence of volcanogenic sandstones and intermediate-composition flows, tuffs, and breccias (Irwin, 1977, 1981; Watkins, 1985; Miller and Harwood, 1990). Fossils in interbedded strata are late Early Carboniferous to latest Carboniferous or Early Permian (Watkins, 1985). The Baird Formation is conformably overlain by the Lower Permian McCloud Limestone, which in turn is overlain by volcanic and sedimentary rocks of Late Permian through Middle Jurassic age (Irwin, 1977, 1981; Watkins, 1985; Miller and Harwood, 1990).

## RESULTS

For this study, detrital zircon samples were collected from sandstones of the Upper Devonian(?)–Lower Carboniferous Bragdon Formation, the Upper Carboniferous–Lower Permian Baird Formation, and the Middle-Upper Triassic Pit Formation. For each unit, ~20 kg was collected from a narrow stratigraphic interval at a single locality. Zircons were extracted from the samples using standard mineral-separation techniques (Gehrels, this volume, Introduction). Because our sample from the Pit Formation did not yield zircons larger than ~60 µm, the sample was not processed further. The larger grains from each of the other two samples were separated into color and morphology groups, and then representatives from each group were abraded, cleaned in dilute $HNO_3$, and selected for analysis. Chemical separations and thermal ionization mass spectrometry were conducted following procedures outlined by Gehrels (this volume, Introduction), and the resulting data were reduced and plotted (Figs. 4 and 5) using programs of Ludwig (1991a, 1991b).

### *Bragdon Formation*

Our sample was collected from a roadcut near McCloud Reservoir, at the same locality as sample M86-11 of Miller and Saleeby (1991). We sampled a coarse-grained sandstone near the base of a 2-m-thick graded bed in a sequence of sandstone-siltstone-shale turbidites. The sandstone is a chert-rich quartz-lithic sandstone, according to Dickinson and Gehrels (this volume, Chapter 11).

We recovered ~1.6 gm of zircon from our sample; the largest grains are ~200 µm in length. All of the >125 µm grains in the sample were divided into groups based on color (colorless, light pink, and medium pink) and morphology (moderately rounded and/or spherical versus euhedral). The smaller size fractions were examined for grains with different characteristics, although none were found. The sample contained approximately equal proportions of colorless and pinkish grains, and moderately rounded grains were much more abundant than euhedral grains.

The 31 grains analyzed yield mainly concordant to slightly discordant ages, as shown in Table 1 and Figure 4. The interpreted ages shown in Table 1 and plotted in Figure 6 are

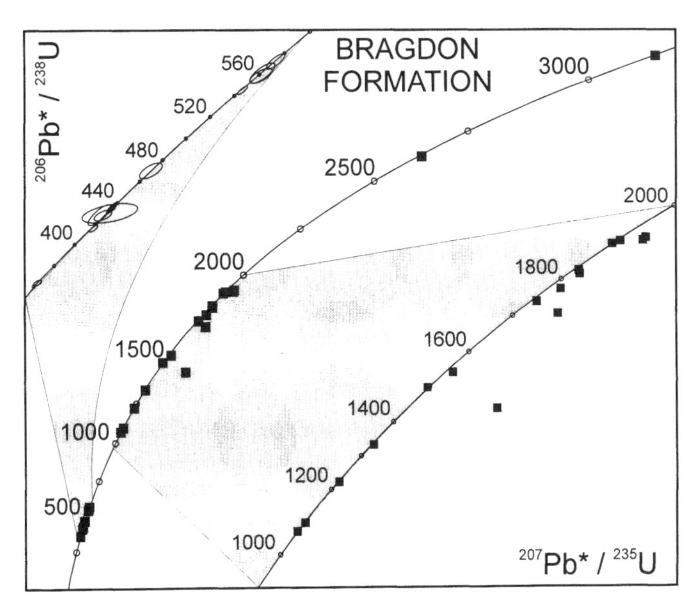


Figure 4. U-Pb concordia diagram of detrital zircons in Bragdon Formation. All analyses are of single zircon grains.

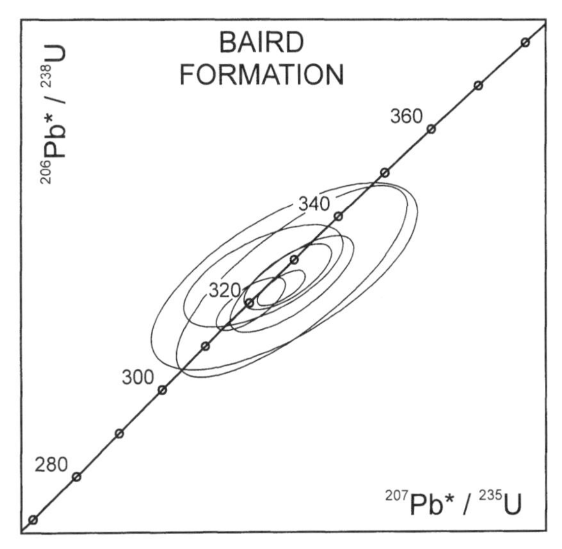


Figure 5. U-Pb concordia diagram of detrital zircons in Baird Formation. All analyses are of single zircon grains, and error ellipses are shown at 95% confidence level.

$^{206}$Pb*/$^{238}$U ages for the <1.0 Ga grains, whereas ages for the older grains are projected from 100 Ma. This assumes that the younger grains have not undergone significant Pb loss and that discordance in the older grains resulted from Pb loss ca. 100 Ma, which is an approximate minimum age of regional metamorphism in the region. Given the small degrees of discordance, these projected ages are generally only a few million years older than the $^{207}$Pb*/$^{206}$Pb* ages (Table 1).

As shown in Figure 6, the grains from the Bragdon Formation yield age clusters of ca. 360–580 Ma and 1760–2000 Ma, with a few scattered ages of 1.07–1.59 Ga and >2.6 Ga. The younger grains have ages predominantly of 415–437 and 545–572 Ma, with additional ages of ca. 363 and 470 Ma. The 415–572 Ma ages are an excellent match with the ages of igneous rocks in the Trinity terrane, and suggest that at least some of the detritus in the Bragdon Formation was derived from the Trinity complex. The ca. 363 Ma grain may have been derived from latest Devonian or earliest Mississippian (Harland et al., 1990; Tucker et al., 1998) volcanism during or slightly preceding deposition of the Bragdon Formation.

The older grains in our sample originated in a continental region containing widespread >1.0 Ga rocks. The grains need not have been derived directly from a continental region, however, as several assemblages or terranes in northern California–western Nevada contain detrital zircons of the appropriate age. These potential sources include lower Paleozoic strata of the Antelope Mountain Quartzite in the Yreka terrane, the Shoo Fly Complex in the northern Sierra terrane, and the Roberts Mountains allochthon (Figs. 2 and 6). If older zircons were recycled from the Yreka terrane, all of the detritus in the Bragdon Formation could have been derived from the eastern Klamath terrane and its underlying basement.

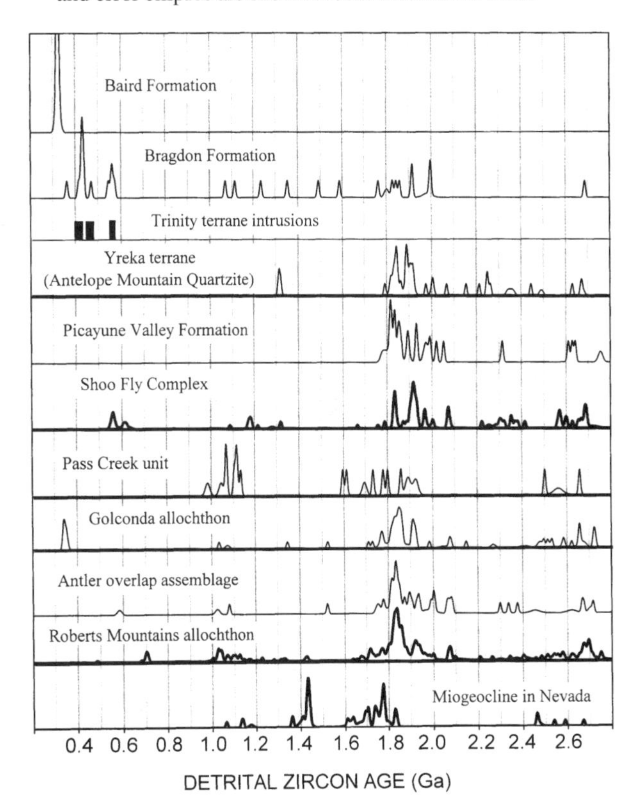


Figure 6. Relative age-probability plots comparing ages from Baird and Bragdon Formations with age spectra from miogeoclinal strata in Nevada (Gehrels and Dickinson, 1995), Roberts Mountains allochthon (Gehrels et al., this volume, Chapter 1) and its overlap assemblage (Gehrels and Dickinson, this volume), Golconda allochthon (Riley et al., this volume), Carboniferous strata of Pass Creek unit in Black Rock terrane (Darby et al., this volume), Shoo Fly complex (Harding et al., this volume) and overlying Carboniferous strata of Picayune Valley Formation (Spurlin et al., this volume), Yreka terrane (Wallin et al., this volume), and Trinity terrane (Wallin et al., 1988, 1995; Metcalf et al., 1998). Note that each curve represents all detrital zircon ages from sample or set of samples, and that curves are normalized according to number of constituent grains.

## TABLE 1. U-Pb ISOTOPIC DATA AND AGES

| Grain type | Grain wt. (μg) | $Pb_c$ (pg) | U (ppm) | $^{206}Pb_m/^{204}Pb$ | $^{206}Pb_c/^{208}Pb$ | Apparent ages (Ma) $^{206}Pb^*/^{238}U$ | $^{207}Pb^*/^{235}U$ | $^{207}Pb^*/^{206}Pb^*$ | Interpreted age (Ma) |
|---|---|---|---|---|---|---|---|---|---|
| Bragdon Formation (41°09'45"N, 122°06'11"W) | | | | | | | | | |
| **CR** | **10** | **4** | **260** | **2160** | **6.6** | **363 ± 2** | **364 ± 4** | **367 ± 20** | **363 ± 10** |
| **CE** | **9** | **14** | **450** | **1250** | **6.8** | **415 ± 3** | **418 ± 4** | **432 ± 13** | **415 ± 10** |
| **LR** | **17** | **7** | **69** | **720** | **6.4** | **428 ± 8** | **423 ± 11** | **395 ± 33** | **428 ± 10** |
| **CR** | **8** | **22** | **293** | **485** | **3.4** | **428 ± 4** | **426 ± 8** | **420 ± 35** | **428 ± 10** |
| CR | 10 | 39 | 122 | 153 | 3.1 | 430 ± 7 | 433 ± 25 | 448 ± 120 | 430 ± 20 |
| **CR** | **18** | **4** | **131** | **2500** | **6.8** | **433 ± 3** | **433 ± 4** | **434 ± 11** | **433 ± 10** |
| **LR** | **10** | **4** | **275** | **2700** | **6.6** | **435 ± 3** | **435 ± 4** | **435 ± 14** | **435 ± 10** |
| **CR** | **12** | **8** | **450** | **3080** | **14.5** | **437 ± 3** | **436 ± 4** | **431 ± 10** | **437 ± 10** |
| **CR** | **8** | **19** | **206** | **433** | **5.2** | **470 ± 6** | **470 ± 10** | **470 ± 36** | **470 ± 10** |
| **LR** | **14** | **30** | **990** | **2590** | **5.2** | **545 ± 4** | **546 ± 4** | **552 ± 9** | **545 ± 10** |
| **CR** | **10** | **19** | **172** | **540** | **4.1** | **559 ± 6** | **561 ± 9** | **572 ± 26** | **559 ± 10** |
| **CR** | **9** | **12** | **131** | **580** | **6.6** | **561 ± 8** | **562 ± 11** | **566 ± 27** | **561 ± 10** |
| **LR** | **19** | **4** | **49** | **1170** | **4.6** | **572 ± 6** | **573 ± 8** | **576 ± 17** | **572 ± 10** |
| **CE** | **8** | **13** | **188** | **1270** | **7.8** | **1072 ± 8** | **1071 ± 10** | **1070 ± 10** | **1070 ± 10** |
| LR | 13 | 8 | 125 | 2240 | 8.4 | 1098 ± 7 | 1103 ± 10 | 1113 ± 12 | 1114 ± 10 |
| CE | 8 | 9 | 163 | 1870 | 4.8 | 1222 ± 9 | 1227 ± 11 | 1234 ± 9 | 1235 ± 10 |
| LR | 9 | 33 | 186 | 705 | 6.9 | 1332 ± 10 | 1340 ± 15 | 1352 ± 14 | 1353 ± 10 |
| **CR** | **10** | **4** | **44** | **2800** | **6.9** | **1498 ± 14** | **1495 ± 15** | **1491 ± 6** | **1491 ± 10** |
| LR | 13 | 4 | 24 | 1230 | 3.9 | 1542 ± 17 | 1560 ± 21 | 1586 ± 13 | 1587 ± 10 |
| CR | 14 | 5 | 77 | 4100 | 9.5 | 1739 ± 21 | 1751 ± 23 | 1764 ± 5 | 1765 ± 10 |
| MR | 14 | 50 | 452 | 2335 | 5.6 | 1774 ± 11 | 1799 ± 13 | 1827 ± 5 | 1828 ± 10 |
| CE | 8 | 52 | 495 | 1435 | 6.2 | 1825 ± 11 | 1834 ± 13 | 1843 ± 6 | 1844 ± 10 |
| CE | 8 | 11 | 91 | 1290 | 4.1 | 1816 ± 14 | 1836 ± 15 | 1858 ± 6 | 1860 ± 10 |
| LR | 8 | 13 | 139 | 1580 | 4.2 | 1706 ± 11 | 1793 ± 15 | 1896 ± 9 | 1903 ± 20 |
| CR | 12 | 6 | 67 | 2930 | 4.1 | 1905 ± 15 | 1909 ± 17 | 1914 ± 6 | 1915 ± 10 |
| CR | 16 | 4 | 91 | 7200 | 5.5 | 1897 ± 8 | 1905 ± 9 | 1915 ± 4 | 1916 ± 10 |
| *MR* | *9* | *20* | *185* | *1530* | *3.6* | *1439 ± 11* | *1665 ± 15* | *1962 ± 8* | *1984 ± 40* |
| MR | 9 | 9 | 137 | 2860 | 8.2 | 1906 ± 14 | 1948 ± 15 | 1993 ± 4 | 1996 ± 20 |
| MR | 10 | 10 | 189 | 4030 | 4.4 | 1914 ± 13 | 1954 ± 15 | 1995 ± 6 | 1997 ± 10 |
| LR | 9 | 10 | 145 | 3680 | 6.1 | 2633 ± 14 | 2633 ± 17 | 2687 ± 5 | 2688 ± 10 |
| **LR** | **14** | **2** | **43** | **12550** | **8.2** | **3115 ± 13** | **3115 ± 16** | **3115 ± 5** | **3115 ± 10** |
| Baird Formation (41°08'09"N, 122°06'51"W) | | | | | | | | | |
| **YE** | **16** | **13** | **253** | **1030** | **8.4** | **320 ± 5** | **320 ± 7** | **324 ± 36** | **320 ± 10** |
| **BE** | **16** | **8** | **233** | **1580** | **7.1** | **323 ± 4** | **327 ± 6** | **355 ± 25** | **323 ± 10** |
| **CR** | **14** | **6** | **73** | **590** | **9.9** | **324 ± 6** | **330 ± 12** | **366 ± 60** | **324 ± 10** |
| **CE** | **16** | **10** | **62** | **182** | **8.2** | **325 ± 18** | **330 ± 24** | **362 ± 95** | **325 ± 20** |
| **YE** | **22** | **18** | **89** | **380** | **6.4** | **326 ± 9** | **323 ± 15** | **300 ± 95** | **326 ± 10** |
| **YE** | **18** | **8** | **172** | **790** | **6.6** | **326 ± 4** | **329 ± 7** | **347 ± 45** | **326 ± 10** |

*Note:*

Grain with $^{206}Pb^*/^{238}U$ age within 15% of $^{207}Pb^*/^{206}Pb^*$ age in plain type.

Grain with $^{206}Pb^*/^{238}U$ age not within 15% of $^{207}Pb^*/^{206}Pb^*$ age in italics.

Grain with concordant age and moderate to high precision analysis in bold.

* Radiogenic Pb.

Grain type: L—light pink, M—medium pink, C—colorless, Y—yellow, B—brownish; R—slightly rounded, E—euhedral.

Unless otherwise noted, all grains are clear and/or translucent and abraded with air abrasion.

$^{206}Pb/^{204}Pb$ is measured ratio, uncorrected for blank, spike, or fractionation.

$^{206}Pb/^{208}Pb$ is corrected for blank, spike, and fractionation.

Most concentrations have an uncertainty of 25% due to uncertainty in weight of grain.

Constants used: $^{238}U/^{235}U$ = 137.88. Decay constant for $^{235}U = 9.8485 \times 10^{-10}$. Decay constant for $^{238}U = 1.55125 \times 10^{-10}$.

All uncertainties are at the 95% confidence level.

Pb blank ranged from 2 to 10 pg. U blank was consistently <1 pg.

Interpreted ages for concordant grains are $^{206}Pb^*/^{238}U$ ages if <1.0 Ga and $^{207}Pb^*/^{206}Pb^*$ ages if >1.0 Ga.

Interpreted ages for discordant grains are projected from 100 Ma.

All analyses conducted using conventional isotope dilution and thermal ionization mass spectrometry, as described by Gehrels (this volume, Introduction).

The fact that generally similar age groups of >1.0 Ga detrital zircons occur in the Bragdon Formation and in Carboniferous strata of the northern Sierra terrane (Picayune Valley Formation), Golconda allochthon, Black Rock terrane (Pass Creek unit), and Antler overlap assemblage, however, suggests instead that the Bragdon Formation was part of a broad submarine fan system. According to Harwood and Murchey (1990), the Roberts Mountains allochthon was the most likely source of siliciclastic detritus in this submarine fan system. If this is correct, the Bragdon Formation consists of detritus shed from an active arc and older arc rocks within the Klamath Mountains as well as the Antler highlands along the Cordilleran margin.

### *Baird Formation*

One sample was collected from near Battle Creek, at the east end of McCloud Reservoir. We collected ~15 kg of medium-grained, magnetite-rich sandstone from a 1-m-thick bed in a sequence of feldspathic sandstones interlayered with siltstone and mudstone. The magnetite-rich layer that we sampled is interpreted to have accumulated in a shoreline environment.

The sample yielded ~630 mg of zircon grains, most of which are <100 μm in sieve size. There are ~125 grains in the 100–145 μm size range; no grains are >145 μm. The larger grains were divided into three groups: yellowish-brown euhedral grains with length:width of ~2:1, elongate (~4:1) euhedral grains that are colorless to yellowish, and a small number of brownish euhedral grains with abundant dark inclusions. No grains with other colors or morphologies were identified in the <100 μm size fractions.

Six grains were analyzed, all of which yielded apparently concordant (but low precision) ages between 320 and 326 Ma (Table 1; Fig. 5). The similarity of this source age and the depositional age of the Baird Formation suggests that most detritus in this sample was shed from active volcanic centers within or near the eastern Klamath terrane. According to the time scale of Harland et al. (1990), this volcanism occurred during latest Mississippian–earliest Pennsylvanian time.

## CONCLUSIONS

U-Pb ages of detrital zircons in Carboniferous–Early Permian(?) sandstones of the eastern Klamath terrane reflect derivation from three fundamentally different sources. The youngest grains yield ages of 320–326 Ma (Baird Formation) and ca. 363 Ma (Bragdon Formation), which overlap with the ages of deposition of the two units. These young ages, combined with the volcanic-rich composition of sandstones in both formations, require deposition in proximity to active latest Devonian and latest Mississippian–earliest Pennsylvanian volcanic centers within or near the eastern Klamath terrane.

In contrast, grains in the Bragdon Formation with ages between 415 and 572 Ma apparently came from slightly older magmatic arc assemblages that were being dissected during Carboniferous time. Igneous rocks of appropriate ages to provide 415–437, ca. 470, and 545–572 Ma detrital grains are present within the immediately adjacent Trinity terrane, which is consistent with previous suggestions (Charvet et al., 1989, 1990; Metcalf et al., 1998) that the Redding section was deposited upon or accumulated near the Trinity complex. The lack of significant plutonic or ultramafic debris in the Bragdon Formation (Miller and Cui, 1989) suggests, however, that the volume of detritus derived from the Trinity terrane was minor.

The oldest grains in our samples, ca. 1.07–3.12 Ga, must have originated in continental rocks. Rather than being eroded directly from continental source rocks, however, these grains were most likely recycled from lower Paleozoic sandstones of northern California or western Nevada that contain predominantly Proterozoic and Archean detrital grains. These grains may have been recycled from the Ordovician and/or Silurian Antelope Mountain Quartzite in the Yreka terrane of the eastern Klamaths (Wallin et al., this volume, Chapter 9) or the lower Paleozoic Shoo Fly Complex of the northern Sierra terrane (Harding et al., this volume, Chapter 2), because both of these assemblages yield detrital grains of the appropriate ages.

It is also possible that the >1.0 Ga grains were shed from the Roberts Mountains allochthon, which formed an orogenic highland during Carboniferous time and was the source of large volumes of sand containing detrital zircons of the appropriate age (Gehrels et al., this volume, Chapter 1). Because the Roberts Mountains allochthon was emplaced onto the continental margin during Late Devonian–Mississippian time, a provenance link with the Roberts Mountains allochthon would require the eastern Klamath terrane to have been in close proximity to the southwest Cordilleran margin during deposition of the Bragdon Formation.

Our data cannot distinguish between these two possibilities, however, and therefore additional studies will be required to resolve the paleogeographic setting of the Redding section.

## ACKNOWLEDGMENTS

This research was supported by National Science Foundation grant EAR-9416933. We acknowledge the help of Hallie Humphrey in processing our two samples. Richard Schweickert and Jason Saleeby provided careful and instructive reviews.

## REFERENCES CITED

Belasky, P., and Runnegar, B., 1994, Permian longitudes of Wrangellia, Stikinia, and eastern Klamath terranes based on coral biogeography: Geology, v. 22, p. 1095–1098.

Boudier, F., LeSueur, E., Nicolaus, A., 1989, Structure of an atypical ophiolite; the Trinity complex, eastern Klamath Mountains, California: Geological Society of America Bulletin, v. 101, p. 820–833.

Brouxel, M., and LaPierre, H., 1988, Geochemical study of an early Paleozoic island-arc–back-arc system. Part 1: The Trinity ophiolite (northern California): Geological Society of America Bulletin, v. 100, p. 1111–1119.

Brouxel, M., LaPierre, H., Michard, A., and Albarede, F., 1988, Geochemical study of an early Paleozoic island-arc–back-arc system. Part 2: Eastern Klamath, early to middle Paleozoic island-arc volcanic rocks (northern California):

Geological Society of America Bulletin, v. 100, p. 1120–1130.

Burchfiel, B.C., and Davis, G.A., 1972, Structural framework and evolution of the southern part of the Cordilleran orogen, western United States: American Journal of Science, v. 272, p. 97–118.

Charvet, J., LaPierre, H., and Campos, C., 1989, Les effets de la phase Antler (Devonian Superieur-Carbonifere inferieur) dans les Klamath orientales (Californie, U.S.A.): Implications geodynamiques: Paris, Comptes Rendus, Académie des Sciences ser. II, p. 1629–1635.

Charvet, J., Lapierre, H., Rouer, O., Coulon, C., Campos, C., Martin, P., and Lecuyer, C., 1990, Tectono-magmatic evolution of Paleozoic and early Mesozoic rocks in the eastern Klamath Mountains, California, and the Blue Mountains, eastern Oregon–Western Idaho, *in* Harwood, D.S., and Miller, M.M., eds., Paleozoic and early Mesozoic paleogeographic relations; Sierra Nevada, Klamath Mountains, and related terranes: Geological Society of America Special Paper 255, p. 255–276.

Dickinson, W.R., 1977, Paleozoic plate tectonics and the evolution of the Cordilleran continental margin, *in* Stewart, J.H., et al., eds., Paleozoic paleogeography of the western United States: Pacific Section, Society of Economic Paleontologists and Mineralogists, Pacific Coast Paleogeography Symposium I, p. 137–155.

Diller, J.S., 1905, The Bragdon Formation: American Journal of Science, ser. 4 p. 379–387.

Gehrels, G.E., and Dickinson, W.R., 1995, Detrital zircon provenance of Cambrian to Triassic miogeoclinal and eugeoclinal strata in Nevada: American Journal of Science, v. 295, p. 18–48.

Hacker, B.R., and Goodge, J.W., 1990, Comparison of early Mesozoic high-pressure rocks in the Klamath Mountains and Sierra Nevada, *in* Harwood, D.S., and Miller, M.M., eds., Paleozoic and early Mesozoic paleogeographic relations; Sierra Nevada, Klamath Mountains, and related terranes: Geological Society of America Special Paper 255, p. 277–295.

Harland, W.B., Armstrong, R.L., Cox, A.V., Craig, L.E., Smith, A.G., and Smith, D.G., 1990, A geologic time scale: Cambridge, Cambridge University Press, 263 p.

Harwood, D.S., and Murchey, B.L., 1990, Biostratigraphic, tectonic, and paleogeographic ties between upper Paleozoic volcanic and basinal rocks in the northern Sierra terrane, California, and the Havallah Sequence, Nevada, *in* Harwood, D.S., and Miller, M.M., eds., Paleozoic and early Mesozoic paleogeographic relations; Sierra Nevada, Klamath Mountains, and related terranes: Geological Society of America Special Paper 255, p. 157–173.

Hoffman, P.F., 1989, Precambrian geology and tectonic history of North America, *in* Bally, A.W., and Palmer, A.R., eds., The geology of North America—An overview: Boulder, Colorado, Geological Society of America, Geology of North America, v. A, p. 447–512.

Irwin, W.P., 1966, Geology of the Klamath Mountains province: California Division of Mines and Geology Bulletin 190, p. 19–38.

Irwin, W.P., 1977, Review of Paleozoic rocks of the Klamath Mountains, *in* Stewart, J.H., et al., eds., Paleozoic paleogeography of the western United States: Pacific Section, Society of Economic Paleontologists and Mineralogists, Pacific Coast Paleogeography Symposium I, p. 441–454.

Irwin, W.P., 1981, Tectonic accretion of the Klamath Mountains, *in* Ernst, W.G., ed., The geotectonic evolution of California (Rubey Volume I): Englewood Cliffs, New Jersey, Prentice-Hall, p. 29–49.

Jacobsen, S.B., Quick, J.E., and Wasserburg, G.J., 1984, A Nd and Sr isotopic study of the Trinity peridotite; implications for mantle evolution: Earth and Planetary Science Letters, v. 68, p. 361–378.

Kinkel, A.R., Jr., Hall, W.E., and Albers, J.P., 1956, Geology and base-metal deposits of west Shasta Copper-Zinc district, Shasta County, California: U.S. Geological Survey Professional Paper 285, 156 p.

Kistler, R.W., 1991, Chemical and isotopic characteristics of plutons in the Great Basin, *in* Raines, G.L., et al., eds., Geology and ore deposits of the Great Basin: Reno, Geological Society of Nevada, p. 107–109.

Kistler, R.W., and Peterman, Z.E., 1973, Variations in Sr, Rb, K, Na, and initial 87Sr/86Sr in Mesozoic granitic rocks and intruded wall rocks in central California: Geological Society of America Bulletin, v. 84, p. 3489–3512.

Ludwig, K.R., 1991a, A computer program for processing Pb-U-Th isotopic data: U.S. Geological Survey Open-File Report 88-542.

Ludwig, K.R., 1991b, A plotting and regression program for radiogenic-isotopic data: U.S. Geological Survey Open-File Report 91-445.

Metcalf, R.V., Wallin, E.T., and Willse, K., 1998, Devonian volcanic rocks of the eastern Klamath Mountains: An island arc or volcanic cover of the Trinity ophiolite: Geological Society of America Abstracts with Programs, v. 30, no. 5, p. 54.

Miller, M.M., 1987, Dispersed remnants of a northeast Pacific fringing arc; upper Paleozoic terranes of Permian McCloud faunal affinity, western U.S.: Tectonics, v. 6, p. 807–830.

Miller, M.M., and Cui, B., 1989, Submarine fan characteristics and dual sediment provenance, Lower Carboniferous Bragdon Formation, eastern Klamath terrane, California: Canadian Journal of Earth Sciences, v. 26, p. 927–940.

Miller, M.M., and Harwood, D.S., 1990, Paleogeographic setting of upper Paleozoic rocks in the northern Sierra and eastern Klamath terranes, northern California, *in* Harwood, D.S., and Miller, M.M., eds., Paleozoic and early Mesozoic paleogeographic relations; Sierra Nevada, Klamath Mountains, and related terranes: Geological Society of America Special Paper 255, p. 175–192.

Miller, M.M., and Saleeby, J.B., 1991, Continental detrital zircon in Carboniferous ensimatic arc rocks, Bragdon Formation, eastern Klamath terrane, northern California: Geological Society of America Bulletin, v. 103, p. 268–276.

Oldow, J.S., 1984, Evolution of a late Mesozoic back-arc fold and thrust belt, northwestern Great Basin, U.S.A.: Tectonophysics, v. 102, p. 245–274.

Peacock, S.M., and Norris, P.J., 1989, Metamorphic evolution of the CMB, Klamath province, California: An inverted metamorphic sequence beneath the Trinity peridotite: Journal of Metamorphic Geology, v. 7, p. 191–209.

Potter, A.W., Hotz, P.E., and Rohr, D.M., 1977, Stratigraphy and inferred tectonic framework of lower Paleozoic rocks in the eastern Klamath Mountains, northern California, *in* Stewart, J.H., et al., eds., Paleozoic paleogeography of the western United States: Pacific Section, Society of Economic Paleontologists and Mineralogists, Pacific Coast Paleogeography Symposium 1, p. 421–440.

Potter, A.W., Boucot, A.J., Bergstrom, S.M., Blodgett, R.B., Dean, W.T., Flory, R.A., Ormiston, A.R., Pedder, A.E.H., Rigby, J.K., Rohr, D.M., and Savage, N.M., 1990, Early Paleozoic stratigraphic, paleogeographic, and biogeographic relations of the eastern Klamath belt, northern California, *in* Harwood, D.S., and Miller, M.M., eds., Paleozoic and early Mesozoic paleogeographic relations; Sierra Nevada, Klamath Mountains, and related terranes: Geological Society of America Special Paper 255, p. 57–74.

Quick, J.E., 1981, Petrology and petrogenesis of the Trinity peridotite, an upper mantle diapir in the eastern Klamath Mountains, northern California: Journal of Geophysical Research, v. 86, p. 11837–11863.

Ross, G.M., 1991, Precambrian basement in the Canadian Cordillera: An introduction: Canadian Journal of Earth Sciences, v. 28, p. 1133–1139.

Rubin, C.M., Miller, M.M., and Smith, G.E., 1990, Tectonic development of Cordilleran mid-Paleozoic volcanoplutonic complexes: Evidence for convergent margin tectonism, *in* Harwood, D.S., and Miller, M.M., eds., Paleozoic and early Mesozoic paleogeographic relations; Sierra Nevada, Klamath Mountains, and related terranes: Geological Society of America Special Paper 255, p. 1–16.

Saleeby, J.B., 1992, Petrotectonic and paleogeographic settings of U.S. Cordilleran ophiolites, *in* Burchfiel, B.C., et al., eds., The Cordilleran orogen: Conterminous U.S.: Boulder, Colorado, Geological Society of America, Geology of North America, v. G-3, p. 653–682.

Schweickert, R.A., and Snyder, W.S., 1981, Paleozoic plate tectonics of the Sierra Nevada and adjacent regions, *in* Ernst, W.G., ed., The geotectonic development of California: Englewood Cliffs, New Jersey, Prentice-Hall, p. 183–201.

Silberling, N.J., 1991, Allochthonous terranes of western Nevada: Current status, *in* Raines, G.L., et al., eds; Geology and ore deposits of the Great Basin: Reno, Geological Society of Nevada, p. 101–102.

Silberling, N.J., Jones, D.L., Blake, M.C., Jr., and Howell, D.G., 1987, Lithotectonic

terrane map of the western conterminous United States: U.S. Geological Survey Miscellaneous Field Studies Map MF-1874-C, scale 1:2,500,000.
Speed, R.C., 1979, Collided Paleozoic microplate in the western United States: Journal of Geology, v. 87, p. 279–292.
Stevens, C.H., Yancey, T.E., and Hanger, R.A., 1990, Significance of the provincial signature of Early Permian faunas of the eastern Klamath terrane; *in* Harwood, D.S., and Miller, M.M., eds., Paleozoic and early Mesozoic paleogeographic relations; Sierra Nevada, Klamath Mountains, and related terranes: Geological Society of America Special Paper 255, p. 201–218.
Stewart, J.H., Poole, F.G., Ketner, K.B., Madrid, R.J., Roldan-Quintana, J., and Amaya-Martinez, R., 1990, Tectonics and stratigraphy of the Paleozoic and Triassic southern margin of North America, Sonora, México, *in* Gehrels, G.E., and Spencer, J.E., eds., Geologic excursions through the Sonoran Desert region, Arizona and Sonora: Arizona Geological Survey Special Paper 7, p. 183–202.
Tucker, R.D., Bradley, D.C., Ver Straeten, C.A., Harris, A.G., Ebert, J.R., and McCutcheon, S.R., 1998, New U-Pb zircon ages and the duration and division of Devonian time: Earth and Planetary Science Letters, v. 158, p. 175–186.
Van Schmus, W.R., and 24 others, 1993, Transcontinental Proterozoic provinces, *in* Reed, J.C., et al., eds., Precambrian: Conterminous U.S.: Boulder, Colorado, Geological Society of America, Geology of North America, v. C-2, p. 171–334.
Wallin, E.T., 1989, Provenance of lower Paleozoic sandstones in the eastern Klamath Mountains and the Roberts Mountains allochthon, California and Nevada [Ph.D. Thesis]: Lawrence, University of Kansas, 152 p.
Wallin, E.T., Mattinson, J.M., and Potter, A.W., 1988, Early Paleozoic magmatic events in the eastern Klamath Mountains, northern California: Geology, v. 16, p. 144–148.
Wallin, E.T., Coleman, D.S., Lindsey-Griffin, N., and Potter, A.W., 1995, Silurian plutonism in the Trinity terrane (Neoproterozoic and Ordovician), Klamath Mountains, California, United States: Tectonics, v. 14, p. 1007–1013.
Watkins, R., 1985, Volcaniclastic and carbonate sedimentation in late Paleozoic island-arc deposits, eastern Klamath Mountains, California: Geology, v. 13, p. 709–713.
Wright, J.E., 1982, Permo-Triassic accretionary subduction complex, southwestern Klamath Mountains, northern California: Journal of Geophysical Research, v. 87, p. 3805–3818.

Manuscript Accepted by the Society January 24, 2000

Printed in U.S.A.

Geological Society of America
Special Paper 347
2000

# *Detrital zircon geochronology of Upper Triassic strata in western Nevada*

**Jeffrey D. Manuszak***
*Department of Geosciences, University of Arizona, Tucson, Arizona 85721, USA*
**Joseph I. Satterfield**
*Geology Department, San Jacinto College North, Houston, Texas 77049, USA*
**George E. Gehrels**
*Department of Geosciences, University of Arizona, Tucson, Arizona 85721, USA*

## ABSTRACT

**Several distinct assemblages of basinal, volcanic arc, and shelfal strata in western Nevada are separated from each other and from adjacent assemblages by known and inferred thrust and strike-slip faults. We analyzed 84 detrital zircon grains from Upper Triassic strata in three assemblages to constrain their paleogeography and tectonic history. Basinal strata that probably belong to the Lovelock assemblage yielded grains with ages between 950–1140 Ma, three grains of ca. 1417 Ma, and grains with a scattering of ages between 227 and 683 Ma. Shelfal strata of the Luning assemblage yield grains with ages mainly of 218–229 and 1643–1966 Ma; additional grains have ages of ca. 272, 1058, 1089, 1438, 1444, and 2325 Ma. All grains analyzed from the volcanic-rich Pine Nut assemblage yielded ages of ca. 231 Ma.**

**The detrital zircon grains were apparently derived from sources both to the east and the west. The 1.40–1.45 and 1.63–1.74 Ga grains most likely originated in Precambrian basement of the southwestern United States, which contains igneous rocks of these ages. Grains >1.8 Ga could have been derived from the Golconda or Roberts Mountains allochthons to the east, or the northern Sierra and eastern Klamath terranes to the west. Grains of Paleozoic and Triassic age probably originated in preexisting and active magmatic arcs to the west. These provenance ties are consistent with stratigraphic and regional tectonic arguments suggesting that Triassic assemblages in western Nevada formed in a backarc basin between the Sierra-Klamath arc terranes to the west and the Cordilleran continental margin to the east.**

## INTRODUCTION

Mesozoic terranes or lithotectonic assemblages in western North America are controversial, and diverse and conflicting histories have been proposed (e.g., Burchfiel and Davis, 1972, 1975; Coney et al., 1980; Oldow, 1984; Speed et al., 1989; Schweickert and Lahren, 1990; Burchfiel et al., 1992). Major differences among interpretations concern: the number of terranes or assemblages, the locations of assemblage boundaries, and the amounts and senses of displacement on assemblage-bounding faults.

In most interpretations, Mesozoic assemblages resided in western Nevada, near the North American craton, throughout their history. Triassic and Jurassic age strata of western Nevada are interpreted by most workers to have accumulated in a backarc basin between a west-facing volcanic arc, now preserved in the northern Sierra and eastern Klamath terranes, and primarily nonmarine strata deposited on the North American craton to the east.

*Current address: Department of Earth and Atmospheric Sciences, Purdue University, West Lafayette, Indiana 47906, USA.

Manuszak, J.D., Satterfield, J.I., and Gehrels, G.E., 2000, Detrital zircon geochronology of Upper Triassic strata in western Nevada, *in* Soreghan, M.J., and Gehrels, G.E., eds., Paleozoic and Triassic paleogeography and tectonics of western Nevada and northern California: Boulder, Colorado, Geological Society of America Special Paper 347, p. 109–118.

In an alternative interpretation, several of these assemblages and their ophiolitic basement were accreted to North America during a Middle Jurassic collision with an east-facing volcanic arc (Speed, 1978; Dilek et al., 1988).

### Regional geology

The Mesozoic assemblages of western Nevada consist of coeval or partly coeval rocks with different rock types, depositional environments, and/or structural histories (Speed, 1978; Oldow, 1984; Speed et al., 1988, 1989; Silberling, 1991; Oldow et al., 1993) (Figs. 1 and 2). The Humboldt and Luning assemblages contain siliciclastic and carbonate rocks of Triassic and Jurassic age that were deposited in shallow-marine, intertidal, and deltaic environments. The Pine Nut and Pamlico assemblages contain Triassic–Jurassic volcanic, volcanogenic, and carbonate rocks deposited in or near a volcanic arc in shallow-marine to subaerial environments. The Sand Springs and Black Rock assemblages include Triassic–Jurassic volcanic, volcanogenic, and carbonate rocks deposited primarily in deep-marine environments. The Lovelock assemblage contains Triassic–Jurassic siliciclastic rocks and carbonate rocks deposited primarily in deep-marine environments.

These assemblages are separated by mapped and inferred thrust and strike-slip faults of Jurassic and/or Cretaceous age (Oldow, 1981, 1984; Schweickert and Lahren, 1990; Oldow et al., 1993). The Luning and Fencemaker thrust faults (Fig. 2) are the frontal faults of the Luning-Fencemaker thrust belt. The Humboldt and Gold Range assemblages (Fig. 2), in the footwall of the thrust belt, are parautochthonous and unconformably overlie Paleozoic eugeoclinal rocks that were accreted to the continental margin prior to Late Triassic time. The Luning, Pamlico,

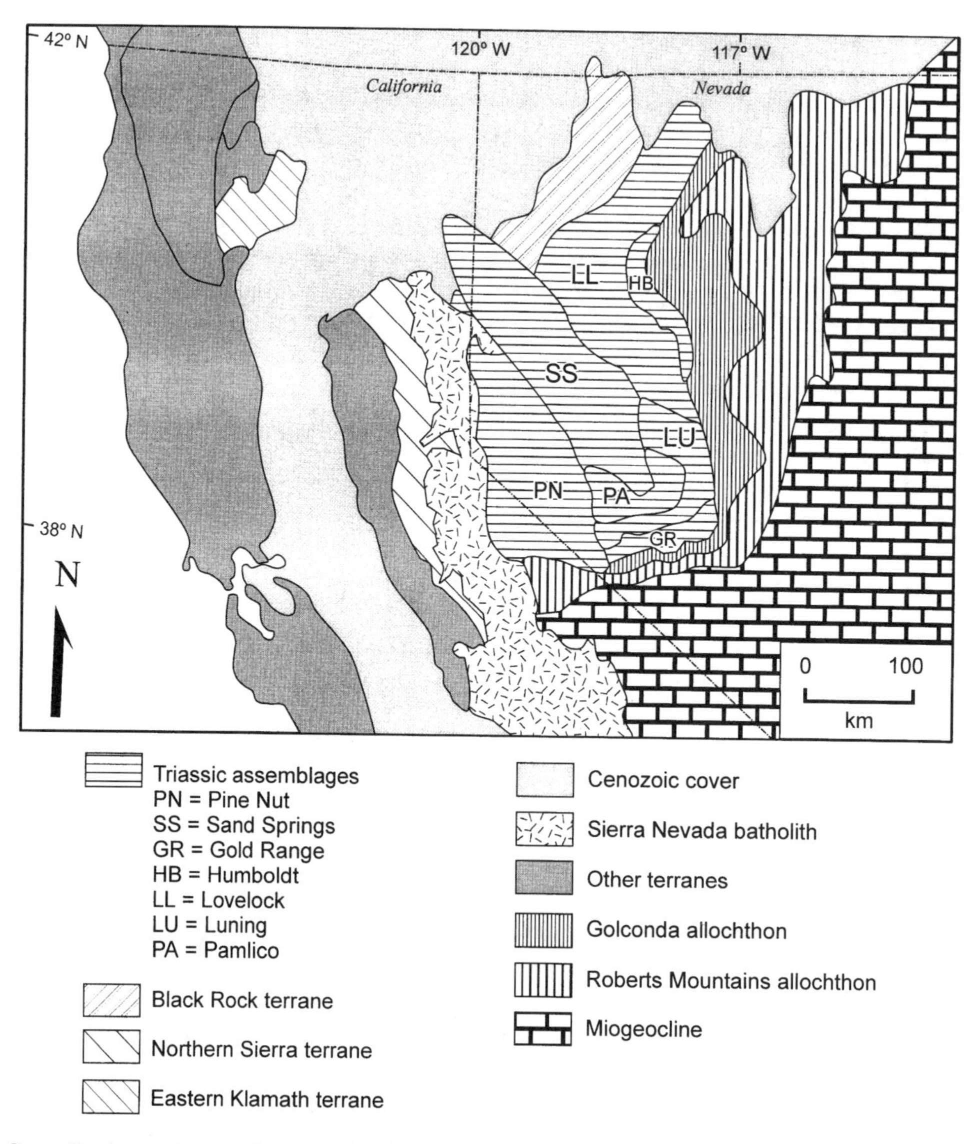


Figure 1. Generalized tectonic map of western Nevada and northern California (adapted from Silberling, 1991; Oldow, 1984).

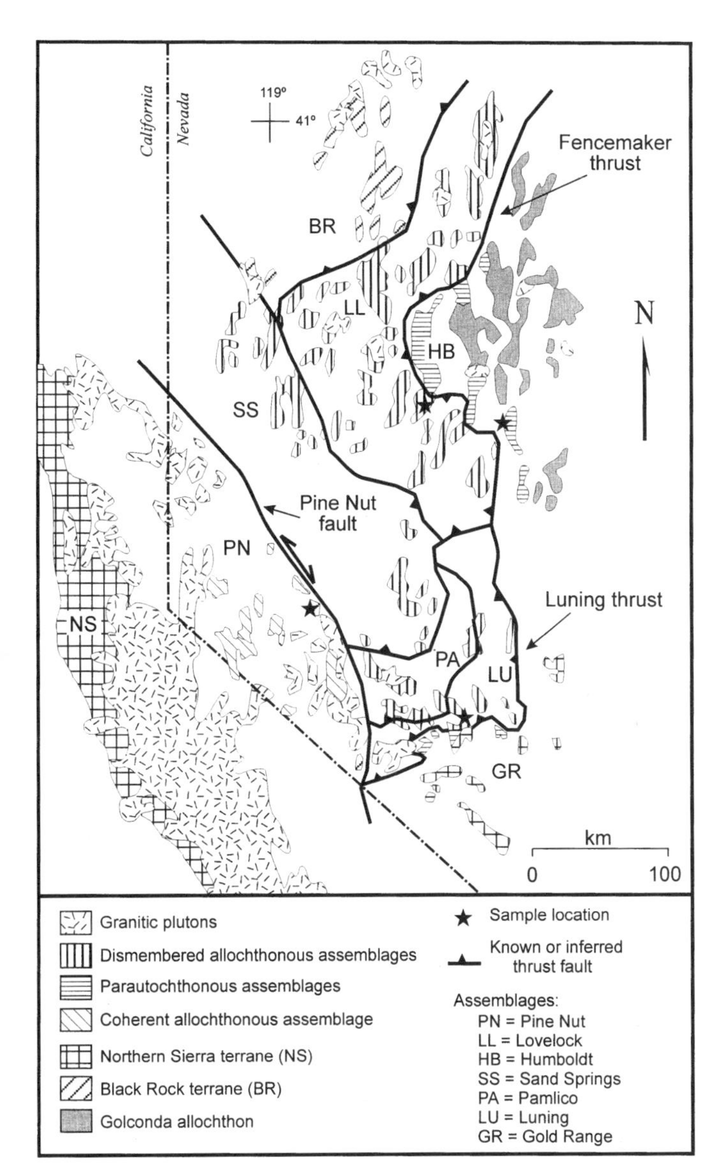


Figure 2. Pre-Cenozoic exposures in western Nevada and northern California showing assemblage boundaries (modified from Silberling, 1991; Oldow et al., 1993).

Sand Springs, Lovelock, and Black Rock assemblages (Fig. 2), all bounded by mapped and inferred thrust faults of the Luning-Fencemaker belt, record several hundred kilometers of northwest-southeast shortening. Individual thrust faults, such as the Luning thrust, have displacements estimated at tens of kilometers. The Pine Nut assemblage is separated from assemblages to the east by a mapped and inferred strike-slip fault, the Pine Nut fault (Fig. 2), which was inferred by Oldow (1984) and Oldow et al. (1993) to have accommodated several hundred kilometers of left-lateral motion during displacement on the Luning-Fencemaker thrust belt. The Pine Nut fault may also be the northern continuation of the Mojave–Snow Lake fault, in which case it would have undergone ~400 km of dextral displacement during Early Cretaceous time (Schweickert and Lahren, 1990).

### *Present study*

The purpose of this study is to use the ages of detrital zircons to determine the provenance of clastic detritus in several of the Triassic assemblages of western Nevada. Samples were collected from the Pine Nut, Lovelock(?), and Luning assemblages (Figs. 2 and 3). An additional sample from the Pamlico assemblage, from the Pamlico Formation in the western part of the Garfield Hills (field trip stop 2-2 of Oldow et al. [1993], 38°28′20″N, 118°27′45″W), was processed but did not yield zircons.

Ages from these samples complement detrital zircon data from Triassic strata of the Black Rock terrane (Darby et al., this volume, Chapter 5) and the Humboldt assemblage (Gehrels and Dickinson, 1995). Collectively these data sets allow comparisons between shelfal, basinal, and arc-type assemblages, and between parautochthonous, coherent allochthonous, and dismembered allochthonous strata.

Each sample was collected from a narrow stratigraphic interval at a single outcrop. Zircons were extracted using conventional mineral separation techniques (Gehrels, this volume, Introduction). The individual zircons were sieved into size fractions, and the larger (>125 μm) grains were separated into populations based on their color (pink, yellowish, and brownish) and degree of textural maturity (rounded versus euhedral). Representatives of each color and/or morphology group were then abraded to ~70% of their original size, cleaned in warm $HNO_3$, and at least 20 grains from each sample were selected for analysis. The U-Pb isotope analyses were conducted utilizing standard isotope dilution-thermal ionization mass spectrometry (Gehrels, this volume, Introduction).

The data are reported in Table 1, plotted on $^{206}Pb^*/^{238}U$ vs. $^{207}Pb^*/^{235}U$ concordia diagrams in Figure 4, and summarized on relative age-probability plots on Figure 5. For concordant analyses, the interpreted ages (Table 1) are determined from $^{206}Pb^*/^{238}U$ ages for zircons younger than ca. 1.0 Ga and $^{207}Pb^*/^{206}Pb^*$ ages for older zircons. For discordant analyses, all of which are >700 Ma, the interpreted age is the upper concordia intercept of a discordia projected from 100 Ma. This results in an age that is slightly older than the $^{207}Pb^*/^{206}Pb^*$ age, and is considered more reliable, because the main phase of regional metamorphism and isotopic disturbance in the region ended prior to ca. 100 Ma (Speed et al., 1988). There is little evidence for recent thermal disturbance, which is implicit in the use of $^{207}Pb^*/^{206}Pb^*$ ages. Because an interpreted age is less reliable with increasing degree of discordance, the ages assigned in Table 1 and used in constructing Figure 5 have uncertainties that increase with degree of discordance.

## RESULTS

### *Pine Nut assemblage*

Mesozoic strata in the northern part of the Wassuk Range comprise one of the most complete and least metamorphosed

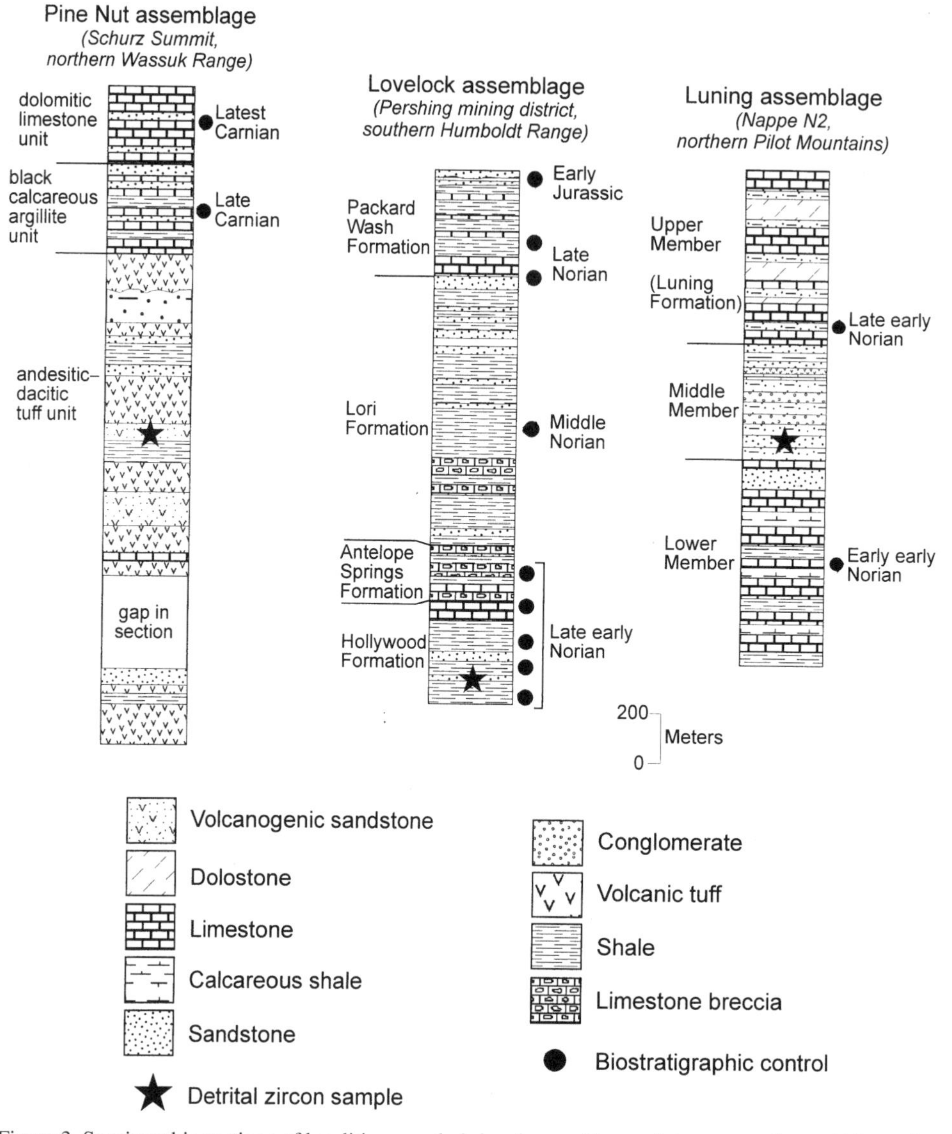


Figure 3. Stratigraphic sections of localities sampled showing positions of zircon samples and biostratigraphic control. Sources of data used to construct columns: Schurz Summit, northern Wassuk Range—Proffett and Dilles (1984), D.E. Cameron (1998, written commun.); Nappe N2, northern Pilot Mountains—Oldow (1981), Reilly et al. (1980), Silberling (1984); Pershing mining district, southern Humboldt Range—Oldow et al. (1990).

sections in the Pine Nut assemblage, and include a continuous 2700-m-thick section divided into three map units (Fig. 3; Proffett and Dilles, 1984; Oldow et al., 1993). The basal unit is composed of interbedded crystal-lithic tuff, volcanogenic sandstone, volcanogenic conglomerate, and minor shale and limestone. A middle unit of black calcareous argillite, known to be of Late Carnian age (Proffett and Dilles, 1984), contains interbedded limestone, shale, and volcanogenic rocks. The upper unit contains mostly limestone and lesser amounts of shale, conglomerate, and volcanogenic sandstone. Similar Upper Triassic sections elsewhere in the Pine Nut assemblage are interpreted to have been deposited in shallow-marine environments (Oldow, 1984; Oldow et al., 1993; Wyld and Wright, 1993).

One sample was collected from near the middle of the lower, volcanic-rich unit of Proffett and Dilles (1984) in the northernmost Wassuk Range (near Schurz Summit; Fig. 2). The sample is a medium- to coarse-grained sandstone composed of ~43% plagioclase and 16%–21% each of volcanic rock fragments, sedimentary rock fragments, and monocrystalline quartz (Dickinson and Gehrels, this volume, Chapter 11).

Zircons recovered from this sample are predominantly colorless to light pink and euhedral. Subordinate populations are yellowish to brownish. No rounded or pinkish grains were found in any of the size fractions. We analyzed 20 individual detrital zircons, all >125 μm in sieve size prior to abrasion (Table 1). All of the grains are apparently concordant, and 17 analyses are of mod-

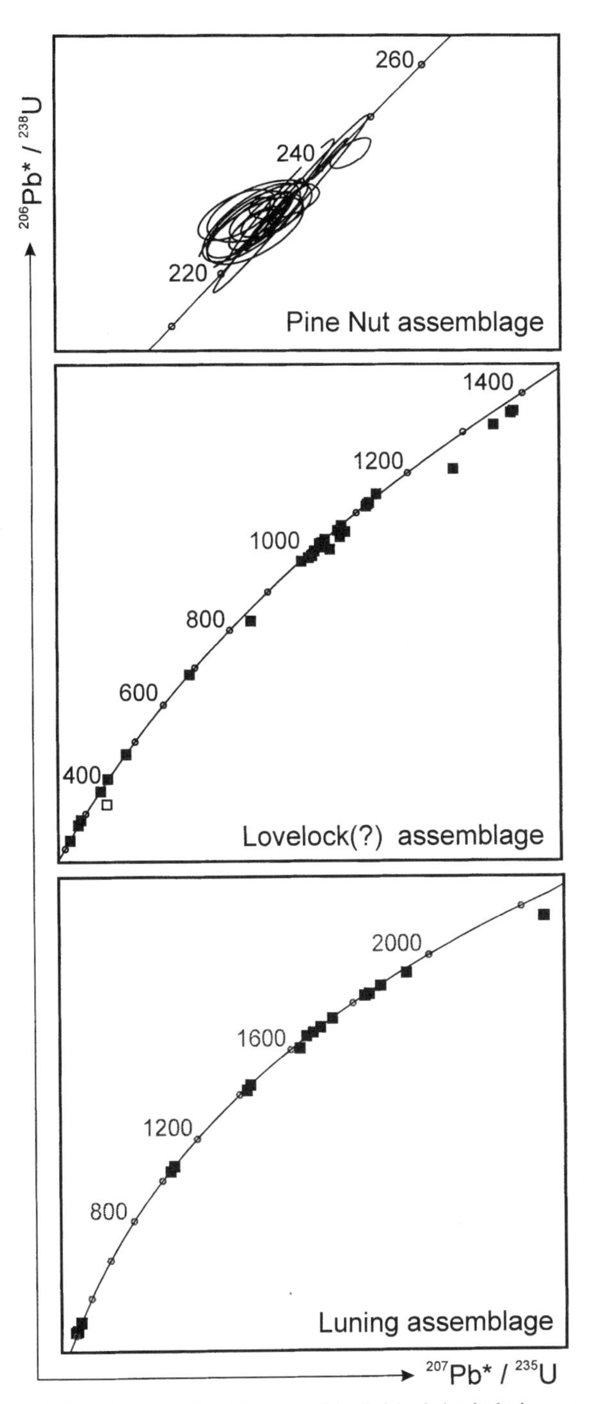


Figure 4. U-Pb concordia diagram of individual detrital zircon grains from Lovelock(?), Luning, and Pine Nut assemblages. Filled boxes are used rather than ellipses where ellipse would not be visible at scale of diagram. Open square in Lovelock(?) sample is sufficiently discordant that it is not considered further. Data reduction and plotting are from Ludwig (1991a, 1991b).

erate to high precision (Fig. 4). The interpreted ages of these grains, based on their $^{206}Pb^{*}/^{238}U$ ages, range from 228 to 243 Ma, and a peak in age probability occurs at 231 Ma (Fig. 5; Table 2). The age of these grains indicates that the sandstone is probably latest Middle to early Late Triassic, because the Middle-Late Triassic boundary is ca. 227 Ma, according to the time scale of Gradstein et al. (1994).

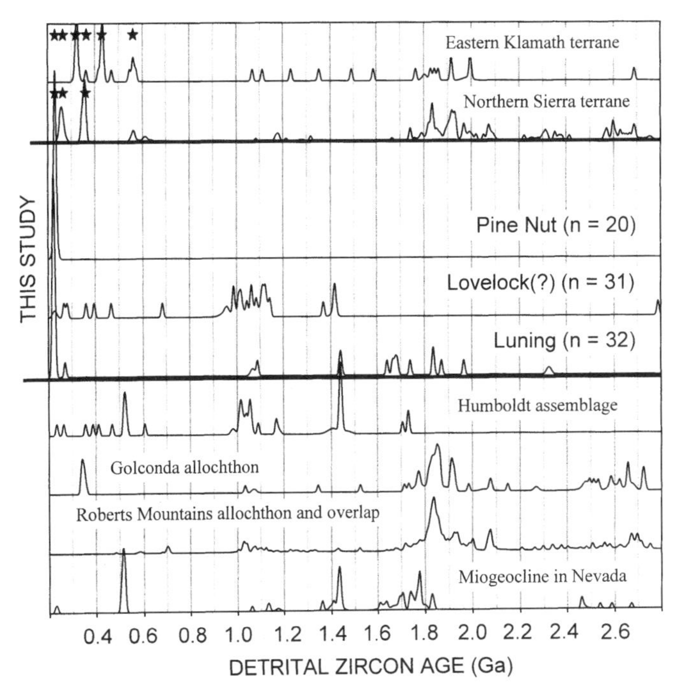


Figure 5. Relative age-probability plots of detrital zircon ages from Pine Nut, Lovelock(?), and Luning assemblages (this study), together with detrital zircon ages from strata in northern Sierra terrane (Harding et al., this volume; Spurlin et al., this volume), eastern Klamath terrane (Gehrels and Miller, this volume), Golconda allochthon (Riley et al., this volume) and overlying Triassic strata of Humboldt assemblage (Gehrels and Dickinson, 1995), Roberts Mountains allochthon (Gehrels et al., this volume, Chapter 1) and overlying strata (Gehrels and Dickinson, this volume), and Cordilleran miogeocline (Gehrels and Dickinson, 1995). Ages of igneous rocks in Sierra Nevada and eastern Klamath terranes, indicated with stars, are from Miller and Harwood (1990) and Saleeby and Busby-Spera (1992).

### *Lovelock(?) assemblage*

The Hollywood Formation, the basal map unit of a continuous 2300 m section, consists of shale, siltstone, and intercalated nonvolcanogenic sandstone, limestone breccia, and minor limestone beds (Fig. 3; Oldow et al., 1990). These strata are interpreted to have been deposited in basinal submarine fan or slope environments (Oldow et al., 1990; Heck and Speed, 1987).

Our sample was obtained from the Hollywood Formation in the southern Humboldt Range, near the Hollywood mine in the Pershing mining district (Wallace et al., 1969; Heck and Speed, 1987; Oldow et al., 1990). Late early Norian ammonites recovered from strata above and below the sample constrain this sample to be late early Norian (Fig. 3; Oldow et al., 1990). Figure 2 shows the sample location in the hanging wall of the Fencemaker thrust, following the interpretation of Oldow et al. (1990). In contrast, the sample location is in the footwall of the Fencemaker thrust on the map of Heck and Speed (1987). Pending more

TABLE 1. U-Pb ISOTOPIC DATA AND AGES

| Grain type | Grain wt. (μg) | $Pb_c$ (pg) | Pb (ppm) | U (ppm) | $^{206}Pb_m/^{204}Pb$ | $^{206}Pb_c/^{204}Pb$ | $^{206}Pb/^{208}Pb$ | $^{206}Pb^*/^{238}U$ | $^{207}Pb^*/^{235}U$ | $^{207}Pb^*/^{206}Pb^*$ | Interpreted age (Ma) |
|---|---|---|---|---|---|---|---|---|---|---|---|
| Pine Nut (Carnian: 39°02'29"N, 118°58'45"W, 4966' elevation) | | | | | | | | | | | |
| LE | 16 | 18 | 9.78 | 246 | 515 | 695 | 5.9 | 228 ± 3 | 227 ± 6 | 219 ± 45 | 228 ± 10 |
| BOE | 4 | 19 | 28.29 | 676 | 340 | 450 | 4.7 | 228 ± 4 | 227 ± 7 | 219 ± 59 | 228 ± 10 |
| LE | 12 | 30 | 12.47 | 285 | 280 | 330 | 4.3 | 228 ± 3 | 228 ± 6 | 223 ± 53 | 228 ± 10 |
| LE | 14 | 13 | 17.43 | 466 | 1140 | 1805 | 7.7 | 228 ± 3 | 229 ± 3 | 231 ± 21 | 228 ± 10 |
| LE | 13 | 22 | 7.05 | 159 | 240 | 300 | 4.2 | 228 ± 5 | 230 ± 10 | 254 ± 79 | 228 ± 10 |
| BOE | 5 | 24 | 23.78 | 555 | 280 | 345 | 4.8 | 229 ± 5 | 230 ± 9 | 245 ± 71 | 229 ± 10 |
| BOE | 7 | 24 | 15.87 | 357 | 265 | 325 | 4.0 | 229 ± 5 | 225 ± 10 | 183 ± 83 | 229 ± 20 |
| LE | 19 | 31 | 14.18 | 343 | 500 | 590 | 5.0 | 230 ± 2 | 229 ± 4 | 215 ± 37 | 230 ± 10 |
| YE | 12 | 23 | 12.4 | 293 | 370 | 460 | 4.7 | 230 ± 3 | 230 ± 6 | 224 ± 53 | 230 ± 10 |
| BOE | 5 | 28 | 20.31 | 428 | 195 | 230 | 3.6 | 230 ± 5 | 230 ± 12 | 230 ± 98 | 230 ± 10 |
| LE | 13 | 24 | 40.43 | 1082 | 1340 | 1680 | 8.2 | 230 ± 5 | 230 ± 6 | 231 ± 19 | 230 ± 10 |
| LE | 15 | 20 | 6.97 | 164 | 310 | 410 | 4.9 | 231 ± 4 | 227 ± 8 | 187 ± 65 | 231 ± 20 |
| LE | 20 | 30 | 7.19 | 163 | 275 | 320 | 4.5 | 231 ± 4 | 230 ± 8 | 219 ± 68 | 231 ± 10 |
| YE | 6 | 33 | 15.3 | 290 | 145 | 165 | 2.9 | 232 ± 5 | 228 ± 13 | 186 ± 120 | 232 ± 20 |
| YE | 5 | 26 | 67.18 | 1701 | 770 | 945 | 6.2 | 232 ± 3 | 233 ± 4 | 242 ± 31 | 232 ± 10 |
| LE | 44 | 29 | 12.1 | 312 | 1110 | 1330 | 7.1 | 233 ± 11 | 234 ± 11 | 241 ± 21 | 233 ± 10 |
| LE | 39 | 29 | 10.81 | 272 | 885 | 1065 | 6.3 | 234 ± 3 | 233 ± 4 | 221 ± 27 | 234 ± 10 |
| ME | 11 | 23 | 13.62 | 323 | 378 | 475 | 5.2 | 234 ± 3 | 233 ± 8 | 232 ± 76 | 234 ± 10 |
| LE | 62 | 27 | 17.09 | 404 | 2260 | 2770 | 4.5 | 241 ± 3 | 241 ± 3 | 241 ± 9 | 241 ± 10 |
| BOE | 4 | 38 | 58.46 | 1152 | 320 | 360 | 2.7 | 243 ± 4 | 244 ± 7 | 252 ± 59 | 243 ± 10 |
| Lovelock(?) (Early Norian: 40°08'47"N, 118°06'32"W, 4576' elevation) | | | | | | | | | | | |
| LE | 18 | 10 | 28.75 | 826 | 3390 | 6720 | 14.2 | 227 ± 24 | 227 ± 24 | 231 ± 10 | 227 ± 20 |
| LE | 17 | 7 | 9.26 | 210 | 1340 | 4190 | 6.6 | 267 ± 3 | 266 ± 4 | 258 ± 19 | 267 ± 10 |
| LE | 10 | 4 | 12.17 | 255 | 2000 | 6195 | 5.4 | 281 ± 3 | 279 ± 3 | 257 ± 19 | 281 ± 10 |
| LE | 14 | 12 | 16.28 | 240 | 1020 | 1695 | 3.4 | 360 ± 10 | 362 ± 10 | 379 ± 21 | 360 ± 10 |
| YE | 22 | 7 | 35.6 | 537 | 6950 | 25730 | 6.1 | 394 ± 3 | 393 ± 3 | 382 ± 5 | 394 ± 10 |
| YE | 20 | 5 | 57.34 | 756 | 13515 | 209930 | 8.4 | 466 ± 3 | 466 ± 3 | 469 ± 5 | 466 ± 10 |
| LS | 9 | 12 | 30.09 | 247 | 1345 | 2290 | 5.4 | 683 ± 6 | 684 ± 8 | 689 ± 13 | 683 ± 10 |
| LE | 12 | 10 | 25.79 | 486 | 1950 | 3890 | 8.7 | 325 ± 9 | 390 ± 11 | 792 ± 9 | 948 ± 40 |
| LS | 9 | 3 | 43.87 | 323 | 7580 | 30430 | 11.6 | 824 ± 5 | 857 ± 6 | 946 ± 7 | 958 ± 20 |
| LE | 17 | 13 | 41.09 | 249 | 3410 | 5570 | 8.8 | 963 ± 5 | 969 ± 7 | 984 ± 9 | 986 ± 10 |
| LS | 10 | 3 | 32.94 | 195 | 5900 | 21220 | 8.2 | 977 ± 5 | 980 ± 5 | 987 ± 6 | 988 ± 10 |
| LE | 18 | 12 | 22.21 | 123 | 1890 | 3130 | 5.8 | 995 ± 6 | 999 ± 8 | 1007 ± 10 | 1008 ± 10 |
| BOE | 6 | 105 | 167.71 | 1158 | 515 | 540 | 6.1 | 782 ± 7 | 838 ± 14 | 989 ± 29 | 1012 ± 40 |
| LE | 20 | 11 | 19.36 | 115 | 2320 | 4230 | 14.4 | 1018 ± 7 | 1018 ± 9 | 1018 ± 11 | 1018 ± 10 |
| LS | 15 | 155 | 83.85 | 361 | 380 | 390 | 2.4 | 1003 ± 6 | 1008 ± 21 | 1020 ± 38 | 1021 ± 20 |
| BOE | 13 | 4 | 59.21 | 339 | 21000 | 45554 | 11.8 | 1039 ± 5 | 1040 ± 5 | 1043 ± 5 | 1044 ± 10 |
| LE | 15 | 18 | 40.21 | 219 | 1980 | 2700 | 7.9 | 1046 ± 6 | 1050 ± 9 | 1060 ± 12 | 1061 ± 10 |
| LE | 15 | 9 | 16.13 | 86 | 1600 | 3440 | 7.5 | 1064 ± 9 | 1064 ± 10 | 1065 ± 11 | 1065 ± 10 |
| BE | 11 | 13 | 18.76 | 91 | 890 | 1435 | 3.9 | 1050 ± 10 | 1056 ± 12 | 1070 ± 16 | 1072 ± 20 |
| LS | 10 | 5 | 20.13 | 105 | 2320 | 65481 | 5.7 | 1055 ± 19 | 1064 ± 20 | 1083 ± 8 | 1085 ± 10 |
| BOE | 13 | 95 | 86.72 | 474 | 705 | 740 | 10.3 | 1034 ± 9 | 1055 ± 15 | 1097 ± 20 | 1104 ± 20 |
| LS | 12 | 6 | 9.85 | 51 | 1180 | 5310 | 9.8 | 1108 ± 19 | 1109 ± 20 | 1111 ± 14 | 1111 ± 20 |
| LE | 14 | 19 | 72.74 | 388 | 3400 | 4595 | 13.9 | 1115 ± 6 | 1114 ± 8 | 1112 ± 9 | 1112 ± 10 |
| BOE | 17 | 8 | 144.19 | 781 | 19000 | 46330 | 21.1 | 1125 ± 13 | 1125 ± 13 | 1124 ± 5 | 1124 ± 10 |
| LE | 21 | 12 | 8.54 | 41 | 840 | 1365 | 6.1 | 1133 ± 12 | 1130 ± 14 | 1126 ± 15 | 1126 ± 20 |
| LS | 13 | 8 | 20.64 | 120 | 2160 | 5840 | 11.1 | 1008 ± 17 | 1044 ± 18 | 1121 ± 7 | 1131 ± 20 |
| LE | 13 | 9 | 58.81 | 321 | 5480 | 12635 | 11.0 | 1074 ± 4 | 1096 ± 5 | 1140 ± 6 | 1145 ± 10 |
| LS | 12 | 14 | 38.61 | 155 | 1810 | 2705 | 5.8 | 1321 ± 9 | 1338 ± 11 | 1367 ± 9 | 1369 ± 10 |
| LS | 10 | 13 | 42.78 | 180 | 1790 | 2830 | 4.6 | 1224 ± 7 | 1290 ± 10 | 1401 ± 9 | 1414 ± 20 |
| LE | 18 | 12 | 53.55 | 218 | 4650 | 7565 | 8.1 | 1359 ± 8 | 1379 ± 11 | 1412 ± 9 | 1415 ± 10 |
| LE | 16 | 7 | 44.91 | 184 | 9030 | 18380 | 8.9 | 1363 ± 7 | 1384 ± 9 | 1417 ± 7 | 1420 ± 10 |
| LE | 15 | 3 | 27.51 | 38 | 6500 | 37180 | 2.8 | 2785 ± 16 | 2787 ± 18 | 2788 ± 5 | 2788 ± 10 |
| Luning (Early Norian: 38°24'33"N, 118°01'42"W, 5668' elevation) | | | | | | | | | | | |
| YE | 7 | 12 | 26.6 | 726 | 915 | 1510 | 6.3 | 218 ± 2 | 217 ± 3 | 208 ± 27 | 218 ± 10 |
| LE | 8 | 3 | 6.09 | 174 | 945 | 3750 | 8.4 | 218 ± 2 | 219 ± 3 | 224 ± 21 | 218 ± 10 |
| LE | 8 | 3 | 13.69 | 367 | 2205 | 13650 | 5.1 | 220 ± 2 | 221 ± 3 | 240 ± 18 | 220 ± 10 |
| LE | 9 | 7 | 18.81 | 508 | 1395 | 4340 | 5.4 | 220 ± 2 | 220 3 | 224 ± 15 | 220 ± 10 |
| LS | 9 | 3 | 5.23 | 145 | 945 | 4480 | 6.9 | 221 ± 4 | 221 ± 5 | 225 ± 21 | 221 ± 10 |
| YE | 5 | 13 | 24.3 | 645 | 550 | 855 | 6.2 | 221 ± 3 | 219 ± 6 | 194 ± 52 | 221 ± 10 |
| BO | 5 | 30 | 43.28 | 1086 | 425 | 505 | 5.1 | 221 ± 2 | 220 ± 6 | 212 ± 58 | 221 ± 10 |

**TABLE 1. U-Pb ISOTOPIC DATA AND AGES (continued)**

| Grain type | Grain wt. (μg) | $Pb_c$ (pg) | Pb (ppm) | U (ppm) | $^{206}Pb_m/^{204}Pb$ | $^{206}Pb_c/^{204}Pb$ | $^{206}Pb/^{208}Pb$ | $^{206}Pb^*/^{238}U$ | $^{207}Pb^*/^{235}U$ | $^{207}Pb^*/^{206}Pb^*$ | Interpreted age (Ma) |
|---|---|---|---|---|---|---|---|---|---|---|---|
| **YE** | **10** | **11** | **24.77** | **675** | **1970** | **5470** | **6.6** | **223 ± 2** | **223 ± 3** | **225 ± 19** | **223 ± 10** |
| **YE** | **13** | **24** | **14.21** | **350** | **450** | **565** | **4.8** | **224 ± 2** | **224 ± 6** | **225 ± 53** | **224 ± 10** |
| **BOE** | **7** | **13** | **24.66** | **632** | **810** | **1325** | **5.0** | **224 ± 2** | **225 ± 3** | **232 ± 30** | **224 ± 10** |
| **YE** | **14** | **17** | **23.57** | **617** | **1165** | **1640** | **5.8** | **225 ± 1** | **224 ± 3** | **214 ± 22** | **225 ± 10** |
| **YE** | **13** | **13** | **22.22** | **578** | **1330** | **2150** | **5.3** | **225 ± 2** | **224 ± 3** | **219 ± 19** | **225 ± 10** |
| **YE** | **9** | **35** | **12.9** | **259** | **170** | **190** | **2.9** | **225 ± 4** | **226 ± 10** | **241 ± 9** | **225 ± 10** |
| **YE** | **15** | **13** | **21.41** | **567** | **1480** | **2355** | **6.5** | **227 ± 2** | **226 ± 3** | **208 ± 27** | **227 ± 10** |
| **LS** | **10** | **4** | **11.19** | **305** | **2040** | **7280** | **7.8** | **228 ± 3** | **228 ± 3** | **228 ± 14** | **228 ± 10** |
| **YE** | **8** | **15** | **19.61** | **489** | **630** | **940** | **4.9** | **228 ± 3** | **228 ± 5** | **229 ± 38** | **228 ± 10** |
| **LE** | **8** | **3** | **11.93** | **302** | **1690** | **6690** | **4.7** | **229 ± 2** | **228 ± 3** | **215 ± 16** | **229 ± 10** |
| **YE** | **14** | **20** | **8.39** | **168** | **340** | **445** | **5.0** | **272 ± 4** | **271 ± 8** | **268 ± 56** | **272 ± 10** |
| LS | 7 | 5 | 35 | 173 | 315 | 641 | 5.5 | 1057 ± 35 | 1058 ± 38 | 1058 ± 34 | 1058 ± 10 |
| **LS** | **10** | **16** | **28.18** | **126** | **920** | **1320** | **3.1** | **1084 ± 9** | **1086 ± 22** | **1089 ± 38** | **1089 ± 10** |
| LE | 9 | 6 | 31.65 | 99 | 2290 | 12600 | 2.3 | 1406 ± 13 | 1418 ± 14 | 1437 ± 6 | 1438 ± 10 |
| **LS** | **6** | **22** | **91.55** | **326** | **1345** | **1705** | **5.2** | **1438 ± 8** | **1440 ± 10** | **1444 ± 9** | **1444 ± 10** |
| **LE** | **7** | **2** | **11.74** | **36** | **2085** | **4680** | **4.9** | **1642 ± 13** | **1642 ± 14** | **1643 ± 7** | **1643 ± 10** |
| LS | 9 | 16 | 52.01 | 162 | 1640 | 2350 | 5.6 | 1630 ± 33 | 1645 ± 35 | 1665 ± 9 | 1666 ± 10 |
| **LS** | **12** | **18** | **26.46** | **78** | **950** | **1280** | **5.0** | **1677 ± 23** | **1677 ± 25** | **1677 ± 10** | **1677 ± 10** |
| LE | 8 | 4 | 16.18 | 52 | 1660 | 3480 | 8.6 | 1666 ± 13 | 1674 ± 15 | 1684 ± 7 | 1686 ± 10 |
| **LS** | **7** | **12** | **61.87** | **188** | **2085** | **3450** | **8.5** | **1736 ± 14** | **1738 ± 19** | **1740 ± 13** | **1740 ± 10** |
| LS | 7 | 4 | 19.89 | 51 | 1895 | 5070 | 3.8 | 1820 ± 15 | 1827 ± 16 | 1834 ± 6 | 1835 ± 10 |
| **LS** | **5** | **21** | **43.07** | **101** | **490** | **625** | **3.1** | **1837 ± 19** | **1837 ± 27** | **1838 ± 17** | **1838 ± 10** |
| **LS** | **9** | **3** | **9.33** | **23** | **1390** | **5850** | **3.7** | **1867 ± 23** | **1869 ± 24** | **1871 ± 8** | **1871 ± 10** |
| LS | 8 | 20 | 63.89 | 162 | 1380 | 1800 | 5.9 | 1929 ± 9 | 1946 ± 12 | 1964 ± 6 | 1966 ± 10 |
| LS | 7 | 8 | 54.63 | 120 | 2590 | 6510 | 5.7 | 2160 ± 21 | 2243 ± 23 | 2320 ± 6 | 2325 ± 20 |

*Note:*

Grain with $^{206}Pb^*/^{238}U$ age within 15% of $^{207}Pb^*/^{206}Pb$ age in plain type.

Grain with $^{206}Pb^*/^{238}U$ age not within 15% of $^{207}Pb^*/^{206}Pb$ age in italics.

Grain with concordant age and moderate to high precision analysis in bold.

* Radiogenic Pb

$Pb_c$—Total common Pb in picograms.

Grain type: L—light pink to colorless, M—medium pink, Y—yellowish, O—cloudy/grayish; S—moderate to high sphericity, E—euhedral.

$^{206}Pb_m/^{204}Pb$ is measured ratio, uncorrected for blank, spike, fractionation, or initial Pb.

$^{206}Pb_c/^{204}Pb$ and $^{206}Pb/^{208}Pb$ are corrected for blank, spike, fractionation, and initial Pb.

Pb & U concentrations have uncertainties of 25% due to uncertainty in grain weight

Constants used: $^{238}U/^{235}U$ = 137.88. Decay constant for $^{235}U = 9.8485 \times 10^{-10}$. Decay constant for $^{238}U = 1.55125 \times 10^{-10}$.

All uncertainties are at the 95% confidence level.

Pb blank ranged from 2 to 10 pg. U blank was consistently <1 pg.

Interpreted ages for concordant grains are $^{206}Pb^*/^{238}U$ ages if <1.0 Ga and $^{207}Pb^*/^{206}Pb^*$ ages if >1.0 Ga.

Interpreted ages for discordant grains are projected from 100 Ma, which is a reasonable approximate age for Pb loss in the region.

All analyses conducted using conventional isotope dilution and thermal ionization mass spectrometry, as described by Gehrels (this volume, Introduction).

detailed investigations, we follow the interpretation of Oldow et al. (1990) that the sampled strata are above the Fencemaker thrust. According to this interpretation, the results from our sample apply to the Lovelock assemblage. Should future studies confirm the alternative view, that the sampled strata are below the Fencemaker thrust and belong to the Humboldt assemblage (Heck and Speed, 1987), our results would instead apply to the Humboldt assemblage.

We sampled a single bed of fine- to very fine-grained sandstone. The sample is composed of abundant monocrystalline quartz plus small percentages of sedimentary-metasedimentary fragments and polycrystalline quartz (Dickinson and Gehrels, this volume, Chapter 11). Detrital zircon grains from this sample are predominantly light pink and range from euhedral to well rounded. Subordinate populations of brownish to yellowish grains are also present. We analyzed 32 individual detrital zircons (Table 1); 16 grains yielded ages that are of moderate to high precision and apparently concordant, 15 grains are moderately discordant, and one grain is so highly discordant that it is not considered further (Fig. 4). Reliable ages include groups of 940–1150 Ma (n = 19) and 1405–1435 Ma (n = 3), and single ages of ca. 227, 267, 281, 360, 394, 466, 683, 1369, and 2788 Ma (Fig. 5; Table 2).

### *Luning assemblage*

In the Pilot Mountains, the Luning assemblage is separated from the Gold Range assemblage by the Luning thrust (Fig. 2). Several thrust fault–bounded nappes occur within the upper plate

TABLE 2. DOMINANT DETRITAL ZIRCON AGE GROUPS IN TRIASSIC STRATA OF WESTERN NEVADA

| Age range (Ma) | Peak ages (Ma) | Relative abundance |
|---|---|---|
| Pine Nut | | |
| 210-250 | 231 | 1.00 |
| Lovelock(?) | | |
| 940-1150 | 987, 1016, 1063, 1113* | 0.61 |
| 1405-1435 | 1417 | 0.09 |
| Luning | | |
| 210-235 | 223 | 0.51 |
| 1425-1445 | 1441 | 0.06 |

*Note:* These ages and relative abundances refer to the age-probability plots shown in Figure 5.
* Indicates the peak with the highest probability.

of the Luning thrust (Oldow, 1981). Nappe N2, the unit of interest for this study, comprises a continuous 2200-m-thick section consisting of a lower member of interbedded limestone and mudstone, a middle member of siliciclastic rocks, and an upper member of limestone and dolomite interstratified with sparse mudstone (Fig. 3). The middle member is composed of mudstone and sandy mudstone that contains interbeds of conglomerate, sandstone, and rare limestone (Oldow and Dockery, 1993; Oldow, 1981; Reilly et al., 1980). Reilly et al. (1980) interpreted the middle member of the Luning Formation in Nappe N2 to be delta-front and lower delta-plain deposits of a delta complex. Abundant shallow-marine fossils of early Norian age in carbonate rocks of the upper and lower members indicate shallow-marine deposition (Oldow, 1981; Silberling, 1984).

Our sample was obtained from a single bed of coarse-grained sandstone in the middle member of the Luning Formation from the north flank of Dunlap Canyon in the northern Pilot Mountains (Ferguson and Muller, 1949; Oldow, 1981; Oldow and Dockery, 1993). The sample is composed of polycrystalline quartz plus small amounts of sedimentary rock fragments, monocrystalline quartz, and matrix (Dickinson and Gehrels, this volume, Chapter 11). Zircons are equally divided between yellowish and colorless to light pinkish grains. Most grains are moderately to well rounded, although euhedral grains are also common. A few brownish, cloudy grains are present in each size fraction as well. We analyzed 32 individual detrital zircons from this sample, of which 15 yielded apparently concordant ages of moderate to high precision (Fig. 4). Most ages range from 218 to 229 Ma (n = 17), with a peak in age probability at 223 Ma (Table 2). Two grains have interpreted ages of ca. 1058 and 1089 Ma, two grains are ca. 1441 Ma, and the rest of the grains range in age from 1643 to 2325 Ma (Fig. 5).

## PROVENANCE OF THE DETRITAL ZIRCONS

The detrital zircon grains in our three samples yield ages that in most cases can be linked to recognized source areas in the southwestern United States. Figure 5 shows the ages of our grains together with ages of grains in other assemblages in the region.

A first-order observation is that all three of our samples, plus sandstones in the Humboldt assemblage (Gehrels and Dickinson, 1995), contain detrital zircons of Middle-Late Triassic age. In addition, there is a progressive decrease in the proportion of these Triassic grains from southwest to northeast, from 20 of 20 grains in the Pine Nut assemblage, to 17 of 32 grains in the Luning assemblage, to 1 of 31 grains in the Lovelock (or possibly Humboldt) assemblage, and 1 of 40 grains in the previously studied Humboldt assemblage. These grains were likely derived from arc-type sources in the northern Sierra and eastern Klamath terranes to the west, which contain igneous rocks of Late Triassic age (Fig. 5). Other grains that may have been derived from arc-type sources to the west include the 267–466 Ma grains in the Lovelock(?) and Luning assemblages. As shown in Figure 5, igneous rocks in the Sierra-Klamath terranes match well with the 267–281, 360, 394, and 466 Ma grains.

Grains with Precambrian ages are common in the Lovelock(?) and Luning assemblages (Fig. 5). Grains with ages of ca. 1414–1444 Ma and ca. 1643–1740 Ma in the Luning and Lovelock(?) assemblages most likely originated in the southwestern United States, where there are widespread basement rocks of these ages. This match is best seen in a comparison with the ages of detrital zircons in miogeoclinal strata of central Nevada, as shown on the lower curve of Figure 5.

In addition to grains that probably originated in basement rocks of the southwest, samples from the Luning and Lovelock(?) assemblages both contain >1.8 Ga grains. These are similar in age to detrital zircons recovered from the Roberts Mountains and Golconda allochthons to the east and the northern Sierra terrane to the west (Fig. 5). Sandstones within the Golconda allochthon are a likely source for these grains because the sandstones are locally overlain by Upper Triassic strata of the Humboldt assemblage (Oldow, 1984).

Another group of Precambrian grains in both the Luning and Lovelock(?) samples ranges between ca. 958 and 1145 Ma in age (Fig. 5). These grains could have come directly from the Grenville Province in the southern United States, as is likely for Triassic strata of the miogeocline in Nevada (Gehrels and Dickinson, 1995), or they may have been recycled from off-shelf assemblages such as the Roberts Mountains allochthon, Golconda allochthon, or northern Sierra terrane (Fig. 5). Among the latter assemblages, the most likely source is a sequence of sandstones of the Harmony Formation in the Roberts Mountains allochthon that yields detrital zircon grains of predominantly of 1040–1240 Ma (Gehrels et al., this volume, Chapter 1).

Our interpretations of detrital zircon ages are consistent with provenance interpretations based on sandstone compositions and depositional environments. In the Pine Nut assemblage, where detrital zircons were shed from a nearby active arc, the sandstones are characterized by the presence of abundant plagioclase, volcanic rock fragments, and unstrained volcanic quartz grains (Dickinson and Gehrels, this volume, Chapter 11). Abundant tuff breccia and tuff in the sampled map unit (Fig. 3) also indicate a nearby volcanic source.

In the Lovelock(?) assemblage, our sample from the Hollywood Formation suggests derivation from both arc-type rocks to the west and various assemblages of continental or continental-margin affinity to the east. Stratigraphic relations support derivation of detritus from the east in that the basinal quartz-rich sandstone and mudstone succession is similar in composition to deltaic strata in the Humboldt assemblage, which were fed by rivers to the east and/or southeast (Oldow et al., 1990; Heck and Speed, 1987). In the Clan Alpine Range of the Lovelock assemblage, southeast of our sample location, sparse conglomerate beds contain quartzite and chert clasts derived from immediately to the east (Speed, 1978).

Detrital zircons from our sample of the Luning assemblage also record a western arc-type source and an eastern continental or continental margin source. This is consistent with petrographic evidence for a chert-rich province and a plagioclase–volcanic rock fragment–chert province in Luning strata (Reilly et al., 1980). In the Shoshone Mountains, in the extreme eastern Luning assemblage, Luning sandstones belong to the chert-rich province, indicating that the source area of the chert-rich province likely is to the east (Reilly et al., 1980). In the eastern Garfield Hills, to the west within the Luning assemblage, all sandstones point-counted are exclusively in the plagioclase–volcanic rock fragment–chert province, which is a mixture of the eastern sedimentary source area described here and an andesitic-dacitic volcanic arc source area (Reilly et al., 1980). The increase in volcanic arc–derived components to the west indicates that a volcanic arc was located some distance to the west of the Luning assemblage. Nappe N2 in the Pilot Mountains, the location of our sample, contains sandstones of both provinces. It is accordingly not surprising that our Luning assemblage sample yields a mixture of arc-derived and continent-derived detrital zircons.

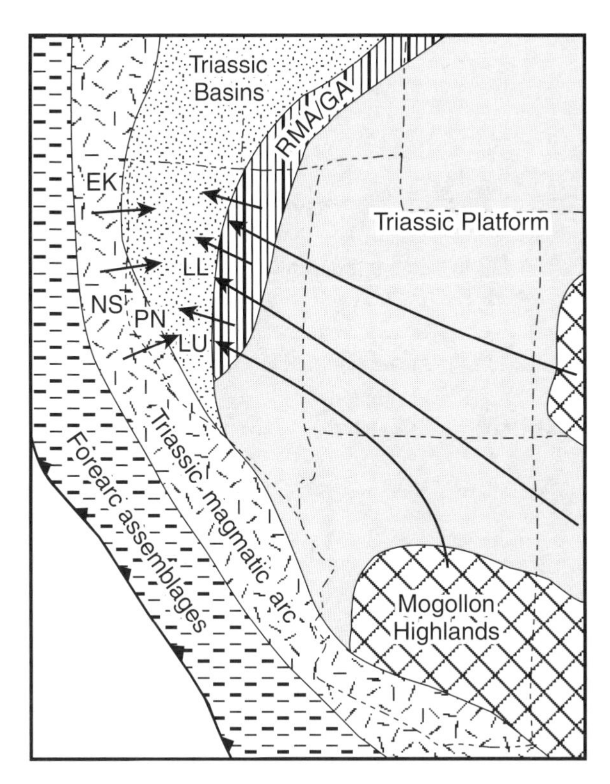


Figure 6. Late Triassic reconstruction of western margin of North America (from Saleeby and Busby-Spera, 1992) showing inferred source areas of detrital zircon grains in Triassic strata of western Nevada. Basement rocks in regions shown as Triassic platform and Mogollon Highlands are mainly 1.63–1.80 Ga with abundant 1.40–1.45 Ga plutons (Van Schmus et al., 1993). EK—eastern Klamath terrane, NS—northern Sierra terrane, PN—Pine Nut assemblage, LU—Luning assemblage, LL—Lovelock assemblage, RMA/GA—Roberts Mountains and Golconda allochthons.

## CONCLUSIONS

Our data are consistent with a relatively simple paleogeography and tectonic setting for the Nevada-California region during Late Triassic time. As discussed by Burchfiel and Davis (1972, 1975), Speed (1978), Oldow (1984), Burchfiel et al. (1992), Saleeby and Busby-Spera (1992), Oldow et al. (1993), and many other workers, Triassic strata of western Nevada likely formed in a backarc basin that separated a magmatic arc to the west from the continental margin to the east. Original sources for nearly all of the grains in our samples can be identified in adjacent regions (Fig. 6), as follows.

The 218–243 Ma grains that decrease in relative abundance northeastward were apparently shed from active arc-type magmatism in the Sierra-Klamath region.

The 267–466 Ma grains were apparently derived from preexisting arc assemblages in the Sierra-Klamath region.

The 958–1145 Ma grains have several possible sources, including the Grenville province to the southeast and south, the Roberts Mountains and/or Golconda allochthons to the east and southeast, and lower Paleozoic strata of the northern Sierra terrane.

The 1414–1444 Ma and 1643–1740 Ma grains were almost certainly derived from basement rocks in southwestern North America (Van Schmus et al., 1993).

The 1835–2788 Ma grains could have been recycled from the Roberts Mountains or Golconda allochthons to the east or from the northern Sierra terrane to the west.

Although our data are consistent with this relatively simple tectonic scenario, we are not able to rule out other possible configurations of the Cordilleran margin during Late Triassic time. Several other terranes along the Cordilleran margin also could have yielded the Phanerozoic detrital grains found in our samples from the Lovelock(?) and Luning assemblages (Coney et al., 1980; Saleeby and Busby-Spera, 1992; Burchfiel et al., 1992). Hence, as emphasized by Oldow et al. (1993), the importance of strike-slip faulting within and between the tectonic elements shown in Figure 6 is difficult to resolve. In particular, abrupt changes of stratigraphy and structural history across the Pine Nut fault and at other assemblage boundaries (Oldow, 1984; Schweickert and Lahren, 1990; Oldow et al., 1993) imply possibly large but difficult to quantify displacements between assemblages which were part of the same Triassic backarc basin.

## ACKNOWLEDGMENTS

This research was supported by National Science Foundation grant EAR-9416933. We thank J. Oldow and W.R. Dickinson for advice in selecting sample sites and W.R. Dickinson, M. Spurlin, B. Darby, J. Harding, B. Lareau, and B. Riley for helpful comments on the manuscript. Reviewed by Richard Schweickert, John Oldow, and Robert Speed.

## REFERENCES CITED

Burchfiel, B.C., and Davis, G.A., 1972, Structural framework and evolution of the southern part of the Cordilleran orogen, western United States: American Journal of Science, v. 272, p. 97–118.

Burchfiel, B.C., and Davis, G.A., 1975, Nature and controls of Cordilleran orogenesis, western United States: Extensions of an earlier synthesis: American Journal of Science, v. 275-A, p. 363–396.

Burchfiel, B.C., Cowan, D.S., and Davis, G.A., 1992, Tectonic overview of the Cordilleran orogen in the western United States, *in* Burchfiel, B.C., et al., eds., The Cordilleran orogen: Conterminous U.S.: Boulder, Colorado, Geological Society of America, Geology of North America, v. G-3, p. 407–480.

Coney, P.J., Jones, D.L., and Monger, J.W.H., 1980, Cordilleran suspect terranes: Nature, v. 288, p. 329–333.

Dilek, Y., Moores, E.M., and Erskine, D.W., 1988, Ophiolitic thrust nappes in western Nevada: Implications for the Cordilleran orogen: Geological Society of London Journal, v. 145, p. 969–975.

Ferguson, H.G., and Muller, S.W., 1949, Structural geology of the Hawthorne and Tonopah quadrangles, Nevada: U.S. Geological Survey Professional Paper 216, 5 p.

Gehrels, G.E., and Dickinson, W.R., 1995, Detrital zircon provenance of Cambrian to Triassic miogeoclinal and eugeoclinal strata in Nevada: American Journal of Science, v. 295, p. 18–48.

Gradstein, F.M., Agterberg, F.P., Ogg, J.G., Hardenbol, J., van Veen, P., Thierry, J. and Huang, Z. 1994, A Mesozoic time scale: Jourual of Geophysical Research, v. 99, no. B12, p. 24,051–24,074.

Heck, F.R., and Speed, R.C., 1987, Triassic olistostrome and shelf-basin transition in the western Great Basin: Paleogeographic implications: Geological Society of America Bulletin, v. 99, p. 539–551.

Ludwig, K.R., 1991a, A computer program for processing Pb-U-Th isotopic data: U.S. Geological Survey Open-File Report 88-542.

Ludwig, K.R., 1991b, A plotting and regression program for radiogenic-isotopic data: U.S. Geological Survey Open-File Report 91-445.

Miller, M.M., and Harwood, D.S., 1990, Paleogeographic setting of upper Paleozoic rocks in the northern Sierra and eastern Klamath terranes, northern California, *in* Harwood, D.S., and Miller, M.M., eds., Paleozoic and early Mesozoic paleogeographic relations; Sierra Nevada, Klamath Mountains, and related terranes: Geological Society of America Special Paper 255, p. 175–192.

Oldow, J.S., 1981, Structure and stratigraphy of the Luning allochthon and the kinematics of allochthon emplacement, Pilot Mountains, west-central Nevada: Geological Society of America Bulletin, v. 92, p. 888–911.

Oldow, J.S., 1984, Evolution of a late Mesozoic back-arc fold and thrust belt, northwestern Great Basin, U.S.A.: Tectonophysics, v. 102, p. 245–274.

Oldow, J.S., and Dockery, H.A., 1993, Geologic map of the Mina quadrangle, Nevada: Nevada Bureau of Mines and Geology Field Studies Map 6, scale 1:24 000.

Oldow, J.S., Bartel, R.L., and Gelber, A.W., 1990, Depositional setting and regional relationships of basinal assemblages: Pershing Ridge Group and Fencemaker Canyon sequence in northwestern Nevada: Geological Society of America Bulletin, v. 102, p. 193–222.

Oldow, J.S., Satterfield, J.I., and Silberling, N.J., 1993, Jurassic to Cretaceous transpressional deformation in the Mesozoic marine province of the northwestern Great Basin, *in* Lahren, M.M., Trexler, J.H., Jr., and Spinosa, C., eds., Crustal evolution of the Great Basin and Sierra Nevada: Reno, Department of Geological Sciences, University of Nevada, p. 129–166.

Proffett, J.M., Jr., and Dilles, J.H., 1984, Geologic map of the Yerington District, Nevada: Nevada Bureau of Mines and Geology Map 77, scale 1:24,000.

Reilly, M.B., Breyer, J.A., Oldow, J.S., 1980, Petrographic provinces and provenance interpretation, Upper Triassic Luning Formation, west-central Nevada: Geological Society of America Bulletin, v. 91, part II, p. 2112–2151.

Saleeby, J.B., and Busby-Spera, C., 1992, Early Mesozoic evolution of the Western Cordillera, *in* Burchfiel, B.C., et al., eds., The Cordilleran orogen: Conterminous U.S.: Boulder, Colorado, Geological Society of America, Geology of North America, v. G-3, p. 107–168.

Schweickert, R.A., and Lahren, M.M., 1990, Speculative reconstruction of the Mojave–Snow Lake fault: Implications for Paleozoic and Mesozoic orogenesis in the western United States: Tectonics, v. 9, p. 1609–1629.

Silberling, N.J., 1984, Map showing localities and correlation of age-diagnostic lower Mesozoic megafossils, Walker Lake 1 × 2 degree quadrangle, Nevada and California: U.S. Geological Survey Miscellaneous Field Studies Map MF-2062, scale 1:250 000.

Silberling, N.J., 1991, Allochthonous terranes of western Nevada: Current status, *in* Raines, G.L., et al., eds., Geology and ore deposits of the Great Basin: Reno, Geological Society of Nevada, p. 101–102.

Speed, R.C., 1978, Paleogeographic and plate-tectonic evolution of the early Mesozoic marine province of the western Great Basin, *in* Howell, D.G., and McDougall, K.A., eds., Mesozoic paleogeography of the western United States: Pacific Section, Society of Economic Paleontologists and Mineralogists, Pacific Paleogeography Symposium 2, p. 253–270.

Speed, R.C., Elison, M.W., and Heck, F.R., 1988, Phanerozoic tectonic evolution of the Great Basin, *in* Ernst, W.G., ed., Metamorphism and crustal evolution of the western United States (Rubey Volume VII): Englewood Cliffs, New Jersey, Prentice-Hall, p. 573–605.

Speed, R.C., Silberling, N.J., Elison, M.W., Nichols, K.M., and Snyder, W.S., 1989, IGC field trip T122: Early Mesozoic tectonics of the western Great Basin, Nevada: International Geological Congress, 28th, Field Trip Guidebook T122: Washington, D.C., American Geophysical Union, 54 p.

Stacey, J.S., and Kramers, J.D., 1975, Approximation of terrestrial lead isotope evolution by a two-stage model: Earth and Planetary Science Letters, v. 26, p. 207–221.

Van Schmus, W.R., and 24 others, 1993, Transcontinental Proterozoic provinces, *in* Reed, J.C., et al., eds., Precambrian: Conterminous U.S.: Boulder, Colorado, Geological Society of America, Geology of North America, v. C-2, p. 171–334.

Wallace, R.E., Silberling, N.J., Irwin, W.P., and Tatlock, D.B., 1969, Geologic map of the Buffalo Mountain quadrangle, Pershing and Churchill Counties, Nevada: U.S. Geological Survey Map GQ-821, scale 1:62 500.

Wyld, S.J., and Wright, J.E., 1993, Mesozoic stratigraphy and structural history of the southern Pine Nut Range, west-central Nevada, in Dunne, G., and McDougall, K., eds., Mesozoic paleogeography of the western United States—II: Pacific Section, Society of Economic Paleontologists and Mineralogists book 71, p. 289–306.

Manuscript Accepted by the Society January 24, 2000

Geological Society of America
Special Paper 347
2000

# *Provenance of the Antelope Mountain Quartzite, Yreka terrane, California: Evidence for large-scale late Paleozoic sinistral displacement along the North American Cordilleran margin and implications for the mid-Paleozoic fringing-arc model*

**E. Timothy Wallin**
**Robert C. Noto**
*Department of Geoscience, University of Nevada, Las Vegas, Nevada 89154-4010, USA*
**George E. Gehrels**
*Department of Geosciences, University of Arizona, Tucson, Arizona 85721, USA*

## ABSTRACT

**The Antelope Mountain Quartzite is a coarse feldspathic siliciclastic unit in the Yreka terrane of the eastern Klamath Mountains. U-Pb geochronology of single detrital zircons from the Antelope Mountain indicates a source to the north of the 1.7–1.8 Ga mobile belt that transects the United States. When compared to cratonic and miogeoclinal detrital zircon ages, the signature of the Antelope Mountain is most compatible with derivation from a "northern British Columbia" source. The manner in which that detritus reached its present position is explained best by catastrophic failure of the continental margin in northern British Columbia during the early Paleozoic, incorporation of olistoliths into melange of the Yreka terrane by an indeterminate amount of subduction-related transport before the Middle Devonian, and southward sinistral offset of the Yreka terrane during the late Paleozoic. This inferred tectonic transport of detritus in the Antelope Mountain Quartzite conflicts with models in which the eastern Klamath province is parautochthonous relative to the Paleozoic United States.**

## NATURE OF THE PROBLEM

Despite great progress in understanding the assembly of the North American Cordillera since the advent of the terrane concept (Irwin, 1972; Helwig, 1974; Coney et al., 1980), many aspects of its coarse structure remain uncertain. In particular, the paleogeography of some Paleozoic crustal fragments is still poorly known. Knowledge of their provenance is important because Paleozoic crustal fragments commonly form the basement of younger terranes, the paleogeography of which may be misunderstood if the accepted paleogeographic positions of their basement rocks are incorrect.

The eastern Klamath Mountains include several Paleozoic crustal fragments that have been interpreted both as parautochthonous (Burchfiel and Davis, 1972; Rubin et al., 1990; Gehrels et al., this volume) and exotic relative to the part of the mid-Paleozoic craton in the United States (Speed, 1979; Wallin, 1993). These interpretations are fundamentally different, and each has important implications for Cordilleran tectonic evolution. According to the parautochthonous school, the quasilinear distribution of mid-Paleozoic igneous rocks in the Cordillera is a result of the accretion of an intraoceanic magmatic arc due to collapse of backarc basins along the continental margin (cf. Rubin et al., 1990, their Fig. 1). Although the simplicity of such a

Wallin, E.T., Noto, R.C., and Gehrels, G.E., 2000, Provenance of the Antelope Mountain Quartzite, Yreka terrane, California: Evidence for large-scale late Paleozoic sinistral displacement along the North American Cordilleran margin and implications for the mid-Paleozoic fringing-arc model, *in* Soreghan, M.J., and Gehrels, G. E., eds., Paleozoic and Triassic paleogeography and tectonics of western Nevada and northern California: Boulder, Colorado, Geological Society of America Special Paper 347, p. 119–131.

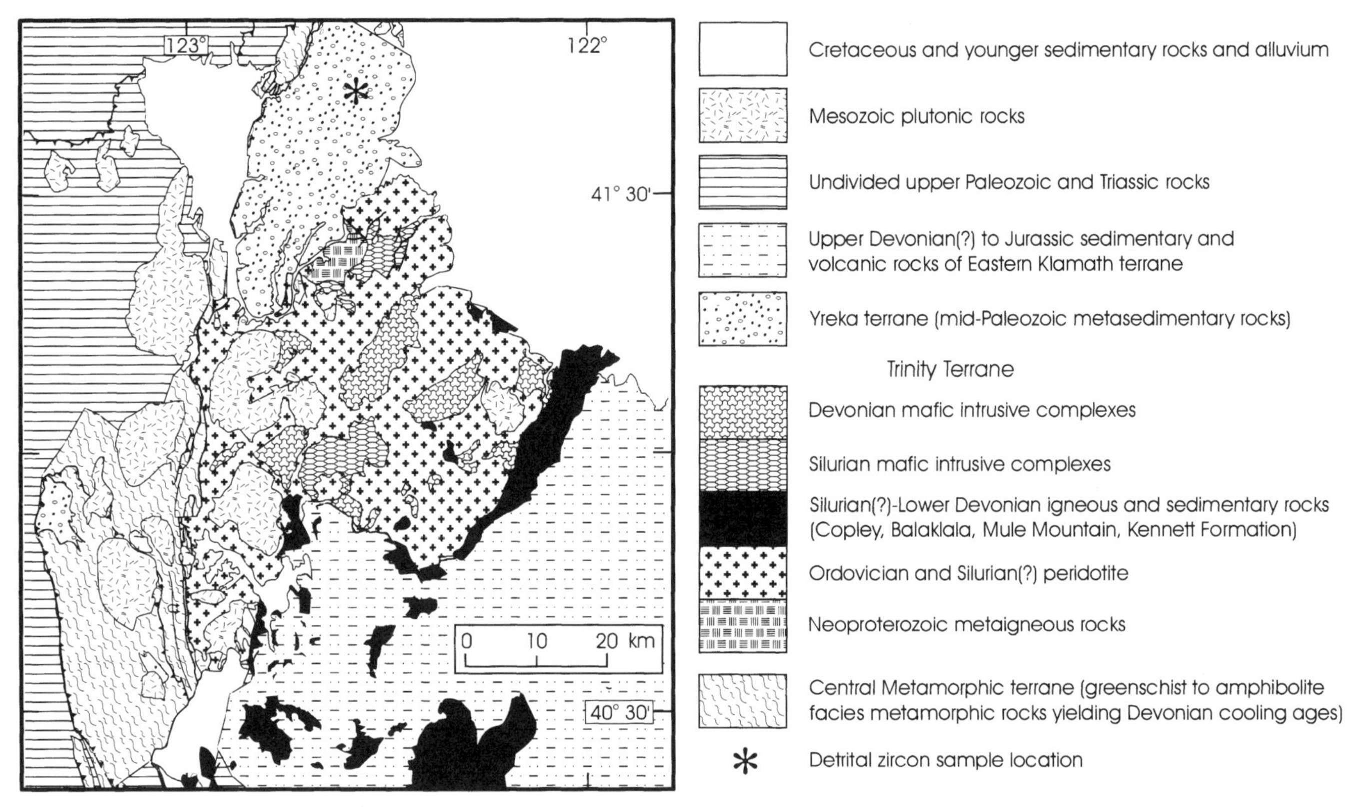


Figure 1. Geologic map of eastern Klamath Mountains (modified after Irwin, 1994).

"fringing volcanic arc" model is appealing, a growing body of evidence indicates that the entire late Paleozoic North American Cordilleran margin was dominated by sinistral strike slip (Eisbacher, 1983; Walker, 1988; Stone and Stevens, 1988; Mahoney et al., 1991; Trexler et al., 1991). This body of evidence calls into question the validity of the fringing arc model. Such margin-parallel sinistral offset would probably have dismembered and/or juxtaposed fragments of any such mid-Paleozoic volcanic arc.

Methods and data typically used to estimate terrane displacements (piercing points, paleomagnetism, paleobiogeography, geologic evidence) have so far been insufficient to discriminate between these end-member models. An alternative approach is to reconstruct the paths and maximum distance of sediment dispersal using sedimentological data, and thus reject hypotheses that require geologically unrealistic distances of sedimentary transport. As the coarsest feldspathic siliciclastic unit in the eastern Klamath Mountains, the Antelope Mountain Quartzite holds the potential to test whether major sinistral strike slip transported eugeoclinal rocks southward along the Cordilleran margin.

A variety of geologic evidence sets the stage for such a test. First, Paleozoic rocks of the eastern Klamaths contain exclusively North American fauna (Boucot et al., 1974; Rohr, 1980; Potter, 1990), thus we can safely assume that the Antelope Mountain Quartzite was derived from North America. Earlier work on multigrain fractions of detrital zircon in the Antelope Mountain Quartzite yielded an isotopic signature indicating the possibility of derivation from Precambrian basement in the western Canadian shield (Wallin, 1989, 1993). Second, aspects of the sedimentology of the Antelope Mountain Quartzite preclude sedimentary transport across long distances (> ~100 km), which means that any substantial apparent offset from its source rocks might approximate the net amount of tectonic transport the unit has undergone since deposition (Wallin, 1993). Recent developments permit isotopic analysis of single detrital zircons, allowing a more rigorous test of the idea that the Antelope Mountain Quartzite was derived from western Canada. In this chapter, single-grain analyses are used to infer the provenance of the Antelope Mountain Quartzite, and are then combined with sedimentological data to estimate the amounts of sedimentary and tectonic transport of that detritus. These new data are most consistent with a source in northern British Columbia, and we conclude that the Antelope Mountain Quartzite is in its current position primarily as a result of tectonic transport that occurred during the late Paleozoic.

## GEOLOGIC CONTEXT

The Klamath Mountains consist of east-dipping lithotectonic belts that record a long history of episodic subduction-related underplating, metamorphism, magmatism, and sedimentation ranging from mid-Paleozoic in the east to Jurassic-Cretaceous in the west (Irwin, 1989). The oldest rocks in the region are in the eastern Klamath Mountains (Fig. 1), which are part of an extensive belt of lower Paleozoic eugeoclinal rocks that crop out discontinuously from northern Mexico to central Alaska (Silberling et al., 1992). Recent work indicates that a transect of a mid-Paleozoic intraoceanic convergent plate margin is preserved in the eastern Klamath Mountains (Wallin and Metcalf, 1998). Elements of this margin include Silurian and Devonian subduction-related igneous rocks (Eastern Klamath and Trinity terranes; Kinkel et al., 1956; Lapierre et al., 1985; Doe et al., 1985; Wallin and Metcalf, 1998), Lower Devonian trench-forearc sedimentary successions (Wallin and Trabert, 1994; Wallin and Gehrels, 1995), and mid-Paleozoic subduction-related metamorphic rocks (Peacock, 1987; Peacock and Norris, 1989; Lindsley-Griffin et al., 1991).

Metasedimentary strata in the Yreka terrane (Fig. 2) occur as nappes above a thin, variably named accretionary prism preserved at the boundary between the Yreka and Trinity terranes (Wallin et al., 1988; Lindsley-Griffin et al., 1991). Concordant Late Silurian U-Pb ages (410–420 Ma) for detrital zircons in basal sedimentary rocks of the Yreka terrane, coupled with Devonian isotopic ages of metamorphism (Duzel Phyllite; Cashman, 1980), reveal that virtually all sedimentation in these nappes occurred during the Early Devonian, rather than throughout much of the early Paleozoic as was previously thought (Wallin and Gehrels, 1995).

### *Sedimentology*

The Antelope Mountain Quartzite is part of the Yreka terrane, and was named by Hotz (1977), who erected an excellent stratigraphic nomenclature for the Yreka terrane that remains largely valid today. The Antelope Mountain Quartzite contains several rock types, including fine-pebble conglomerate, quartzite, quartz arenite, phyllitic argillite, chert, and red, purple, and green argillite (Hotz, 1977; Noto, 2000). Thin-bedded, black radiolarian chert caps the stratigraphic section, and is at least 100 m thick. The Antelope Mountain Quartzite is unfossiliferous save for the sparse occurrence in red argillites of circular traces that resemble soft-body parts recorded in some Ediacaran strata (Lindsley-Griffin, et al., 1989). Thus its age is known only to be pre-Devonian because it is tightly folded along with the Mallethead thrust system, as detailed in the following discussion.

Coarse quartzofeldspathic detritus in the Antelope Mountain Quartzite is compositionally uniform and is manifested principally as quartzite and quartz arenite. Many beds contain abundant fine pebbles in both pebbly sandstone and clast-supported conglomerate. Petrographic analysis of the Antelope Mountain Quartzite revealed that it was derived from a continental block provenance, with sparse rock fragments consisting mainly of altered tectonite grains and chert (Bond and Devay, 1980).

The sedimentary processes and depositional environment responsible for deposition of the Antelope Mountain are enigmatic, mainly because of the dearth of physical and biogenic sedimentary structures. Aspects of the sedimentology of the Antelope Mountain bear on our interpretation, thus a synopsis of the salient points is necessary here. Hotz (1977), noting this dearth of sedimentary structures, suggested a shallow-water paleoenvironment of deposition. In contrast, Klanderman (1978) argued that the structureless character of the Antelope Mountain is best explained by redeposition of well-sorted sediment as turbidites.

One of us (Gehrels) prefers a turbidite origin for the Antelope Mountain, an interpretation based mainly on its regional tectonostratigraphic context (Gehrels et al., this volume). In contrast, Wallin and Noto observed numerous examples of straight-crested, symmetrical ripple marks in medium sandstone (7 superbly preserved symmetrical ripple marks; 10 probable marks; Noto, 2000), and believe that they require deposition by oscillatory flow in shallow water. These ripple marks all have spacings of <5 cm, a vertical form index (wavelength/height) of 6–7, and ripple symmetry indices (stoss projection/lee projection) of one. Application of either empirical considerations (Tanner, 1971) or theoretical analysis based on Airy wave theory (Clifton and Dingler, 1984) indicates that ripples with spacings of <5 cm occur only in very shallow water (centimeter-scale depth), where wave height and period are limited. The short spacings and high symmetry of these ripple marks also indicate that they are not reversing-current ripples formed by long-period oscillatory flows in deep water.

These sedimentological considerations do not augur well for interpretations involving a deep-sea depositional environment for the Antelope Mountain Quartzite. If the environment of deposition was shallow, then it is not credible that such detritus could have been transported more than 1500 km southward on the shelf by myriad traction currents without greater sorting or contamination by sediment mixing. However, it seems virtually impossible to produce the sedimentary structures in the Antelope Mountain via long-distance transport by turbidites in deep water. Wallin and Noto maintain that the apparent shallow-water environment of deposition must be explained differently if the Antelope Mountain Quartzite is to be interpreted as part of a far-traveled turbidite system.

### *Structural context*

Hotz interpreted the Antelope Mountain Quartzite as a formation in the local stratigraphy, but field mapping conducted as part of this study revealed that it is a very large block in the subjacent melange. First, we outline Hotz's interpretations, then describe how our detailed mapping and stratigraphic analysis result in considerable simplification of the local geology. Hotz drew structural contacts between the Antelope Mountain Quartzite, the Schulmeyer Gulch melange, and the Duzel Phyllite as thrusts, with the Antelope Mountain in the hanging wall (Fig. 3A). In contrast, our field mapping and structural analysis of the Antelope Mountain Quartzite reveal that it is a large tectonic block in the Schulmeyer Gulch melange that is in the footwall of the Mallethead thrust (Fig. 3B). The Antelope Mountain Quartzite was incorporated into the melange by the Early to Middle Devonian because its relict stratigraphy is tightly folded along with the floor thrust of the Mallethead thrust

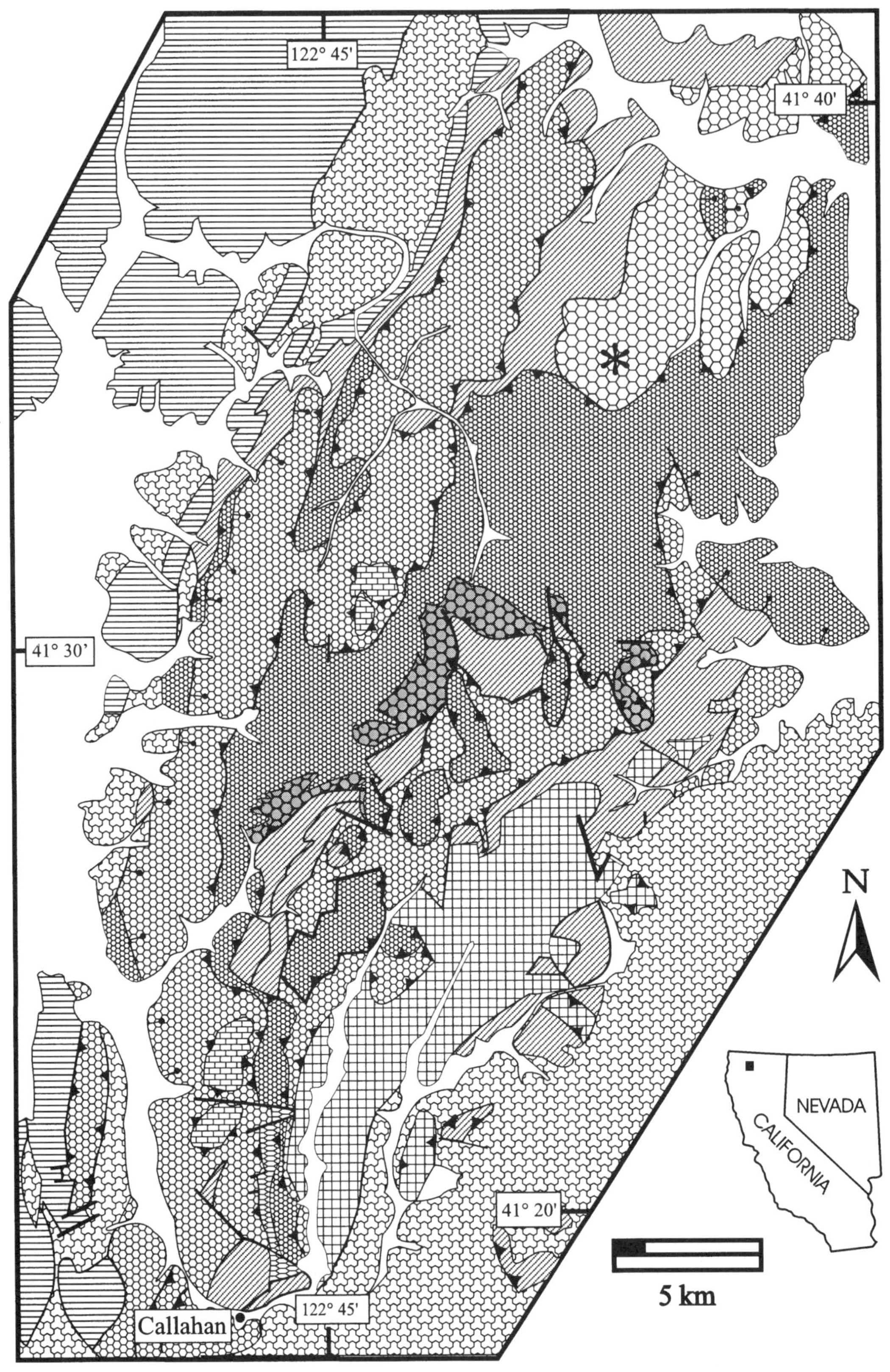


Figure 2. (on this and opposite page). Geologic map of Yreka terrane (modified after Hotz, 1978; Lindsley-Griffin and Griffin, 1983).

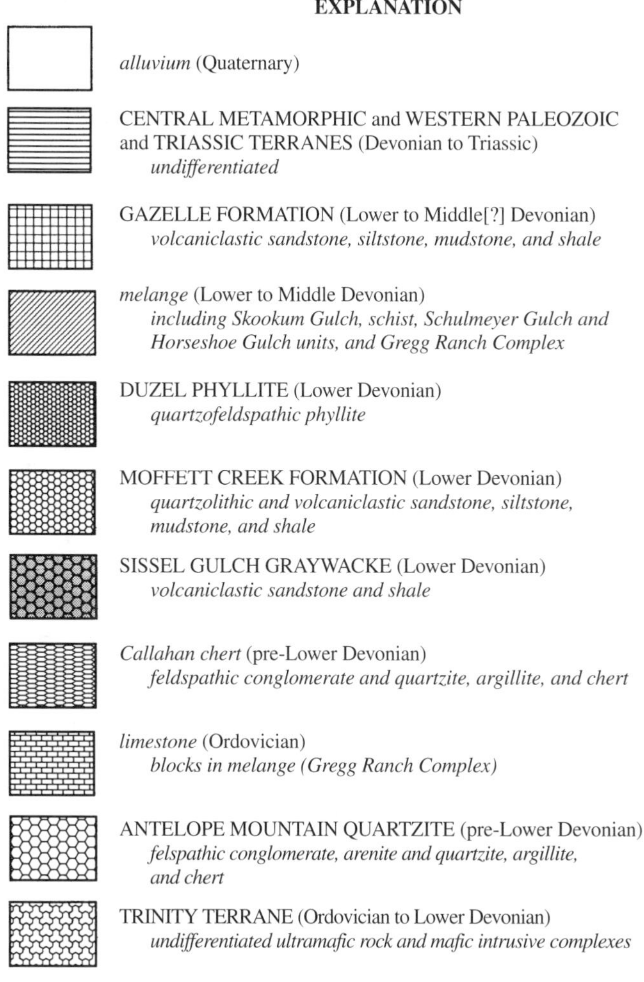


system (Fig. 3B). The timing of motion of the Mallethead thrust system is known to be Devonian for two reasons. First, the Duzel conformably overlies the Sissel Gulch Graywacke, which contains 410 Ma detrital zircon (Wallin and Gehrels, 1995). Second, work at the southern terminus of the Yreka terrane has tightly bracketed the age of east-vergent thrusting there to be Early to Middle Devonian and these thrusts are demonstrably correlative with those to the north (Eschelbacher and Wallin, 1998). The Mallethead thrust has not undergone any subsequent deformation, save the gentle Jurassic warping recognized by Zdanowicz (1971); thus all of the compressional deformation in the Yreka terrane appears to be Early to Middle Devonian.

### *Stratigraphic context*

Despite his solid contribution to our understanding of the stratigraphy of the Yreka terrane, the structural and stratigraphic context of the Antelope Mountain Quartzite indicated on his map of the Yreka quadrangle (Hotz, 1977) was incorrect. Hotz's misunderstanding of structural and sedimentological relations in the vicinity of Antelope Mountain was probably caused by misinterpretation of phyllitic argillites in the Antelope Mountain Quartzite. Hotz interpreted the Antelope Mountain as a facies of the Duzel Phyllite because he believed that phyllitic argillites interbedded with quartzite of the Antelope Mountain were tongues of the Duzel Phyllite. This apparent interfingering of the Duzel and the Antelope Mountain led Hotz to draw a number of depositional contacts between the two units (Fig. 3A). Careful examination of each putative depositional contact between the Antelope Mountain and the Duzel reveals a sharp contact at each locality, the origin of which is entirely equivocal. No sedimentological evidence (e.g., gradational contacts, interfingering) exists that requires those sharp contacts to be depositional. Detailed stratigraphic analysis of the lithologically variable succession in the Antelope Mountain reveals subtle but distinct differences in lithology and metamorphic grade between its phyllitic argillites and the higher grade Duzel Phyllite. In short, the Antelope Mountain Quartzite contains phyllitic argillites, whereas the coarser grained Duzel Phyllite exhibits more recrystallization, as well as metamorphic segregations in the axes of both microscopic and mesoscopic folds (Noto, 2000).

Hotz (1977) estimated the thickness of the Antelope Mountain Quartzite to be 700 m; however, he was unaware that it is a block in melange, and that it been affected by extensional faulting. Our stratigraphic and structural analysis indicates that the Antelope Mountain has been folded tightly along with the floor thrust of the Mallethead thrust system (Fig. 3B), and that subsequent extension of the Antelope Mountain Quartzite by 14% occurred along two generations of normal faults (Noto and Wallin, 1998). The total stratigraphic thickness of the Antelope Mountain Quartzite is uncertain because it is a block in melange and its base is not exposed. A minimum thickness of 220 m has been reconstructed by measurement of stratigraphic sections among marker beds in extensional fault blocks (Noto, 2000). The reconstructed dimensions of the original tectonic block of Antelope Mountain Quartzite prior to both folding and subsequent extension are 13 km × 9 km × ≥0.22 km (Noto, 2000).

## PROVENANCE

### *Methods*

Dating of detrital zircon affords an opportunity to provide greater specificity regarding the source of the Antelope Mountain Quartzite than that provided by Bond and Devay (1980) or Wallin (1989). We processed 30 kg of representative feldspathic granule conglomerate from the east face of Antelope Mountain and obtained 140 mg of zircon. The detrital zircon population is quite diverse in terms of size, shape, roundness, clarity, and color. In selecting grains for analysis, every effort was made to select grains representative of the population's diversity. Grains were collected, selected, and prepared for iso-

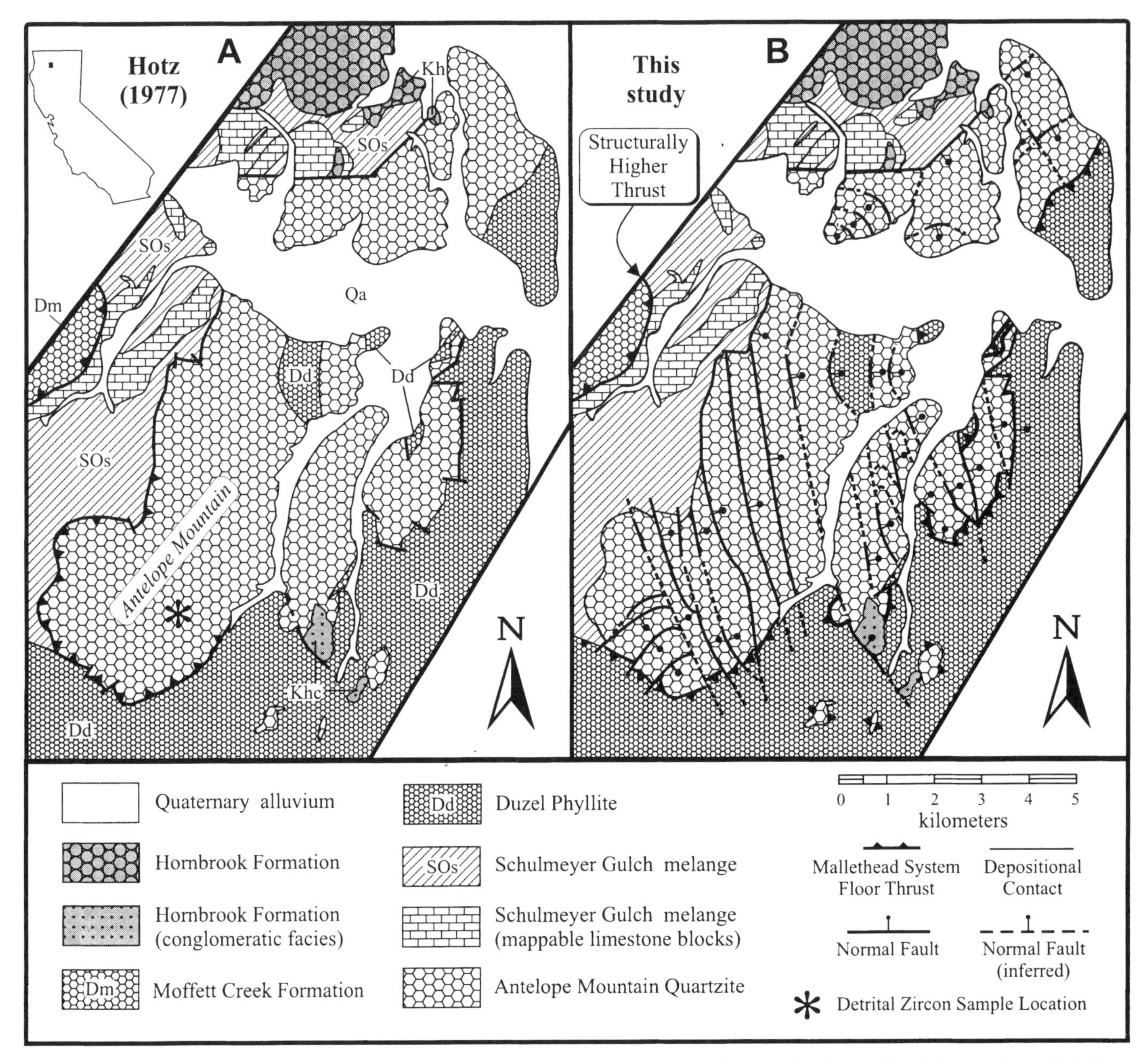


Figure 3. Differing interpretations of structural and stratigraphic context of Antelope Mountain Quartzite. A: Antelope Mountain Quartzite is intercalated depositionally with Duzel Phyllite and is locally thrust over it and Schulmeyer Gulch melange (Hotz, 1977). B: New interpretation in which Antelope Mountain Quartzite is tectonic block in Schulmeyer Gulch melange, which is beneath Mallethead thrust. Antelope Mountain is older than Duzel Phyllite, and is not facies of it. Area containing Antelope Mountain has been extended by several generations of normal faults that are at least in part Cretaceous in age.

topic analysis by Wallin at the University of Nevada, and were analyzed isotopically primarily by Gehrels at the University of Arizona. Grains analyzed isotopically by Wallin at the University of Kansas are so indicated on Table 1. The methods employed in the isotopic analyses are detailed by Gehrels (this volume). U-Pb isotopic analysis of 46 apparently core-free, inclusion-free detrital zircons from the Antelope Mountain Quartzite yielded a broad spectrum of concordant and slightly discordant Precambrian ages (Figs. 4 and 5; Table 1).

### *Provenance*

All of the detrital zircons we analyzed are Precambrian, with ages ranging from 1.32 to 3.0 Ga, consistent with derivation from either the North American craton or its sedimentary cover. Precambrian subprovinces of the craton exhibit a general southward decrease in age (Reed et al., 1993). Baseline study of the detrital zircon content of the Cordilleran miogeocline along a north-south transect indicated that the shelf faithfully reflects this

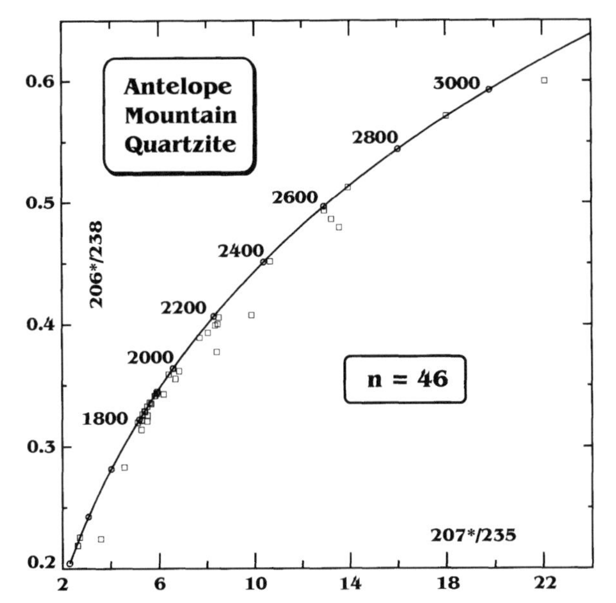


Figure 4. Concordia diagram showing U-Pb isotopic data for detrital zircons from Antelope Mountain Quartzite.

gradient, the provenance of its zircons being divisible into northern and southern Cordilleran source regions (Gehrels et al., 1995). Most of our concordant ages cluster between 1.8 and 1.9 Ga, with a subordinate number of older Early Proterozoic and Archean grains. These ages permit derivation from rocks anywhere from Nevada to Alaska. Overall, however, the spectrum of ages is most consistent with derivation from the northern Cordilleran source region (Fig. 5). Negative evidence also supports this interpretation, in that we found no evidence of grains derived from the 1.7–1.5 Ga belts that transect the central United States in an east-northeast–west-southwest direction. The lack of grains from these younger belts indicates a source well to the north of them because the source of the Antelope Mountain contained compositionally mature sedimentary rocks, which would have sampled the basement there. Although the number of grains analyzed is large for a study of this type, it is worth noting that when sample sizes are so small relative to the size of the population, the limitations of negative evidence should be borne in mind. Nevertheless, we believe that the absence of such grains is not an artifact of sparse sampling, because empirically, few samples yield additional ages once more than about 20 grains representative of the grain diversity have been analyzed (Gehrels, this volume).

Perhaps the most difficult data to explain according to this interpretation are those for the three grains derived from 1.3 Ga rocks. Rocks of this age occur mainly in an east-southeast–west-southwest–trending belt in the central United States and, in most places, intrude rocks ranging from 1.7 to 1.4 Ga. Thus, assuming that the data are geologically representative, the occurrence of 1.3 Ga grains without grains from the 1.7–1.4 Ga country rock is difficult to reconcile. This paradox indicates that one of two interpretations is likely. First, 1.3 Ga rocks may exist in the subsurface of western Canada and simply have not yet been documented. Alternatively, these 1.3 Ga grains were well rounded and exhibited high sphericity; thus they may have been derived by recycling of a minor sedimentary unit capping the crystalline rocks that constituted the principal source of the Antelope Mountain Quartzite. Despite the presence of these grains in the Antelope Mountain Quartzite, the best interpretation of its provenance, given our current understanding of the distribution of zircon ages in both crystalline and sedimentary rocks, is that it was derived from northern British Columbia.

## DISPLACEMENT HISTORY OF THE ANTELOPE MOUNTAIN QUARTZITE

The detrital zircon signature of the Antelope Mountain Quartzite is virtually identical to that of lower Paleozoic rocks in the northern Sierra terrane (Harding et al., this volume), which indicates that they had a common ultimate, if not proximate, provenance in northern British Columbia. The manner in

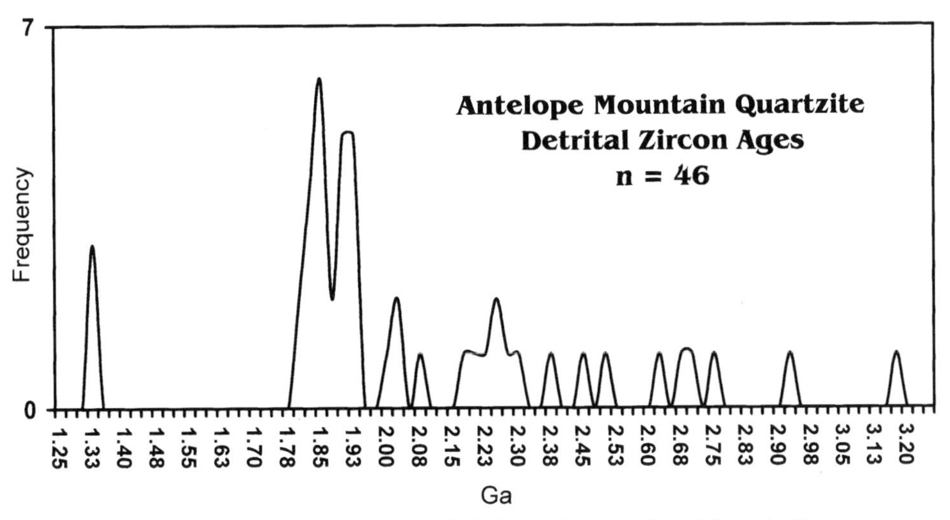


Figure 5. Spectrum of U-Pb ages for detrital zircons from Antelope Mountain Quartzite.

TABLE 1. U-Pb ISTOPIC DATA AND APPARENT AGES–
DETRITAL ZIRCONS FROM THE ANTELOPE MOUNTAIN QUARTZITE

| Grain description | Sample weight (mcg) | $^{206}Pb_m$ / $^{204}Pb$ | %err | $^{206}Pb_c$ / $^{207}Pb$ | %err | $^{206}Pb_c$ / $^{208}Pb$ | Ages$_d$ $^{206}Pb^*$ / $^{238}U$ | $^{207}Pb^*$ / $^{235}U$ | $^{207}Pb^*$ / $^{206}Pb^*$ | 2σ error |
|---|---|---|---|---|---|---|---|---|---|---|
| AMQ-60 nm(0)1.7A clr/rnd | 17 | 4100 | 10 | 11.38 | 0.3 | 5.7 | 1277 | 1289 | 1309 | ± 10 |
| AMQ-59 nm(0)1.7A clr/rnd | 13 | 265 | 1 | 7.29 | 0.5 | 3.1 | 1312 | 1313 | 1314 | ± 31 |
| AMQ-52 nm(0)1.7A pk/rnd | 12 | 2200 | 5 | 10.99 | 0.3 | 7.9 | 1275 | 1292 | 1320 | ± 10 |
| AMQ-1 nm(0)1.7A eu/pk+ | 17 | 1236 | 0.5 | 8.44 | 0.1 | 2.8 | 1684 | 1727 | 1778 | ± 2 |
| **AMQ-8 nm(0)1.7A eu/pk** | **38** | **1140** | **3** | **8.27** | **0.3** | **2.9** | **1788** | **1790** | **1793** | **± 9** |
| AMQ-2 nm(0)1.7A eu/pk+ | 10 | 1154 | 0.9 | 8.31 | 0.1 | 3.3 | 1792 | 1802 | 1812 | ± 3 |
| AMQ-69 m(0)1.7A eu/pk | 5 | 1500 | 5 | 8.37 | 0.3 | 5.3 | 1607 | 1699 | 1814 | ± 10 |
| *AMQ-14 nm(0)1.7A eu/pk* | *13* | *870* | *3* | *7.93* | *0.5* | *4.8* | *1305* | *1513* | *1818* | *± 14* |
| AMQ-15 nm(0)1.7A eu/pk | 6 | 990 | 3 | 8.05 | 1 | 2.2 | 1821 | 1820 | 1819 | ± 22 |
| **AMQ-67 nm(0)1.7A sub/pk** | **13** | **5440** | **10** | **8.76** | **0.2** | **6.8** | **1834** | **1833** | **1831** | **± 6** |
| **AMQ-41 nm(0)1.7A eu/pk** | **18** | **7200** | **10** | **8.79** | **0.4** | **5.8** | **1834** | **1834** | **1835** | **± 8** |
| AMQ-29 nm(0)1.7A eu/pk | 15 | 1530 | 5 | 8.27 | 0.5 | 4.4 | 1795 | 1816 | 1840 | ± 12 |
| AMQ-30 nm(0)1.7A eu/pk | 13 | 2120 | 5 | 8.43 | 0.4 | 48 | 1798 | 1820 | 1844 | ± 9 |
| AMQ-24 nm(0)1.7A eu/pk | 21 | 2600 | 5 | 8.5 | 0.4 | 4.4 | 1830 | 1837 | 1846 | ± 9 |
| **AMQ-65 nm(0)1.7A sub/pk** | **8** | **2920** | **5** | **8.54** | **0.5** | **4.8** | **1852** | **1850** | **1848** | **± 10** |
| **AMQ-7 nm(0)1.7A eu/pk** | **19** | **1210** | **4** | **8.04** | **0.2** | **3.6** | **1868** | **1866** | **1863** | **± 9** |
| AMQ-4 nm(0)1.7A eu/pk+ | 8 | 243.4 | 1.9 | 6.08 | 0.3 | 3.7 | 1869 | 1871 | 1873 | ± 10 |
| AMQ-49 m (0)1.4A pur/rnd | 7 | 3160 | 5 | 8.43 | 0.3 | 4.8 | 1759 | 1814 | 1877 | ± 7 |
| AMQ-46 m(0)1.4A pur/rnd | 7 | 3800 | 5 | 8.44 | 0.3 | 5.1 | 1862 | 1874 | 1886 | ± 6 |
| AMQ-17 nm(0)1.7A eu/pk | 12 | 2580 | 5 | 8.31 | 0.3 | 4.9 | 1817 | 1851 | 1889 | ± 7 |
| AMQ-53 nm(0)1.7A clr/rnd | 10 | 2600 | 5 | 8.31 | 0.3 | 5.9 | 1862 | 1875 | 1890 | ± 7 |
| **AMQ-58 nm(0)1.7A clr/rnd** | **16** | **2520** | **5** | **8.27** | **0.4** | **6.2** | **1893** | **1896** | **1899** | **± 9** |
| **AMQ-48 m(0)1.4A pur/rnd** | **5** | **3970** | **5** | **8.36** | **0.3** | **5.45** | **1895** | **1900** | **1906** | **± 6** |
| **AMQ-55 nm(0)1.7A clr/rnd** | **10** | **1790** | **3** | **8.08** | **0.3** | **3.6** | **1905** | **1907** | **1908** | **±7** |
| **AMQ-57 nm(0)1.7A clr/rnd** | **20** | **4900** | **10** | **8.35** | **0.3** | **2.6** | **1910** | **1913** | **1916** | **± 7** |
| AMQ-6 nm(0)1.7A eu/pk | 14 | 1290 | 4 | 7.86 | 0.2 | 3.1 | 1902 | 1909 | 1917 | ± 8 |
| AMQ-28 nm(0)1.7A eu/pk | 15 | 3200 | 5 | 8.24 | 0.4 | 4.7 | 1793 | 1852 | 1919 | ± 8 |
| **AMQ-44 nm(0)1.7A sub/pk** | **11** | **11300** | **10** | **8.19** | **0.2** | **20.6** | **1977** | **1976** | **1975** | **± 4** |
| AMQ-37 nm(0)1.7A eu/pk | 23 | 4600 | 10 | 7.96 | 0.4 | 10.6 | 1900 | 1949 | 2002 | ± 9 |
| **AMQ-31 nm(0)1.7A eu/pk** | **14** | **2370** | **5** | **7.78** | **0.5** | **12.2** | **1999** | **2002** | **2004** | **±10** |
| AMQ-40 nm(0)1.7A eu/pk | 13 | 4400 | 10 | 7.68 | 0.3 | 5.1 | 1960 | 2013 | 2067 | ± 7 |
| AMQ-22 nm(0)1.7A eu/pk | 11 | 2430 | 5 | 7.2 | 0.5 | 6.5 | 2119 | 2136 | 2152 | ± 10 |
| *AMQ-3 nm(0)1.7A eu/pk* | *6* | *187.9* | *2* | *5.19* | *0.4* | *3.95* | *1670* | *1912* | *2185* | *± 16* |

which these lower Paleozoic quartzose strata in the Klamath-Sierran region reached their present position is important because that detritus is 1000–2000 km from its apparent source. For accurate paleogeographic reconstruction, it is critical to try to determine whether that detritus reached its present position as a result of sedimentary or tectonic transport. It has been suggested that the southward transport occurred tectonically during the late Paleozoic along a margin-parallel sinistral strike-slip system (Wallin, 1993), an idea that is consistent with the aforementioned widespread occurrence of late Paleozoic sinistral offset documented in various places along the Cordilleran margin. Both Wallin and Noto prefer that interpretation, which is further detailed later in this paper. Alternatively, Gehrels envisions long-distance transport of the detritus by sediment gravity flows during the early Paleozoic (Gehrels et al., this volume). Here we assess the relative merits of these two very different ideas.

Any model of the origin and displacement history of the Antelope Mountain must explain the following aspects of its stratigraphy and sedimentology: (1) the occurrence in the Antelope Mountain of symmetrical ripples formed by oscillatory flow; (2) the occurrence of coarse detritus in the Antelope Mountain (fine-pebble conglomerate, grit, and coarse detrital mica); and (3) the well-preserved relict stratigraphy, which includes intercalated quartzites, quartz arenites, and argillite horizons, capped by a thick radiolarian chert.

Several aspects of the sedimentology of the Antelope Mountain contraindicate the long-distance sedimentary transport envisioned by Gehrels et al. (this volume). First and least important, it contains coarse grains of detrital mica, which are difficult to reconcile with long distances of sedimentary transport. Given the vastly different hydrodynamic behavior of coarse mica relative to more equant grains, they are rapidly winnowed by or elutriated from turbulent flows. Second and more important, although it has been shown that turbidites can travel long distances, it is highly unlikely that sediment of that caliber (fine pebbles) can be transported such long distances by typical sediment gravity flows. For example, data from the modern Astoria fan off Oregon indicate that fine pebbles in cohesionless sediment are transported only 45 km downslope before deposition (Nelson and Nilsen, 1974). It is unlikely that typical sedi-

**TABLE 1. U-Pb ISTOPIC DATA AND APPARENT AGES–
DETRITAL ZIRCONS FROM THE ANTELOPE MOUNTAIN QUARTZITE (continued)**

| Grain description | Sample weight (mcg) | $^{206}Pb_m/^{204}Pb$ | %err | $^{206}Pb_c/^{207}Pb$ | %err | $^{206}Pb_c/^{208}Pb$ | Ages$_d$ $^{206}Pb^*/^{238}U$ | $^{207}Pb^*/^{235}U$ | $^{207}Pb^*/^{206}Pb^*$ | 2σ error |
|---|---|---|---|---|---|---|---|---|---|---|
| AMQ-25 nm(0)1.7A eu/pk | 12 | 990 | 3 | 6.62 | 0.5 | 6.2 | 2138 | 2175 | 2210 | ± 11 |
| AMQ-47 m(0)1.4A pur/rnd | 7 | 6350 | 10 | 6.99 | 0.2 | 17.5 | 2165 | 2206 | 2245 | ± 5 |
| AMQ-32 nm(0)1.7A eu/pk | 12 | 3480 | 10 | 6.9 | 0.5 | 5.5 | 2196 | 2222 | 2247 | ± 10 |
| AMQ-13 nm(0)1.7A eu/pk | 18 | 4480 | 10 | 6.88 | 0.3 | 6.1 | 2171 | 2218 | 2261 | ± 7 |
| *AMQ-5 nm(0)1.7A eu/pk+* | *11* | *2041* | *0.6* | *6.69* | *0.1* | *6.6* | *1988* | *2136* | *2281* | ± 2 |
| *AMQ-54 nm(0)1.7A clr/rnd* | *16* | *540* | *2* | *5.78* | *0.3* | *6.2* | *2063* | *2213* | *2356* | ± 9 |
| AMQ-43 nm(0)1.7A pk/rnd | 9 | 4000 | 10 | 6.19 | 0.3 | 5.1 | 2402 | 2424 | 2442 | ± 6 |
| *AMQ-20 nm(0)1.7A eu/pk* | *7* | *880* | *3* | *5.69* | *0.5* | *6.2* | *2203* | *2352* | *2485* | ± 11 |
| AMQ-10 nm(0)1.7A pk/rnd | 39 | 5100 | 5 | 5.63 | 0.3 | 2.5 | 2585 | 2601 | 2613 | ± 5 |
| **AMQ-45 m(-1)1.4A pur/rnd** | **10** | **3020** | **5** | **5.39** | **0.3** | **6.4** | **2667** | **2669** | **2671** | ± 6 |
| AMQ-50 m(0)1.4A pur/rnd | 6 | 4050 | 5 | 5.41 | 0.3 | 7.4 | 2554 | 2622 | 2676 | ± 5 |
| AMQ-19 nm(0)1.7A eu/pk | 19 | 3200 | 5 | 5.19 | 0.5 | 6.3 | 2525 | 2645 | 2738 | ± 9 |
| **AMQ-9 nm(0)1.7A pk/rnd** | **26** | **2900** | **4** | **4.67** | **0.3** | **5.1** | **2912** | **2911** | **2911** | ± 5 |
| AMQ-34 nm(0)1.7A eu/pk | 20 | 4100 | 10 | 4.04 | 0.4 | 5.1 | 3029 | 3105 | 3155 | ± 7 |

*Note:*
Grain with $^{206}Pb^*/^{238}U$ age within 5% of $^{207}Pb/^{206}Pb$ age in plain type.
Grain with concordant age and moderate– to high–precision analysis in bold.
Grain with $^{206}Pb^*/^{238}U$ age not within 5% of $^{207}Pb/^{206}Pb$ age in italics.
All analysis were conducted using conventional isotope dilution and thermal ionization mass spectrometry.
Most grains analyzed by Gehrels at the University of Arizona according to procedures in Gehrels (this volume).
For grain descriptions containing (+), isotopic ratios were measured by Wallin using peak switching on the Daly collector of a VG Sector mass spectrometer at the University of Kansas.
Mass fractionation corrections of 0.18% and 0.3% bias per mass unit were applied to Pb and U data, respectively.
* Radiogenic Pb: analytical blanks for Pb range between 2 to 10 pg, those for U are consistently <1 pg.
eu—euhedral, sub—subhedral, pk—pink, pur—purple, clr—colorless, rnd—rounded; c—U and Pb corrected for analytical blank; m—measured; d—Pb corrected for blank and nonradiogenic Pb, U corrected for blank.
Uncertainties are based on precision of measured ratios, uncertainty in mass fractionation correction, uncertainties in concentration and isotopic composition of spikes, and uncertainty in isotopic composition of nonradiogenic Pb.
The isotopic composition of nonradiogenic Pb used to calculate radiogenic $^{207}Pb$ and $^{206}Pb$ was estimated iteratively using Stacey and Kramers (1975) Pb evolution model and the calculated $^{207}Pb/^{206}Pb$ age.
Blank Pb values used were 18.66 ± 1:15 ± 0.1: 38.39 ± 1. Data reduced using Ludwig (1991).
U decay constants are from Steiger and Jäger (1977). Uncertainties are reported at 2 σ level.

ment gravity flows were more competent during the Paleozoic. Third, it is very difficult to reconcile the occurrence of symmetrical ripple marks with a deep-water turbidite origin. Fourth, one cannot invoke resedimentation of sediment carried southward on the shelf by longshore transport because the Antelope Mountain Quartzite is both texturally and mineralogically less mature than the Ordovician miogeoclinal rocks that have a northern British Columbia provenance signature.

In our judgment, the only way the Antelope Mountain could have been transported by deep-water sedimentary processes is through catastrophic failure of stratified, lithified, continental margin, creating a laminar flow of exceptional magnitude, comparable to those that occur in modern oceans (e.g., Jacobi, 1976; Moore et al., 1989, 1994). Such failures transport both unconsolidated sediment and blocks of lithified strata with preserved internal stratigraphy onto the abyssal plain, where they are blanketed by subsequent sedimentation. Such an event, coupled with some amount of tectonic transport on subducting oceanic crust, can explain its current position and geologic context. Furthermore, the reconstructed dimensions of the Antelope Mountain Quartzite (13 km × 9 km × ≥0.22 km) approximate the mean slide thickness/slide width value for 28 documented modern submarine slides (Woodcock, 1979). Given that such slides occur orthogonally with respect to the topographic and/or bathymetric strike of their sources, the Antelope Mountain Quartzite is probably in its current position largely as a consequence of tectonic, rather than sedimentary, transport.

A conceptual model of the sequence of events that can explain these disparate observations is shown in Figure 6. This combination of events conveniently explains the four considerations noted herein. First, initial deposition of the Antelope Mountain formation in a shelfal or platformal setting near its proximate source easily explains the occurrence of shallow-water oscillation ripples. Second, the occurrence of the thick radiolarian chert that overlies the Antelope Mountain Quartzite is difficult to explain in terms of a shallow-water environment of deposition. It is much easier to explain by invoking resedimentation of blocks of miogeoclinal strata into deeper water, and subsequent blanketing by pelagic sedimentation. These considerations lead us (Wallin and Noto) to conclude that detritus represented by the Antelope

Mountain Quartzite was transported southward by a combination of catastrophic collapse of the continental margin off British Columbia during the early Paleozoic, and tectonic transport during the middle and late Paleozoic. That the Antelope Mountain Quartzite is a block in melange does not change that interpretation, except that it decouples the Antelope Mountain from the rest of the Yreka terrane during part of its transport history.

In summary, any model chosen to explain the assembly of the Cordillera must explain the present position of the Antelope Mountain Quartzite, which is not easy to do using a fringing-arc model in which the eastern Klamath Mountains are parautochthonous relative to the mid-Paleozoic United States. Although the scenario in Figure 6 is nonunique, only a few combinations of plate motion vectors, trench orientations, and subduction polarities can explain Cordilleran geology satisfactorily, and it appears that any such combination requires southward tectonic transport of the Yreka terrane. With regard to the timing of such tectonic transport, paleomagnetic data indicate that the Klamath Mountains have been in their current paleolatitudinal position relative to cratonic North America since the Late Permian (Mankinen et al., 1989). Given that paleomagnetic study typically entails uncertainties on the order of several hundred kilometers, this means that large-scale sinistral displacement of the eastern Klamath Mountains relative to the North American craton could have occurred only after their formation in the mid-Paleozoic, and prior to the Late Permian.

## IMPLICATIONS FOR CORDILLERAN TECTONICS

One of the fundamental questions in tectonics is the extent to which the distribution of terranes in accretionary orogens is controlled by either accordion tectonics (repetitive formation and contraction of backarc basins orthogonal to the continental margin) or dispersion tectonics (displacement dominated by strike slip). If our model is correct, it implies that the distribution of Paleozoic terranes along the Cordilleran margin has been influenced at least as much by dispersion tectonics as by accordion tectonics. According to this model, the mid-Paleozoic magmatic arc is fringing (Burchfiel and Davis, 1972; Rubin et al., 1990) in the sense that it is outboard of the craton (e.g., Kay, 1951), but it is not separated from the craton by backarc basins during most of its evolution. Our interpretation differs from the fringing-arc model, in that it explains the following four geologic considerations.

1. There is a vast distance between the Antelope Mountain Quartzite and its apparent source in northern British Columbia. If long-distance sedimentary transport of the detritus in the Antelope Mountain can be discounted, as we suggest, then a substantial amount of southward tectonic transport appears to be required to explain the present location of the Antelope Mountain Quartzite.

2. There is a dearth of volcaniclastic detritus in the Roberts Mountains allochthon which, according to the fringing-arc model, formed in a backarc basin. The allochthon contains only sparse, thin, usually aphyric felsic tuffs, and does not contain the thick volcaniclastic successions characteristic of arc-related basins (e.g., Marsaglia, 1995).

3. The Mississippian geologic record in the eastern Klamath and northern Sierra terranes lacks evidence of volumetrically significant magmatism (Gehrels and Miller, this volume, notwithstanding), which is difficult to reconcile with the idea of an active Mississippian volcanic island arc fringing the North American Cordilleran margin. The record is arguably more consistent with formation in either a transpressive or transtensional tectonic setting.

4. The Klamath Mountains have undergone ~106° of clockwise vertical-axis rotation since the Permian-Triassic (Mankinen et al., 1989), which means that the north-south–trending mid-Paleozoic convergent margin preserved in the eastern Klamaths was oriented orthogonally to the continental margin prior to that rotation. Although the pre-Permian rotational history of the Klamaths is unknown, the paleomagnetic data are not consistent with a simple fringing-arc model, and they must be explained by any model of tectonic evolution.

Figure 6 (on opposite page). Sequence of conceptual diagrams illustrating Paleozoic transport history of Antelope Mountain Quartzite. Most Cordilleran terranes are not present until after Permian, but are shown for geographic reference. Plate motion vectors are inferred from geological evolution. A: Nonunique initial orientation of subduction zone is based on intraoceanic setting, North American faunas, and subsequent Devonian intraoceanic and continental arc magmatism. Large-scale failure of continental margin in British Columbia transports Antelope Mountain Quartzite downslope in laminar flow of exceptional magnitude. This translation onto abyssal plain permits blanketing of Antelope Mountain Quartzite by thick radiolarian chert. TT=proto-arc represented by Trinity terrane. B: Indentation of overriding oceanic plate by Laurentia causes reversal of subduction polarity at northern margin of Laurentia, resulting in Devonian magmatism of Ellesmere orogen and refraction of east-facing intraoceanic protoarc. Subduction of cratonic detritus produces Devonian magmatism with upper crustal component. Sedimentation in trench and forearc creates metasedimentary units of Yreka terrane (EK), some of which are derived from southwestern United States. Black arrow indicates long-distance transport of this fine-grained quartzolithic sediment into trench (Duzel Phyllite and Moffett Creek Formation) in manner analogous to that in modern Aleutian trench. C: Magmatism continues in Ellesmere orogen. Intraoceanic margin-parallel transform begins to form at northwestern corner of Laurentia (where buoyancy of subducting plate changes abruptly), causing cessation of magmatism in Silurian-Devonian protoarc (including Klamath Mountains). Roberts Mountain allochthon and coeval chert-rich assemblages are emplaced by some combination of trench rollback and transpression along transform. D: Continued late Paleozoic movement of inactive arc edifice on this transform transports eastern Klamath Mountains southward at least 1000 km. E: Early Pennsylvanian decrease in obliquity of plate motions begins convergence and intraoceanic subduction-related magmatism through Permian-Triassic along west-facing arc. Collision of this active margin and craton results in emplacement of Sonomia north of California salient in intial Sr. 0.706 contour, and sinistral truncation of continental margin occurs south of that salient (Walker, 1988). As Sonomia is emplaced, slivers of it are shaved off its western margin by sinistral shear and transported at least 400 km farther southward. Dextral slip on Lower Cretaceous Mojave–Snow Lake fault (Lahren et al., 1990) requires that these slivers were transported 400 km northward during Early Cretaceous. Eastern Klamath Mountains underwent ~106° of post-Triassic clockwise rotation (Irwin and Mankinen, 1998).

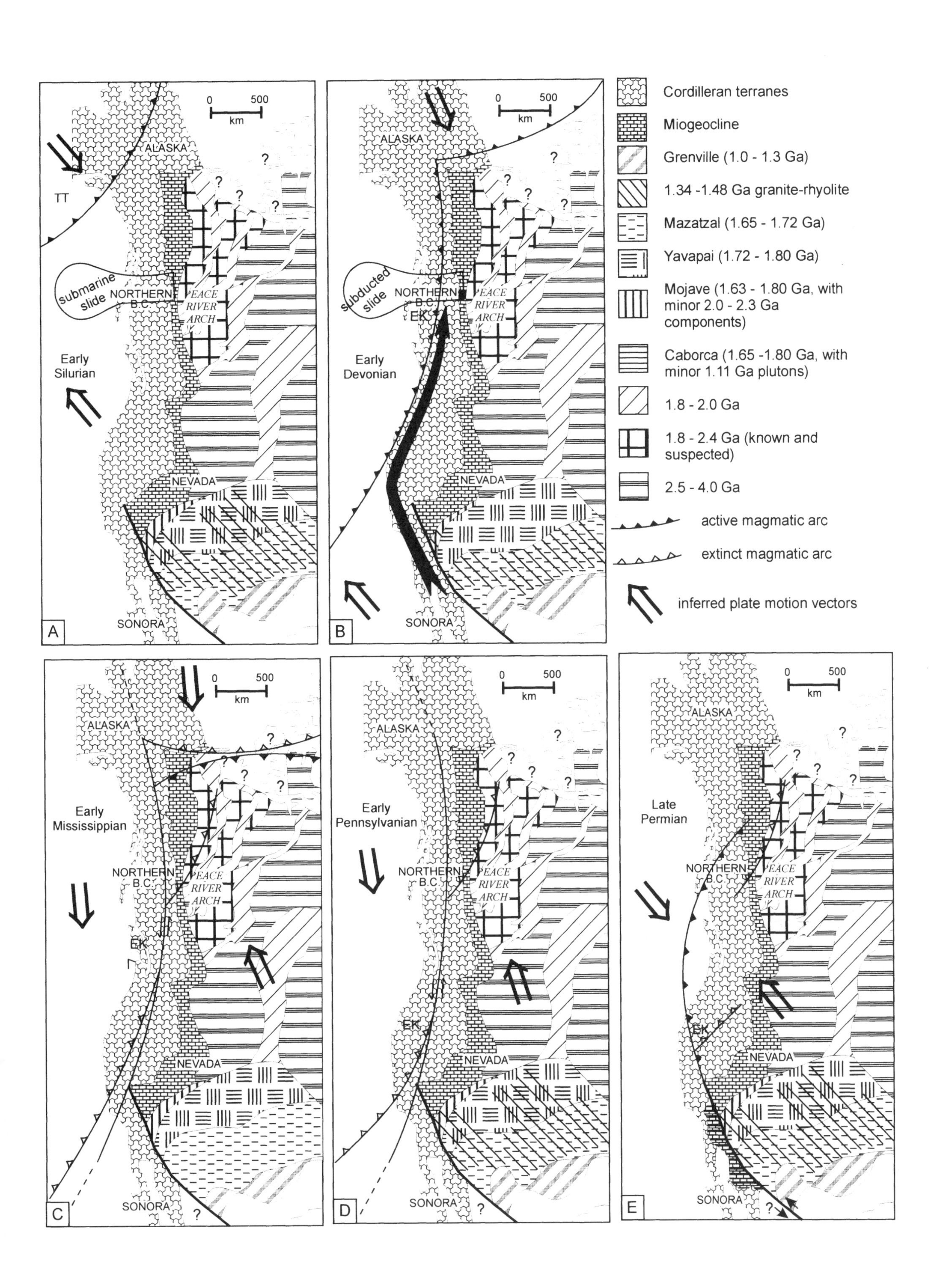

Cordilleran terranes
Miogeocline
Grenville (1.0 - 1.3 Ga)
1.34 -1.48 Ga granite-rhyolite
Mazatzal (1.65 - 1.72 Ga)
Yavapai (1.72 - 1.80 Ga)
Mojave (1.63 - 1.80 Ga, with minor 2.0 - 2.3 Ga components)
Caborca (1.65 -1.80 Ga, with minor 1.11 Ga plutons)
1.8 - 2.0 Ga
1.8 - 2.4 Ga (known and suspected)
2.5 - 4.0 Ga
active magmatic arc
extinct magmatic arc
inferred plate motion vectors
A
Early Silurian
TT
submarine slide
ALASKA
NORTHERN B.C.
PEACE RIVER ARCH
NEVADA
SONORA
B
Early Devonian
subducted slide
EK
C
Early Mississippian
D
Early Pennsylvanian
E
Late Permian
0 500 km

A number of striking tectonostratigraphic similarities exist between the northern Sierra terrane and the eastern Klamath Mountains, indicating that they may have been part of a single mid-Paleozoic convergent margin (Davis, 1969; Schweickert and Snyder, 1981; Wallin et al., 1988). Although the nature and sequence of Paleozoic events in both terranes is broadly similar, there are significant differences in the timing of those events in each terrane. One important question is whether those differences are simply the result of the inherent lateral variation in both the basement and magmatic history of volcanic arcs. An explanation of these differences may be in the Permian-Triassic and Mesozoic structural evolution of the Cordillera. According to the model presented here, Sonomia is emplaced north of the basement salient represented by the $^{87/86}Sr_i$ 0.706 contour as envisioned by Speed (1979), but given the inferred plate motion vectors, it is likely that tectonic slivers of Sonomia were shaved off its western edge and transported hundreds of kilometers farther southward. Such slivering appears to be required to explain the current location of the Eastern Klamath and northern Sierra terranes, because Paleozoic rocks of the Eastern Klamath and northern Sierra terranes are outboard of the Mojave–Snow Lake fault in the Sierra Nevada, which accommodated 400 km of dextral displacement during the Cretaceous (Lahren et al., 1990). Such a combination of late Paleozoic sinistral offset and Mesozoic dextral displacement (e.g., Stevens et al., 1992) may have juxtaposed different segments of a single diachronous volcanic arc that were not adjacent during its formation.

## CONCLUSIONS

1. The Antelope Mountain Quartzite occurs as a large tectonic block in melange, rather than as a facies of the metasedimentary strata in overlying nappes, as was previously believed.

2. U-Pb ages for detrital zircon from the Antelope Mountain are most compatible with derivation from source rocks in northern British Columbia.

3. The mechanism by which detritus in the Antelope Mountain reached its current position is explained best by catastrophic failure of the continental margin in northern British Columbia during the early Paleozoic, followed by an indeterminate amount of subduction-related transport prior to incorporation into the Yreka terrane by the Early Devonian, followed by as much as 1000 km of southward sinistral offset of the Yreka terrane during the late Paleozoic.

## ACKNOWLEDGMENTS

Research was supported by National Science Foundation grant 94-17954 to Wallin. Wallin thanks W.R. Van Schmus and J.D. Walker for use of the Isotope Geochemistry Laboratory at the University of Kansas, Preston Hotz for his groundbreaking field work on the Yreka terrane, and J.W. Eschelbacher and G.M. Gin for discussions. R.A. Schweickert, C.H. Stevens, and W.S. Snyder provided incisive, constructive reviews of the manuscript. We thank M. Soreghan for his editorial efforts on behalf of this volume.

## REFERENCES CITED

Bond, G.C., and Devay, J.C., 1980, Pre-Upper Devonian quartzose sandstones in the Shoo Fly Formation, northern California—Petrology, provenance, and implications for regional tectonics: Journal of Geology, v. 88, p. 285–308.

Boucot, A.J., Dunkle, D.H., Potter, A., Savage, N.M., and Rohr, D., 1974, Middle Devonian orogeny in western North America?: A fish and other fossils: Journal of Geology, v. 82, p. 691–708.

Burchfiel, B.C., and Davis, G.A., 1972, Structural framework and evolution of the southern part of the Cordilleran orogen, western United States: American Journal of Science, v. 272, p. 97–118.

Cashman, S.M., 1980, Devonian metamorphic event in the northeastern Klamath Mountains, California: Geological Society of America Bulletin, v. 91, p. 453-459.

Clifton, H.E., and Dingler, J.R., 1984, Wave-formed structures and paleoenvironmental reconstruction: Marine Geology, v. 60, p. 165–198.

Coney, P.J., Jones, D.L., and Monger, J.W.H., 1980, Cordilleran suspect terranes: Nature, v. 288, p. 329–333.

Davis, G.A., 1969, Tectonic correlations, Klamath Mountains and western Sierra Nevada, California: Geological Society of America Bulletin, v. 80, p. 1095–1108.

Doe, B.R., Delevaux, M.H., and Albers, J.P., 1985, The plumbotectonics of the West Shasta mining district, eastern Klamath Mountains, California: Economic Geology, v. 80, p. 2136–2148.

Eisbacher, G.H., 1983, Devonian-Mississippian sinistral transcurrent faulting along the cratonic margin of western North America: Geology, v. 11, p. 7–10.

Eschelbacher, J.W., and Wallin, E.T., 1998, Assembly of a subduction complex during the Early to Middle Devonian, Yreka terrane, California: Geological Society of America Abstracts with Programs, v. 30, no 5, p. 13.

Gehrels, G.E., Dickinson W.R., Ross, G.M., Stewart, J.H., and Howell, D.G., 1995, Detrital zircon reference for Cambrian to Triassic miogeoclinal strata of western North America: Geology, v. 23, p. 831–834.

Helwig, J., 1974, Eugeosynclinal basement and a collage concept of orogenic belts, *in* eds., Modern and ancient geosynclinal sedimentation: Society of Economic Paleontologists and Mineralogists Special Publication 19, p. 359–376.

Hotz, P.E., 1977, Geology of the Yreka quadrangle, Siskiyou County, California: U.S. Geological Survey Bulletin, v. 1436, 72 p.

Hotz, P.E., 1978, Geologic map of the Yreka quadrangle and parts of the Fort Jones, Etna, and China Mountain quadrangles, California: U.S. Geological Survey Open-File Report 78–12, scale 1:62 500.

Irwin, W.P., 1972, Terranes of the western Paleozoic and Triassic belt in the southern Klamath Mountains, California: U.S. Geological Survey Professional Paper 800-C, p. 103–111.

Irwin, W.P., 1989, Terranes of the Klamath Mountains, California and Oregon *in* Blake, M.C., and Harwood, D.S., eds., Tectonic evolution of northern California: International Geological Congress, 28th, Field Trip Guidebook T108, p. 19–32.

Irwin, W.P., 1994, Geologic map of the Klamath Mountains, California and Oregon: U.S. Geological Survey Miscellaneous Investigations Series Map I-2148, scale 1:500 000.

Irwin, W.P., and Mankinen, E.A., 1998, Rotational and accretionary evolution of the Klamath Mountains, California and Oregon, from Devonian to present time: U.S. Geological Survey Open-File Report 98-114.

Jacobi, R.D., 1976, Sediment slides on the northwestern continental margin of Africa: Marine Geology, v. 22, p. 157–173.

Kay, M., 1951, North American geosynclines: Geological Society of America Memoir 48, 143 p.

Kinkel, A.R., Hall, W.E., and Albers, J.P., 1956, Geology and base-metal deposits of West Shasta copper-zinc district, Shasta County, California: U.S. Geological Survey Professional Paper 285, 156 p.

Klanderman, D., 1978, Geology of the Antelope Mountain area, Yreka quadrangle, California [M.S. thesis]: Oregon State University, 101 p.

Lahren, M.M., Schweickert, R.A., Mattinson, J.M., and Walker, J.D., 1990, Evidence of uppermost Proterozoic to Lower Cambrian miogeoclinal rocks and the Mojave–Snow Lake fault: Snow Lake pendant, central Sierra Nevada, California: Tectonics, v. 9, p. 1585–1608.

Lapierre, H., Albarede, F., Albers, J.P., Cabanis, B., and Coulon, C., 1985, Early Devonian volcanism in the eastern Klamath Mountains, California: Evidence for an immature island arc: Canadian Journal of Earth Sciences, v. 22, p. 214–227.

Lindsley-Griffin, N., and Griffin, J.R., 1983, The Trinity terrane—An early Paleozoic microplate assemblage, *in* Stevens, C.H., ed., Pre-Jurassic rocks in western North American suspect terranes: Pacific Section, Society of Economic Paleontologists and Mineralogists, p. 63–76.

Lindsley-Griffin, N., Griffin, J.R., and de la Fuente, J., 1989, Unidentified fossil-like objects (UFOs) from the Antelope Mountain Quartzite, lower Paleozoic Yreka terrane, Klamath Mountains, California: Geological Society of America Abstracts with Programs, v. 21, no. 5, p. 107.

Lindsley-Griffin, N., Griffin, J.R., and Wallin, E.T., 1991, Redefinition of the Gazelle Formation of the Yreka terrane, Klamath Mountains, California: Paleogeographic implications, *in* Cooper, J.D. and Stevens, C.H., eds., Paleozoic paleogeography of the western United States: Pacific Section, SEPM (Society for Sedimentary Geology), p. 609–624.

Ludwig, K.R., 1991, ISOPLOT: A plotting and regression program for radiogenic isotope data (version 2.53): U.S. Geological Survey Open-File Report, v. 91-445, 39 p.

Mahoney, J.B., Link, P.K., Burton, B.R., Geslin, J.K., and O'Brien, J.P., 1991, Pennsylvanian and Permian Sun Valley Group, Wood River Basin, south-central Idaho, *in* Cooper, J.D. and Stevens, C.H., eds., Paleozoic paleogeography of the western United States: Pacific Section, SEPM (Society for Sedimentary Geology), p. 551–571.

Mankinen, E.A., Irwin, W.P., and Gromme, C.S., 1989, Paleomagnetic study of the eastern Klamath terrane, California, and implications for the tectonic history of the Klamath Mountains Province: Journal of Geophysical Research, v. 94, p. 10444–10472.

Marsaglia, K.M., 1995, Interarc and backarc basins, *in* Tectonics of sedimentary basins: Busby C.J., and Ingersoll, R.V., eds., Cambridge, Massachusetts, Blackwell Science, p. 299–330.

Moore, J.G., Clague, D.A., Holcomb, R.T., Lipman, P.W., Normark, W.R., and Torresan, M.E., 1989, Prodigious submarine landslides on the Hawaiian Ridge: Journal of Geophysical Research, v. 94, 17465–17484.

Moore, J.G., Normark, W.R., and Holcomb, R.T, 1994, Giant Hawaiian landslides: Annual Reviews of Earth and Planetary Science, v. 22, p. 119–144.

Nelson, C.H., and Nilsen, T.H., 1974, Depositional trends of modern and ancient deep-sea fans, *in* Dott, R.H., Jr., and Shaver, R.H., eds., Modern and ancient geosynclinal sedimentation: Society of Economic Paleontologists and Mineralogists Special Publication, 19, p. 69–91.

Noto, R.C., 2000, Structure and stratigraphy of the Antelope Mountain Quartzite, Yreka terrane, California [M.S. thesis]: Las Vegas, University of Nevada, 150 p.

Noto, R.C., and Wallin, E.T., 1998, Cretaceous intra-arc extension in the north-eastern Yreka terrane, California: Geological Society of America Abstracts with Programs, v. 30, no. 5, p. 56.

Peacock, S.M., 1987, Serpentinization and infiltration metasomatism in the Trinity peridotite, Klamath province, northern California: Implications for subduction zones: Contributions to Mineralogy and Petrology, v. 95, p. 55–70.

Peacock, S.M., and Norris, P.J., 1989, Metamorphic evolution of the Central Metamorphic Belt, Klamath Province, California: An inverted metamorphic gradient beneath the Trinity peridotite: Journal of Metamorphic Geology, v. 7, p. 191–209.

Potter, A.W., 1990, Middle and Late Ordovician brachipods from the eastern Klamath Mountains, northern California: Palaeontographica, v. 212, p. 31–158.

Reed, J.C., Jr., et al., 1993, Precambrian: Conterminous U.S.: Boulder, Colorado, Geological Society of America, The Geology of North America, v. C-2, p. 1–657.

Rohr, D.M., 1980, Ordovician-Devonian Gastropoda from the Klamath Mountains, California: Palaeontographica, v. 171, p. 141–199.

Rubin, C.M., Miller, M.M., and Smith, G.M., 1990, Tectonic development of Cordilleran mid-Paleozoic volcano-plutonic complexes; evidence for convergent margin tectonism: in Harwood, D.S., and Miller, M.M., eds., Paleozoic and early Mesozoic paleogeographic relations: Sierra Nevada, Klamath Mountains, and related terranes: Geological Society of America Special Paper 255, p. 1–16.

Schweickert, R.A., and Snyder, W.S., 1981, Paleozoic plate tectonics of the Sierra Nevada and adjacent regions, *in* Ernst, W.G., eds., The geotectonic development of California: Englewood Cliffs, New Jersey; Prentice Hall, p. 182–201.

Silberling, N.J., Jones, D.L., Monger, J.W.H., and Coney, P.J., 1992, Lithotectonic terrane map of the North American Cordillera: U.S. Geological Survey Miscellaneous Investigation Series Map I-2176, scale 1:5 000 000.

Speed, R.C., 1979, Collided Paleozoic microplate in the western United States: Journal of Geology, v. 87, p. 279–292.

Stacey, J.S., and Kramers, J.D., 1975, Approximation of terrestrial lead isotope evolution by a two-stage model: Earth and Planetary Science Letters, v. 26, p. 207-221.

Steiger, R.H. and Jager, E., 1977, Subcommission on geochronology: Convention on the use of decay constants in geo- and cosmochronology: Earth and Planetary Science Letters, v. 28, p. 359-362.

Stevens, C.H., 1991, Permian paleogeography of the western United States, *in* Cooper, J.D., and Stevens, C.H., eds., Paleozoic paleogeography of the western United States: Pacific Section, SEPM (Society for Sedimentary Geology), p. 149–166.

Stevens, C.H., Stone, P., and Kistler, R.W., 1992, A speculative reconstruction of the middle Paleozoic continental margin of southwestern North America: Tectonics, v. 11, p. 405–419.

Stone, P., and Stevens, C.H., 1988, Pennsylvanian and Early Permian paleogeography of east-central California: Implications for the shape of the continental margin and the timing of continental truncation: Geology, v. 16, p. 330–333.

Tanner, W.F., 1971, Numerical estimates of ancient waves, water depth, and fetch: Sedimentology, v. 16, p. 71–88.

Trexler, J.H., Jr., Snyder, W.S., Cashman, P.H., Gallegos, D.M., and Spinosa, C., 1991, Mississippian through Permian orogenesis in eastern Nevada: Post-Antler, pre-Sonoma tectonics of the western Cordillera, *in* Cooper, J.D., and Stevens, C.H., eds., Paleozoic paleogeography of the western United States, Pacific Section, SEPM (Society for Sedimentary Geology), p. 317–329.

Walker, J.D., 1988, Permian and Triassic rocks of the Mojave Desert and their implications for timing and mechanisms of continental truncation: Tectonics, v. 7, p. 685–709.

Wallin, E.T., 1989, Provenance of lower Paleozoic sandstones in the eastern Klamath Mountains and the Roberts Mountains allochthon, California and Nevada [Ph.D. thesis]: Lawrence, University of Kansas, 152 p.

Wallin, E.T., 1993, Sonomia revisited: Evidence for a western Canadian provenance of the eastern Klamath and northern Sierra Nevada terranes, California: Geological Society of America Abstracts with Programs, v. 25, no. p. 173.

Wallin, E.T., and Gehrels, G.E., 1995, Sedimentary record of Silurian volcanism reveals denudation of the Trinity terrane and the timing of siliciclastic sedimentation in the Yreka terrane, northern California: Geological Society of America Abstracts with Programs, v. 27, no. p. 83.

Wallin, E.T., and Metcalf, R.V., 1998, Supra-subduction zone ophiolite formed in an extensional forearc: Trinity terrane, California: Journal of Geology, v. 106, p. 591–608.

Wallin, E.T., and Trabert, D.W., 1994, Eruption-controlled epiclastic sedimentation in a Devonian trench-slope basin: Evidence from sandstone petrofacies, Klamath Mountains, California: Journal of Sedimentary Research, v. A64, p. 373–385.

Wallin, E.T., Mattinson, J.M., and Potter, A.W., 1988, Early Paleozoic magmatic events in the eastern Klamath Mountains, northern California: Geology, v. 16, p. 144–148.

Woodcock, N.H., 1979, Sizes of submarine slides and their significance: Journal of Structural Geology, v. 1, p. 137–142.

Zdanowicz, T.A., 1971, Stratigraphy and structure of the Horseshoe Gulch area, Etna and China Mountain quadrangles, California [M.S. thesis]: Oregon State University, 88 p.

Manuscript Accepted by the Society January 24, 2000

Printed in U.S.A.

Geological Society of America
Special Paper 347
2000

# Tectonic implications of detrital zircon data from Paleozoic and Triassic strata in western Nevada and northern California

**George E. Gehrels, William R. Dickinson, Brian J. Darby*, James P. Harding, Jeffrey D. Manuszak*, Brook C.D. Riley*, Matthew S. Spurlin**
*Department of Geosciences, University of Arizona, Tucson, Arizona 85721, USA*
**Stanley C. Finney**
*Department of Geological Sciences, California State University, Long Beach, California 90840, USA*
**Gary H. Girty**
*Department of Geological Sciences, San Diego State University, San Diego, California 92182, USA*
**David S. Harwood**
*U.S. Geological Survey, 345 Middlefield Road, Menlo Park, California 94025, USA*
**M. Meghan Miller**
*Department of Geology, Central Washington University, Ellensburg, Washington 98926, USA*
**Joseph I. Satterfield**
*Geology Department, San Jacinto College North, Houston, Texas 77049, USA*
**Moira T. Smith**
*Tech Exploration Ltd., #350 – 272 Victoria Street, Kamloops, British Columbia V2C 2A2, Canada*
**Walter S. Snyder**
*Department of Geological Sciences, Boise State University, Boise, Idaho 83725, USA*
**E. Timothy Wallin**
*Department of Geoscience, University of Nevada, Las Vegas, Nevada 89154, USA*
**Sandra J. Wyld**
*Department of Geology, University of Georgia, Athens, Georgia 30602, USA*

## ABSTRACT

**U-Pb analyses of detrital zircons from various allochthonous assemblages of Paleozoic and early Mesozoic age in western Nevada and northern California yield new constraints on the sediment dispersal patterns and tectonic evolution of western North America. During early Paleozoic time, a large submarine fan system formed in slope, rise, basinal, and perhaps trench settings near the continental margin, west of continental shelf deposits of the Cordilleran miogeocline. Our detrital zircon data suggest that most of the detritus in this fan system along the western U.S. segment of the margin was derived from the Peace River Arch region of northwestern Canada, and some detritus was shed from basement rocks of the southwestern United States or western Mexico. In most cases, the detritus in the allochthonous assemblages was recycled through platformal and/or miogeoclinal sedimentary units prior to accumulating in offshelf environments. Lower Paleozoic rocks of the Roberts Mountains allochthon, Shoo Fly**

*Current addresses: Darby, Department of Earth Sciences, University of Southern California, Los Angeles, California 90089, USA; Manuszak, Department of Earth Sciences, Purdue University, West Lafayette, Indiana 47907, USA; Riley, Department of Geological Sciences, University of Texas, Austin, Texas 78712, USA.

Gehrels, G.E., et al., 2000, Tectonic implications of detrital zircon data from Paleozoic and Triassic strata in western Nevada and northern California, *in* Soreghan, M.J., and Gehrels, G.E., eds., Paleozoic and Triassic paleogeography and tectonics of western Nevada and northern California: Boulder, Colorado, Geological Society of America Special Paper 347, p. 133–150.

**Complex, and Yreka terrane are interpreted to have been parts of this fan complex that accumulated along the central U.S. segment of the continental margin, probably within 1000 km of the miogeocline.**

**During the mid-Paleozoic Antler orogeny, parts of the lower Paleozoic fan complex were deformed and uplifted, and strata of the Roberts Mountains allochthon were tectonically emplaced onto the continental margin. This orogeny was apparently driven at least in part by convergence of the Sierra-Klamath arc with the continental margin, as has been proposed by many previous workers, because these arc terranes are overlain by Mississippian clastic strata derived from the Roberts Mountains allochthon. Our data are not sufficient, however, to determine the polarity of the arc, or whether the arc formed along the continental margin or was exotic to western North America.**

**Detrital zircon data indicate that following the Antler orogeny, clastic sediments derived from the Roberts Mountains allochthon were deposited both on the continental margin to the east and within intra-arc and backarc basins to the west. The occurrence of this detritus in terranes of western Nevada and northern California indicates that they were proximal to each other and to the continental margin during late Paleozoic time.**

**The presence of upper Paleozoic volcanic and plutonic rocks and arc-derived detrital zircons in strata of the northern Sierra, eastern Klamath, and Black Rock terranes records the existence of a west-facing magmatic arc near the continental margin during late Paleozoic time. Our data are not supportive of scenarios in which these arc terranes were located farther north or thousands of kilometers offshore of the Nevada continental margin during late Paleozoic time.**

**Following a second phase of uplift, erosion, and allochthon emplacement during the Permian–Early Triassic Sonoma orogeny, Middle and Upper Triassic strata now preserved in west-central Nevada accumulated in a backarc basin. Our data indicate that the basinal assemblages contain detritus from arc terranes to the west as well as the craton to the east.**

## INTRODUCTION

As described in the preceding chapters of this volume, U-Pb ages have been determined for 648 detrital zircon grains from sandstones of western Nevada and northern California. The ages, which have been determined by isotope dilution-thermal ionization mass spectrometry, are in most cases concordant to slightly discordant and of moderate to high precision. The resulting ages are considered to be reliable indicators of crystallization age, with an average uncertainty of 10–20 m.y. (at the 95% confidence level). Details of our analytical procedures are described by Gehrels (this volume, Introduction).

From each sandstone sampled, ~20 grains were analyzed. Rather than selecting the grains at random, we have analyzed representatives of all of the different colors, morphologies, and shapes of grains present in each sample. In theory, this selection procedure should increase the likelihood that all of the main age groups in each sample have been recognized. In practice, we find that new age groups are rarely identified after the first 20 grains are analyzed. Hence, we are confident that the sets of ages reported for each sample are an accurate reflection of the ages of the main zircon-bearing sources that contributed detritus to each sandstone.

The ages of detrital zircons in these units provide a powerful tool for reconstructing the geologic and tectonic history of the host sandstones. Two of the main applications used in this study are (1) reconstructing the dispersal history of detritus that accumulated along the continental margin and in outboard terranes during Paleozoic–early Mesozoic time, and (2) establishing provenance links between the clastic units and their source regions. This information is used herein to help reconstruct the Paleozoic–early Mesozoic paleogeography and tectonic history of parts of western Nevada and northern California.

## GEOLOGIC AND TECTONIC FRAMEWORK

The sandstones analyzed in this study belong to terranes and overlap assemblages that are located outboard of the Cordilleran miogeocline in western Nevada and northern California (Fig. 1). The distribution of these terranes and assemblages is shown in Figure 2, and critical aspects of their stratigraphy are shown in Figure 3. They can be divided into three types.

1. Arc-type terranes contain abundant volcanic and volcaniclastic rocks of late Paleozoic and early Mesozoic age. These include the eastern Klamath and Black Rock terranes, the Pine Nut assemblage, and the upper part of the northern Sierra terrane.

2. Ocean-floor or basinal assemblages consist mainly of Paleozoic chert, argillite, sandstone, and subordinate mafic volcanic rocks and carbonate. These include the Roberts Mountains

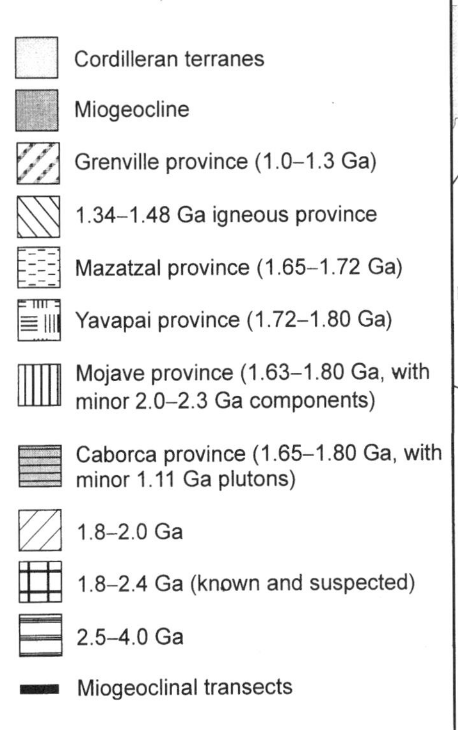


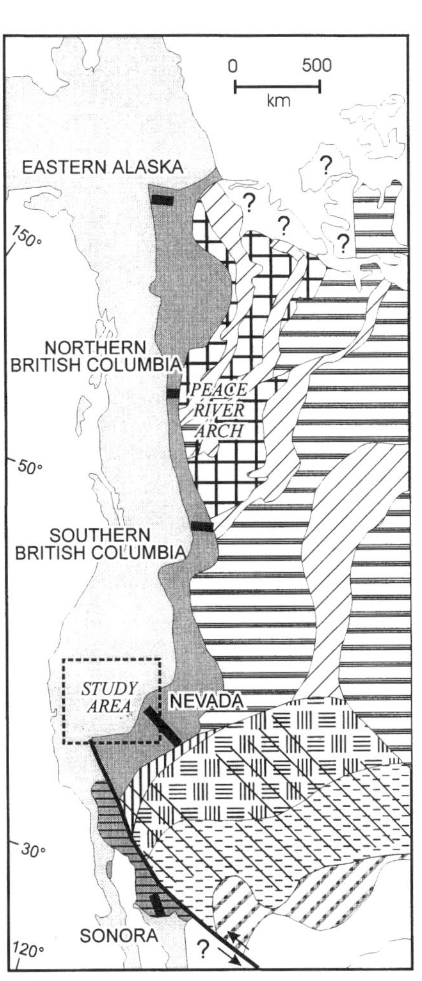


Figure 1. Schematic map showing location of study area in relation to Cordilleran miogeocline and outboard terranes, and first-order basement provinces of western North America. Also shown are locations of five transects that yield detrital zircon data for miogeoclinal strata (Gehrels, this volume). Basement provinces are simplified from Hoffman (1989) for cratonal interior, Ross (1991) and Villeneuve et al. (1993) for western Canadian shield, Van Schmus et al. (1993) for southwestern United States, and Stewart et al. (1990) for northwestern Mexico.

allochthon, the Golconda allochthon, the lower part of the northern Sierra terrane, and the Yreka terrane.

3. The third type comprises Triassic sedimentary overlap assemblages that are known or presumed to overlie older terranes.

The location of these terranes outboard of the Cordilleran miogeocline (Fig. 1) was recognized many years ago (Schuchert, 1923; Kay, 1951), and the tectonic mobility of outboard tectonic elements was established with the initial applications of plate tectonic principles (Wilson, 1968; Hamilton, 1969; Moores, 1970; Burchfiel and Davis, 1972; Helwig, 1974). One of the fundamental questions in Cordilleran tectonics concerns the degree of this tectonic mobility—whether the terranes in Nevada-California formed and evolved in off-shelf settings near their present position, or whether they originated far from where they are found today. These issues remain controversial largely because of the long and complicated orogenic history of the region, which has alternated between periods of convergence, such as the Antler and Sonoma orogenies, and phases of crustal extension. Strike-slip faults with large-scale, orogen-parallel displacement may also have been active intermittently during Paleozoic and Mesozoic time. The end product of this long process is a series of crustal fragments with disparate geologic records that are bounded by faults and tectonic breaks of uncertain total displacement (Helwig, 1974; Coney et al., 1980) (Fig. 2).

Five main models have been proposed for the displacement history of the terranes in western Nevada and northern California (Fig. 4). The first model portrays the Sierra–Klamath–Black Rock terranes as west-facing magmatic arcs that were separated from western North America by a series of backarc basins (Fig. 4A). These basins collapsed and were partially obducted onto the continental margin during the Antler and Sonoma orogenies due to changes in plate motions within a consistently west-facing subduction system. Only one long-lived magmatic arc is involved in this model. This noncollisional view was first proposed by Burchfiel and Davis (1972) and has been favored in many subsequent tectonic syntheses (e.g., Churkin, 1974; Burchfiel and Davis, 1975; Miller et al., 1984, 1992; Miller, 1987; Harwood and Murchey, 1990; Burchfiel et al., 1992).

A second model, favored by Schweickert and Snyder (1981), has an east-facing exotic arc collide with the continental margin during the mid-Paleozoic Antler orogeny, accreting arc-type materials that formed far from western North America as well as a large subduction complex composed of ocean-floor, basinal, and trench-fill sediments that accumulated along the Cordilleran margin

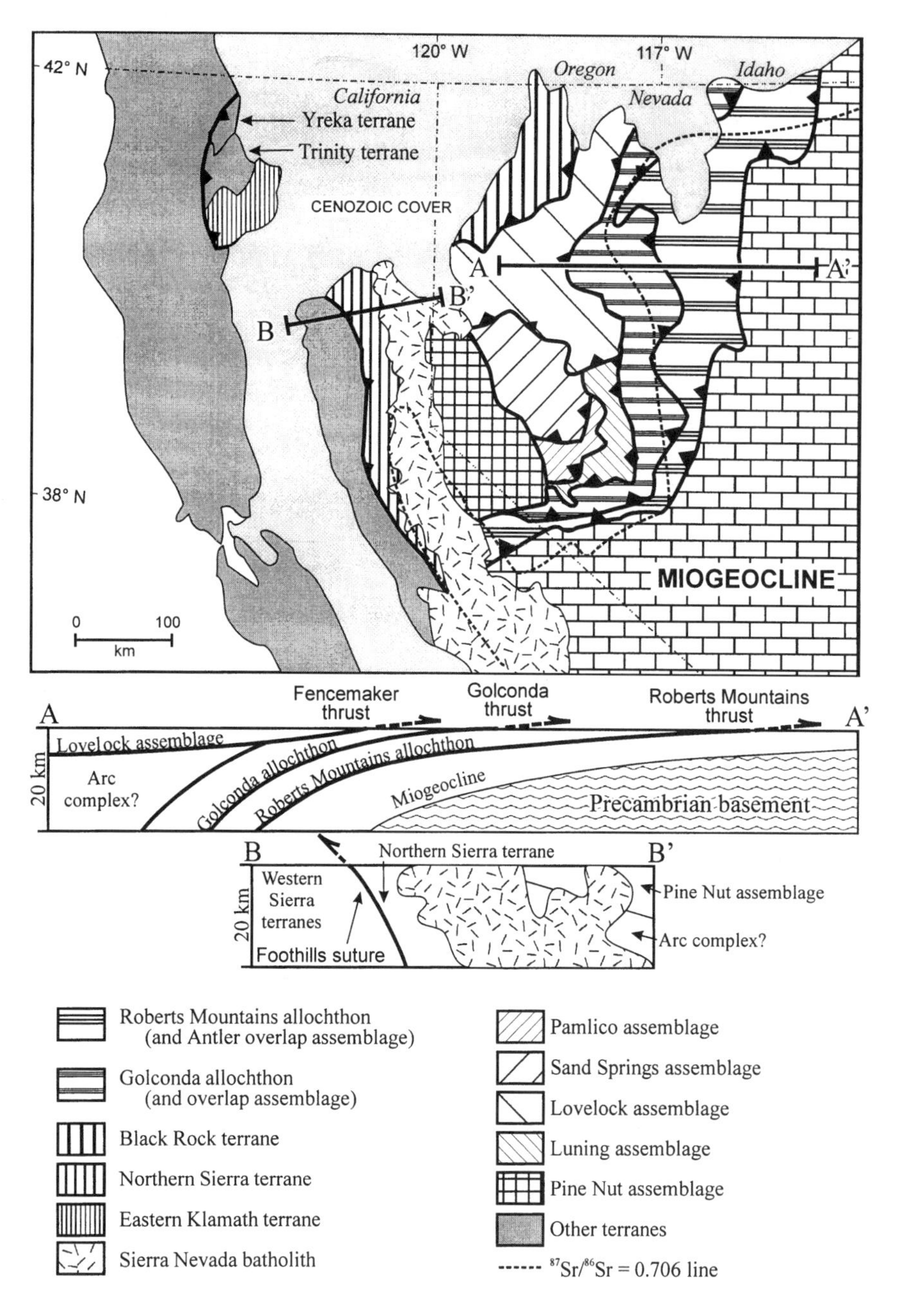


Figure 2. Schematic map and cross sections of main terranes and/or assemblages and their bounding faults in western Nevada and northern California. Map configuration of terranes and assemblages is adapted primarily from Oldow (1984), Silberling (1991), and Silberling et al. (1987), and cross sections are highly simplified from Blake et al. (1985) and Saleeby (1986). Location of $^{87}Sr/^{86}Sr_i = 0.706$ line is from Kistler and Peterman (1973) and Kistler (1991). Note that widespread Cretaceous and younger rocks and structures are omitted in map and sections in effort to emphasize configuration of pre-Cretaceous features.

(Fig. 4B). The sediments are now preserved in the Roberts Mountains allochthon, Shoo Fly Complex, and Yreka terrane, whereas the early Paleozoic part of the magmatic arc may have been the Alexander terrane of southeast Alaska. Following the Antler orogeny, arc polarity switches, the arc rifts away, and a backarc basin opens behind the west-facing magmatic arc, forming the strata of the Golconda allochthon. During the Permian–Early Triassic Sonoma orogeny, a brief phase of east-dipping subduction closed the backarc basin and thrust the backarc basin strata (Golconda allochthon) onto the continental margin. Following the Sonoma orogeny, much of the early Paleozoic arc system was removed by strike-slip faulting, and then west-facing subduction resumed on the west edge of the accreted Sierra-Klamath magmatic arc.

A third model envisions most of the Paleozoic rocks in western Nevada and northern California as belonging to an arc-type crustal fragment that first arrived along the continental margin during the Permian-Triassic Sonoma orogeny (Dickinson, 1977; Speed, 1979; Speed and Sleep, 1982; Dickinson et al., 1983) (Fig. 4C). This fragment, commonly referred to as "Sonomia" (Speed, 1979), consists of upper Paleozoic assemblages in the Black Rock terrane and the northern Sierra and eastern Klamath Mountains. The polarity of subduction in this arc system was eastward, as with model 2. Emplacement of the Roberts Mountains allochthon during the Antler orogeny would have resulted from collision of an early Paleozoic arc system (the Antler arc) that is no longer exposed along the Nevada-California margin,

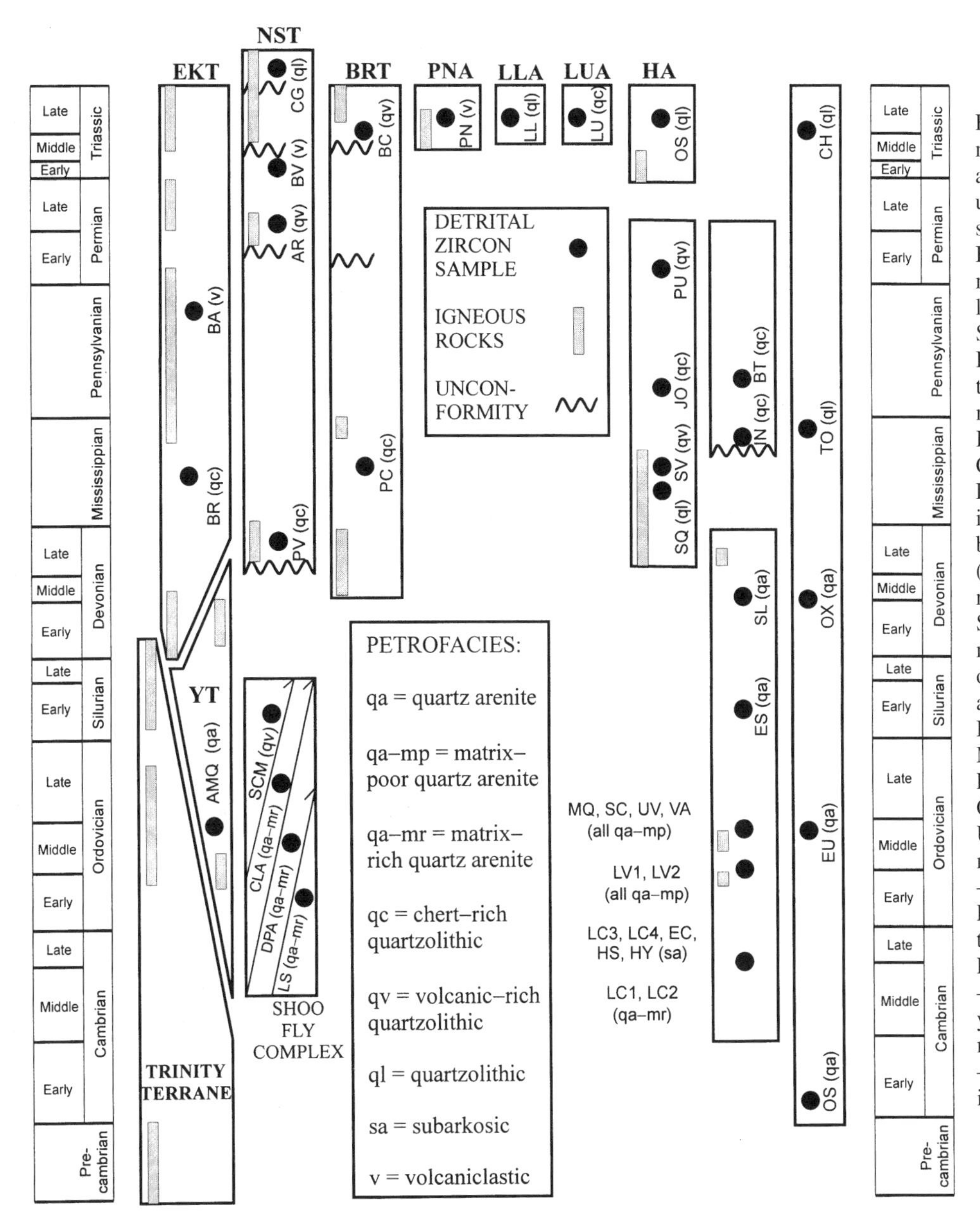


Figure 3. Simplified columns showing name, age, and petrofacies of each sample analyzed and ages of igneous rocks and unconformities in various assemblages studied. EKT (eastern Klamath terrane): BA—Baird Formation, BR—Bragdon Formation. YT (Yreka terrane): AMQ—Antelope Mountain Quartzite. NST (northern Sierra terrane): CG—Cisco Grove unit, BV—Big Valley Formation, AR—Arlington Formation, PV—Picayune Valley Formation. BRT (Black Rock terrane): BC—Bishop Creek formation, PC—Pass Creek unit. PNA—Pine Nut assemblage. LLA—Lovelock assemblage. LUA—Luning assemblage. HA (Humboldt assemblage): OS—Osobb Formation. GA (Golconda allochthon): PU—Pumpernickel Formation, JO—Jory Sandstone, SV—volcanic facies of Schoonover Formation, SQ—quartz-rich phase of Schoonover Formation. AOA (Antler overlap assemblage): BT—Battle Formation, IN—Inskip Formation. RMA (Roberts Mountains allochthon): SL—Slaven Chert, ES—Elder Sandstone, MQ—McAfee Quartzite, SC—Snow Canyon Formation, UV—upper sandstone unit in Vinini Formation, VA—Valmy Formation, LV1—lower sandstone in Vinini Formation #1, LV2—lower sandstone in Vinini Formation #2, LC1-4—Harmony Formation from Little Cottonwood Canyon (1–4), EC—Harmony Formation from Elbow Canyon. MIO (miogeocline): CH—Chinle Formation, TO—Tonka Formation, OX—Oxyoke Sandstone, EU—Eureka Quartzite, OS—Osgood Mountains Quartzite.

either because it lies beneath Sonomia or has been tectonically removed from the region. This scenario is supported by biogeographic analyses of Permian fauna in the eastern Klamath terrane, which Stevens et al. (1990) suggested formed at least 5000 km outboard of the Cordilleran margin.

A fourth model calls on changes in subduction geometry to explain the interplay between arc magmatism and accretion of ocean-floor assemblages (Fig. 4D) (Burchfiel and Royden, 1991; Burchfiel et al., 1992). In this model, the Antler and Sonoma orogenies record brief phases of east-facing (west-dipping) subduction beneath an island arc that was generally west facing. Backarc basins inboard of the main Sierra–Klamath–Black Rock arc are interpreted to have closed during the Antler and Sonoma orogenies due to brief periods of subduction along the inboard (eastern) margin of the arc system. The Golconda and Roberts Mountains allochthons would have developed as subduction complexes associated with the trench inboard of the arc. Slab rollback during this phase of backarc subduction (Fig. 4D) led to continentward migration of the subduction complexes and simultaneous extension farther west. Allochthon emplacement is interpreted to have occurred when the migrating subduction complexes arrived at the continental margin.

A fifth alternative, not shown in Figure 4, includes the possibility that some terranes in the region may have formed farther north or south along the margin (Oldow, 1984; Schweickert and Lahren, 1990; Wallin, 1990a, 1993; Wyld and Wright, 1997). Wallin (1993) and Wallin et al. (this volume, Chapter 9) argue, largely on the basis of detrital zircon ages, that the northern Sierra and eastern Klamath terranes were located along the Canadian Cordilleran margin during early Paleozoic time, whereas Wallin

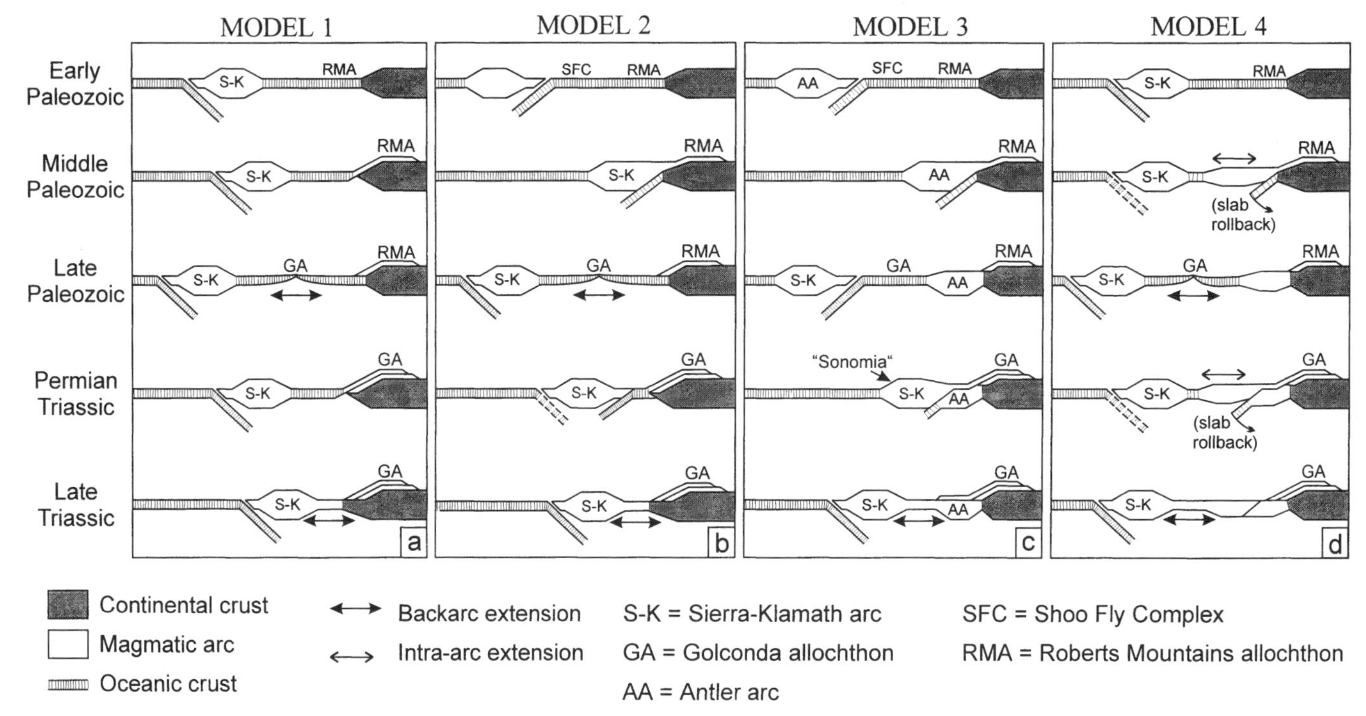


Figure 4. Summary of existing tectonic models for Cordilleran margin. A, Consistently west facing convergent margin with arc separated by series of marginal basins (Burchfiel and Davis, 1972, 1975; Churkin, 1974; Miller et al., 1984, 1992; Harwood and Murchey, 1990; Burchfiel et al., 1992). B, Collision of exotic arc during Antler orogeny followed by mainly west facing convergent margin (Schweickert and Snyder, 1981). C, Collision of two east-facing arc systems during Antler and Sonoma orogenies (Dickinson, 1977; Speed, 1979; Speed and Sleep, 1982; Dickinson et al., 1983). D, Generally west facing convergent margin with brief phases of east-facing subduction and extensional arc magmatism during Antler and Sonoma orogenies (Burchfiel and Royden, 1991; Burchfiel et al., 1992).

(1990a) suggested that the Harmony Formation of the Roberts Mountains allochthon may have originated near coastal Mexico. Oldow (1984), Schweickert and Lahren (1990), and Wyld and Wright (1997) argued for margin-parallel strike-slip displacement of outboard terranes during the Jurassic and Cretaceous. These models do not, however, invoke as much displacement of terranes as the models of Wallin (1993) and Wallin et al. (this volume).

## PRESENT STUDY

Our detrital zircon analyses were conducted in an attempt to test models for the paleogeography and tectonic evolution of western Nevada–northern California during Paleozoic–early Mesozoic time. U-Pb ages of zircons provide a powerful tool for reconstructing this history because the western part of the North American craton consists of Precambrian provinces with distinct ages (Hoffman, 1989; Ross, 1991; Villeneuve et al., 1993; Van Schmus et al., 1993; Stewart et al., 1990). Specific cratonal sources of detrital zircons in Cordilleran terranes can therefore be identified in many cases, establishing provenance ties. U-Pb ages of zircons are a robust tool for constructing provenance links because crystallization ages are preserved in most cases through moderate and even high grades of metamorphism, and because zircon grains are extremely durable in sedimentary and weathering environments.

The stratigraphic positions of the samples that we have analyzed are shown in Figure 3, and the interpreted ages for the grains are summarized on relative age-probability plots. All of our interpreted provenance links are listed in Table 1, are shown graphically on probability plots in Figures 5, 7, and 10, and are shown in map view in Figures 6, 8, and 11 (see also Fig. 9).

## EARLY PALEOZOIC PALEOGEOGRAPHY

Our data indicate that lower Paleozoic strata of western Nevada and northern California received detritus from several different regions of the North American craton, with only minor input from an outboard magmatic arc. The dominant source for sandstones of the Roberts Mountains allochthon (Gehrels et al., this volume, Chapter 1), the Shoo Fly Complex (Harding et al., this volume, Chapter 2), and at least part of the Yreka terrane (Wallin et al., this volume, Chapter 9) consisted of >1.8 Ga basement rocks and overlying platformal strata in the Peace River arch region of northwestern Canada (provenance links 2, 4, 5, and 8 in Figs. 5 and 6 and Table 1). This detritus appears in Cambrian(?), Ordovician, Silurian, and Devonian strata of the Roberts Mountains allochthon and in Ordovician(?) strata of the northern Sierra and Yreka terranes. Feldspathic sands (e.g., most of the Harmony Formation [Roberts Mountains allochthon] and part of the Shoo Fly Complex [northern Sierra terrane] and Antelope

TABLE 1. DETRITAL ZIRCON PROVENANCE LINKS FOR STRATA OF WESTERN NEVADA AND NORTHERN CALIFORNIA

| Provenance link number | Sandstone | Source | Distinctive ages |
|---|---|---|---|
| Early Paleozoic provenance links | | | |
| 1 | RMA: Harmony A (Cambrian?) | Mojave-Sonora craton | ca. 0.7, 1.0-1.35 Ga |
| 2 | RMA: Harmony B (Cambrian?) | Southern British Columbia craton | 1.76-2.7 Ga |
| 3 | RMA: Lower Vinini (early Middle Ordovician) | Southwestern U.S. craton | ca. 1.43, 1.6-1.8 Ga |
| 4 | RMA: late Middle Ordovician, Silurian, and Devonian units | Northern British Columbia craton (Peace River arch) | ca. 1.03, 1.8-2.0, ca. 2.07 Ga |
| 5 | NST: Lang, Duncan Peak, Culbertson Lake allochthons (Ordovician?) | Northern British Columbia craton (Peace River arch) | 1.8-2.0, ca. 2.07, 2.2-2.4, 2.5-2.7 Ga |
| 6 | NST: Sierra City melange (Silurian-Devonian?) | Mojave-Sonora craton | 1.0-1.35 Ga |
| 7 | NST: Sierra City melange (Silurian-Devonian?) | Trinity terrane igneous rocks | ca. 560 Ma |
| 8 | YT: Antelope Mountain Quartzite (Ordovician?) | Northern British Columbia craton (Peace River arch) | 1.8-1.95, 2.2-2.4, 2.6-2.7 Ga |
| Late Paleozoic provenance links | | | |
| 9 | Antler overlap assemblage (Mississippian and Penn units) | Roberts Mountains allochthon | 1.8-2.1, 2.3-2.8 Ga |
| 10 | GA: Mississippian, Penn, and Permian units | Roberts Mountains allochthon | 1.8-2.1, 2.4-2.8 Ga |
| 11 | GA: Permian strata | Southwestern U.S. craton | 1.7-1.8 Ga |
| 12 | GA: Mississippian and Permian strata | Igneous rocks in northern Sierra terrane | 338-358 Ma |
| 13 | BRT: Mississippian strata | Roberts Mountains allochthon | 1.0-1.2 and 1.8-2.0 Ga |
| 14 | BRT: Mississippian strata | Salmon River arch region? | ca. 1.6, 1.6-1.8 Ga |
| 15 | NST: Permian strata | Igneous rocks in northern Sierra terrane | 344-362 Ma |
| 16 | NST: Devonian-Mississippian strata | Roberts Mountains allochthon and/or Shoo Fly Complex | 1.8-2.1, 2.3-2.8 Ga |
| 17 | EKT: Devonian-Mississippian strata | Yreka terrane and/or Roberts Mountains allochthon | 1.8-2.0 Ga |
| 18 | EKT: Devonian-Mississippian strata | Igneous rocks in Trinity terrane | 398-437, ca. 470, 545-572 Ma |
| 19 | EKT: Devonian-Mississippian and Penn-Perm strata | Volcanism in/near eastern Klamath terrane | 320-326 and ca. 363 Ma |
| Early Mesozoic provenance links | | | |
| 20 | Humboldt assemblage | Miogeocline in Nevada | ca. 1.45 and 1.6-1.8 Ga |
| 21 | TA: Lovelock assemblage | Miogeocline in Nevada | ca. 1.45 Ga |
| 22 | TA: Lovelock assemblage | Igneous rocks in EKT, NST, BRT, and Pine Nut assemblage | 227-394 Ma |
| 23 | TA: Luning assemblage | Miogeocline in Nevada | ca. 1.45 Ga and 1.6-1.75 Ga |
| 24 | TA: Luning assemblage | Roberts Mountains or Golconda allochthons | 1.0-1.2 and 1.8-2.0 Ga |
| 25 | TA: Luning assemblage | Igneous rocks in EKT, NST, BRT, and Pine Nut assemblage | 218-272 Ma |
| 26 | TA: Pine Nut assemblage | Igneous rocks in Black Rock terrane | 228-243 Ma |
| 27 | BRT: Bishop Canyon | Roberts Mountains or Golconda allochthons | >1.8 Ga |
| 28 | BRT: Bishop Canyon | Salmon River arch? | ca. 1.6 Ga |
| 29 | BRT: Bishop Canyon | Igneous rocks in EKT, NST, and BRT | 268-441 Ma |
| 30 | NST: Big Valley | Igneous rocks in NST or EKT | 254-270 Ma |

Notes: RMA—Roberts Mountains allochthon; NST—northern Sierra terrane; YT—Yreka terrane; GA—Golconda allochthon; BRT—Black Rock terrane; TA—Triassic assemblages in western Nevada; EKT—eastern Klamath terrane.

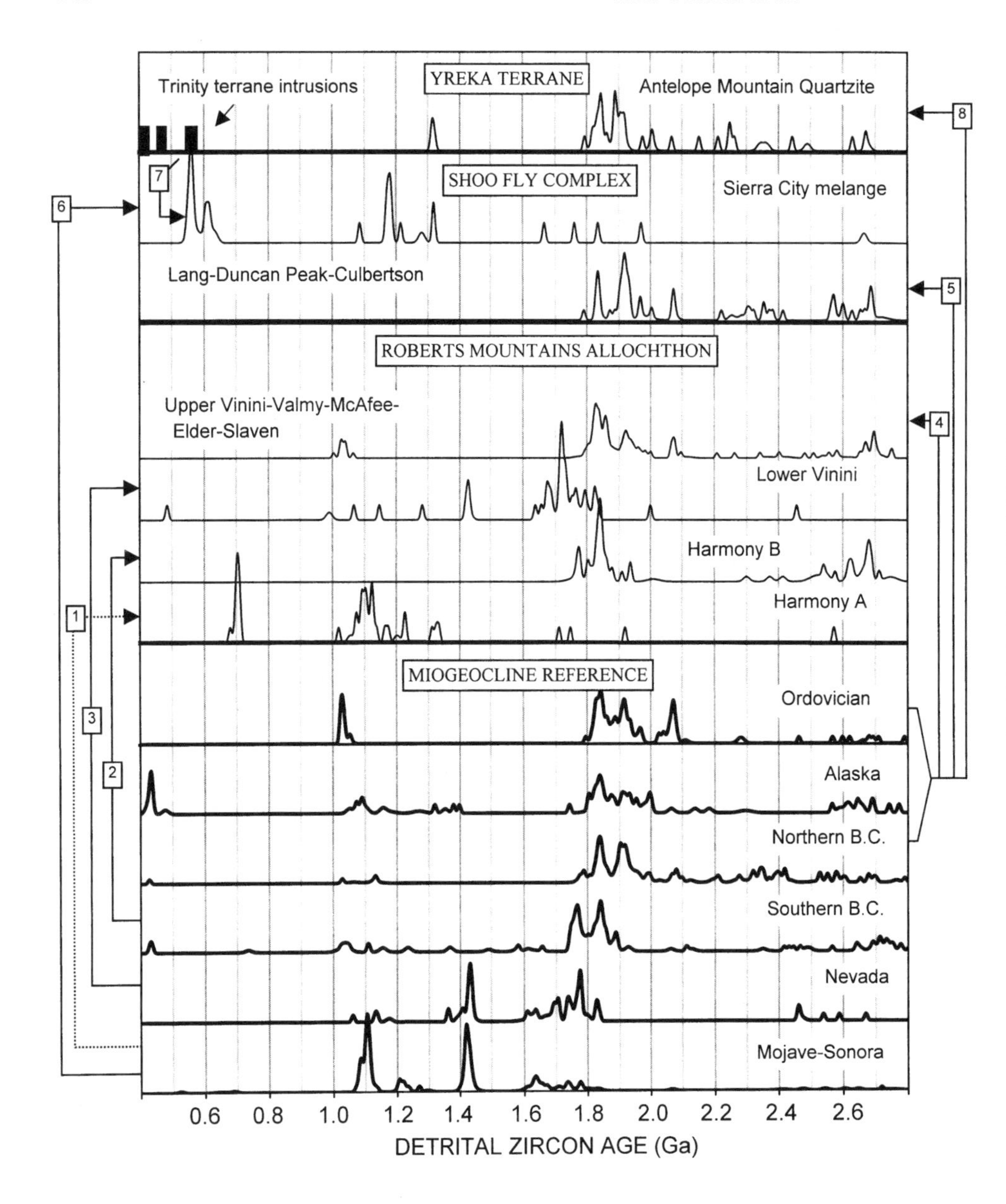


Figure 5. Relative age-probability curves for Paleozoic strata of Cordilleran miogeocline (Gehrels, this volume) and lower Paleozoic strata of Roberts Mountains allochthon (Gehrels et al., this volume, Chapter 1), Shoo Fly Complex (Harding et al., this volume), and Yreka terrane (Wallin et al., this volume). Ages of Trinity terrane intrusive rocks are from Wallin et al. (1995). B.C.—British Columbia. Numbers in boxes refer to provenance links listed in Table 1.

Mountain Quartzite [Yreka terrane]) may have been derived directly from basement rocks exposed along the continental margin, whereas more mature quartz sands were probably recycled from miogeoclinal and/or platformal strata in the Peace River arch region.

Ordovician strata of the Roberts Mountains allochthon also contain 1.8–1.4 Ga grains that originated in basement rocks of the southwestern United States (provenance link 3). As concluded by Finney and Perry (1991) on the basis of biostratigraphic and facies relations, there is a strong link between lower Middle Ordovician rocks of the Vinini Formation and coeval strata of the miogeocline directly to the east. Derivation of sediment from nearby continental sources apparently resulted from a low sea-level stand, during which sand was eroded from either exposed basement rocks or their platformal cover and transported across the miogeocline via channels carved into the continental shelf (Finney and Perry, 1991).

Units in the Roberts Mountains allochthon and Shoo Fly Complex also contain detrital zircons of 1.0–1.3 and ca. 0.7 Ga, which were most likely shed from basement rocks exposed along the southern continental margin (Wallin, 1990a) (provenance links 1 and 6). The dominant source may have been a westward continuation of 1.0–1.3 Ga rocks of the Grenville province, perhaps now preserved in the Oaxaca terrane of Mexico (Coney and Campa, 1987; Ruiz et al., 1988). The 0.7 Ga grains in the Harmony Formation were originally reported to have an uncertain provenance (Wallin, 1990a; Smith and Gehrels, 1994), but the unique occurrence of a 0.7 Ga grain in

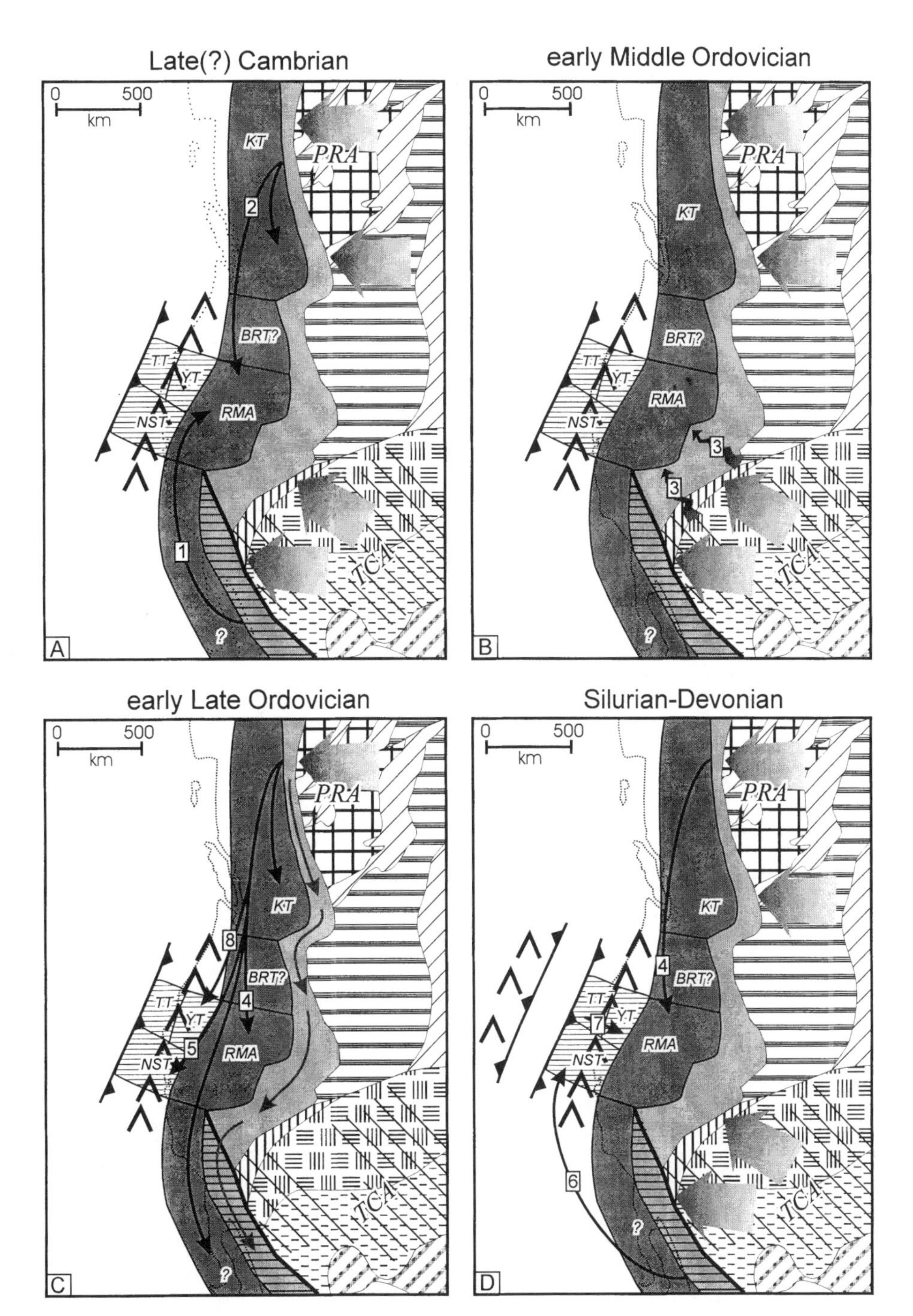


Figure 6. Schematic maps of western North America showing interpreted provenance links and paleogeography for Cordilleran margin during early Paleozoic time. Cratonal provinces and miogeoclinal strata are as shown in Figure 1. Darker gray pattern represents proposed distribution of off-shelf assemblages such as Roberts Mountains allochthon, whereas horizontal ruled pattern outboard of miogeocline represents convergent margin assemblages such as Yreka terrane. The inverted V pattern represents possible trace of a magmatic arc outboard of Cordilleran margin. West-facing arc may have existed along margin during much or all of early Paleozoic time, or east-facing arc may have approached the margin during Devonian time. Black arrows represent interpreted dispersal patterns of sand in offshore settings, with numbers that are keyed to provenance links listed in Table 1. Gray arrows reflect general transport of sand that accumulated within miogeoclinal strata (Gehrels, this volume). Provenance of detritus in Kootenay terrane is from Smith and Gehrels (1991). Provenance of detritus in eugeoclinal strata of Mexico is from Gehrels and Stewart (1998). PRA—Peace River arch, TCA—Trans-continental arch, KT—Kootenay terrane, BRT—Black Rock terrane, TT—Trinity terrane, YT—Yreka terrane, RMA—Roberts Mountains allochthon, NST—northern Sierra terrane.

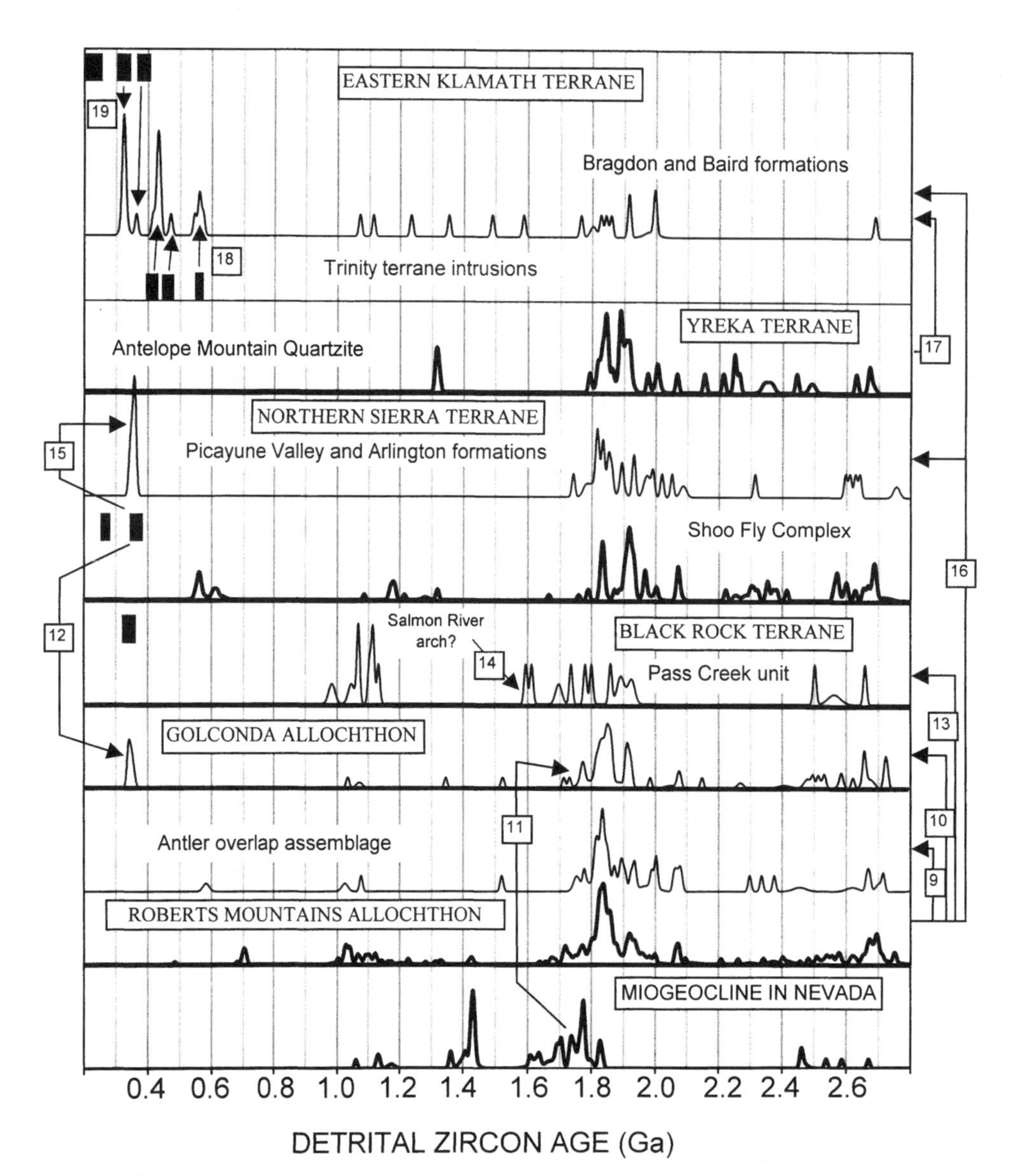


Figure 7. Relative age-probability curves for Paleozoic strata of miogeocline in Nevada (Gehrels, this volume), lower Paleozoic strata of Roberts Mountains allochthon (Gehrels et al., this volume, Chapter 1) and overlying strata (Gehrels and Dickinson, this volume), Golconda allochthon (Riley et al., this volume), Black Rock terrane (Darby et al., this volume), lower Paleozoic rocks of Shoo Fly Complex (Harding et al., this volume) and overlying strata (Spurlin et al., this volume), Ordovician(?) Antelope Mountain Quartzite of Yreka terrane (Wallin et al., this volume), and upper Paleozoic strata of eastern Klamath terrane (Gehrels and Miller, this volume). Ages of igneous rocks in eastern Klamath and Black Rock terranes are from Miller and Harwood (1990) and Wallin et al. (1995). Numbers in boxes refer to provenance links listed in Table 1.

the Wood Canyon Formation of the Mojave Desert suggests that these grains may also have come from the southern Cordilleran margin (provenance link 1).

The Sierra City melange, which is the structurally highest component of the Shoo Fly Complex (Schweickert et al., 1984; Girty et al., 1990), contains grains with ages of 540–640 Ma. These grains were likely shed from igneous rocks in an arc that is now represented only by small crustal fragments in the Trinity terrane (Wallin et al., 1988; Wallin and Metcalf, 1998; provenance link 7) and by granitoid blocks in the Sierra City melange (Saleeby, 1990). A link between this hypothetical arc system and continental rocks is provided by igneous rocks of the Yreka terrane that contain xenocrystic zircons of sufficient age (>1.0 Ga) that they must have a continental source (Wallin, 1990b).

Collectively, these provenance ties indicate that most rocks of the Roberts Mountains allochthon, Shoo Fly Complex (northern Sierra terrane), and Yreka and Trinity terranes (eastern Klamath Mountains) formed in proximity to western North America, prob-

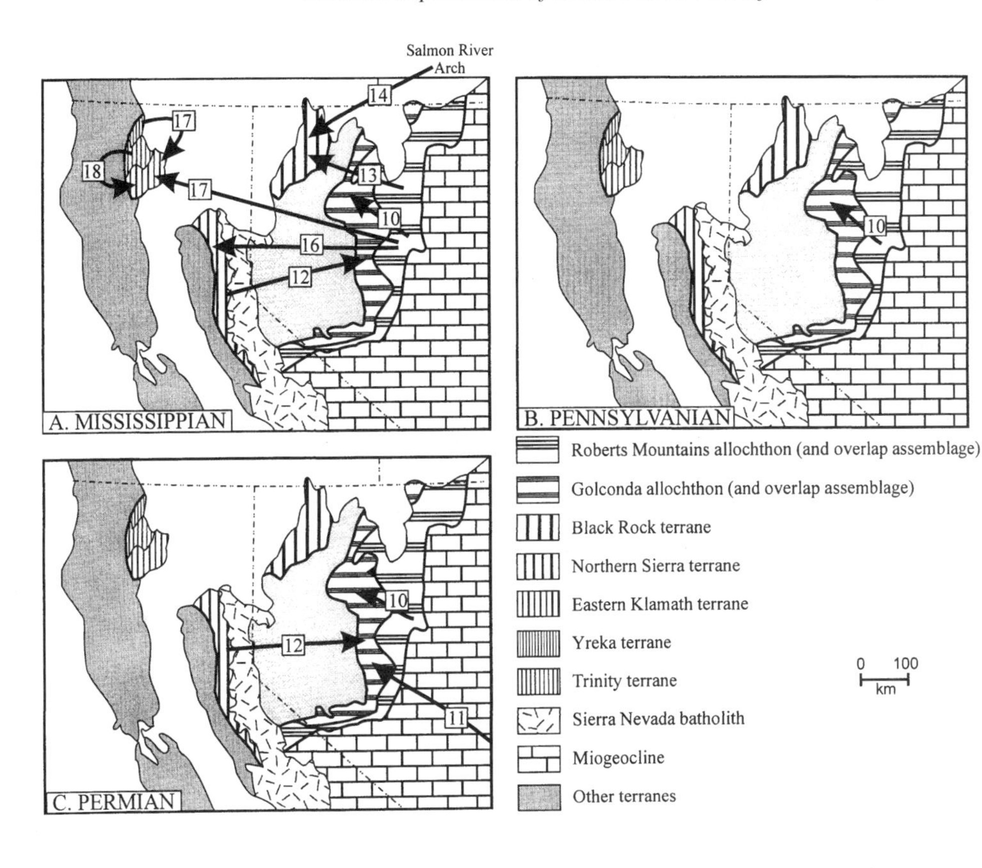


Figure 8. Schematic maps of western Nevada–northern California showing detrital zircon provenance links during late Paleozoic time. Numbers are keyed to list in Table 1. Cratonal provinces and miogeoclinal strata are as shown in Figure 2.

ably within ~1000 km of their present positions along the Cordilleran margin. As shown in Figure 6, we envision the clastic units as having formed as part of a continuous sedimentary apron along and outboard of the Cordilleran margin, with sand arriving from the craton to the east, the Peace River arch region to the north, and Mexico to the south. The sand was apparently transported largely by turbidity currents, mixing sediment of different provenances in coalescing submarine fans. Extensional basins, as described by Miller et al. (1992), may have been important in channeling sand parallel to the continental margin for inboard assemblages such as the Roberts Mountains allochthon. In outboard assemblages, such as the Shoo Fly Complex and Yreka terrane, sand may have been channeled within a subduction-related trench offshore from the continental margin (Schweickert and Snyder, 1981; Girty et al., 1990; Wallin et al., this volume, Chapter 9).

In terms of tectonic models for the early Paleozoic setting of the Cordilleran margin (Fig. 4), the links described here demonstrate that rocks in the northern Sierra, Yreka, and Black Rock terranes and the Roberts Mountains allochthon were in proximity to the Cordilleran margin long before the Sonoma orogeny. This conclusion precludes model 3, which would have these terranes at a considerable distance from the margin until Permian-Triassic time. Unfortunately, the other early Paleozoic models are difficult to test with our data. The primary differences between models 1, 2, and 4 are the facing direction of the Sierra-Klamath arc system, and the existence of an exotic magmatic arc during early Paleozoic time. As shown in Figure 6, craton-derived sediment could have accumulated in proximity to an arc system that faced to the east or west. If the arc faced westward (model 1), the detritus would have accumulated in both backarc (Roberts Mountains allochthon) and forearc (Shoo Fly Complex and Yreka terrane) settings. Given evidence for extension in the Klamath arc (Wallin and Metcalf, 1998), transport of detritus across a low-lying arc construct is plausible. Conversely, the strata could have accumulated entirely inboard of an exotic arc that faced eastward for all of early Paleozoic time (Fig. 4B), or within an extensional arc that faced eastward for just a short period during the mid-Paleozoic (Fig. 4D). Hence, in Figure 6 we show alternative geometries of a west-facing arc existing along the Cordilleran margin for all of early Paleozoic time (Fig. 6, A–D) versus an east-facing arc approaching the continental margin just prior to the Antler orogeny (Fig. 6D).

The possibility that some of the lower Paleozoic assemblages formed farther north or south along the margin can be partially evaluated with our data. Our interpretations that the Roberts Mountains allochthon and the Shoo Fly Complex formed near the Nevada segment of the margin rests mainly on (1) the tie between lower Middle Ordovician strata of the Roberts Mountains allochthon and the adjacent craton (provenance link 3), and (2) the presence in the Shoo Fly Complex and the Roberts Mountains allochthon of detritus derived from both the north (Peace River arch) and south (Mojave-Sonora craton) (Harding et al., this volume, Chapter 2; Gehrels et al., this volume, Chapter 1). Given the scale of dispersal systems envisioned along the continental margin (Fig. 6), paleoposition along the margin using these constraints can be resolved to only ~1000 km. Because lower Paleozoic strata of the Yreka terrane (Antelope Mountain

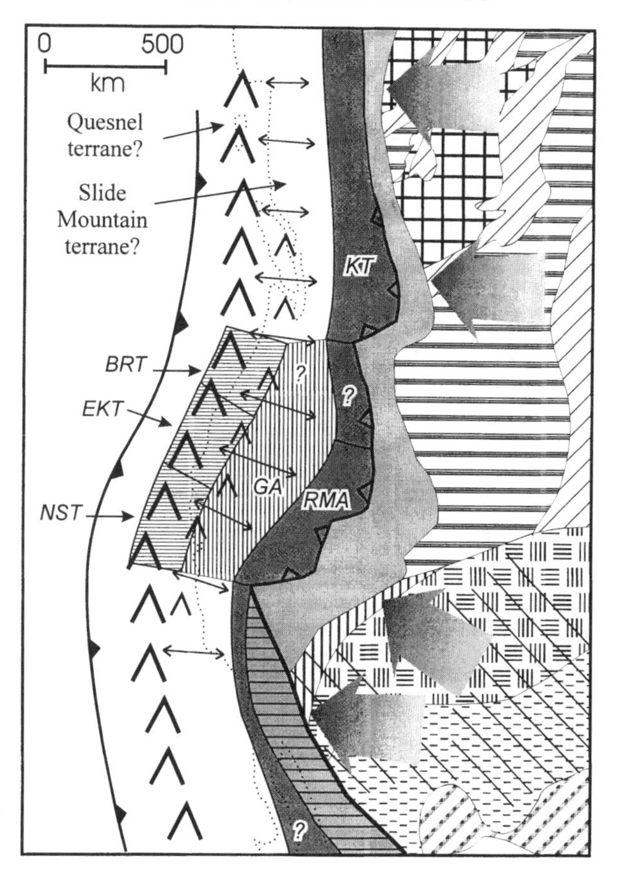


Figure 9. Schematic map of western North America showing our preferred paleogeography for Cordilleran margin during late Paleozoic time. Cratonal provinces and miogeoclinal strata are as shown in Figure 1. Horizontal ruled region represents arc-type terranes such as eastern Klamath terrane. Vertical ruled region represents basinal assemblages, such as Golconda allochthon, that formed in backarc basin setting. Small inverted V pattern represents extensional, east-facing arc active during emplacement of Roberts Mountains allochthon (following Burchfiel and Royden, 1991). Large inverted V pattern represents west-facing magmatic arc that is interpreted to have been active after Antler orogeny. Small black arrows show the inferred directions of crustal extension within this arc system. KT—Kootenay terrane, BRT—Black Rock terrane, GA—Golconda allochthon, EKT—eastern Klamath terrane, RMA—Roberts Mountains allochthon, NST—northern Sierra terrane. Large gray arrows reflect the general transport of sand that accumulated within miogeoclinal strata (Gehrels, Introduction).

Quartzite) are apparently lacking in detritus shed from the southern craton margin, it is possible that this terrane formed farther north along the margin (Wallin et al., this volume, Chapter 9).

## MID-PALEOZOIC PALEOGEOGRAPHY

Beginning in Devonian time, the lower Paleozoic assemblages described here were variably deformed and metamorphosed. Strata of the Roberts Mountains allochthon were imbricated along east-vergent thrust faults and displaced over shelf facies strata for a distance of ~200 km (Roberts et al., 1958; Madrid, 1987; Miller et al., 1992; Burchfiel et al., 1992). This deformation, which is referred to as the Antler orogeny, involves rocks as young as Fammenian (latest Devonian), and was completed by Late Mississippian time (Poole, 1974; Dickinson et al., 1983; Madrid, 1987; Miller et al., 1992).

Rocks of the Shoo Fly Complex are unconformably overlain by Fammenian and younger strata, which requires that imbrication and internal deformation of the Shoo Fly assemblages was completed significantly prior to onset of the Antler orogeny (Schweickert et al., 1984; Harwood, 1992). Likewise, lower Paleozoic rocks of the Yreka terrane were deformed before Early-Middle Devonian time, which considerably predates emplacement of the Roberts Mountains allochthon. Both of these assemblages are known or inferred to be overlain by Devonian-Mississippian rocks that record arc-type magmatism prior to and during emplacement of the Roberts Mountains allochthon.

The lack of an obvious relationship between tectonic events in the Sierra-Klamath region and emplacement of the Roberts Mountains allochthon together with uncertainties about the facing direction of the Sierra-Klamath arc (cf. Girty et al., 1990; Schweickert and Snyder, 1981; Dickinson, this volume, Chapter 14) have led to a variety of models for mid-Paleozoic paleogeography of the Cordilleran margin (Fig. 4). As noted herein, our data suggest that model 3 is not viable, but are not sufficient to determine the facing direction of arcs involved in the Antler orogeny. Hence, we are not able to discriminate between models 1, 2, and 4 during mid-Paleozoic time.

## LATE PALEOZOIC PALEOGEOGRAPHY

All of the lower Paleozoic rocks described herein are known or inferred to be overlain unconformably by upper Paleozoic strata that have been analyzed as part of this project. The upper Paleozoic sequences in the northern Sierra and eastern Klamath terranes are volcanic rich, reflecting development of an offshore magmatic arc, but they also contain some units rich in siliciclastic and carbonate strata (summarized in Miller and Harwood, 1990). In contrast, the Roberts Mountains allochthon is overlain by a thick sequence of clastic and carbonate strata of the Antler overlap assemblage (Roberts et al., 1958; Dickinson et al., 1983). Miogeoclinal units to the east of the Roberts Mountains allochthon are overlain by Carboniferous foreland basin strata derived from the Antler orogen, and by Permian clastic and carbonate rocks that signal a return to passive margin sedimentation (Poole, 1974; Dickinson et al., 1983).

Other upper Paleozoic successions, the Golconda allochthon and Black Rock terrane, are located between the Sierra-Klamath arc and the Roberts Mountains allochthon (Fig. 2). The Golconda allochthon consists of an ocean-basin assemblage of chert, argillite, basalt, sandstone, and limestone (Miller et al., 1984, 1992). The Black Rock terrane consists of a mixture of siliciclastic, carbonate, and volcanogenic rocks similar to those in the Sierra-Klamath terranes, but with less abundant volcanic strata (Wyld, 1990).

Clastic sedimentary rocks in the Antler overlap and foreland basin assemblages were derived primarily from erosion of the Roberts Mountains allochthon, which formed a subaerial high-

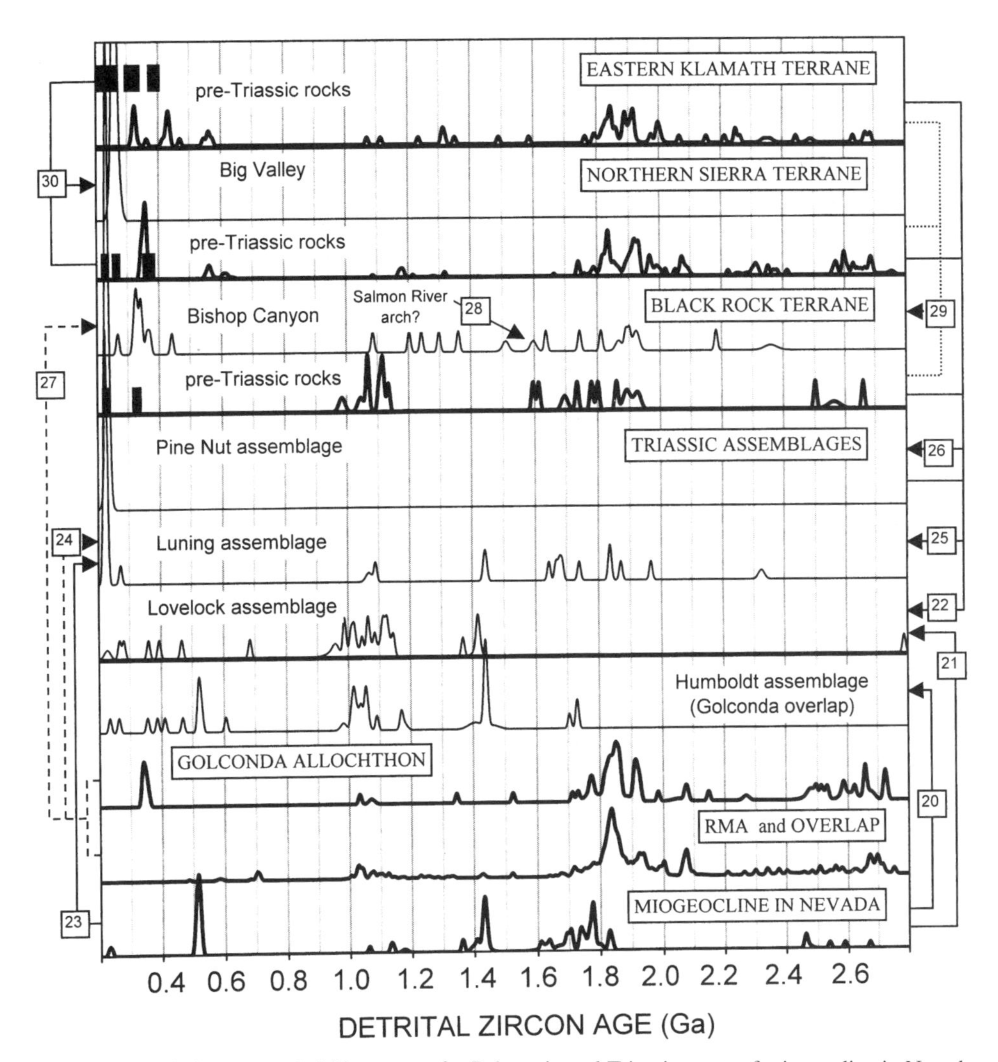


Figure 10. Relative age-probability curves for Paleozoic and Triassic strata of miogeocline in Nevada (Gehrels, this volume), lower Paleozoic rocks of Roberts Mountains allochthon (Gehrels et al., this volume, Chapter 1) and overlying strata (Gehrels and Dickinson, this volume), Golconda allochthon (Riley et al., this volume) and overlying strata (Gehrels et al., 1995), Triassic assemblages in western Nevada (Manuszak et al., this volume), Black Rock terrane (Darby et al., this volume), northern Sierra terrane (Harding et al., this volume; Spurlin et al., this volume), and eastern Klamath terrane (Gehrels and Miller, this volume). Ages of igneous rocks (shown in filled boxes) are from Miller and Harwood (1990) for northern Sierra and eastern Klamath terranes and Wyld (1990) for Black Rock terrane. Numbers in boxes refer to provenance links listed in Table 1.

land along the continental margin during Carboniferous time (Roberts et al., 1958; Poole, 1974; Dickinson et al., 1983; Miller et al., 1992). Previous provenance arguments were based largely on facies relations, paleocurrent studies, and petrographic similarities. Our detrital zircon data strongly support these provenance links (link 9 in Table 1 and Fig. 7), and provide a means of specifying which units in the allochthon sourced most of the detritus. As described by Gehrels and Dickinson (this volume, Chapter 3), comparison of the age-probability curve for overlap and foreland basin strata with the curves for various units within the Roberts Mountains allochthon indicates that Ordovician through Devonian quartzites and part of the Harmony Formation (Harmony B) were the dominant sources of detritus, whereas other strata in the Harmony (Harmony A) and Lower Vinini sandstones were minor sources. These proportions are generally similar to the present-day areal distribution of the various units in the Roberts Mountains allochthon.

Outboard of Antler orogen, the Golconda allochthon (Fig. 2) also contains sandstones that were apparently derived from the Roberts Mountains allochthon (Schweickert and Snyder, 1981; Dickinson et al., 1983; Miller et al., 1984, 1992; Harwood and Murchey, 1990). As shown in Figure 7, detrital zircon ages from Golconda strata (Riley et al., this volume, Chapter 4) are very similar to the ages from strata in the Roberts Mountains allochthon and from strata in the Antler overlap assemblage (provenance link 10). Hence, as described by Schweickert and Snyder (1981),

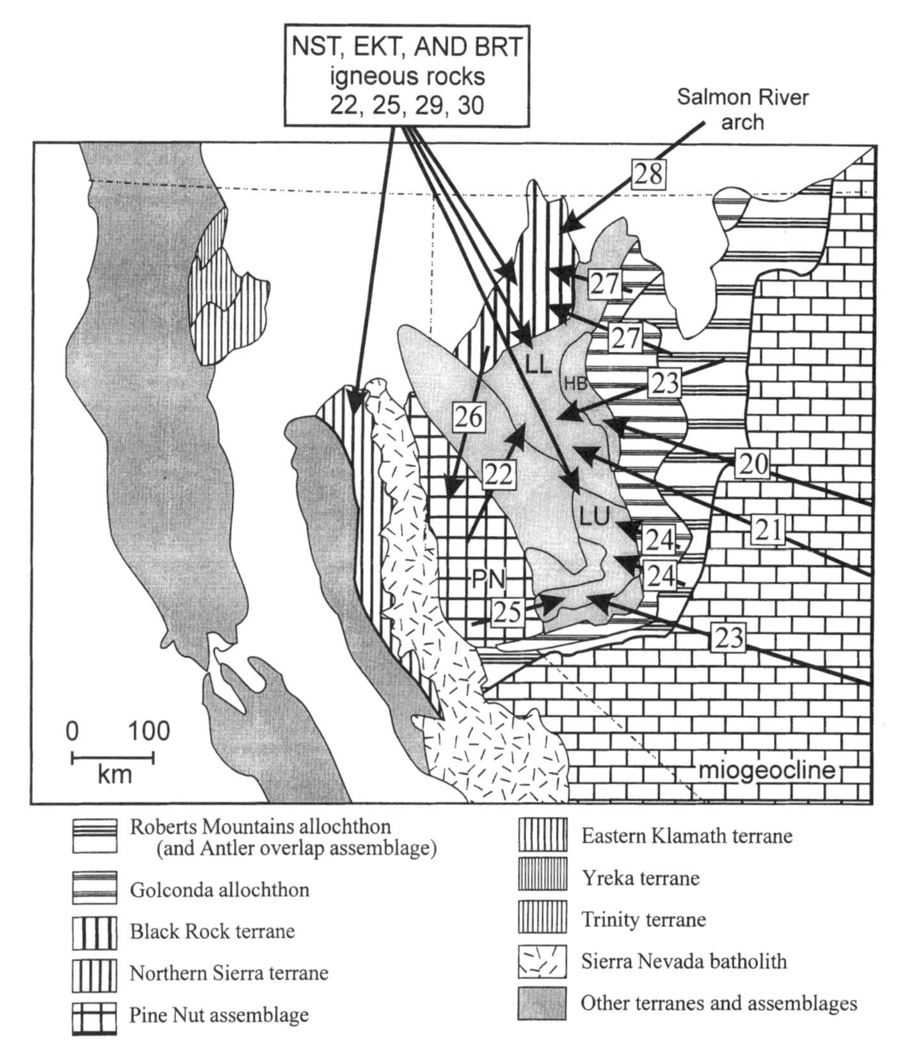


Figure 11. Schematic map of western Nevada–northern California showing detrital zircon provenance links during Triassic time. Numbers are keyed to list in Table 1. Cratonal provinces and miogeoclinal strata are as shown in Figure 2. LL—Lovelock assemblage, HB—Humboldt assemblage, LU—Luning assemblage, PN—Pine Nut assemblage.

Dickinson et al. (1983), Miller et al. (1984, 1992), and Harwood and Murchey (1990), strata in the Golconda allochthon likely were deposited immediately west of, and received detritus primarily from, the Antler orogen. During Permian time, strata in the Golconda allochthon also incorporated detritus derived originally from basement rocks of the southwestern United States (Riley et al., this volume, Chapter 4; provenance link 11). This is consistent with stratigraphic evidence, which suggests that the Antler highland subsided and was overlapped by marine strata during Permian time, thereby allowing craton-derived detritus to reach the offshore basins (Poole, 1974; Miller et al., 1992).

Sandstones in the Golconda allochthon also contain 338–358 Ma detrital grains that were most likely shed from the Sierran and/or Klamath arc systems to the west (Riley et al., this volume; provenance link 12). This apparent provenance link supports the long-held view that sandstones within the Golconda allochthon were deposited in a backarc basin separating the Sierran-Klamath arc from the passive continental margin (Burchfiel and Davis, 1972; Schweickert and Snyder, 1981; Dickinson et al., 1983; Miller et al., 1984, 1992; Miller, 1987; Harwood and Murchey, 1990).

Upper Paleozoic strata of the Black Rock terrane consist largely of detritus derived from the Roberts Mountains allochthon and from cratonal rocks to the east (Darby et al., this volume, Chapter 5; provenance link 13). Cratonal sources apparently included the Salmon River arch (provenance link 14), where ca. 1.6 Ga grains could have originated, whereas 1.6–1.8 Ga and perhaps 1.0–1.2 Ga grains could have been recycled from miogeoclinal and/or platformal strata. A reasonable scenario, as shown in Figures 6 and 9, is that the Black Rock terrane may have been a northern continuation of the Sierra-Klamath arc system, with detritus coming mainly from the northern part of the Roberts Mountains allochthon and from underlying basement, platformal, and/or miogeoclinal rocks. This interpretation is consistent with suggestions by Wyld (1990) and Wyld and Wright (1997) that the Black Rock terrane may have formed between the Sierra-Klamath terrane and arc-type terranes that now reside in Oregon and Washington.

Upper Paleozoic units in the northern Sierra and eastern Klamath terranes contain abundant Pennsylvanian-Permian volcanic and volcaniclastic rocks (Schweickert, 1981; Schweickert et al.,

1984; Harwood and Murchey, 1990; Harwood, 1992), as well as Mississippian quartz-rich clastic strata. Detrital zircons in Mississippian sandstones indicate recycling from either the Roberts Mountains allochthon or underlying lower Paleozoic rocks (i.e., Shoo Fly Complex and Antelope Mountain Quartzite; provenance links 16 and 17), with little input from volcanic sources (Spurlin et al., this volume, Chapter 6; Gehrels and Miller, this volume, Chapter 7). Harwood and Murchey (1990) and Harwood et al. (1991) argued that the Roberts Mountains allochthon is a more likely source for the detritus in upper Paleozoic strata of the northern Sierra terrane on the basis of facies, conglomerate petrology, and thickness relations. In contrast, Pennsylvanian and Permian epiclastic strata were derived from older and coeval arc rocks. The ages of these younger grains are an excellent match for the ages of igneous rocks in the eastern Klamath and northern Sierra terranes, which is consistent with a local or intraterrane origin for much of the detritus (provenance links 15, 18, and 19).

Our preferred model for late Paleozoic paleogeography (Fig. 9) is generally similar to that presented originally by Burchfiel and Davis (1972). As portrayed in models 1, 2, and 4 (Fig. 4), a marginal basin is interpreted to have formed outboard of the continental margin, separating the Antler highlands and its possible northern and southern continuations from an outboard magmatic arc. Deposits of the marginal basin are preserved largely in the Golconda allochthon and its possible northern continuation, the Slide Mountain terrane (Monger et al., 1992), whereas the Black Rock, northern Sierra, and eastern Klamath terranes are interpreted to have formed in and along the eastern margin of the outboard arc. This arc system faced westward, away from the continental margin, and may have been continuous with arc rocks in the Quesnel terrane (Monger et al., 1992) to the north.

The provenance links described herein are inconsistent with biogeographic relations suggesting that the Klamath and related arc terranes were located ~5000 km to the west of the Cordilleran margin during Permian time (Stevens et al., 1990). Although we cannot specify distances from our provenance ties, it is unlikely that detritus from the Roberts Mountains allochthon reached ~5000 km out into the paleo-Pacific basin, particularly given the several hundred meter– to several kilometer–thick sections of siliciclastic strata present in the eastern Klamath and northern Sierra terranes.

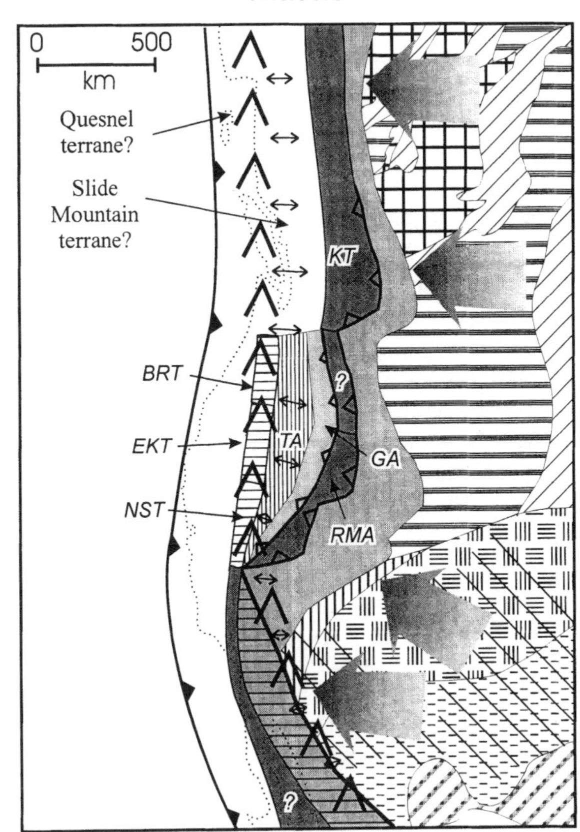


Figure 12. Schematic map of western North America showing our preferred paleogeography for Cordilleran margin during Triassic time. Cratonal provinces and miogeoclinal strata are as shown in Figure 1. Horizontal ruled region represents arc-type terranes such as eastern Klamath terrane. Vertical ruled region represents basinal assemblages in central Nevada that formed in backarc basin setting. Inverted V pattern represents trace of west-facing magmatic arc outboard of Cordilleran margin. Small black arrows show inferred direction of crustal extension behind this arc system. Large gray arrows reflect the general transport of sand that accumulated within miogeoclinal strata (Gehrels, this volume, Introduction). KT—Kootenay terrane, BRT—Black Rock terrane, TT—Trinity terrane, YT—Yreka terrane, GA—Golconda allochthon, EKT—eastern Klamath terrane, RMA—Roberts Mountains allochthon, NST—northern Sierra terrane, TA—Triassic assemblages.

## LATEST PALEOZOIC–EARLIEST MESOZOIC PALEOGEOGRAPHY

The main tectonic event during this time period was the Sonoma orogeny, during which the Golconda allochthon was emplaced onto the Roberts Mountains allochthon and the Antler overlap assemblage (Gabrielse et al., 1983; Miller et al., 1992). Our provenance constraints from various Paleozoic units, as described herein, indicate that the Golconda allochthon consists of ocean-floor rocks that formed near the western margin of the Roberts Mountains allochthon, rather than far from the continental margin. A west-facing magmatic arc, preserved primarily in the eastern Klamath and northern Sierra terranes, is interpreted to have existed outboard of the Golconda allochthon. Emplacement of the allochthon during continuous arc magmatism may have been accomplished by a brief phase of west-dipping subduction along the inboard margin of the arc system (following Schweickert and Snyder, 1981; Burchfiel et al., 1992).

## MIDDLE AND LATE TRIASSIC PALEOGEOGRAPHY

By Middle Triassic time, arc-related magmatism is known to have occurred within cratonal rocks of the Mojave Desert region of California (Walker, 1988; Saleeby and Busby-Spera,

1992). This continental margin arc system apparently continued northward into an offshore island arc now preserved in the northern Sierra, eastern Klamath, and Black Rock terranes and the Pine Nut assemblage (summarized in Saleeby and Busby-Spera, 1992), and possibly southward into Mexico (Fig. 12) (Stewart et al., 1986; Sedlock et al., 1993). Triassic marine strata in central Nevada (Pamlico, Sand Springs, Lovelock, and Luning assemblages of Fig. 2) are interpreted by most workers to have been deposited within a backarc basin to shelf setting within this west-facing convergent margin (Speed, 1978; Oldow, 1984; Saleeby and Busby-Spera, 1992). Our data are entirely consistent with this paleogeography in that western assemblages are dominated by zircons derived from active arc and arc-basement sources (provenance links 22, 25, 26, 29, and 30), whereas assemblages closer to the continental margin are dominated by continent-derived detritus (provenance links 20, 21, 23, 24, 27, and 28) (Fig. 10).

## SUMMARY

Our data are consistent with relatively simple and tectonically conservative models for the Paleozoic–early Mesozoic evolution of the U.S. Cordilleran margin (models 1, 2, and 4 of Fig. 4). Basinal strata that we have studied generally yield detrital zircons that were shed from basement rocks of the North American craton, whereas more outboard volcanic-rich assemblages generally contain a mix of craton-derived and arc-derived detritus. The provenance links between cratonal, basinal, and arc-type assemblages are consistent with models in which west-facing, fringing island arcs formed in proximity to the Cordilleran margin, and were separated from the continent by a series of marginal basins. These basinal assemblages were progressively emplaced onto the continental margin during the Antler and Sonoma orogenies as a result of changes in subduction dynamics, and perhaps brief phases of east-dipping subduction.

Although consistent with these relatively conservative models, our data are also consistent with the collision of an exotic, east-facing arc along the margin during the Antler orogeny. Likewise, the suggestion that the Yreka terrane may have formed farther north along the margin during early Paleozoic time (Wallin, 1993; Wallin et al., this volume) is not precluded by our data.

Our data also reveal interesting patterns of sediment transport during Paleozoic and early Mesozoic time. The dominant pattern during early Paleozoic time involved transport of detritus long distances along the continental margin, both from the north and south. This detritus accumulated in a large submarine fan complex on the slope, rise, and abyssal plain, and probably extended for at least several hundred kilometers out into the proto-Pacific ocean basin. This dominance of margin-parallel sedimentary transport during early Paleozoic time may have resulted from relatively high sea-level stands coupled with the presence of the Peace River arch and Transcontinental arch near the northern and southern ends of the Cordilleran margin (Fig. 6).

Beginning in mid-Paleozoic time, the accreted ocean-floor sediments were uplifted and eroded, shedding detritus eastward onto the craton and westward into backarc basins along the continental margin. Farther west, the recycled detritus was mixed with arc-derived clastic debris from active and formerly active magmatic arcs. The upper Paleozoic basinal assemblages, which were in turn deformed, uplifted, and eroded during Permian-Triassic time, were sources of much detritus that was recycled into Triassic marginal basins along the continental margin.

## ACKNOWLEDGMENTS

This research has been supported by National Science Foundation grant EAR-9416933. We acknowledge critical input from Clark Burchfiel, Greg Davis, John Oldow, Richard Schweickert, and Jim Wright. Reviewed by John H. Stewart and Richard Schweickert.

## REFERENCES CITED

Blake, M.C., Jr., Bruhn, R.L., Miller, E.L., Moores, E.M., Smithson, S.B., and Speed, R.C., 1985, C-1 Menodocino triple junction to North American craton: Geological Society of America Centennial Continent-Ocean Transect #12.

Burchfiel, B.C., and Davis, G.A., 1972, Structural framework and evolution of the southern part of the Cordilleran orogen, western United States: American Journal of Science, v. 272, p. 97–118.

Burchfiel, B.C., and Davis, G.A., 1975, Nature and controls of Cordilleran orogenesis, western United States: Extensions of an earlier synthesis: American Journal of Science, v. 275-A, p. 363–396.

Burchfiel, B.C., and Royden, L.H., 1991, Antler orogeny: A Mediterranean-type orogeny: Geology, v. 19, p. 66–69.

Burchfiel, B.C., Cowan, D.S., and Davis, G.A., 1992, Tectonic overview of the Cordilleran orogen in the western United States, *in* Burchfiel, B.C., et al., eds., The Cordilleran orogen: Conterminous U.S.: Boulder, Colorado, Geological Society of America, Geology of North America, v. G-3, p. 407–480.

Churkin, M., Jr., 1974, Paleozoic marginal ocean basin-volcanic arc in the Cordilleran foldbelt, *in* Dott, R.H., and Shaver, R.H., eds., Modern and ancient geosynclinal sedimentation: Society of Economic Paleontologists and Mineralogists Special Publication 19, p. 174–192.

Coney, P.J., and Campa, M.F., 1987, Lithotectonic terrane map of Mexico: U.S. Geological Survey Miscellaneous Geologic Investigations Map MF-1874-D, scale 1:2,500,000.

Coney, P.J., Jones, D.L., and Monger, J.W.H., 1980, Cordilleran suspect terranes: Nature, v. 288, p. 329–333.

Dickinson, W.R., 1977, Paleozoic plate tectonics and the evolution of the Cordilleran continental margin, *in* Stewart, J.H., et al., eds., Paleozoic paleogeography of the western United States: Pacific Section, Society of Economic Paleontologists and Mineralogists, Pacific Coast Paleogeography Symposium I: p. 137–155.

Dickinson, W.R., Harbaugh, D.W., Saller, A.H., Heller, P.L., and Snyder, W.S., 1983, Detrital modes of upper Paleozoic sandstones derived from Antler orogen in Nevada: Implications for the nature of the Antler orogeny: American Journal of Science, v. 282, p. 481–509.

Finney, S.C., and Perry, B.D., 1991, Depositional setting and paleogeography of Ordovician Vinini Formation, central Nevada, *in* Cooper, J.D., and Stevens, C.H., eds., Paleozoic paleography of the Western United States—II, Volume 2: Pacific Section, SEPM Society for Sedimentary Geology book 67, p. 747–766.

Gabrielse, H., Snyder, W.S., and Stewart, J.H., 1983, Sonoma orogeny and Permian to Triassic tectonism in western North America (Penrose Conference Report): Geology, v. 11, p. 484–486.

Gehrels, G.E., and Stewart, J.H., 1998, Detrital zircon geochronology of Cambrian to Triassic miogeoclinal and eugeoclinal strata of Sonora, Mexico: Journal of Geophysical Research, v. 103, no. B2, p. 2471–2487.

Gehrels, G.E., Dickinson, W.R., Ross, G.M., Stewart, J.H., and Howell, D.G., 1995, Detrital zircon reference for Cambrian to Triassic miogeoclinal strata of western North America: Geology, v. 23, p. 831–834.

Girty, G.H., Gester, K.C., and Turner, J.B., 1990, Pre-Late Devonian geochemical, stratigraphic, sedimentologic, and structural patterns, Shoo Fly Complex, northern Sierra Nevada, California, *in* Harwood, D.S., and Miller, M.M., eds., Paleozoic and early Mesozoic paleogeographic relations; Sierra Nevada, Klamath Mountains, and related terranes: Geological Society of America Special Paper 255, p. 43–56.

Hamilton, W.B., 1969, Mesozoic California and the underflow of Pacific mantle: Geological Society of America Bulletin, v. 80, p. 2409–2430.

Harwood, D.S., 1992, Stratigraphy of Paleozoic and lower Mesozoic rocks in the northern Sierra terrane, California: U.S. Geological Survey Bulletin 1957, 78 p.

Harwood, D.S., and Murchey, B.L., 1990, Biostratigraphic, tectonic, and paleogeographic ties between upper Paleozoic volcanic and basinal rocks in the northern Sierra terrane, California, and the Havallah Sequence, Nevada, *in* Harwood, D.S., and Miller, M.M., eds., Paleozoic and early Mesozoic paleogeographic relations; Sierra Nevada, Klamath Mountains, and related terranes: Geological Society of America Special Paper 255, p. 157–173.

Harwood, D.S., Yount, J.C., and Seiders, V.M., 1991, Upper Devonian and Lower Mississippian island-arc and back-arc deposits in the northern Sierra Nevada, California, *in* Cooper, J.D., and Stevens, C.H., eds., Paleozoic paleogeography of the western United States—II: Pacific Section, Society of Economic Paleontologists and Mineralogists, book 67, p. 717–733.

Helwig, J., 1974, Eugeosynclinal basement and a collage concept of orogenic belts, *in* Dott R.H., and Shaver, R.H., eds., Modern and ancient geosynclinal sedimentation: Society of Economic Paleontologists and Mineralogists Special Publication 19, p. 359–376.

Hoffman, P.F., 1989, Precambrian geology and tectonic history of North America, *in* Bally, A.W., and Palmer, A.R., eds., The geology of North America—An overview: Boullder, Colorado, Geological Society of America, Geology of North America, v. A, p. 447–512.

Kay, M., 1951, North American geosynclines: Geological Society of America Memoir 48, 143 p.

Kistler, R.W., 1991, Chemical and isotopic characteristics of plutons in the Great Basin, *in* Raines, G.L., et al., eds., Geology and ore deposits of the Great Basin: Reno, Geological Society of Nevada, p. 107–109.

Kistler, R.W., and Peterman, Z.E., 1973, Variations in Sr, Rb, K, Na, and initial $^{87}Sr/^{86}Sr$ in Mesozoic granitic rocks and intruded wall rocks in central California: Geological Society of America Bulletin, v. 84, p. 3489–3512.

Madrid, R.J., 1987, Stratigraphy of the Roberts Mountains allochthon in north-central Nevada [Ph.D. thesis]: Stanford, California, Stanford University, 336 p.

Miller, E.L., Holdsworth, B.K., Whiteford, W.B., and Rodgers, D., 1984, Stratigraphy and structure of the Schoonover sequence, northeastern Nevada: Implications for Paleozoic plate-margin tectonics: Geological Society of America Bulletin, v. 95, p. 1063–1076.

Miller, E.L., Miller, M.M., Stevens, C.H., Wright, J.E., and Madrid, R., 1992, Late Paleozoic paleographic and tectonic evolution of the western U.S. Cordillera, *in* Burchfiel, B.C., et al., eds., The Cordilleran orogen: Conterminous U.S.: Boulder, Colorado, Geological Society of America, Geology of North America, v. G-3, p. 57–106.

Miller, M.M., 1987, Dispersed remnants of a northeast Pacific fringing arc: Upper Paleozoic terranes of Permian McCloud faunal affinity, western U.S.: Tectonics, v. 6, p. 807–830.

Miller, M.M., and Harwood, D.S., 1990, Paleogeographic setting of upper Paleozoic rocks in the northern Sierra and eastern Klamath terranes, northern California, *in* Harwood, D.S., and Miller, M.M., eds., Paleozoic and early Mesozoic paleogeographic relations; Sierra Nevada, Klamath Mountains, and related terranes: Geological Society of America Special Paper 255, p. 175–192.

Monger, J.W.H., Wheeler, J.O., Tipper, H.W., Gabrielse, H., Harms, T., Struik, L.C., Campbell, R.B., Dodds, C.J., Gehrels, G.E., and O'Brien, J., 1992, Part B, Cordilleran terranes, *in* Gabrielse, H., and Yorath, C.J., eds., Geology of the Cordilleran orogen in Canada: Boulder, Colorado, Geological Society of America, Geology of North America, v. G-2, p. 281–327.

Moores, E., 1970, Ultramafics and orogeny, with models of the U.S. Cordillera and the Tethys: Nature, v. 228, p. 837–842.

Oldow, J.S., 1984, Evolution of a late Mesozoic back-arc fold and thrust belt, northwestern Great Basin, U.S.A.: Tectonophysics, v. 102, p. 245–274.

Poole, F.G., 1974, Flysch deposits of Antler foreland basin, western United States: *in* Dickinson, W.R., ed., Tectonics and Sedimentation: Society of Economic Paleontologists and Mineralogists Special Publication 22, p. 58–82.

Roberts, R.J., Hotz, P.E., Gilluly, J., and Ferguson, H.G., 1958, Paleozoic rocks of north-central Nevada: American Association of Petroleum Geologists Bulletin, v. 42, p. 2813–2857.

Ross, G.M., 1991, Precambrian basement in the Canadian Cordillera: An introduction: Canadian Journal of Earth Sciences, v. 28, p. 1133–1139.

Ruiz, J., Patchett, P.J., and Ortega-Gutierrez, F., 1988, Proterozoic and Phanerozoic basement terranes of Mexico from Nd isotopic studies: Geological Society of America Bulletin, v. 100, p. 274–281.

Saleeby, J.B., 1986, C-2 Central California offshore to Colorado Plateau: Geological Society of America Centennial Continent-Ocean Transect #10.

Saleeby, J.B., 1990, Geochronologic and tectonostratigraphic framework of Sierran-Klamath ophiolitic assemblages, *in* Harwood, D.S., and Miller, M.M., eds., Paleozoic and early Mesozoic paleogeographic relations; Sierra Nevada, Klamath Mountains, and related terranes: Geological Society of America Special Paper 255, p. 93–114.

Saleeby, J.B., and Busby-Spera, C., 1992, Early Mesozoic tectonic evolution of the western U.S. Cordillera, *in* Burchfiel, B.C., et al., eds., The Cordilleran orogen: Conterminous U.S.: Boulder, Colorado, Geological Society of America, Geology of North America, v. G-3, p. 107–168.

Schuchert, C., 1923, Sites and nature of the North American geosynclines: Geological Society of America Bulletin, v. 34, p. 151–230.

Schweickert, R.A., 1981, Tectonic evolution of the Sierra Nevada, *in* Ernst, W.D., ed., The geotectonic development of California (Rubey Volume I): Englewood Cliffs, New Jersey, Prentice-Hall, p. 88–131.

Schweickert, R.A., and Lahren, M.M., 1990, Speculative reconstruction of the Mojave–Snow Lake fault: Implications for Paleozoic and Mesozoic orogenesis in the western United States: Tectonics, v. 9, p. 1609–1629.

Schweickert, R.A., and Snyder, W.S., 1981, Paleozoic plate tectonics of the Sierra Nevada and adjacent regions, *in* Ernst, W.G., ed., The geotectonic development of California: Prentice-Hall, Englewood Cliffs New Jersey, p. 183–201.

Schweickert, R.A., Harwood, D.S., Girty, G.H., and Hanson, R.E., 1984, Tectonic development of the northern Sierra terrane: An accreted late Paleozoic island arc and its basement: *in* Lintz, J., Jr., ed., Western geological excursions, Geological Society of America 1984 Annual Meeting Field Trip Guide volume 4: Reno, University of Nevada Mackay School of Mines, p. 1–65. 50 p.

Sedlock, R.L., Ortega-Gutierrez, F., and Speed, R.C., 1993, Tectonostratigraphic terranes and tectonic evolution of Mexico: Geological Society of America Special Paper 278, 153 p.

Silberling, N.J., 1991, Allochthonous terranes of western Nevada: Current status, *in* Raines, G.L., et al., eds., Geology and ore deposits of the Great Basin: Reno, Geological Society of Nevada, p. 101–102.

Silberling, N.J., Jones, D.L., Blake, M.C., Jr., and Howell, D.G., 1987, Lithotectonic terrane map of the western conterminous United States: U.S. Geological Survey Miscellaneous Field Studies Map MF-1874-C, scale 1:2,500,000.

Smith, M.T., and Gehrels, G.E., 1991, Detrital zircon geochronology of Upper Proterozoic to lower Paleozoic continental margin strata of the Kootenay Arc; implications for the early Paleozoic tectonic development of the eastern Canadian Cordillera: Canadian Journal of Earth Sciences, v. 28,

p. 1271–1284.
Smith, M.T., and Gehrels, G.E., 1994, Detrital zircon geochronology and the provenance of the Harmony and Valmy Formations, Roberts Mountains allochthon, Nevada: Geological Society of America Bulletin, v. 106, p. 968–979.
Speed, R.C., 1978, Paleogeographic and plate-tectonic evolution of the early Mesozoic marine province of the western Great Basin, *in* Howell, D.G., and McDougall, K.A., eds., Mesozoic paleogeography of the western United States: Pacific Section, Society of Economic Paleontologists and Mineralogists, Pacific Paleogeography Symposium 2, p. 253–270.
Speed, R.C., 1979, Collided Paleozoic microplate in the western United States: Journal of Geology, v. 87, p. 279–292.
Speed, R.C., and Sleep, N.H., 1982, Antler orogeny and foreland basin: A model: Geological Society of America Bulletin, v. 93, p. 815–828.
Stevens, C.H., Yancey, T.E., and Hanger, R.A., 1990, Significance of the provincial signature of Early Permian faunas of the eastern Klamath terrane, *in* Harwood, D.S., and Miller, M.M., eds., Paleozoic and early Mesozoic paleogeographic relations; Sierra Nevada, Klamath Mountains, and related terranes: Geological Society of America Special Paper 255, p. 201–218.
Stewart, J.H., Anderson, T.H., Haxel, G.B., Silver, L.T., and Wright, J.E., 1986, Late Triassic paleogeography of the southern Cordillera: The problem of a source for voluminous volcanic detritus in the Chinle Formation of the Colorado Plateau region: Geology, v. 14, p. 567–570.
Stewart, J.H., Poole, F.G., Ketner, K.B., Madrid, R.J., Roldan-Quintana, J., and Amaya-Martinez, R., 1990, Tectonics and stratigraphy of the Paleozoic and Triassic southern margin of North America, Sonora, Mexico, *in* Gehrels, G.E., and Spencer, J.E., eds., Geologic excursions through the Sonoran Desert region, Arizona and Sonora: Arizona Geological Survey Special Paper 7, p. 183–202.
Van Schmus, W.R., and 24 others, 1993, Transcontinental Proterozoic Provinces, *in* Reed, J.C., et al., eds., Precambrian: Conterminous U.S.: Boulder, Colorada, Geological Society of America, Geology of North America, v. C-2, p. 171–334.
Villeneuve, M.E., Ross, G.M., Theriault, R.J., Miles, W., Parrish, R.R., and Broome, J., 1993, Tectonic subdivision and U-Pb geochronology of the crystalline basement of the Alberta basin, western Canada: Geological Survey of Canada Bulletin 447, 86 p.
Walker, J.D., 1988, Permian and Triassic rocks of the Mojave Desert and their implications for the timing and mechanisms of continental truncation: Tectonics, v. 7, p. 685–709.
Wallin, E.T., 1990a, Provenance of selected lower Paleozoic siliciclastic rocks in the Roberts Mountains allochthon, Nevada, *in* Harwood, D.S., and Miller, M.M., eds., Paleozoic and early Mesozoic paleogeographic relations; Sierra Nevada, Klamath Mountains, and related terranes: Geological Society of America Special Paper 255, p. 17–32.
Wallin, E.T., 1990b, Petrogenetic and tectonic significance of xenocrystic Precambrian zircon in Lower Cambrian tonalite, eastern Klamath Mountains, California: Geology, v. 18, p. 1057–1060.
Wallin, E.T., 1993, Sonomia revisited: Evidence for a western Canadian provenance of the eastern Klamath and northern Sierra Nevada terranes: Geological Society of America Abstracts with Programs, v. 25, no. 6, p. A-173.
Wallin, E.T., and Metcalf, R.V., 1998, Supra-subduction zone ophiolite formed in an extensional forearc: Trinity terane, Klamath Mountains, California: Journal of Geology, v. 106, p. 591–608.
Wallin, E.T., Mattinson, J.M., and Potter, A.W., 1988, Early Paleozoic magmatic events in the eastern Klamath Mountains, northern California: Geology, v. 16, p. 144–148.
Wallin, E.T., Coleman, D.S., Lindsey-Griffin, N., and Potter, A.W., 1995, Silurian plutonism in the Trinity terrane (Neoproterozoic and Ordovician), Klamath Mountains, California, United States: Tectonics, v. 14, p. 1007–1013.
Wilson, J.T., 1968, Static or mobile earth, the current scientific revolution: American Philosophical Society Proceedings, v. 112, p. 309–320.
Wyld, S.J., 1990, Paleozoic and Mesozoic rocks of the Pine Forest Range, northwest Nevada, and their relation to volcanic arc assemblages of the western U.S. Cordillera, *in* Harwood, D.S., and Miller, M.M., eds., Paleozoic and early Mesozoic paleogeographic relations; Sierra Nevada, Klamath Mountains, and related terranes: Geological Society of America Special Paper 255, p. 219–237.
Wyld, S.J., and Wright, J.E., 1997, Triassic-Jurassic tectonism and magmatism in the Mesozoic continental arc of Nevada: Classic relations and new developments, *in* Link, P.K., and Kowallis, B.J., eds., Geological Society of America field trip guide book, Proterozoic to recent stratigraphy, tectonics, and volcanology, Utah, Nevada, southern Idaho and central Mexico: Brigham Young University Geology Studies, v. 42, p. 197–224.

Manuscript Accepted by the Society January 24, 2000

Printed in U.S.A.

Geological Society of America
Special Paper 347
2000

# *Sandstone petrofacies of detrital zircon samples from Paleozoic and Triassic strata in suspect terranes of northern Nevada and California*

**Willliam R. Dickinson and George E. Gehrels**
*Department of Geosciences, University of Arizona, Tucson, Arizona 85721, USA*

## ABSTRACT

**We collected 35 sandstones and metasandstones for analysis of detrital zircons from Paleozoic and Triassic strata in suspect terranes of northern Nevada and northern California that represent five generic petrofacies: subarkose eroded from uplifted continental basement, quartzarenite (matrix-poor and matrix-rich subfacies) derived from cratonal sources, quartzolithic (quartz-rich, chert-rich, and quartzolithic-volcaniclastic subfacies) derived from recycled orogenic sources, volcaniclastic from volcanic arc structures, and volcanoplutonic from a dissected magmatic arc. Precambrian ages of dominant detrital zircons in the subarkose petrofacies (exclusively Harmony Formation of Cambrian age) and in most samples of the lower to middle Paleozoic quartzarenite petrofacies in the Roberts Mountains allochthon, the Shoo Fly Complex, and the Yreka terrane reflect a provenance in northern Laurentia or a continental block other than Laurentia. Samples from the lower part of the Vinini Formation in central Nevada are the only examples of the quartzarenite petrofacies yielding zircon populations that could have been derived from adjacent southwestern segments of Laurentia. Dominantly Grenville-age zircons in subordinate quartzose sandstones from the mainly subarkosic Harmony Formation suggest derivation from farther south along the Cordilleran continental margin. Zircons in the quartzolithic petrofacies were largely recycled into varied Paleozoic-Mesozoic overlap assemblages and successor basins, including the Havallah sequence of the Golconda allochthon, from older sedimentary sources in lower to middle Paleozoic strata of underlying or nearby terranes. Phanerozoic zircons in the upper Paleozoic to lower Mesozoic volcaniclastic petrofacies of the Sierran-Klamath belt reflect sources in coeval or older volcanic arc assemblages, with homogeneous age populations in each sample. Latest Precambrian and Grenville-age zircons in the volcanoplutonic petrofacies (one sample from the Sierra City mélange) reflect sources in a magmatic arc that was not part of the Cordilleran miogeoclinal margin. The close correlations between petrofacies compositions and the age spectra of detrital zircon populations provide significant and consistent constraints for interpretations of regional tectonic history, the two datasets jointly offering more interpretive insight than either affords alone.**

## INTRODUCTION

Petrographic study of sandstone and metasandstone samples, collected for detrital zircon analysis from Paleozoic and Triassic strata of so-called suspect terranes in northern Nevada and northern California, was undertaken to test for correlations between the overall petrography of zircon-bearing sands and the age populations of detrital zircons. Figure 1 shows the areal distribution of the samples. Petrography establishes systematic correlations, with a few notable exceptions, between sandstone petrology and detrital zircon ages. Petrographic results are presented here in terms of generic petrofacies (Dickinson and Rich, 1972; Inger-

Dickinson, W.R., and Gehrels, G.E., 2000, Sandstone petrofacies of detrital zircon samples from Paleozoic and Triassic strata in suspect terranes of northern Nevada and California, *in* Soreghan, M.J., and Gehrels, G.E., eds., Paleozoic and Triassic paleogeography and tectonics of western Nevada and northern California: Boulder, Colorado, Geological Society of America Special Paper 347, p. 151–171.

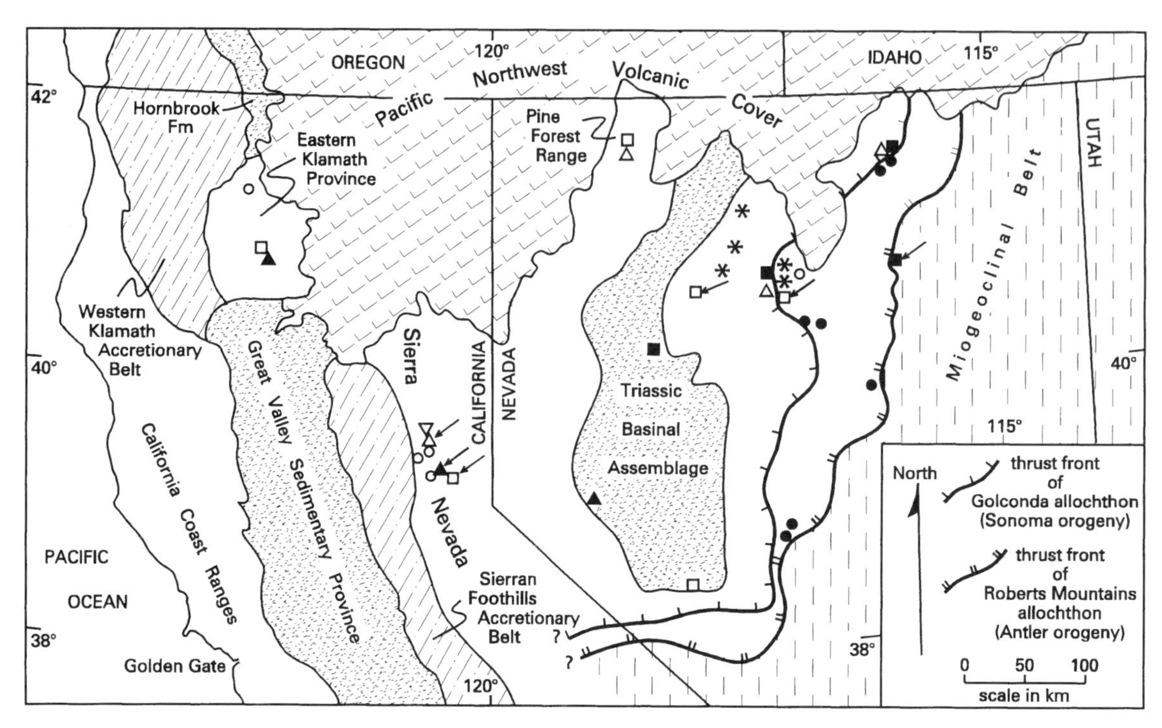


Figure 1. Distribution of detrital zircon samples in northern Nevada and northern California (samples are plotted as petrofacies symbols defined in Fig. 2 and 3; inverted triangle represents single sample of petrofacies E, not plotted in Fig. 3). Trends of Nevada thrust fronts (after Dickinson et al., 1983b) into California (after Schweickert and Lahren, 1987) are queried where truncated by eastern side of Sierra Nevada batholith. Arrows point in Nevada to samples from Antler overlap sequence, and in California to samples from varied overlap successions younger than Shoo Fly Complex of northern Sierra Nevada and Yreka terrane of eastern Klamath Mountains (see Fig. 2 for stratigraphic relations).

soll, 1978, 1983) that have distinctively different kinds of detrital modes (Dickinson, 1970).

Our grouping of the samples into different petrofacies is a means of classifying the detrital zircon samples in terms of sedimentary petrology, but the array of petrofacies that we describe is not necessarily a full census of all the petrofacies that may be present within the terranes collected for detrital zircon analysis. The purpose of our petrographic study restricts our attention to the samples for which age spectra of detrital zircons have been determined. All of our samples were selected, however, to be characteristic of the terranes sampled, and the sedimentary petrology of the samples is broadly representative of the various formations and assemblages from which they were collected. The detrital modes reported here stem from point counts of individual samples from which detrital zircons were separated; no attempt was made to calculate average petrofacies compositions.

Our provenance discussions for each petrofacies are abbreviated and are meant only to set broad constraints for provenance interpretations, which are treated more fully in other chapters of this volume dealing separately with each tectonostratigraphic assemblage of the study region.

## DETRITAL MODES

Just as modal mineralogy pertains to the minerals present in a rock, the detrital mode of a sandstone refers to the population of sand grain types that make up the detrital framework, exclusive of interstitial cement and matrix. Where advanced diagenesis or incipient metamorphism has modified frameworks of sandstones, either mineralogically or texturally, determination of detrital modes requires varying degrees of interpretation. Characteristic mineralogical replacements, such as albite pseudomorphs of detrital plagioclase or chloritic alteration of the groundmasses of volcanic rock fragments, present no fundamental difficulties for determining detrital modes. However, extensive metasomatic changes (McBride, 1985) or wholesale textural transformations render inferred detrital modes only approximate (Dickinson and Ingersoll, 1990), as we indicate for some of our samples.

In detail, sands may include monocrystalline grains of any rock-forming mineral and polycrystalline fragments of any kind of aphanitic rock composed of internal crystals smaller than sand in size. In practice, however, crystals of quartz and feldspar are typically far and away the most abundant types of mineral grains in sandstones. Subordinate micas are diagnostic components of some frameworks, and placering concentrates pyriboles (pyroxenes + amphiboles) and opaque oxides to significant abundance in selected instances. Aphanitic lithic fragments form a wide spectrum of grain types, but can be grouped into a limited number of generic categories based on mineralogical and textural criteria. Consequently, different sandstone petrofacies can be defined on the basis of the relative proportions of standard grain types that serve as generic parameters significant for sand provenance.

## GRAIN TYPES

Table 1 lists the categories of framework grain types used to define sandstone petrofacies in this study. To minimize differential effects of grain size on detrital modes, the Gazzi-Dickinson convention (Ingersoll et al., 1984) was used in calculating the percentages of grain types from the results of point counts. By this procedure, points falling within phenocrysts or intragrain clastic particles larger than silt size (>0.0625 mm) are assigned to monomineralic grain types (Qm, K, P), even though they occur within rock fragments. This convention assures that lithic grain types represent exclusively aphanitic materials, and that the reported percentages of monocrystalline and polycrystalline grain types do not vary solely as a function of the degree of disintegration of source rock as the mean size of sandy detritus varies. The convention does not, however, alter any actual differences in the proportions of different mineral grains, such as quartz and feldspar, as grain size varies (Odom et al., 1976).

Monocrystalline quartz grains (Qm) include grains polygonized into a mosaic of multiple strained domains during compactive diagenesis or incipient metamorphism; provided the resultant domains are individually as large as sand size. Polycrystalline quartz grains (Qp) thus represent chert, metachert, quartzite, or vein quartz in which individual crystals are smaller than sand size. Rocks metamorphosed severely enough to generate analogous polycrystalline domains from monocrystalline detrital quartz grains were not encountered in our study. Because of severe diagenetic modification of quartz grains from compaction and related processes, however, intragrain variations in polycrystallinity, or in the intensity and prevalence of undulatory extinction (Basu et al., 1975), are not reliable guides to the varieties of source rocks that contributed quartz grains to the sands.

Feldspar grains (F) were counted as either K-feldspar (K), including sanidine and perhaps anorthoclase, or plagioclase (P), including albite. Selected rocks containing abundant feldspars were stained to confirm identifications, although much of the feldspar present is albite, which remains unstained. Because albite grains are assumed uniformly to be pseudomorphs of detrital plagioclase, our methods cannot detect possible replacement of K-feldspar by albite (Dickinson et al., 1982; Walker, 1984), but we observed no evidence for that less common replacement process in any of the rocks studied. As our provenance arguments do not depend to any significant degree on distinctions between different kinds of feldspar, the diagenetic histories of constituent feldspars were not investigated in detail. If intrastratal solution of feldspar (McBride, 1987) has occurred on a significant scale, we could not detect it by our methods, but we observed no textural evidence for dissolution (Shanmugam, 1985), nor did we observe any clay domains that could have formed from replacement of feldspars (Helmold, 1985).

**TABLE 1. CATEGORIES OF SAND GRAIN TYPES (QFL POPULATION)***

|  |
|---|
| A. Quartzose grains (Q) |
| Monocrystalline quartz mineral grains: Qm |
| Polycrystalline quartzose lithic fragments: Qp |
| B. Monocrystalline feldspar mineral grains (F) |
| K-feldspar : K |
| Plagioclase : P |
| C. Labile polycrystalline aphanitic lithic fragments (L) |
| Sedimentary-metasedimentary: Lsm |
| Volcanic-metavolcanic: Lvm |

*Note:* *Derivative triangular compositional diagrams (Figure 3): QFL and QmFLt, where Lt = L + Qp.

### *Lithic fragments*

Lithic fragments are classified initially into monomineralic and polymineralic groups: (1) polycrystalline but monomineralic quartzose fragments (Qp), interpreted as being dominantly of sedimentary or metasedimentary origin, but probably including some grains derived from silicified volcanic rocks; and (2) polymineralic as well as polycrystalline fragments (L), the so-called labile or unstable lithic fragments derived from rock types less resistant to weathering than the source rocks for quartzose lithic fragments. The labile lithic fragments are further subdivided into two main categories based on combined mineralogical and textural criteria (Dickinson, 1970; Wolf, 1971; Graham et al., 1976; Dickinson et al., 1979; Ingersoll and Suczek, 1979; Suczek and Ingersoll, 1985): (1) sedimentary and metasedimentary (Lsm), and (2) volcanic and metavolcanic (Lvm).

Distinctions between sedimentary or volcanic lithic fragments and their metamorphosed counterparts are not feasible because advanced diagenesis or incipient metamorphism has modified in place the original mineralogy and fabric of lithic fragments in nearly all the rocks studied. Consequently, except where prominent textural or fabric modifications accompanied metamorphism of source rocks, distinctions between intrastratal alteration and prior alterations in the provenance cannot be made with confidence. Criteria for positive identification of metamorphic lithic fragments are limited to foliation, which is visible only where fragments are appropriately oriented within thin sections, or the presence (despite severe diagenesis) of characteristic high-temperature metamorphic mineralogy, which is difficult to detect readily within individual sand grains.

During point counting, volcanic lithic fragments (Lvm) with different groundmass textures (felsitic, microlitic, lathlike) were recorded separately, but observed variations did not prove significant for provenance inferences at a level important for our conclusions, and are not discussed further here. Sedimentary-metasedimentary lithic fragments (Lsm) were routinely counted as argillitic (clay-rich mudstone-shale-argillite), tectonitic (foliated slate-phyllite), or microgranular (silt-rich clastic, hornfels), and these distinctions come into play for some provenance determinations. Microgranular grains in some units also include aplitic or hypabyssal igneous grains, and the overall provenance implications of microgranular lithic grains are thus ambiguous to some extent in selected rocks. In general, however, we assume that quartz-rich microgranular grains were derived primarily from sedimentary protoliths, whereas feldspar-rich microgranular grains were derived from igneous protoliths.

Gradations from argillitic to microgranular or tectonitic lithic fragments, as well as from argillitic to cherty lithic fragments, are common in all the sample suites. These ambiguities were resolved by restricting quartzose lithic fragments (Qp) to those wholly lacking nonquartzose constituents, requiring tectonitic lithic fragments to display oriented micaceous minerals visible as discrete crystals (leaving shaly fragments displaying mass extinction among the argillitic grains), and restricting clastic microgranular grains to lithic fragments in which discrete silt clasts are dominant over clayey matrix. Even so, multiple operators might obtain somewhat different modal percentages for these partly overlapping categories of grain types. As no significant provenance distinctions are based here on different proportions of the grain types in question, the problem of reproducible discrimination is not a major issue with respect to our conclusions.

## MODAL SUMMARIES

Modal percentages of sandstone constituents are recalculated in three different ways for tables presenting modal sand compositions: (1) QFL% refers to percentages of quartz grains, feldspar grains, and lithic fragments, forming the so-called QFL population of the framework, as summed jointly to 100%; (2) fmwk% refers to percentages of the total framework including micas and heavy minerals, and is used to report abundances of the latter types of grains apart from the QFL grain population; (3) WR% refers to percentages of whole rocks, and is used to report proportions of interstitial constituents (matrix and cement), as opposed to framework grains. By convention, a visible diameter of 0.0625 mm, the lower limit of the sand size range, was used to separate framework sand grains from smaller grains within matrix of either detrital or secondary origin.

Identifiable pseudomatrix (Dickinson, 1970), derived from the compactive deformation of lithic fragments between more rigid framework grains, was interpreted as part of the sand framework, with assignment to the appropriate category of lithic fragment based on relict texture, fabric, and mineralogy preserved within pseudomatrix domains. The dominant matrix materials present in all studied rocks are epimatrix (Dickinson, 1970), formed by uncertain combinations of diagenetic pore filling and recrystallization of detrital matrix (orthomatrix). Because the proportion of matrix does not exceed ~15% in any of the rocks studied, there is no reason to suppose that contributions to secondary matrix from alteration of framework grains (Cox and Lowe, 1996) have significantly modified any reported detrital modes.

Modal percentages are based on grid counts of 400 points in each QFL population, together with the associated framework and interstitial points needed to reach that partial total. The figure of 400 points was chosen to keep standard deviations of counting error for all QFL% values below 2.5 percentage points (van der Plas and Tobi, 1965), and below 1.5 and 1.0 percentage points for constituents present in amounts of <10% and <5%, respectively (standard deviation of counting error: $[p(100-p)]^{1/2}$, where p is the counted percentage of a given grain type).

## PROVENANCE TYPES

Detailed discussions of provenance for each detrital zircon sample and discussion of the tectonic implications of specific provenance determinations are presented elsewhere in this volume. We note here only the general kinds of provenances reflected by each petrofacies. For that purpose, the following general catalogue of provenance types is sufficient (Dickinson and Suczek, 1979; Dickinson et al., 1983a; Mack, 1984; Dickinson, 1985): (1) uplifted continental basement provenances that yield immature first-cycle quartzofeldspathic detritus from granitic and gneissic bedrock; (2) cratonal provenances that yield mature quartzose detritus from deeply weathered continental basement or through multicyclic transport of basement-derived detritus; (3) recycled orogenic provenances that yield quartzolithic detritus from the erosion of older strata deformed within fold-thrust belts or stripped from uplifted continental basement; and (4) magmatic arc provenances that yield volcaniclastic to volcanoplutonic (Dickinson, 1982) detritus from the erosion of volcanic chains and their crustal underpinnings.

The correlation of petrofacies with provenance types depends partly on the scale of sediment dispersal systems (Potter, 1986; Ingersoll, 1990), and this effect influences interpretations of petrofacies (Ingersoll et al., 1993; Potter, 1994). The more extensive the dispersal system, the greater the chance that detritus from multiple provenance types will be mingled prior to final deposition (Potter, 1978). In our provenance interpretations, we have tried to account for the variable scales of dispersal systems (Dickinson, 1988) that may have affected the region of interest.

## PETROFACIES AND SUBFACIES

The areal and stratigraphic distributions of the following petrofacies, designated by letters for clarity of discussion, are shown by Figures 1 and 2, respectively.

A is a subarkose petrofacies (derived from continental basement sources) including five samples, all from the Upper Cambrian Harmony Formation of the Roberts Mountains allochthon in Nevada;

B is a quartzarenite petrofacies (derived from cratonal sources) including 12 samples of lower Paleozoic rocks mainly from the Roberts Mountains allochthon in Nevada (matrix-poor subfacies B1; n = 7) and the related Shoo Fly Complex of the northern Sierra Nevada in California (matrix-rich subfacies B2; n = 3), but also including analogous sandstones from the lower Paleozoic Antelope Mountain Quartzite (n = 1) of the Yreka ter-

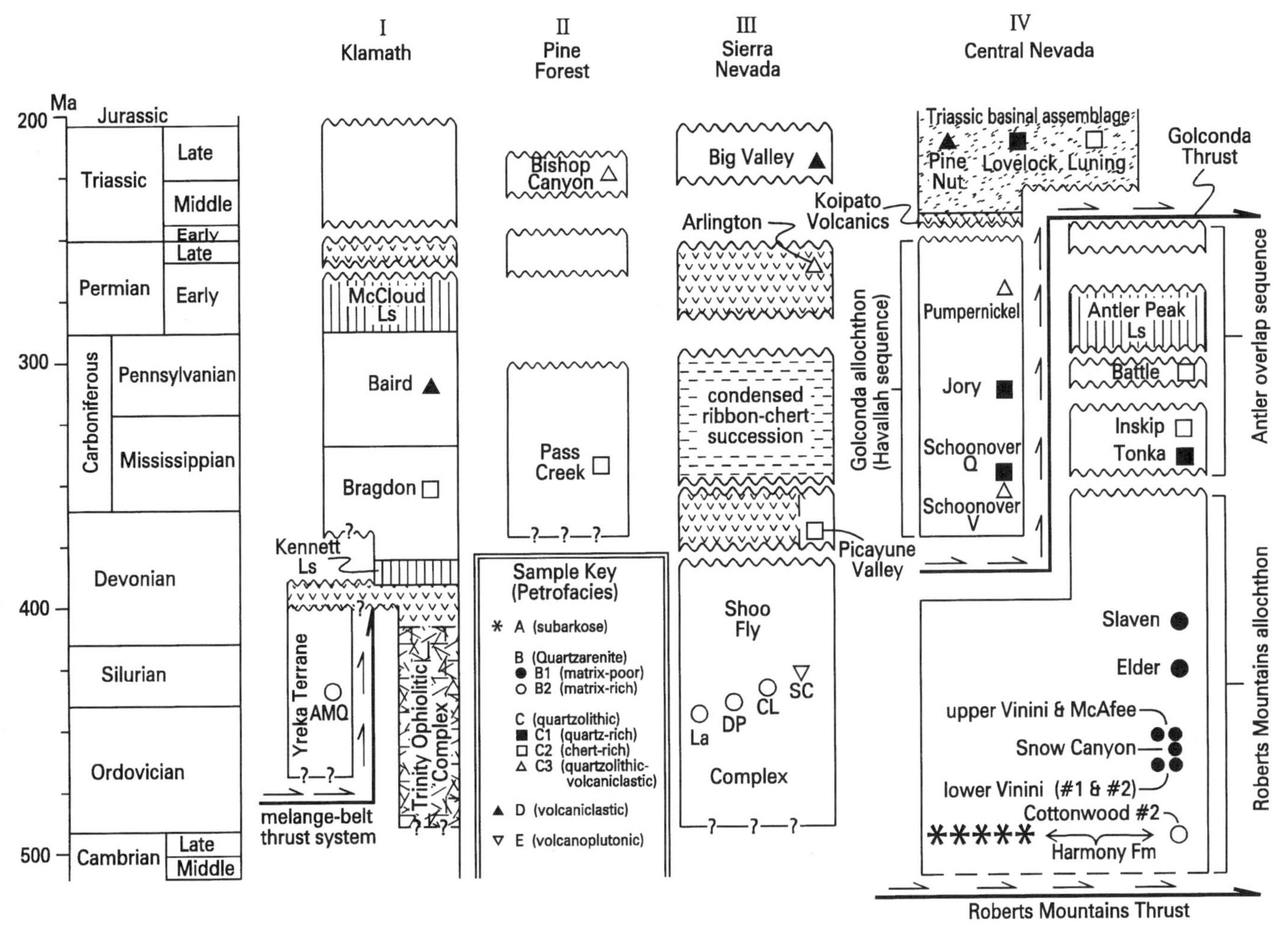


Figure 2. Regional stratigraphic context of detrital zircon samples from eastern Klamath province (I), Pine Forest Range (II), northern Sierra Nevada (III), and central (to west-central) Nevada (IV). V symbols denote principal Devonian and Permian volcanic-arc successions (diachronous) of Klamath Mountains and northern Sierra Nevada, and Triassic Koipato Volcanics of central Nevada. See Figure 1 for areal relations and Figure 3 for petrofacies compositions. Stratigraphic ages of lower Paleozoic samples from Yreka terrane and Shoo Fly Complex are inherently uncertain, but lower Paleozoic samples from Roberts Mountains allochthon and upper Paleozoic samples from Golconda allochthon are plotted at appropriate age horizons. Sample names are keyed to Figure 4 and Tables 2–8 (AMQ—Antelope Mountain Quartzite of Yreka terrane; samples from Shoo Fly Complex are plotted in order of structural stacking of internal thrust panels: La—Lang sequence; DP—Duncan Peak allochthon; CL—Culbertson Lake allochthon; SC—Sierra City mélange; see Table 2 for names of subarkose samples from Cambrian Harmony Formation of Roberts Mountains allochthon). Time scale after Dickinson (this volume).

rane in the Klamath Mountains and the Harmony Formation (n = 1) in Nevada (both of matrix-rich subfacies B2, the matrix-poor subfacies containing $<\sim 5\%$ interstitial matrix and the matrix-rich subfacies containing $>\sim 10\%$ interstitial matrix);

C is a quartzolithic petrofacies (derived from recycled orogenic sources) including 14 samples from upper Paleozoic rocks of the Golconda allochthon in Nevada (n = 4), and from varied upper Paleozoic and Triassic strata capping older and more deformed rocks in Nevada (n = 7) and California (n = 3);

D is a volcaniclastic petrofacies (derived from arc sources) including three samples from upper Paleozoic and Triassic strata in northern California and the adjacent western fringe of Nevada;

E is a volcanoplutonic petrofacies (derived from dissected arc sources) represented by one sample from the Sierra City mélange, a component of the lower Paleozoic Shoo Fly Complex of the northern Sierra Nevada in California.

The quartzolithic petrofacies is further subdivided into three related subfacies: quartz-rich (petrofacies C1; n = 4), chert-rich (petrofacies C2; n = 6), and quartzolithic-volcaniclastic (petrofacies C3; n = 4), the latter containing prominent, though subordinate, admixtures of volcanic detritus of arc derivation. The quartz-rich and chert-rich subfacies jointly display a gradational spectrum of compositions separated according to whether monocrystalline quartz grains or polycrystalline lithic fragments are most abundant in the QFL grain population. The third subfacies, containing minor admixed volcaniclastic detritus, represents mingling of detritus similar to the chert-rich end member of the petrofacies with arc-derived plagioclase grains and volcanic lithic fragments.

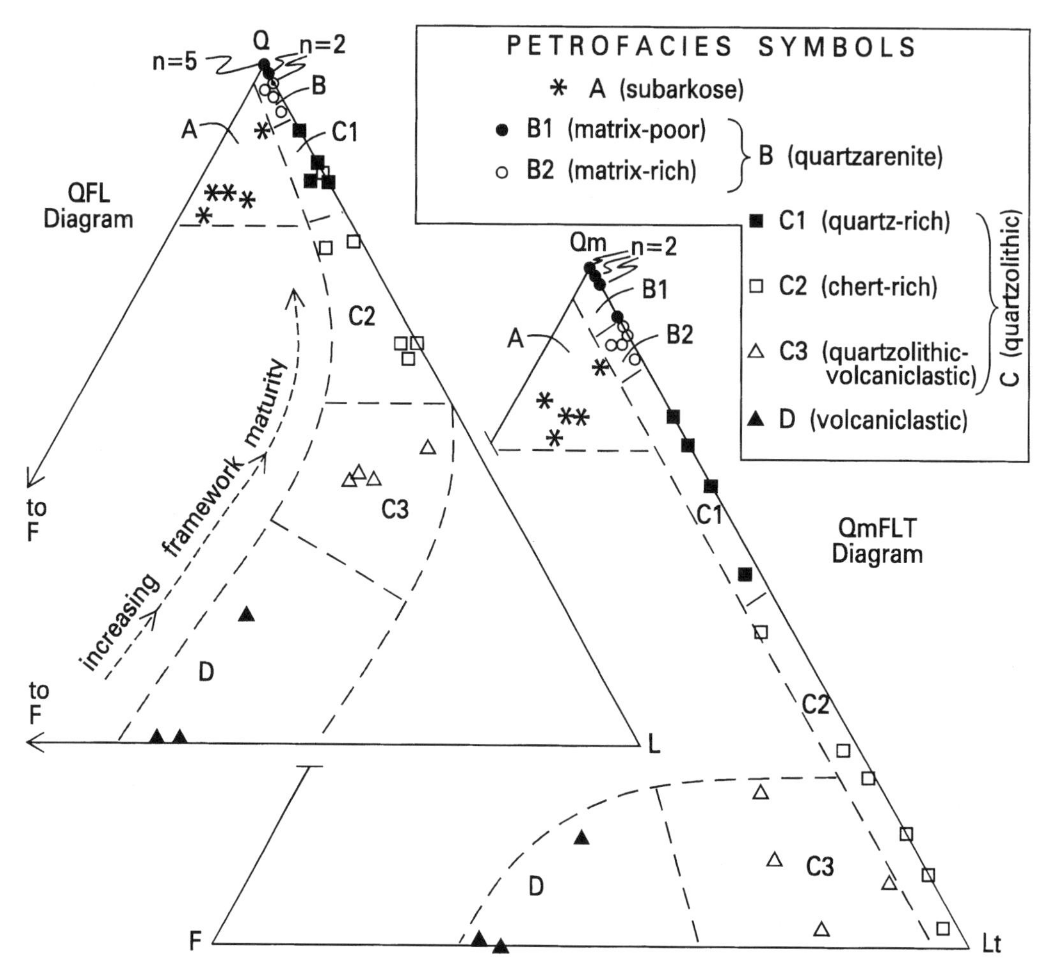


Figure 3. Standard QFL and QmFLt triangular diagrams showing compositions of detrital zircon samples as grouped into generic petrofacies (Tables 2–8). See Figure 1 for areal and Figure 2 for stratigraphic distribution of samples.

Figure 3 shows the positions of the various petrofacies and subfacies in standard QFL and QmFLt space (Table 1), and Tables 2–8 summarize the detrital modes of samples from each petrofacies. With one exception (see footnote to Table 4), the detrital modes reported and plotted here derive entirely from point counts of the samples collected for zircon analysis, and are not necessarily indicative of the mean compositions of any of the petrofacies they represent. Figure 4 indicates the age spectra of the zircon populations recovered from each sample.

## PETROFACIES ASSOCIATIONS

Provenance interpretations depend partly upon the following tectonic associations of the different petrofacies within various stratigraphic assemblages of the study region.

1. The Roberts Mountains allochthon (Fig. 2, IV), composed of strongly deformed lower Paleozoic eugeoclinal strata thrust eastward to southeastward across coeval miogeoclinal strata of Nevada during the Late Devonian to Early Mississippian Antler orogeny (Stewart, 1980; Dickinson et al., 1983b), yielded sandstones of both the subarkose (A; n = 5) and the quartzarenite (matrix-poor subfacies B1; n = 7) petrofacies (Tables 2–3). Samples of the subarkose petrofacies were collected entirely from the Cambrian Harmony Formation, whereas the quartzarenite petrofacies occurs widely in strata ranging in age from Ordovician to Devonian. The Harmony Formation also contains minor quartzose sandstone (n = 1) belonging to the quartzarenite petrofacies (matrix-rich subfacies B2).

2. The Shoo Fly Complex (Fig. 2, III), composed of strongly deformed lower Paleozoic strata forming thrust panels imbricated southward in the northern Sierra Nevada of California (D'Allura et al., 1977, Fig. 10; Schweickert and Snyder, 1981; Harwood, 1992), yielded sandstones of both the quartzarenite (subfacies B2; n = 3) and the volcanoplutonic (E; n = 1) petrofacies, the latter from a tectonic block within the Sierra City mélange. The Antelope Mountain Quartzite of the Yreka terrane in the Klamath Mountains (Fig. 2, I) contains similar quartzarenite petrofacies (Table 4). The single sample of the volcanoplutonic petrofacies was collected from the highest fault-bounded panel of the Shoo Fly thrust stack, which also includes, in descending structural order, the Culbertson Lake and Duncan Peak allochthons and the Lang sequence (Harwood, 1988, 1992; Girty et al., 1990), each of which yielded one of the other zircon samples (Table 4).

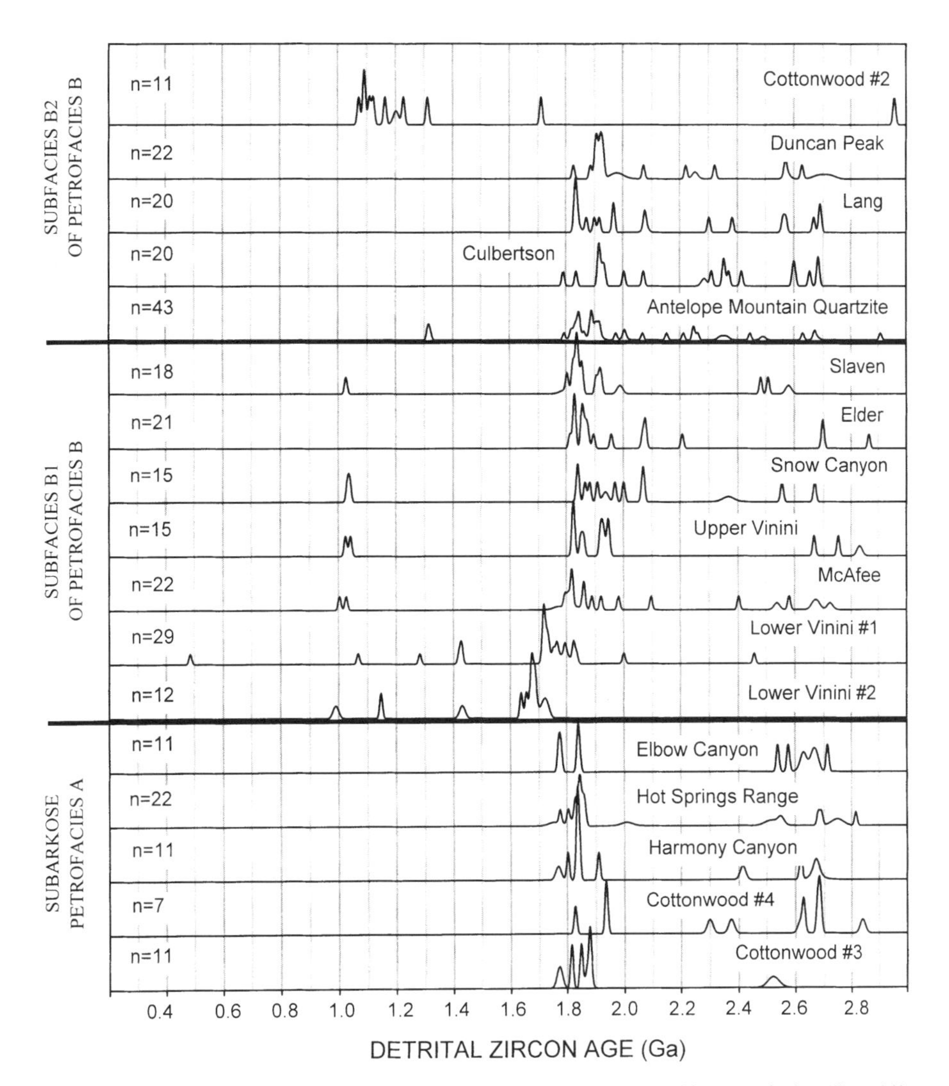


Figure 4 (on this and facing page). Detrital zircon age spectra grouped by petrofacies (Ga = billion years), where n is number of detrital zircon grains from each sample for which concordant or nearly concordant U-Pb ages were obtained (total n = 700). Curves are cumulative probability plots (Gehrels, this volume) of single-crystal zircon ages, normalized to number of grains plotted for each sample (areas under curves constant). Smallest peaks on age spectra represent individual detrital zircon crystals of uncertain significance for provenance.

3. Carboniferous strata of the Antler overlap sequence of central Nevada (Fig. 2, IV), that unconformably overlie eroded vestiges of the Roberts Mountains allochthon emplaced during the Antler orogeny (Johnson and Pendergast, 1981; Speed and Sleep, 1982; Dickinson et al., 1983b), yielded sandstones of both the quartz-rich (subfacies C1; n = 1) and chert-rich (subfacies C2; n = 2) variants of the quartzolithic petrofacies (Tables 5–6). Similar detrital modes occur in sandstones deposited along the western fringe (Inskip, Table 6), within the center (Battle, Table 6), and along the eastern fringe (Tonka, Table 5) of the Antler orogen.

4. The Golconda allochthon (Fig. 2, IV), composed of strongly deformed Carboniferous to mid-Permian turbidites and associated strata of the Havallah sequence (Miller et al., 1982, 1984; Brueckner and Snyder, 1985; Stewart et al., 1986; Murchey, 1990; Jones, 1991), which wasthrust eastward to southeastward across the Antler overlap sequence and the underlying Antler orogen during the Late Permian to Early Triassic Sonoma orogeny (Speed, 1979; Stewart, 1980; Dickinson et al., 1983b), yielded sandstones of both the quartz-rich (subfacies C1; n = 2) and quartzolithic-volcaniclastic (subfacies C3; n = 2) variants of the quartzolithic petrofacies (Tables 5 and 7). The two subfacies appear to be intermingled, both areally and stratigraphically, within the allochthon.

5. Samples from upper Paleozoic and Triassic cover successions capping more deformed rocks of the Klamath-Sierran and Antler-Sonoma orogens (Fig. 1) include the following: In the Klamath Mountains (Fig. 2, I), Carboniferous strata (n = 2) of the Eastern Klamath succession (Miller, 1989) depositionally overlie the Trinity ophiolitic complex (Miller, 1989). In the Pine Forest Range

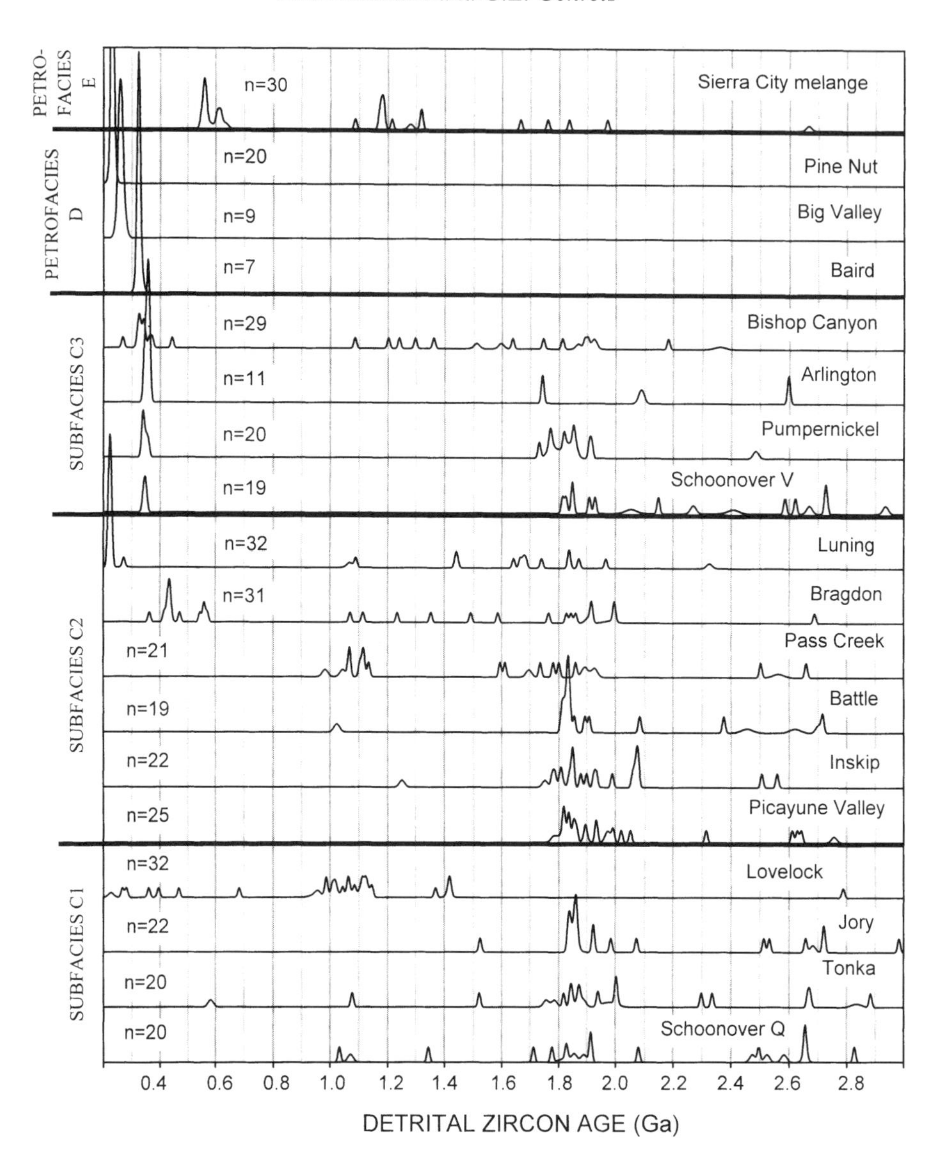


(Wyld, 1990) of northwestern Nevada, there are Carboniferous and Triassic analogues (n = 2) of the Eastern Klamath succession (Fig. 2, II). In the northern Sierra Nevada, Devonian strata (n = 1) unconformably overlie the Shoo Fly Complex and Permian-Triassic strata (n = 2) that depositionally overlie volcanogenic successions that cap the Shoo Fly Complex (Fig. 2, III). Upper Triassic strata (n = 3) of a thick basinal succession are exposed west of the Antler and Sonoma orogens (Fig. 2, IV) in western Nevada (Burke and Silberling, 1973; Speed, 1977; Oldow, 1984; Elison and Speed, 1988).

The varied overlap units of the cover successions yielded sandstones of both the chert-rich (subfacies C2; n = 4) and the quartzolithic-volcaniclastic (subfacies C3; n = 2) variants of the quartzolithic petrofacies (Tables 6–7), together with one sandstone representative (Table 5) of the quartz-rich subfacies (C1; n = 1) and all the examples (n = 3) of volcaniclastic petrofacies D (Table 8). The petrographic variability of the overlap assemblages reflects diverse and mixed sources in a region of complex paleogeography, influenced locally by Mesozoic arc magmatism, located in the western part of the study region during late Paleozoic and early Mesozoic time.

In the following sections, the compositions and provenance relationships of each of the petrofacies and subfacies are discussed in order (petrofacies A–E), with reference to salient petrofacies associations where appropriate.

## SUBARKOSE PETROFACIES

The moderately sorted, fine-grained to very coarse-grained, subrounded to subangular sand in subarkose petrofacies A (Table 2; Fig. 5, A–D) was evidently derived from homogeneous uplifted continental basement of restricted plutonic composition, with at least moderate relief to suppress weathering loss of feldspar in derivative subarkosic detritus. The typical Qm/(Qm + F) ratio of 0.85, approximately twice the characteristic value of 0.40–0.45 for California Cenozoic arkoses derived from strongly uplifted blocks (Dickinson, 1985, Table 3), suggests that the

TABLE 2. DETRITAL MODES OF SUBARKOSE PETROFACIES A

| Grain type | Cottonwood #3 Galena Range | Cottonwood #4 Galena Range | Harmony Canyon Sonoma Range* | Hot Springs Range | Elbow Canyon Sonoma Range* |
|---|---|---|---|---|---|
| Qm | 74 | 78 | 78 | 80 | 86 |
| Qp | 4 | 3 | 2 | 1 | 4 |
| F† | 18 | 14 | 12 | 16 | 5 |
| L§ | 4 | 5 | 8 | 3 | 5 |
| Mica(fmwk%)# | 2 | 1 | 3 | 2 | 2 |
| Matrix (WR%) | 5 | 6 | 8 | 8 | 8 |

*Note:* All samples from Harmony Formation of Roberts Mountains allochthon in Nevada; QFL grain populations summed to 100% (see Table 1 for grain type symbols).
* Sonoma Range (Gilluly, 1967), northeast (Harmony Canyon) and southwest (Elbow Canyon).
† Subequal proportions of K and P (Table 1), within counting error.
§ Dominantly microgranular grains, of both igneous (hypabyssal) and hornfelsic origin.
# Detrital muscovite and biotite both present in varying proportions.

source region had either less relative relief or was more humid than modern ranges exposing continental basement in southern California. The dominance of microgranular hypabyssal and hornfelsic grains among the minor lithic fragments implies that little or no volcanic or sedimentary cover was present in the provenance. Granitic or gneissic sources yielded subequal proportions of K-feldspar and plagioclase, and included both muscovite and biotite as varietal or accessory minerals. The Harmony Formation samples are slightly more quartzose and less feldspathic than the mean composition of sandstones in the formation as a whole, but are within the known spectrum of variation (Suczek, 1977).

The generally uniform lithofacies of sandy midfan turbidites (Stewart and Suczek, 1977) forming the Cambrian Harmony Formation, the only unit of the Roberts Mountains allochthon containing examples of the subarkose petrofacies, suggests deposition on a large subsea fan derived from an extensive basement provenance. The thickness (1.2–1.5 km) and preserved extent (3500–4000 km$^2$) of the Harmony Formation imply a volume of at least $5 \times 10^3$ km$^3$ of plutoniclastic detritus within the unit, and the original extent of the subsea fan system may have been much greater than its present remnants document. Judging from the limited abrasion of quartz and feldspar grains, the turbidite dispersal system of the Harmony Formation was probably fed directly by a fluviodeltaic transport system draining basement highlands, with little modification of basement detritus by reworking along strandline beaches or marine shelves.

The dominant detrital zircons (1.75–1.95 Ga, 2.55–2.75 Ga) in the subarkose petrofacies of the Harmony Formation (Fig. 4) could not have been derived from subsequently buried Precambrian sources anywhere on the adjacent North American craton south of the Wyoming Archean province (Gehrels et al., this volume, Chapter 1, 10). Nor could they have been derived from the Salmon River arch of Idaho, suggested as the source of Harmony Formation sand by others on the basis of combined biogeographic and paleogeographic arguments (Rowell et al., 1979). Sources in the northern reaches of Laurentia, blocks rifted from northern Laurentia, or some separate continental block seemingly must be inferred for subarkosic Harmony detritus. As the Harmony Formation forms the structurally highest thrust sheet within the Roberts Mountains allochthon (Schweickert and Snyder, 1981), and is restricted to that thrust sheet, its provenance has uncertain significance for the allochthon as a whole.

## QUARTZARENITE PETROFACIES

The sands of quartzarenite petrofacies B include two distinct variants: (1) well sorted, fine-grained to medium-grained quartzose sand of subfacies B1 (Table 3; Fig. 6, A and B) collected from Ordovician (Vinini, Snow Canyon, McAfee), Silurian (Elder), and Devonian (Slaven) quartzites of the Roberts Mountains allochthon in Nevada, and (2) poorly sorted, very fine-grained to coarse-grained quartzose sand of subfacies B2; (Table 4; Fig. 6, C and D) collected from thrust slices of generally coeval strata (Varga and Moores, 1981; Harwood, 1983) within the Shoo Fly Complex of the northern Sierra Nevada in California, from the undated but almost surely lower Paleozoic Antelope Mountain Quartzite of the Yreka terrane in the Klamath Mountains (Potter et al., 1977, 1990), and from anomalous feldspar-poor strata within the Harmony Formation of the Roberts Mountains allochthon in Nevada. Interstitial matrix is minor (<6%) or absent in subfacies B1, but abundant (11–15 WR%) in subfacies B2, QFL% Qm is consistently somewhat higher (92%–100%) in subfacies B1 than in subfacies B2 (87%–91%); and subfacies B2 contains minor (3–6 WR%) labile lithic fragments that are nearly absent from subfacies B1.

The slight differences in modal composition could reflect a difference in provenance for the two subfacies, but alternately might reflect somewhat different transport histories and depositional settings for detritus derived from the same provenance. The progressive loss of minor Qp and Lsm grains from craton-derived detritus is expected with increasing textural and compositional maturity. Each subfacies was probably derived from a broad continental craton of subdued relief to allow most grains less stable or durable than quartz to be eliminated by weathering and abrasion during exposure and prolonged terrestrial transport. On bal-

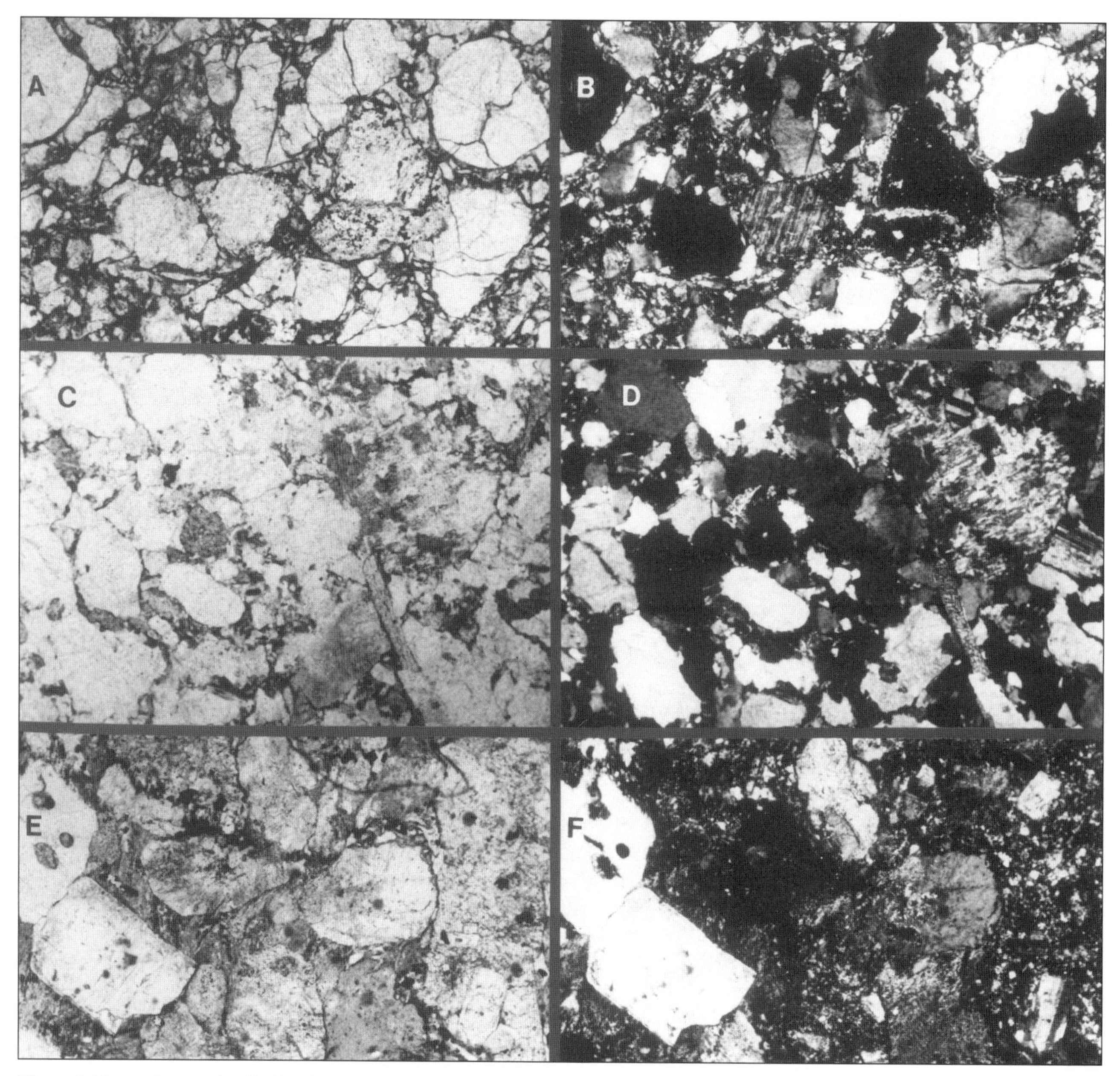


Figure 5. Photomicrographs (fields of view 2.25 mm × 3.5 mm, with plane light on left and crossed nicols on right) of petrofacies A (subarkose, frames A–D) and petrofacies D (volcaniclastic, frames E–F): A–B, Harmony Canyon (Table 2), with polysynthetically twinned plagioclase grain near center; C–D, Cottonwood #3 (Table 2), with perthite grain in upper right and detrital mica flake below it; E–F, Pine Nut (Table 8), with dominantly plagioclase feldspar grains and lithic fragments visible.

ance, the petrographic nature of the two subfacies neither requires the Roberts Mountains allochthon and the Shoo Fly Complex to be parts of the same tectonic assemblage nor precludes them from being so.

The samples from the Shoo Fly Complex (subfacies B2) are petrologically comparable to the bulk of the turbidites studied previously from that assemblage (Bond and Devay, 1980; Girty and Wardlaw, 1985; Girty et al., 1996), although subordinate strata containing appreciable feldspar grains are unrepresented in our collections. Perhaps attempts to collect the least weathered strata for zircon analysis biased our collections toward quartz-rich compositions. Our sample of the Silurian Elder Sandstone (subfacies B1) in the Roberts Mountains allochthon also represents the quartz-rich end member of a spectrum that includes both more feldspathic and more lithic frameworks (Girty et al., 1985). In the case of the Elder Sandstone, however, the feldspar-bearing rocks are generally too fine grained to yield detrital zircons of a size (>100 μm) suitable for single-grain analysis (G.H. Girty, 1996, personal commun.)

**TABLE 3. DETRITAL MODES OF MATRIX-POOR SUBFACIES B1 OF QUARTZARENITE PETROFACIES B**

| Grain type | Lower Vinini #2 | McAfee Quartzite | Lower Vinini #1 | Upper Vinini | Snow Canyon | Elder Sandstone | Slaven Chert |
|---|---|---|---|---|---|---|---|
| Qm | 100* | 100 | 99 | 98 | 98 | 97 | 92 |
| Qp | 0 | 0 | 1 | 1 | 1 | 3 | 8 |
| Lsm | 0 | 0 | 0 | 1 | 1 | 0 | 0 |
| Cement (WR%) | 30[†] | 2[§] | 17[†] | 1[§] | 1[§] | 17[†], 1[§] | 16[#] |
| Matrix (WR%) | 1 | 3 | 0 | 2 | 6 | 0 | 0 |

*Note:* All samples from quartzites of Roberts Mountains allochthon in Nevada; QFL grain populations summed to 100% (see Table 1 for grain type symbols).
* Terrigenous fraction of hybrid sandstone in which 30% of framework grains are calcareous debris
† Calcareous cement
§ Quartz overgrowths
# Chalcedonic cement

The dominant detrital zircons in most samples (n = 5) of subfacies B1 (1.8–2.0 Ga), and the broadly comparable but somewhat more diffuse age range of detrital zircons from most samples (n = 4) of subfacies B2 (1.8–2.7 Ga), could not have been derived from parts of the adjacent North American craton south of the Wyoming Archean province (Fig. 4). The spectrum of detrital zircon ages previously reported (Smith and Gehrels, 1994) for a quartzite sample from the Valmy Formation of the Roberts Mountains allochthon in central Nevada is indistinguishable from those for the associated subfacies B1 samples reported here. Although the sands of subfacies B1 and B2 could have slightly different provenances, with zircons >2.0 Ga in age more common in the latter, the zircon age spectra of the two subfacies are similar; including, for example, a common minor age spike near 2.05–2.1 Ga (Fig. 4), also present in the Valmy age spectrum. Derivation from adjacent areas of the same cratonal block is the most parsimonious hypothesis for the sources of the two subfacies. Moreover, both may have been derived from the same basement block that had previously contributed distinctly more immature basement detritus to subarkose petrofacies A of the older Harmony Formation. Zircon age spectra are generally comparable (Fig. 4), although Harmony Formation zircons encompass the age range 1.75–1.8 Ga, which is not well represented in either of the two quartzarenite subfacies.

The similarity of the detrital zircon populations in the two quartzarenite subfacies suggests that the coeval Roberts Mountains allochthon of Nevada and the Shoo Fly Complex of California may have been contiguous parts of the same tectonic belt prior to intrusion of the Sierra Nevada batholith, and to structural disruption both before and after batholith emplacement. Moreover, the detrital zircon population of the Antelope Mountain Quartzite in the Yreka terrane is essentially indistinguishable from Shoo Fly Complex samples (Fig. 4), and framework detrital modes are also indistinguishable (Bond and Devay, 1980). A widespread early Paleozoic dispersal system may have been responsible for sedimentation of nearly all the lower Paleozoic strata we have studied.

In the lower part of the Vinini Formation (n = 2) of central Nevada, the dominant detrital zircons (1.65–1.75 Ga; n = 2) contrast with the zircon populations of the other petrofacies B samples (Fig. 4), and unlike the others could have been derived from the immediately adjacent southwestern part of the North American craton. The Vinini zircon suite is closely matched by detrital zircon populations in quartzites of the structurally underlying miogeoclinal belt deposited along the continental margin in Nevada (Gehrels and Dickinson, 1995). The Vinini Formation data are understandable in light of the interpretation that the Vinini depositional system was a continental slope and rise assemblage deposited just off the edge of the miogeocline (Finney and Perry, 1991). More westerly and structurally higher thrust slices within the Roberts Mountains allochthon that yield zircon populations exotic to the adjacent craton are typically more complexly deformed, and are commonly interleaved structurally with pillow basalts and associated radiolarian cherts, apparently representing deformed sea-floor assemblages deposited farther from the continental margin (Stanley et al., 1977). The dispersal system responsible for transport of sediment along and near the edge of the continental block in Nevada during the time of deposition of the lower Vinini Formation evidently did not transport sand far enough offshore to influence sedimentation in more distal oceanic settings farther from the continental margin during deposition of older (Cambrian Harmony Formation) and younger (Ordovician-Devonian) turbidites.

Provenance evaluations for other quartzite units in central Nevada are complicated by the dominance of 1.8–2.1 Ga detrital zircon grains in the Ordovician Eureka Quartzite of the miogeoclinal succession in eastern Nevada (Gehrels and Dickinson, 1995). Sands of Eureka Quartzite age apparently spread southward, in modern coordinates, along the Paleozoic continental shelf from sources associated with the Peace River arch in northern Laurentia. Similar sands may have been carried off the shelf into deeper water by local turbidity currents, accounting for some of the sands in the quartzarenite petrofacies of more western assemblages (Miller and Larue, 1983). There are no notable dif-

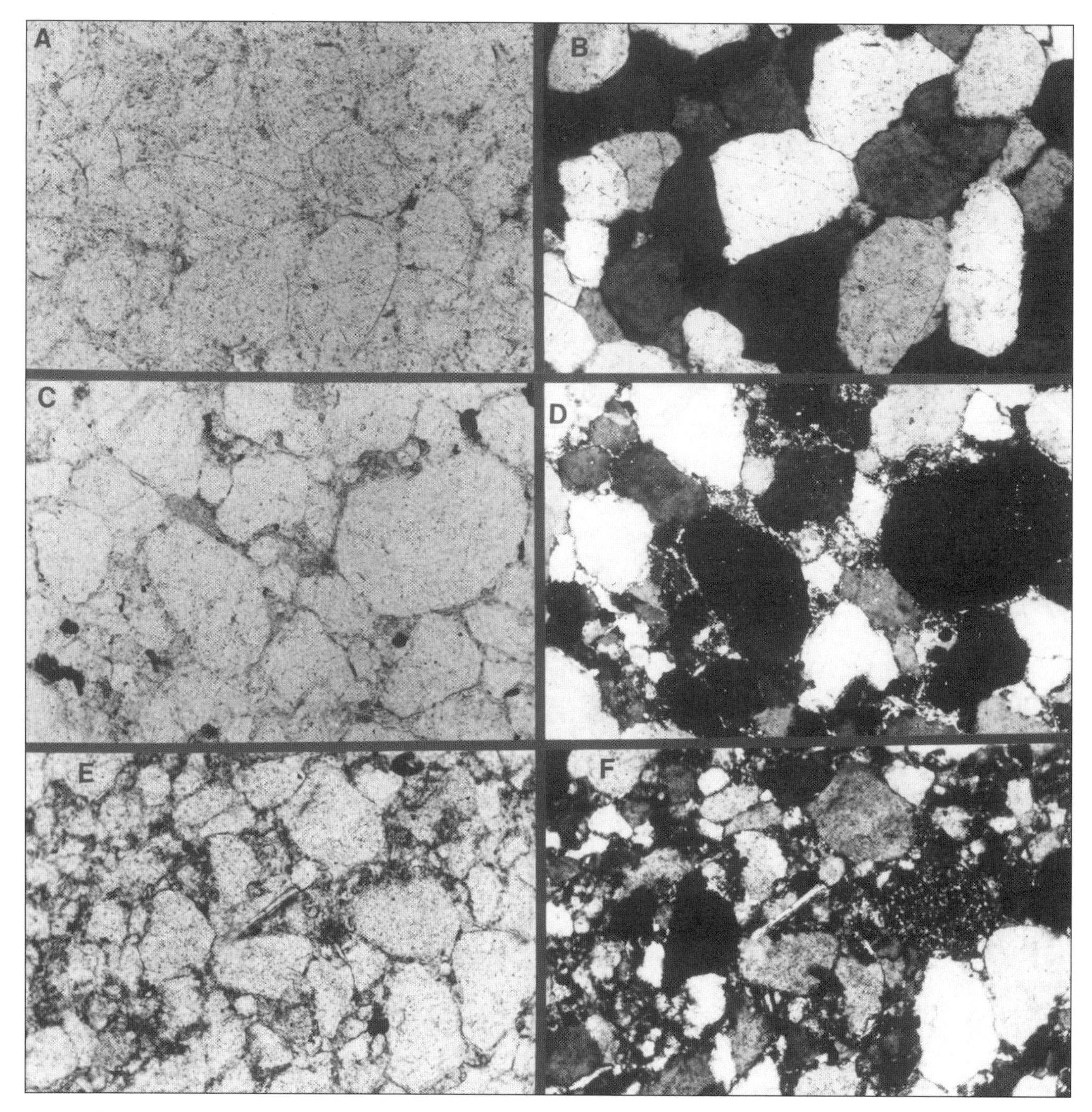


Figure 6. Photomicrographs (fields of view 0.9 mm × 1.4 mm, with plane light on left and crossed nicols on right) of quartzarenite petrofacies B (matrix-poor subfacies B1, frames A–B; matrix-rich subfacies B2, frames C–F): A–B, Upper Vinini (Table 3), with prominent quartz overgrowths visible on rounded detrital quartz grains; C–D, Culbertson Lake (Table 4), with birefringent matrix visible between somewhat less rounded quartz grains; E–F, Cottonwood #2 (Table 4), with detrital mica flake near center and polycrystalline chert grain to right of mica flake.

ferences, however, among the zircon age populations of post-lower Vinini Formation Ordovician, Silurian, and Devonian samples from quartzarenite petrofacies B1 (Figs. 2 and 4), and the nature of the younger sands cannot be ascribed to the Eureka Quartzite dispersal system.

In sum, the dominance of detrital zircons exotic to the adjacent craton in lower Paleozoic tectonic assemblages west of the miogeoclinal belt in Nevada can be attributed alternately or in combination to any of three processes: (1) longitudinal transport of sedimentary detritus by dispersal systems moving subparallel to the continental margin from sources farther north but still within Laurentia; (2) transverse transport of sedimentary detritus back toward the continental margin from an offshore continental block (Madrid, 1987); or (3) tectonic transport of imbricated

**TABLE 4. DETRITAL MODES OF MATRIX-RICH SUBFACIES B2 OF QUARTZARENITE PETROFACIES B**

| Grain type | Antelope Mountain Quartzite*[†] | Culbertson Lake allochthon[§] | Lang sequence[§] | Duncan Peak allochthon[§] | Cottonwood #2 (Harmony Formation)* |
|---|---|---|---|---|---|
| Qm | 89 | 91 | 90 | 89 | 87 |
| Qp | 4 | 6 | 7 | 6 | 6 |
| F | 2 | 0 | 0 | 1 | 1 |
| Lsm[#] | 5 | 3 | 3 | 4 | 6 |
| Matrix (WR%) | 11 | 13 | 12 | 15 | 14 |

*Note:* QFL grain populations summed to 100% (see Table 1 for grain type symbols).
* Also includes 1 Fmwk% mica.
[†] Yreka terrane; petrofacies sample not from same outcrop as detrital zircon sample.
[§] Structural components of Shoo Fly Complex in northern Sierra Nevada
[#] Dominantly quartz-mica metapelite, possibly formed by in situ metamorphism.

**TABLE 5. DETRITAL MODES OF QUARTZ-RICH SUBFACIES C1 OF QUARTZOLITHIC PETROFACIES**

| Grain type | Lovelock* | Schoonover Q | Tonka | Jory[†] |
|---|---|---|---|---|
| Qm | 78 | 74 | 68 | 55 |
| Qp | 5 | 11 | 22 | 28 |
| Lsm | 16[§] | 15[#] | 10[#] | 13** |
| Lvm | 1 | 0 | 0 | 2 |
| Matrix (WR%) | 6 | 9 | 7 | 7 |

*Note:* Column headings are local formation or assemblage names (Figure 2); QFL grain populations summed to 100% (see Table 1 for grain type symbols).
* Also includes 2 fmwk% detrital muscovite.
[†] Also includes 2 QFL% feldspar (F).
[§] Dominantly quartz-mica metapelite, possibly formed by in situ metamorphism.
[#] Dominantly argillite grains with subordinate microgranular siltstone grains.
** Sum of 9% argillite grains and 4% microgranular siltstone grains.

allochthons that were assembled by deformation of sea-floor sediment at a subduction zone either well offshore or farther north.

The sample of quartzarenite petrofacies B (Table 4; Fig. 6, E and F) in the Cambrian Harmony Formation of the Galena Range contains an anomalous zircon population (Fig. 4) dominated by grains of Grenville age (1.1–1.3 Ga), and a similar zircon assemblage was previously reported from a Harmony sample collected nearby (Smith and Gehrels, 1994). Outcrop reconnaissance indicates that the quartzose Harmony Formation turbidites form a thick succession depositionally beneath more typical Harmony Formation turbidites of subarkose petrofacies along a sharp but concordant contact, suggesting that the subarkoses represent a different subsea fan that prograded over a slightly older subsea fan composed of more quartzose detritus. The zircon data indicate that the source of the more quartzose sand in the Harmony Formation was distinctly different from the sources of both subarkose petrofacies A, dominant in the Harmony Formation, and the remainder of petrofacies B. The prevalence of Grenville-age zircons suggests a source farther south along the Cordilleran continental margin in Mexico (Gehrels et al., 1995, Fig. 2). The Harmony Formation quartzarenite petrofacies shows the utility of detrital zircon analysis for detecting provenance differences among supermature sands with uniformly quartzose detrital modes.

## QUARTZOLITHIC PETROFACIES

In sandstones of quartzolithic petrofacies C, the joint dominance of quartz grains (Qm) and sedimentary-metasedimentary lithic fragments (Qp and Lsm) is indicative of recycling from older stratal successions (Dickinson, 1985). By inference, the quartz grains were derived mainly from older sandstones, and the lithic fragments were derived from associated finer grained rocks (e.g., chert, argillite, shale, slate, phyllite). The paucity of feldspar grains, coupled with the abundance of aphanitic lithic fragments, indicates that plutonic basement was not a significant part of the provenance, or feldspar mineral grains would appear in the detrital aggregate (Ingersoll and Suczek, 1979; Suczek and Ingersoll, 1985).

The quartzolithic petrofacies embodies a spectrum of frameworks, from a quartz-rich subfacies (C1; Table 5; Fig. 7, A and B)

TABLE 6. DETRITAL MODES OF CHERT-RICH SUBFACIES C2 OF QUARTZOLITHIC PETROFACIES

| Grain type | Inskip* | Picayune Valley† | Battle§ | Pass Creek# | Luning | Bragdon** |
|---|---|---|---|---|---|---|
| Qm | 46 | 29 | 25 | 17 | 11 | 3 |
| Qp | 27 | 28 | 49 | 42 | 73 | 56 |
| F | 5 | 2 | 1 | 0 | 0 | 2 |
| Lsm†† | 21 | 41 | 20 | 41 | 16 | 35 |
| Lvm | 1 | 0 | 5 | 0 | 0 | 4 |
| Matrix (WR%) | 16 | 9 | 6 | ? | 7§§ | 13 |

*Note:* Column headings are local formation or assemblage names (Figure 3); QFL grain population summed to 100% (see Table 1 for grain type symbols).

* Approximate detrital mode for semischist with newgrown biotite and muscovite along undulating cleavage surfaces bounding lozenge-shaped framework grains.

† Foliated phyllitic semischist in chlorite zone with detrital grains still clearly visible.

§ Sample of Gehrels and Dickinson (1995).

\# Approximate detrital mode for nonfoliated hornfelsic metasandstone in biotite zone.

** Contains gradational population of argillaceous chert and cherty argillite grains.

†† Argillitic/pelitic grains form 75%–85% of Lsm population, together with subordinate microgranular grains (siltstone/hornfels), but gradational varieties are present as well.

§§ Interstitial materials also include 2 WR% pore-filling calcite cement.

to a chert-rich subfacies (C2; Table 6; Fig. 7, C and D), the compositional division between the two being taken as the position along the gradational spectrum where the ratio Qm/(Qp + Lsm) is unity (Fig. 3). Where admixtures of volcaniclastic detritus are present in subfacies C3 (Table 7; Fig. 7, E and F), the joint occurrence of feldspar mineral grains and volcanic-metavolcanic lithic fragments (Lvm) as significant components of the framework is a robust petrographic signal of volcaniclastic debris. Previous petrographic studies confirm that the dominant sandstones of the key tectonic assemblages from which we collected examples of the quartzolithic petrofacies are petrologically similar to our samples (Dickinson et al., 1983b; Miller and Cui, 1989; Whiteford, 1990; Jones, 1991; McLean, 1995).

### *Quartz-rich subfacies*

The relatively quartz rich upper Paleozoic and Upper Triassic sandstones of subfacies C1 (Table 5), all from Nevada (Fig. 1), in which Qm > (Qp + Lsm) and grains are subrounded to subangular, range from well sorted, very fine grained to fine-grained (Lovelock) and fine-grained to medium-grained (Schoonover Q) frameworks to only moderately sorted, fine-grained to coarse-grained (Tonka) and fine-grained to very coarse grained (Jory) frameworks, with a matrix content in the range of 6–9 WR%.

An inverse correlation of Qm content with Qp/Lsm ratio, together with a quasiconstant Qm/Lsm ratio, suggests that the quartz grains were recycled from quartzose clastic successions, of consistent lithology (sandstone-argillite), which were associated with nonclastic chert packets within a thrust-fold belt or subduction complex (Dickinson, 1985). Generally sympathetic variations in Qp/Qm and Qp/Lsm ratios suggest further that the chert in the provenance was admixed with the clastic strata in varying proportions, whether by sedimentary interbedding, tectonic interleaving, or differential weathering of more resistant chert and less resistant clastic sequences.

Because the Roberts Mountains allochthon is composed dominantly of associated chert-argillite sequences and turbidite sandstones of quartzarenite facies, derivation of Carboniferous-Permian and Upper Triassic quartzolithic sands in Nevada from sources within the tectonically or depositionally underlying Roberts Mountains allochthon is an attractive hypothesis (Fig. 2, IV). The dominant detrital zircons (1.8–2.0 Ga) in Carboniferous quartz-rich subfacies C1 of the Antler overlap sequence (Tonka) and the Havallah sequence of the Golconda allochthon (Schoonover Q, Jory) are compatible with recycling in such fashion (Fig. 4), but the subfacies C1 sample from the Lovelock assemblage (Oldow, 1984) of the Upper Triassic basinal assemblage in western Nevada includes abundant Grenville-age (0.95–1.15 Ga) zircons that were perhaps recycled from quartzose Harmony Formation sandstones of the Roberts Mountains allochthon in which zircons of Grenville age are dominant (Fig. 4).

The occurrence of similar zircon age populations in quartzolithic strata of the Antler overlap sequence and the Havallah sequence of the Golconda allochthon implies that detritus from the Antler orogen was shed westward into the oceanic region from which the Golconda allochthon was later thrust, as well as into the Antler foreland basin to the east (Dickinson et al., 1983b). One Havallah sample (Jory) includes a minor admixture of volcaniclastic detritus (F, Lvm) similar to the more abundant admixtures present in quartzolithic-volcaniclastic subfacies C3 (Table 5).

### *Chert-rich subfacies*

The relatively chert rich Carboniferous and Upper Triassic sandstones of subfacies C2 (Table 6), widely distributed in both Nevada and California (Fig. 1), are all poorly sorted, with a full range of sand sizes represented by subrounded to subangular grains and matrix contents in the range of 7–13 WR%. The strata sampled represent various local stratigraphic units (Fig. 2), including: (1) arc-related Carboniferous basins of the Klamath Mountains (Bragdon; Watkins, 1986) and Pine Forest Range (Pass Creek; Wyld, 1990); (2) Devonian-Mississippian beds (Picayune Valley; Harwood, 1992) unconformably overlying the Shoo Fly Complex near Lake Tahoe in the northern Sierra Nevada; (3) Mississippian (Inskip; Whitebread, 1994) and Pennsylvanian (Battle; Saller and Dickinson, 1982) strata of the Antler overlap sequence in Nevada; and (4) the Luning assemblage (Oldow, 1984) of the Upper Triassic basinal assemblage in western Nevada (Luning; Oldow, 1981).

The Roberts Mountains allochthon, the Shoo Fly Complex, or some analogous oceanic assemblage uplifted as a subduction

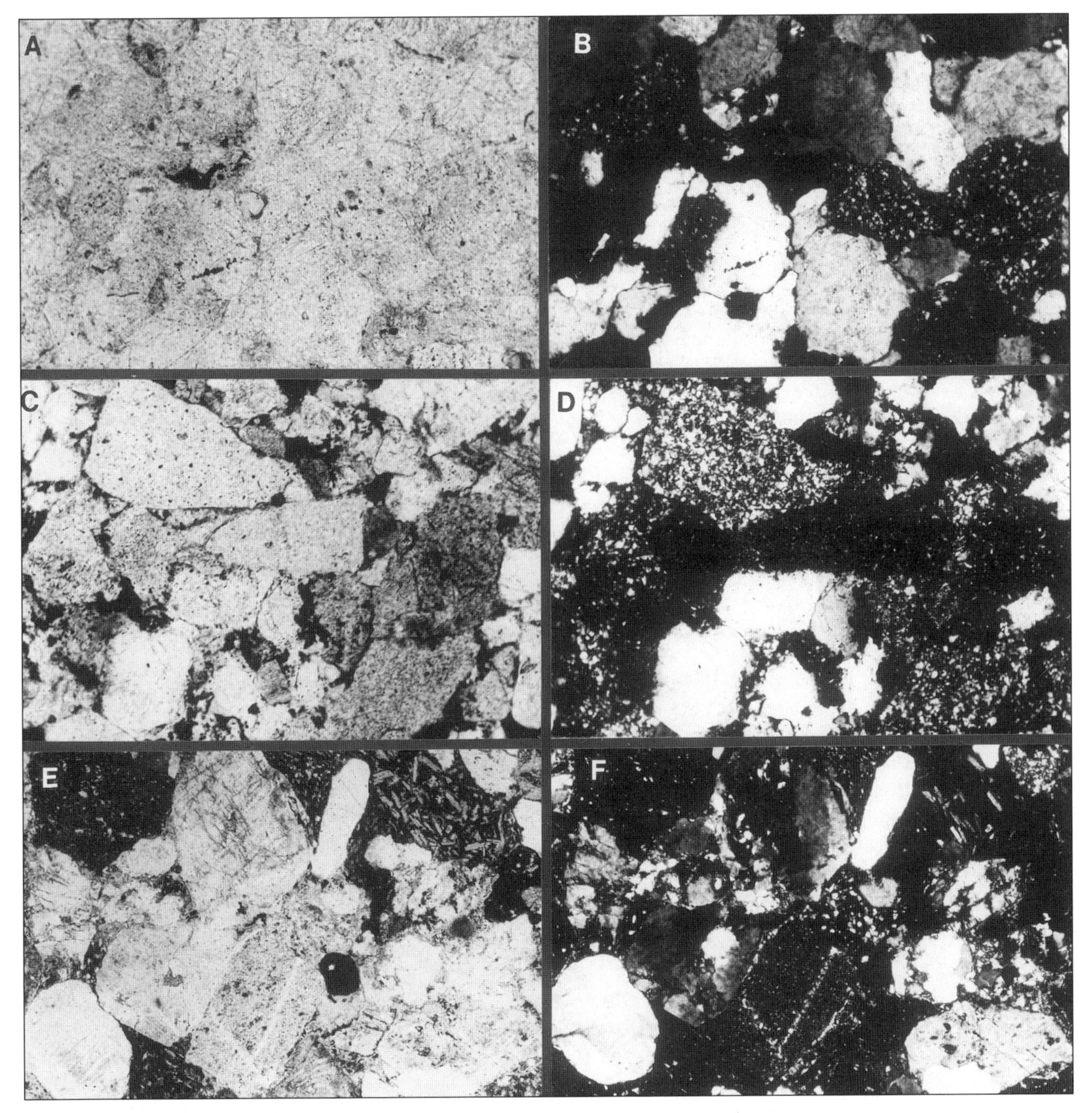


Figure 7. Photomicrographs (fields of view 0.9 mm × 1.4 mm for A–D and 2.25 mm × 3.5 mm for E–F, with plane light on left and crossed nicols on right) of petrofacies C (quartzolithic): A–B, Tonka (Table 5), quartz-rich subfacies C1 with Qm > Qp; C–D, Battle (Table 6), chert-rich subfacies C2 with Qp > Qm; E–F, Pumpernickel (Table 7), volcaniclastic-quartzolithic subfacies C3 (note microlitic volcanic lithic fragment in upper right).

TABLE 7. DETRITAL MODES OF QUARTZOLITHIC-VOLCANICLASTIC SUBFACIES C3 OF QUARTZOLITHIC PETROFACIES

| Grain type | Bishop Canyon* | Pumpernickel | Schoonover V | Arlington |
|---|---|---|---|---|
| Qm | 13 | 10 | 23 | 3 |
| Qp | 26 | 34 | 16 | 36 |
| F | 19 | 6 | 16 | 18 |
| Lsm† | 35 | 26 | 28 | 21 |
| Lvm | 7 | 24 | 17 | 22 |
| Matrix (WR%) | 10 | 6 | 0§ | 7# |

*Note:* Column headings are local formation or assemblage names (Figure 3); QFL grain population summed to 100% (see Table 1 for grain type symbols).

* Approximate detrital mode because of extensive alteration.

† Argillitic grains form 75%–80% of Lsm population, with the remainder microgranular.

§ Interstitial material is 12 WR% calcareous cement, and calcite replaces 9 fmwk% of framework grains, but grain replacement does not appear to be preferential.

# Interstitial materials also include 2 WR% calcareous cement and replacive calcite.

complex or lithologically similar overthrust mass is attractive as a potential source for sands especially rich in detrital chert grains (Dickinson et al., 1979). Either the Roberts Mountains allochthon or the Shoo Fly Complex has been suggested as the provenance of quartzolithic detritus in the Picayune Valley Formation of the Sierra Nevada (Harwood et al., 1991), and the Roberts Mountains allochthon was clearly the source of quartzolithic detritus in the overlying Battle Formation of the Antler overlap sequence (Dickinson et al., 1983b).

Most ($n = 5$) of the subfacies C2 samples yielded detrital zircon age populations with a dominant (Inskip, Battle, Picayune Valley) or significant (Bragdon, Pass Creek) component of the 1.8–2.1 Ga grains so common in petrofacies B sandstones from the Roberts Mountains allochthon, the Shoo Fly Complex, and the Yreka terrane (Fig. 4). Recycling from one or more of those lower Paleozoic terranes is a feasible source for that characteristic fraction of the zircon assemblage. The appreciable (5 QFL%) feldspar content of the Inskip Formation (Whitebread, 1994), and the volcanic lithic fragments (Lvm) present in similar abundance in the Battle Formation (Saller and Dickinson, 1982) may also reflect recycling, respectively, from petrofacies A (Harmony Formation) subarkosic debris and greenstone units within the Roberts Mountains allochthon.

The Devonian-Mississippian Bragdon Formation of the Klamath Mountains yielded an even higher proportion, however, of 415–440 Ma (Silurian) and 545–570 Ma (latest Precambrian) zircon grains probably derived from the Trinity ophiolitic complex (Wallin et al., 1991), which underlies the Bragdon Formation (Fig. 2, I). The most abundant zircon grains, of Grenville age (1.05–1.15 Ga), in correlative strata (Pass Creek) of the Pine Forest Range in northwestern Nevada (Fig. 2, II), are of uncertain derivation and are anomalous with respect to the other subfacies C2 samples. The dominant zircon grains in the Upper Triassic Lovelock assemblage (Fig. 2, IV), collected directly to the south (Fig. 1), also anomalous for the subfacies C1 samples, are also of Grenville age (0.95–1.15 Ga) and may point, together with the Pass Creek zircon assemblage, to recycling of Grenville-age zircons from quartzose Harmony Formation sandstones within the Roberts Mountains allochthon of central Nevada.

The sample from the Luning assemblage of the Upper Triassic basinal assemblage in western Nevada yielded few 1.8–2.1 Ga zircons (Fig. 4), and the dominant zircon grains are Late Triassic in age (218–229 Ma). Derivation from the Sierran arc terrane to the west seems most likely for the Triassic zircons, but we could detect no volcaniclastic detritus in the sample. Previous workers have shown, however, that the Luning sequence as a whole includes both chert-rich quartzolithic and feldspar-bearing volcaniclastic sandstones of mixed provenance, and have suggested that the chert-rich sands may have developed through loss of feldspar and volcanic rock fragments during reworking in shallow-marine environments (Reilly et al., 1980). If so, arc-derived Triassic zircons may have survived destruction while less resistant arc-derived grains were removed.

### *Quartzolithic-volcaniclastic subfacies*

The upper Paleozoic and Upper Triassic sandstones of quartzolithic-volcaniclastic subfacies C3 (Table 7), also widely distributed in both California and Nevada (Fig. 1), are all well sorted, with subrounded to subangular grains, but include fine-grained to medium-grained (Bishop Canyon), medium-grained to coarse-grained (Arlington, Schoonover V), and coarse-grained to very coarse-grained (Pumpernickel) aggregates. Feldspar grains are largely albite, formed dominantly if not entirely by intrastratal alteration of volcanic plagioclase. The presence of feldspar grains, together with varying subordinate proportions of volcanic-metavolcanic lithic fragments (Lvm), reflects admixture of volcaniclastic arc detritus with recycled orogenic detritus, otherwise similar to the frameworks of chert-rich subfacies C2 (Fig. 3). The rocks containing sands of subfacies C3 were collected from varied stratal assemblages (Fig. 2) including: (1) the Carboniferous-Permian Havallah sequence of the Golconda allochthon in Nevada (Pumpernickel, Schoonover V); (2) the mid-Permian epiclastic cover (Arlington) of the Lower Permian volcanogenic succession in the northern

Sierra Nevada (Harwood et al., 1995); and (3) Upper Triassic beds (Bishop Canyon) that unconformably overlie deformed upper Paleozoic strata (Pass Creek) in the Pine Forest Range of northwestern Nevada (Wyld, 1990).

The presence of Precambrian (1.75–1.95 Ga) zircon grains in significant proportions in all but one (Arlington) of the subfacies C3 samples implies that the accompanying cherty detritus was probably derived from the same recycled orogenic sources as the sands of chert-rich subfacies C2. Chert grains in the Arlington Formation of the northern Sierra Nevada may have been derived instead from Carboniferous cherts, which unconformably overlie nearby Devonian-Mississippian volcanic rocks (Fig. 2, III).

Despite the ranges in the depositional ages and geographic distributions of the strata representative of subfacies C3, the only abundant Phanerozoic detrital zircons in any of the samples are 320–370 Ma (largely 340–360 Ma or Mississippian), and grains within that age range form a significant proportion of the zircon populations in all four samples. Their occurrence suggests that the volcaniclastic detritus throughout the areal distribution of subfacies C3 was derived from arc sources of dominantly Mississippian age. This inference serves as both an important guideline and an important challenge for regional tectonic interpretations, because the possible locations of Mississippian volcanogenic sources, apart from the northern Sierra Nevada, are unclear from past work within the region (Fig. 2). From the ages of accompanying detrital zircons, the volcaniclastic detritus in the Arlington Formation of the northern Sierra Nevada was apparently derived from nearby but considerably older Mississippian volcanic rocks, rather than from immediately subjacent Permian volcanic rocks (Fig. 2, III).

## VOLCANICLASTIC PETROFACIES

The dominant grain types in quartz-poor, arc-derived sands of volcaniclastic petrofacies D (Table 8; Fig. 5, E and F) are volcanic lithic fragments, plagioclase grains (now dominantly albite), and minor proportions of clear unstrained volcanic quartz grains (commonly of bipyramidal habit), together with varied types and proportions of accessory heavy minerals (subordinate except in placer beach laminae of the Baird sample). The subrounded to subangular grains of the volcaniclastic frameworks are moderately well sorted, but include very fine-grained to medium-grained (Baird) and medium-grained to very coarse-grained (Pine Nut) aggregates. Comparatively high ratios of plagioclase feldspar grains to volcanic lithic fragments for volcaniclastic sands (Dickinson, 1985) suggest placer concentration of mineral grains relative to rock fragments during transport and sedimentation (Dickinson et al., 1979). In the Pine Nut sample, the content of quartz (Qm) and the presence of both Qp and Lsm grains, indicate a significant admixture of nonvolcanic detritus. The Baird sample is similar to typical rocks of that formation (Murray and Condie, 1973).

Detrital zircon ages (225–325 Ma) imply derivation of the volcaniclastic detritus from contemporaneous or subjacent upper Paleozoic and lower Mesozoic arc terranes of the Sierra Nevada and the Klamath Mountains (Fig. 2), but details of provenance relationships vary from case to case. For example, the Pennsylvanian Baird Formation (Watkins, 1985) of the Klamath Mountains contains coeval zircons (320–325 Ma) presumably derived from a nearby active arc. By contrast, the Pine Nut assemblage (Oldow, 1984) of the Upper Triassic basinal sequence in western Nevada contains zircons (228–243 Ma) apparently derived from a slightly older Middle Triassic segment of the associated magmatic arc, and Upper Triassic conglomerate at Big Valley in the northern Sierra Nevada (Harwood et al., 1995) contains zircons (255–270 Ma) evidently derived entirely from the Late Permian arc terrane that underlies the Triassic strata (Fig. 2, III). Zircon grains of other ages are absent in each case. Textural and compositional aspects of the samples do not reveal whether the volcaniclastic detritus within each represents neovolcanic (syndepositional) or paleovolcanic (predepositional) debris, with reference to the time of volcanism as opposed to the time of sedimentation (Zuffa, 1985; Critelli and Ingersoll, 1995), but the coherent zircon geochronology permits unambiguous distinctions to be drawn between the two alternatives in each case.

## VOLCANOPLUTONIC PETROFACIES

Volcanoplutonic petrofacies E is represented by a single detrital zircon sample from the Sierra City mélange of the Shoo Fly Complex in the northern Sierra Nevada of California (Figs. 1 and 2). The sample is too foliated, and contains too much pseudomatrix (unidentified deformed lithic fragments) and metamatrix (neomorphic phyllosilicates) for a reliable determination of its detrital mode. Unlike other samples from the Shoo Fly Complex,

**TABLE 8. DETRITAL MODES OF VOLCANICLASTIC PETROFACIES D**

| Grain type | Big Valley* | Baird† | PineNut§ |
|---|---|---|---|
| Qm | 0 | 1 | 16 |
| Qp | 1 | 0 | 3 |
| F | 62 | 64 | 43 |
| Lsm | 0 | 0 | 17 |
| Lvm | 37 | 35 | 21 |
| Matrix (WR%) | 14 | 0 | 7 |
| Cement (WR%)# | 0 | 14 | 0 |

*Note:* Column headings are local formation or assemblage names (Figure 2); QFL grain population summed to 100% (see Table 1 for grain type symbols).

*Approximate detrital mode for hornfelsed rock in biotite zone with newgrown interstitial and replacive biotite (as microscopic inclusions within framework grains).

†Contains prominent heavy-mineral laminae (opaque iron oxides, 14 fmwk%; partly chloritized pyriboles, 8 fmwk%) from beach placering of volcanic sand.

§Static recrystallization in biotite zone makes distinctions ambiguous between different kinds of lithic fragments.

#Clear green interstitial pore-filling chloritic cement (common in volcaniclastic sandstone).

however, the overall quartz (Qm) content is only moderate, feldspar grains are present, and discernible rock fragments include coarse-grained plutonic or hypabyssal grains (as well as Lvm and Lsm populations in uncertain proportions). Derivation from a magmatic arc dissected deep enough to expose plutonic-metamorphic roots and perhaps mélange flanks seems most likely (Dickinson, 1982). Previous petrographic work (Girty and Pardini, 1987) has shown that two generic classes of metasandstone blocks occur within the Sierra City mélange: (1) feldspar-rich varieties derived from a dissected magmatic arc, and (2) feldspar-poor varieties of recycled orogenic provenance. Our sample seemingly represents the first type, although both may be gradational variants of the same sand suite forming a continuous compositional spectrum.

The dominant detrital zircons (550–635 Ma, 1.1–1.3 Ga) imply sources in a latest Precambrian magmatic arc, perhaps built on basement rocks of Grenville age. No such arc could have existed along the passive continental margin represented by the Precambrian-Paleozoic miogeocline of Nevada and Utah, but the offshore distance of the parent arc, and its polarity, are indeterminate from the petrofacies data and the zircon geochronology.

## PETROFACIES-PROVENANCE SUMMARY

Regional relationships between sandstone petrofacies and provenance, as inferred from zircon geochronology and sedimentary petrology, define the following patterns.

1. All samples (n = 5) of the subarkose petrofacies and most samples (n = 9) of the quartzarenite petrofacies, both confined to strongly deformed lower Paleozoic assemblages (Roberts Mountains allochthon, Shoo Fly Complex, Yreka terrane), were probably derived from northern Laurentia or a closely related continental block distant from present exposures of the strata containing either petrofacies. Only two samples of the quartzarenite petrofacies, both from a restricted stratigraphic horizon in central Nevada and representing a particular paleotectonic context, were apparently derived from adjacent southwestern Laurentia. A single sample of the quartzarenite petrofacies, from a seemingly unique stratigraphic context, was probably derived from farther south, along the Cordilleran margin within Mexico.

2. Most samples (n = 12) of the quartzolithic petrofacies, collected from varied upper Paleozoic and Upper Triassic overlap sequences and successor basins, were apparently recycled in whole or in part from deformed lower Paleozoic sedimentary assemblages exposed within the study region. Of those 12 samples, 3 also contain admixtures of Mississsippian volcaniclastic detritus, which evidently contributed most of the zircons present in an additional sample of the quartzolithic petrofacies, and another of those 12 samples apparently acquired most of its zircons from analogous admixtures of Triassic arc detritus. One of the 12 samples also contains a significant component of Grenville-age zircons of uncertain origin, and zircons of identical anomalous age are dominant in another additional sample of the quartzolithic petrofacies.

3. All samples (n = 3) of the volcaniclastic petrofacies (Carboniferous and Triassic) are dominated by zircons of severely restricted, though slightly different, ages that were undoubtedly derived from sources in magmatic arcs located along the Klamath-Sierran belt of the Cordilleran continental margin.

4. The single sample of volcanoplutonic facies from a mélange block in the lower Paleozoic Shoo Fly Complex of the northern Sierra Nevada was apparently derived from a Neoproterozoic magmatic arc not directly related to the Cordilleran continental margin.

## DISCUSSION

Detrital modes from the samples collected for detrital zircon analysis establish patterns of regional and stratigraphic petrofacies that are generally consistent with detrital zircon ages. The correlation of the two data sets, based on independent methodology, increases confidence in the value of each for provenance constraints. Accordingly, joint analyses of the sedimentary petrology and the zircon geochronology should provide an improved understanding of the Paleozoic and earliest Mesozoic paleogeography and tectonics of the study region. Coupling petrofacies and zircon indices of provenance offers the potential for revised paleotectonic hypotheses to explain the depositional systems of suspect terranes that have been severely disrupted by multiple episodes of structural deformation. For the most part, lithofacies patterns are obscure and paleocurrent indicators are poorly preserved, with original orientations that are ambiguous because of superposed structural features too complex geometrically to restore with confidence. With petrofacies as a guide to the generic nature of provenance terranes and zircon geochronology as a guide to the ages of respective source rocks, paleogeographic and paleotectonic interpretations can be sharpened beyond the equivocal inferences possible in the past.

Our results also confirm the intrinsic value of detrital zircon geochronology for provenance studies. Selected sandstones of several petrofacies contain anomalous zircon populations that differ from those in more prevalent sandstones of the same petrofacies, and other sandstones with closely comparable zircon populations belong to different petrofacies. Petrofacies type and zircon content are counterpoints that amplify the information provided by each data set, and jointly provide more scope for interpretionthan either data set considered alone.

## ACKNOWLEDGMENTS

We appreciate guidance to sampling localities and discussions of the sedimentary petrology of the samples with S.C. Finney, G.H. Girty, D.S. Harwood, M.M. Miller, J.I. Satterfield, W.S. Snyder, C.A. Suczek, and S.J. Wyld, and the yeoman work of undergraduate students at the University of Arizona in collecting and processing the samples. Jim Abbott of SciGraphics prepared the figures. Reviews by R.V. Ingersoll, S.Y. Johnson, and F.L. Schwab markedly improved our data presentation.

## REFERENCES CITED


Basu, A., Young, S.W., Suttner, L.J., James, W.C., and Mack, G.H., 1975, Re-evaluation of the use of undulatory extinction and polycrystallinity in detrital quartz for provenance interpretation: Journal of Sedimentary Petrology, v. 45, p. 873–882.

Bond, G.C., and Devay, J.C., 1980, Pre-Upper Devonian quartzose sandstones in the Shoo Fly Formation, northern California—Petrology, provenance, and implications for regional tectonics: Journal of Geology, v. 88, p. 285–308.

Brueckner, H.K., and Snyder, W.S., 1985, Structure of the Havallah sequence, Golconda allochthon, Nevada: Evidence for prolonged evolution in an accretionary prism: Geological Society of America Bulletin, v. 96, p. 1113–1130.

Burke, D.B., and Silberling, N.J., 1973, The Auld Lang Syne Group of Late Triassic and Jurassic(?) age, north-central Nevada: U.S. Geological Survey Bulletin 1394-E, p. E1–E14.

Cox, R., and Lowe, D.R., 1996, Quantification of the effects of secondary matrix on the analysis of sandstone composition and a petrographic-chemical technique for retrieving original framework grain modes of altered sandstones: Journal of Sedimentary Research, v. 66, p. 548–558.

Critelli, S., and Ingersoll, R.V., 1995, Interpretation of neovolcanic versus palaeovolcanic sand grains: An example from Miocene deep-marine sandstone of the Topanga Group (southern California): Sedimentology, v. 42, p. 783–804.

D'Allura, J.A., Moores, E.M., and Robinson, L., 1977, Paleozoic rocks of the northern Sierra Nevada: Their structural and paleogeographic implications, *in* Stewart, J.H., et al., eds., Paleozoic paleogeography of the western United States: Pacific Section, Society of Economic Paleontologists and Mineralogists Pacific Paleogeography Symposium 1, p. 395–408.

Dickinson, W. R., 1970, Interpreting detrital modes of graywacke and arkose: Journal of Sedimentary Petrology, v. 40, p. 695–707.

Dickinson, W.R., 1982, Compositions of sandstones in circum-Pacific subduction complexes and fore-arc basins: American Association of Petroleum Geologists Bulletin, v. 66, p. 121–137.

Dickinson, W.R., 1985, Interpreting provenance relations from detrital modes of sandstones, *in* Zuffa, G.G., ed., Provenance of arenites: Dordrecht, Reidel, p. 333–361.

Dickinson, W.R., 1988, Provenance and sediment dispersal in relation to paleotectonics and paleogeography of sedimentary basins, *in* Kleinspehn, K.L., and Paola, C., eds., New perspectives in basin analysis: New York, Springer-Verlag, p. 3–25.

Dickinson, W.R., and Ingersoll, R.V., 1990, Physiographic controls on the composition of sediments derived from volcanic and sedimentary terrains on Barro Colorado Island, Panama: Discussion: Journal of Sedimentary Petrology, v. 60, p. 797–798.

Dickinson, W.R., and Rich, E.I., 1972, Petrologic intervals and petrofacies in the Great Valley sequence, Sacramento Valley, California: Geological Society of America Bulletin, v. 83, p. 3007–3024.

Dickinson, W.R., and Suczek, C.A., 1979, Plate tectonics and sandstone compositions: American Association of Petroleum Geologists Bulletin, v. 63, p. 2164–2182.

Dickinson, W.R., Helmold, K.P., and Stein, J.A., 1979, Mesozoic lithic sandstones in central Oregon: Journal of Sedimentary Petrology, v. 49, p. 501–516.

Dickinson, W.R., Ingersoll, R.V., Cowan, D.S., Helmold, K.P., and Suczek, C.A., 1982, Provenance of Franciscan graywackes in coastal California: Geological Society of America Bulletin, v. 93, p. 95–107.

Dickinson, W.R., Beard, L.S., Brakenridge, G.R., Erjavec, J.L., Ferguson, R.C., Inman, K.F., Knepp, R.A., Lindberg, F.A., and Ryberg, P.T., 1983a, Provenance of North American Phanerozoic sandstones in relation to tectonic setting: Geological Society of America Bulletin, v. 94, p. 222–235.

Dickinson, W.R., Harbaugh, D.W., Saller, A.H., Heller, P.L., and Snyder, W.S., 1983b, Detrital modes of upper Paleozoic sandstones derived from Antler orogen in Nevada: Implications for nature of Antler orogeny: American Journal of Science, v. 283, p. 481–509.

Elison, M.W., and Speed, R.C., 1988, Triassic flysch of the Fencemaker allochthon, East Range, Nevada: Fan facies and provenance: Geological Society of America Bulletin, v. 100, p. 185–199.

Finney, S.C., and Perry, B.D., 1991, Depositional setting and paleogeography of Ordovician Vinini Formation, central Nevada, *in* Cooper, J.D., and Stevens, C.H., eds., Paleozoic paleogeography of the western United States—II: Pacific Section, SEPM (Society for Sedimentary Geology) book 67, p. 747–766.

Gehrels, G.E., and Dickinson, W.R., 1995, Detrital zircon provenance of Cambrian to Triassic miogeoclinal and eugeoclinal strata in Nevada: American Journal of Science, v. 295, p. 18–48.

Gehrels, G.E., Dickinson, W.R., Ross, G.M., Stewart, J.H., and Howell, D.G., 1995, Detrital zircon reference for Cambrian to Triassic miogeoclinal strata of western North America: Geology, v. 23, p. 831–834.

Gilluly, J., 1967, Geologic map of the Winnemucca quadrangle, Pershing and Humboldt Counties, Nevada: U.S. Geological Survey Geologic Quadrangle Map GQ-656, scale 1:62 500.

Girty, G.H., and Pardini, C.H., 1987, Provenance of sandstone inclusions in the Paleozoic Sierra City mélange, Sierra Nevada, California: Geological Society of America Bulletin, v. 98, p. 176–181.

Girty, G.H., and Wardlaw, M.S., 1985, Petrology and provenance of pre-Late Devonian sandstones, Shoo Fly complex, northern Sierra Nevada, California: Geological Society of America Bulletin, v. 96, p. 516–521.

Girty, G.H., Reiland, D.N., and Wardlaw, M.S., 1985, Provenance of the Silurian Elder Sandstone, north-central Nevada: Geological Society of America Bulletin, v. 96, p. 925–930.

Girty, G.H., Gester, K.C., and Turner, J.B., 1990, Pre-Late Devonian geochemical, stratigraphic, sedimentologic, and structural patterns, Shoo Fly Complex, northern Sierra Nevada, California, *in* Harwood, D.S., and Miller, M.M., eds., Paleozoic and early Mesozoic paleogeographic relations; Sierra Nevada, Klamath Mountains, and related terranes: Geological Society of America Special Paper 255, p. 43–56.

Girty, G.H., Lawrence, J., Burke, T., Fortin, A., Gallarano, C.S., Wirths, T.A., Lewis, J.G., Peterson, M.M., Ridge, D.L., Knack, C., and Johnson, D., 1996, The Shoo Fly complex: Its origin and tectonic significance, *in* Girty, G.H., et al., eds., The northern Sierra terrane and associated Mezozoic magmatic units: Implications for the tectonic history of the western Cordillera: Pacific Section, SEPM (Society for Sedimentary Geology) book 81, p. 1–23.

Graham, S.A., Ingersoll, R.V., and Dickinson, W.R., 1976, Common provenance for lithic grains in Carboniferous sandstones from Ouachita Mountains and Black Warrior basin: Journal of Sedimentary Petrology, v. 46, p. 620–632.

Harwood, D.S., 1983, Stratigraphy of upper Paleozoic volcanic rocks and regional unconformities in part of the northern Sierra terrane, California: Geological Society of America Bulletin, v. 94, p. 413–422.

Harwood, D.S., 1988, Tectonism and metamorphism in the northern Sierra terrane, northern California, *in* Ernst, W.G., ed., Metamorphism and crustal evolution of the western United States: Englewood Cliffs, New Jersey, Prentice-Hall, p. 764–788.

Harwood, D.S., 1992, Stratigraphy of Paleozoic and lower Mesozoic rocks in the northern Sierra terrane, California: U.S. Geological Survey Bulletin 1957, 78 p.

Harwood, D.S., Yount, J.C., and Seiders, V.M., 1991, Upper Devonian and Lower Mississippian island-arc and back-arc basin deposits in the northern Sierra Nevada, California, in Cooper, J.D., and Stevens, C.H., eds., Paleozoic paleogeography of the western United States—II: Pacific Section, SEPM (Society for Sedimentary Geology) book 67, p. 717–733.

Harwood, D.S., Fisher, G.R., Jr., and Waugh, B.J., 1995, Geologic map of the Duncan Peak and southern part of the Cisco Grove quadrangles, Placer and Nevada Counties, California: U.S. Geological Survey Miscellaneous Investigations Series Map I-2341, scale 1:24 000.

Helmold, K.P., 1985, Provenance of feldspathic sandstones—The effect of diagenesis on provenance interpretations: A review, *in* Zuffa, G.G., ed., Provenance of arenites: Dordrecht, Reidel, p. 139–163.

Ingersoll, R.V., 1978, Petrofacies and petrologic evolution of the Late Cretaceous fore-arc basin, northern and central California: Journal of Geology, v. 86, p. 335–352.

Ingersoll, R.V., 1983, Petrofacies and provenance of late Mesozoic forearc basin, northern and central California: American Association of Petroleum Geologists Bulletin, v. 67, p. 1125–1142.
Ingersoll, R.V., 1990, Actualistic sandstone petrofacies: Discriminating modern and ancient source rocks: Geology, v. 18, p. 733–736.
Ingersoll, R.V., and Suczek, C.A., 1979, Petrology of Neogene sand from Nicobar and Bengal Fans, DSDP sites 211 and 218: Journal of Sedimentary Petrology, v. 49, p. 1217–1228.
Ingersoll, R.V., Bullard, T.F., Ford, R.L., Grimm, J.P., Pickle, J.D., and Sares, S.W., 1984, The effect of grain size on detrital modes: A test of the Gazzi-Dickinson point-counting method: Journal of Sedimentary Petrology, v. 54, p. 103–116.
Ingersoll, R.V., Kretchmer, A.G., and Valles, P.K., 1993, The effect of sampling scale on actualistic sandstone petrofacies: Sedimentology, v. 40, p. 937–953.
Johnson, J.G., and Pendergast, A., 1981, Timing and mode of emplacement of the Roberts Mountains allochthon, Antler orogeny: Geological Society of America Bulletin, v. 92, p. 648–658.
Jones, A.E., 1991, Sedimentary rocks of the Golconda terrane: Provenance and paleogeographic implications, *in* Cooper, J.D., and Stevens, C.H., eds., Paleozoic paleogeography of the western United States—II: Pacific Section, SEPM (Society for Sedimentary Geology) book 67, p. 783–800.
Mack, G.H., 1984, Exceptions to the relationship between plate tectonics and sandstone composition: Journal of Sedimentary Petrology, v. 54, p. 212–220.
Madrid, R.A., 1987, Stratigraphy of the Roberts Mountains allochthon in north-central Nevada [Ph.D. thesis]: Stanford, California, Stanford University, 341 p.
McBride, E.F., 1985, Diagenetic processes that affect provenance determinations in sandstone, *in* Zuffa, G.G., ed., Provenance of arenites: Dordrecht, Reidel, p. 95–113.
McBride, E.F., 1987, Diagenesis of the Maxon Sandstone (Early Cretaceous), Marathon region, Texas: A diagenetic quartzarenite: Journal of Sedimentary Petrology, v. 57, p. 98–107.
McLean, H., 1995, Reconnaissance study of Mississippian siliclastic sandstones in eastern Nevada: U.S. Geological Survey Bulletin 1988-I, p. I11–I19.
Miller, E.L., and Larue, D.K., 1983, Ordovician quartzite in the Roberts Mountains allochthon, Nevada: Deep sea fan deposits derived from cratonal North America, *in* Stevens, C.H., ed., Pre-Jurassic rocks in western North American suspect terranes: Los Angeles, California, Pacific Section, Society of Economic Paleontologists and Mineralogists, p. 91–102.
Miller, E.L., Kanter, L.R., Larue, D.K., Turner, R.J., Murchey, B., and Jones, D.L., 1982, Structural fabric of the Paleozoic Golconda allochthon, Antler Peak quadrangle, Nevada: Progressive deformation of an oceanic sedimentary assemblage: Journal of Geophysical Research, v. 87, p. 3795–3804.
Miller, E.L., Holdsworth, B.K., Whiteford, W.R., and Rodgers, D., 1984, Stratigraphy and structure of the Schoonover sequence, northeastern Nevada: Implications for Paleozoic plate-margin tectonics: Geological Society of America Bulletin, v. 95, p. 1063–1076.
Miller, M.M., 1989, Intra-arc sedimentation and tectonism: Late Paleozoic evolution of the eastern Klamath terrane, California: Geological Society of America Bulletin, v. 101, p. 170–187.
Miller, M.M., and Cui, B., 1989, Submarine-fan characteristics and dual sediment provenance, Lower Carboniferous Bragdon Formation, eastern Klamath terrane, California: Canadian Journal of Earth Sciences, v. 26, p. 927–940.
Murchey, B.L., 1990, Age and depositional setting of siliceous sediments in the upper Paleozoic Havallah sequence near Battle Mountain, Nevada: Implications for the paleogeography and structural evolution of the western margin of North America, *in* Harwood, D.S., and Miller, M.M., eds., Paleozoic and early Mesozoic paleogeographic relations; Sierra Nevada, Klamath Mountains, and related terranes: Geological Society of America Special Paper 255, p. 137–155.
Murray, M., and Condie, K.C., 1973, Post-Ordovician to early Mesozoic history of the Eastern Klamath subprovince, northern California: Journal of Sedimentary Petrology, v. 43, p. 505–515.
Odom, I.E., Doe, T.W., and Dott, R.H., Jr., 1976, Nature of feldspar-grain size relations in some quartz-rich sandstones: Journal of Sedimentary Petrology, v. 46, p. 862–870.
Oldow, J.S., 1981, Structure and stratigraphy of the Luning allochthon and the kinematics of thrust emplacement, Pilot Mountains, west-central Nevada: Geological Society of America Bulletin, v. 92, p. 1647–1669.
Oldow, J.S., 1984, Evolution of a late Mesozoic back-arc fold and thrust belt, northwestern Great Basin, U.S.A.: Tectonophysics, v. 102, p. 245–274.
Potter, A.W., Hotz, P.E., and Rohr, D.M., 1977, Stratigraphy and inferred tectonic framework of lower Paleozoic rocks in the eastern Klamath Mountains, northern California, *in* Stewart, J.H., et al., eds., Paleozoic paleogeography of the western United States: Pacific Section, Society of Economic Paleontologists and Mineralogists Pacific Coast Paleogeography Symposium 1, p. 421–440.
Potter, A.W., Boucot, A.J., Bergstrom, S.M., Blodgett, R.B., Dean, W.T., Flory, D.A., Ormiston, A.R., Pedder, A.E.H., Rigby, J.K., Rohr, D.M., and Savage, N.M., 1990, Early Paleozoic stratigraphic, paleogeographic, and biogeographic relations of the eastern Klamath belt, northern California, *in* Harwood, D.S., and Miller, M.M., eds., Paleozoic and early Mesozoic paleogeographic relations; Sierra Nevada, Klamath Mountains, and related terranes: Geological Society of America Special Paper 255, p. 57–74.
Potter, P.E., 1978, Significance and origin of big rivers: Journal of Geology, v. 86, p. 13–33.
Potter, P.E., 1986, South America and a few grains of sand: Part 1—Beach sands: Journal of Geology, v. 94, p. 301–319.
Potter, P.E., 1994, Modern sands of South America: Composition, provenance, and global significance: Geologisches Rundschau, v. 83, p. 212–232.
Reilly, M.B., Breyer, J.A., and Oldow, J.S., 1980, Petrographic provinces and provenance of the Upper Triassic Luning Formation, west-central Nevada: Geological Society of America Bulletin, v. 91, p. 573–575, p. 2112–2151.
Rowell, A.J., Rees, M.N., and Suczek, C.A., 1979, Margin of the North American continent in Nevada during Late Cambrian time: American Journal of Science, v. 279, p. 1–18.
Saller, A.H., and Dickinson, W.R., 1982, Alluvial to marine facies transition in the Antler overlap sequence, Pennsylvanian and Permian of north-central Nevada: Journal of Sedimentary Petrology, v. 52, p. 925–940.
Schweickert, R.A., and Lahren, M.M., 1987, Continuation of Antler and Sonoma orogenic belts to the eastern Sierra Nevada, California, and Late Triassic thrusting in a compressional arc: Geology, v. 15, p. 270–273.
Schweickert, R.A., and Snyder, W.S., 1981, Paleozoic plate tectonics of the Sierra Nevada and adjacent regions, *in* Ernst, W.G., ed., The geotectonic development of California: Englewood Cliffs, New Jersey, Prentice-Hall, p. 182–201.
Shanmugam, G., 1985, Significance of secondary porosity in interpreting sandstone composition: American Association of Petroleum Geologists Bulletin, v. 69, p. 378–384.
Smith, M.T., and Gehrels, G.E., 1994, Detrital zircon geochronology and the provenance of the Harmony and Valmy Formations, Roberts Mountains allochthon, Nevada: Geological Society of America Bulletin, v. 106, p. 968–979.
Speed, R.C., 1977, Basinal terrane of the early Mesozoic marine province of the western Great Basin, *in* Howell, D.G., and McDougall, K.A., eds., Mesozoic paleogeography of the western United States: Pacific Section, Society of Economic Paleontologists and Mineralogists Pacific Coast Paleogeography Symposium 2, p. 237–252.
Speed, R.C., 1979, Collided Paleozoic microplate in the western United States: Journal of Geology, v. 87, p. 279–292.
Speed, R.C., and Sleep, N.H., 1982, Antler orogeny and foreland basin: A model: Geological Society of America Bulletin, v. 93, p. 815–828.
Stanley, K.O., Chamberlain, C.K., and Stewart, J.H., 1977, Depositional setting of some eugeosynclinal Ordovician rocks and structurally interleaved Devonian rocks in the Cordilleran mobile belt, Nevada, *in* Stewart, J.H., et al., eds., Paleozoic paleogeography of the western United States: Pacific Section, Society of Economic Paleontologists and Mineralogists Pacific Coast Paleogeography Symposium 1, p. 259–274.
Stewart, J.H., 1980, Geology of Nevada: Nevada Bureau of Mines and Geology Special Publication 4, 136 p.

Stewart, J.H., and Suczek, C.A., 1977, Cambrian and latest Precambrian paleogeography and tectonics in the western United States, *in* Stewart, J.H., et al., Paleozoic paleogeography of the western United States: Pacific Section, Society of Economic Paleontologists and Mineralogists Pacific Coast Paleogeography Symposium 1, p. 1–17.

Stewart, J.H., Murchey, B.L., Jones, D.L., and Wardlaw, B.R., 1986, Paleontologic evidence for complex tectonic interlayering of Mississippian to Permian deep-water rocks of the Golconda allochthon in Tobin Range, north-central Nevada: Geological Society of America Bulletin, v. 97, p. 1122–1132.

Suczek, C. A., 1977, Tectonic relations of the Harmony Formation, northern Nevada [Ph.D. thesis]: Stanford, California, Stanford University, 96 p.

Suczek, C.A., and Ingersoll, R.V., 1985, Petrology and provenance of Cenozoic sand from the Indus Cone and the Arabian Basin, DSDP sites 221, 222, and 224: Journal of Sedimentary Petrology, v. 55, p. 340–346.

Van der Plas, L., and Tobi, A.C., 1965, A chart for determining the reliability of point counting results: American Journal of Science, v. 263, p. 87–90.

Varga, R.J., and Moores, E.M., 1981, Age, origin, and significance of an unconformity that predates island-arc volcanism in the northern Sierra Nevada: Geology, v. 9, p. 512–518.

Walker, T.R., 1984, Diagenetic albitization of potassium feldspar in arkosic sandstones: Journal of Sedimentary Petrology, v. 54, p. 3–16.

Wallin, E.T., Lindsley-Griffin, N., and Griffin, J.R., 1991, Overview of early Paleozoic magmatism in the eastern Klamath Mountains, California: An isotopic perspective, *in* Cooper, J.D., and Stevens, C.H., eds., Paleozoic paleogeography of the western United States—II: Pacific Section, SEPM (Society for Sedimentary Geology) book 67, p. 581–588.

Watkins, R., 1985, Volcaniclastic and carbonate sedimentation in late Paleozoic island-arc deposits, eastern Klamath Mountains, California: Geology, v. 13, p. 709–713.

Watkins, R., 1986, Late Devonian to Early Carboniferous turbidite facies and basinal development of the eastern Klamath Mountains, California: Sedimentary Geology, v. 49, p. 51–71.

Whitebread, D.H., 1994, Geologic map of the Dun Glen quadrangle, Pershing County, Nevada: U.S. Geological Survey Miscellaneous Investigations Series Map I-2409, scale 1:48 000.

Whiteford, W.B., 1990, Paleogeographic setting of the Schoonover sequence, Nevada, and implications for the late Paleozoic margin of western North America, *in* Harwood, D.S., and Miller, M.M., eds., Paleozoic and early Mesozoic paleogeographic relations; Sierra Nevada, KlamathMountains, and related terranes: Geological Society of America Special Paper 255, p. 115–136.

Wolf, K.H., 1971, Textural and compositional transitional stages between various lithic grain types: Journal of Sedimentary Petrology, v. 41, p. 328–332.

Wyld, S.J., 1990, Paleozoic and Mesozoic rocks of the Pine Forest Range, northwest Nevada, and their relation to volcanic arc assemblages of the western U.S. Cordillera, *in* Harwood, D.S., and Miller, M.M., eds., Paleozoic and early Mesozoic paleogeographic relations; Sierra Nevada, Klamath Mountains, and related terranes: Geological Society of America Special Paper 255, p. 219–237.

Zuffa, G.G., 1985, Optical analysis of arenites: Influence of methodology on compositional results, *in* Zuffa, G.G., ed., Provenance of arenites: Dordrecht, Reidel, p. 165–189.

Manuscript Accepted by the Society January 24, 2000

Printed in U.S.A.

Geological Society of America
Special Paper 347
2000

# *Bootstrap technique and the location of the source of siliciclastic detritus in the lower Paleozoic Shoo Fly Complex, northern Sierra terrane, California*

**Gary H. Girty and Jennifer L. Lawrence**
*Department of Geological Sciences, San Diego State University, San Diego, California 92182*

## ABSTRACT

**Published data unambiguously indicate that the source of clastic materials in the lower southwestern miogeocline is located in the adjacent interior of North America. In contrast, the source of siliciclastic materials in the Shoo Fly Complex has been difficult to locate due to its location outboard of clear-cut aboriginal North American materials. Because the Shoo Fly Complex is west of the southern segment of the North American craton that is dominated by Proterozoic components, we explore in this chapter whether the source of detritus in the Shoo Fly Complex was restricted to this southern segment. Our approach in addressing this question was to collect and chemically analyze a number of mudstone samples from the southwest Cordilleran miogeocline and from the Shoo Fly Complex. As a measure of central tendency we utilized the Aitchison Measure of Location (AML). If mudstones from both areas were derived from the same source, then they should have similar compositions. Hence, under this condition the ratios of $\bar{c}_{1mio}/\bar{c}_{1SFC}, \bar{c}_{2mio}/\bar{c}_{2SFC}, \ldots, \bar{c}_{nmio}/\bar{c}_{nSFC}$ should equal 1, where *mio* refers to a constituent of the miogeoclinal sampling group, subscript SFC refers to a constituent of the Shoo Fly Complex group, and $\bar{c}_1, \bar{c}_2, \ldots, \bar{c}_m$ are AML-based average concentrations of chemical constituents 1 through m. In order to address the inherent statistical uncertainties in these ratios we developed a Visual Basic 6 bootstrap program. The results of 5000 bootstrap replications indicate that 60% of the constituent ratios (i.e., $\bar{c}_{1mio}/\bar{c}_{1SFC}, \bar{c}_{2mio}/\bar{c}_{2SFC}, \ldots, \bar{c}_{nmio}/\bar{c}_{nSFC}$) are statistically different from 1 at the 95% confidence level. We therefore conclude that samples analyzed from the southwest Cordilleran miogeocline and from the Shoo Fly Complex did not share the same source. Where was the source of detritus in the Shoo Fly Complex?**

**Published single-crystal U-Pb detrital zircon data show that the detrital zircon population in a number of samples analyzed from the Shoo Fly Complex contains an Archean component. In addition, the cumulative probability diagram of single-crystal U-Pb detrital zircon ages from the Shoo Fly Complex is similar to that produced by detrital zircon ages derived from miogeoclinal strata, the source of which was located in northwestern North America. Hence, these and other data discussed in this chapter are consistent with the idea that the source of the voluminous siliciclastic detritus in the Shoo Fly Complex was probably located in northwestern rather than southwestern North America.**

Girty, G.H. and Lawrence, J.L., 2000, Bootstrap technique and the location of the source of siliciclastic detritus in the lower Paleozoic Shoo Fly Complex, northern Sierra terrane, California, *in* Soreghan, M.J., and Gehrels, G.E., eds., Paleozoic and Triassic paleogeography and tectonics of western Nevada and northern California: Boulder, Colorado, Geological Society of America Special Paper 347, p. 173–184.

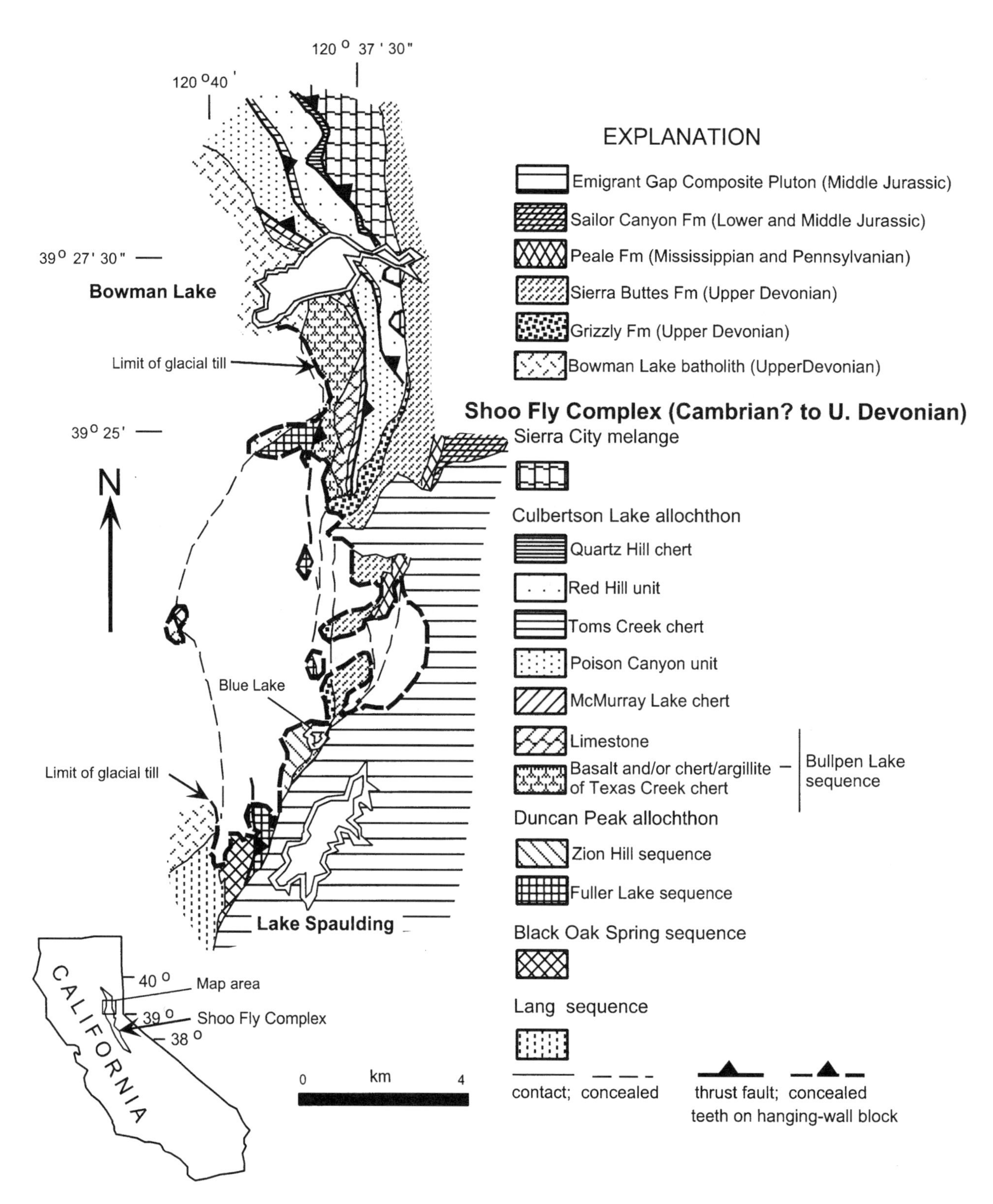


Figure 1. Geologic map depicting major units mapped within the Shoo Fly Complex.

## INTRODUCTION

The Shoo Fly Complex, the most extensive lower Paleozoic terrane containing eugeoclinal or oceanic facies rocks exposed in California, extends from beyond lat 40°N to as far south as 38°N (Fig. 1). It exceeds 30 km in width at its widest extent.

Over the past 20 years detailed geologic mapping indicates that the complex consists of a generally westward verging fold and thrust belt composed of 2 major stratigraphic units and 3 prominent structures (Hannah and Moores, 1986; Girty and Schweickert, 1984; Girty et al., 1990, 1991, 1996; Harwood, 1988, 1992). The stratigraphic units include, from lowest to highest, the Lang and Black Oak spring sequences. Stacked structurally above these two units are, in ascending order, the Duncan Peak allochthon, Culbertson Lake allochthon, and Sierra City melange (Fig. 1). Though the entire complex is generally metamorphosed to chlorite-grade greenschist facies, it is locally upgraded to the hornblende hornfels facies in several well-defined contact-metamorphic aureoles surrounding Middle Jurassic plutons (Girty et al., 1995).

The most thoroughly studied unit in the Shoo Fly Complex is the Culbertson Lake allochthon (e.g., Girty et al., 1996, and references therein). This structure crops out in the glaciated ridges north and south of Bowman Lake, and, in this area, has not been affected by Middle Jurassic contact metamorphism (Fig. 1). Within the Culbertson Lake allochthon primary sedimentological features are moderately to well preserved.

In ascending structural or stratigraphic order, units mapped within the Culbertson Lake allochthon include the Bullpen Lake sequence, McMurray Lake chert, Poison Canyon unit, Toms Creek chert, Red Hill unit, and Quartz Hill chert (Fig. 1). The Poison Canyon and Red Hill units are composed primarily of sandstones derived from subaqueous sediment-gravity flows interstratified with beds and laminations composed of mudstone. Girty and Wardlaw (1985) point-counted 27 sandstone samples from the Poison Canyon unit and 31 from the Red Hill unit. The results of their point counts are plotted on the QFL (Quartz, Feldspar, Aphanitic rock fragments) provenance-discrimination diagram of Dickinson et al. (1983) in Figure 2. Note that samples from both units plot either in the transitional or in the continental interior provenance subfields. Given the quartz-rich character of sandstones from the Lang sequence, Black Oak Spring sequence, and Duncan Peak allochthon, such results appear to be representative of much of the Shoo Fly Complex.

According to McLennan et al. (1993), transitional to continental interior provenances are typically dominated by differentiated old upper continental crust. It is therefore not surprising that most tectonic models that purport to explain the early Paleozoic evolution of the southwest Cordillera portray the source of siliciclastic detritus in the Shoo Fly Complex as coming from somewhere within the interior of the western North American continent (e.g., Burchfiel and Davis, 1975; Schweickert and Snyder, 1981; Burchfiel and Royden, 1991; Burchfiel et al., 1992). Such a conclusion, however, begs the question, Where exactly within the vast interior of western North American was the source of terrigenous debris now found in the Shoo Fly Complex?

## CLUES FROM DETRITAL ZIRCON U-PB GEOCHRONOLOGY

Gehrels and Dickinson (1995) and Gehrels et al. (1996) showed through single-crystal detrital zircon U-Pb geochronology that the sources of sediment dispersed from within the western North American interior during the latest Precambrian and early Paleozoic can be subdivided into southern and northern segments. For example, miogeoclinal sediments in Sonora, Mexico, and Nevada were derived from southern continental interior sources, and are dominated by zircon grains derived from Proterozoic crust ranging from 1.7 to 1.1 Ga (Fig. 3) (Gehrels and Dickinson, 1995; Gehrels et al., 1996). In contrast, miogeoclinal sediments in northern British Columbia and Alaska were derived from northern continental interior source areas, and are dominated by zircon grains derived from Archean and Proterozoic crust ranging from 2.6–2.8 to 2.4–1.8 Ga (Fig. 3). Thus, current data suggest that late Precambrian to early Paleozoic sediment

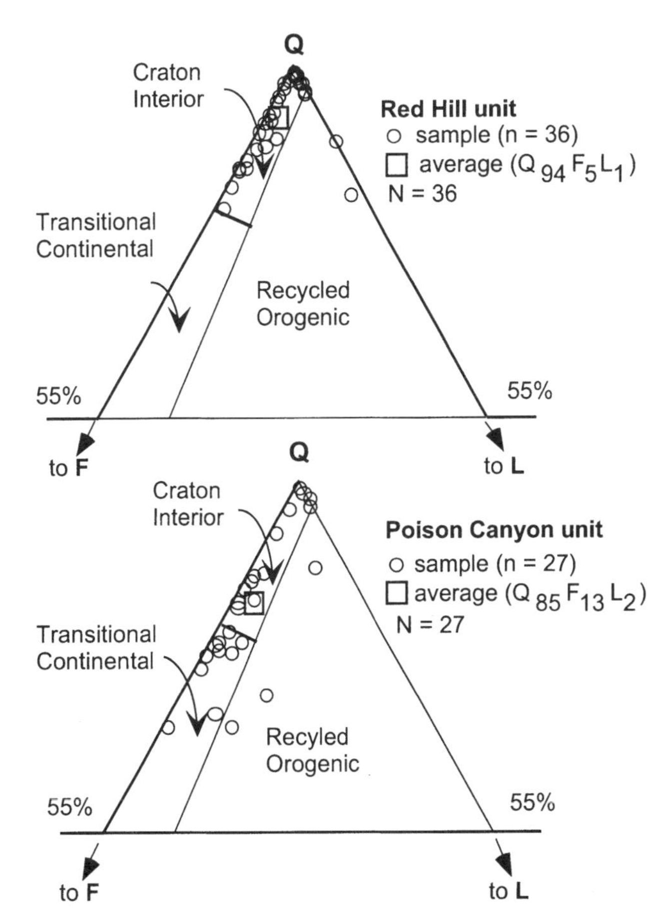


Figure 2. QFL (Quartz, Feldspar, Aphanitic rock fragments) point-count data plotted on provenance-discrimination diagram of Dickinson et al. (1983). Data are from Girty and Wardlaw (1985).

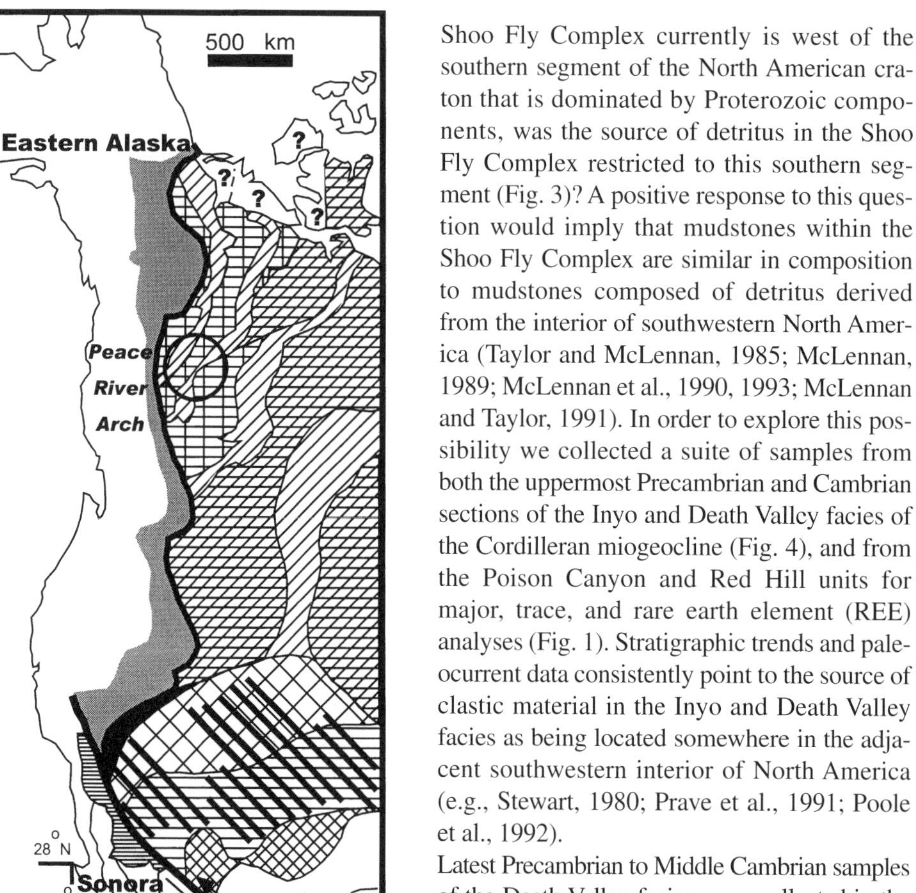


Figure 3. Map depicting locations of major Precambrian basement provinces relative to Cordilleran miogeocline. Basement provinces are from Hoffman (1989) for cratonal interior, various sections of Van Schmus et al. (1993) for southwestern United States, and Stewart et al. (1990) for northwestern Mexico. Map was modified from Figure 10 in Harding et al. (this volume).

derived from the southern interior of North America would be composed primarily of Proterozoic rock debris, while sediment derived from the northern interior of North America would include some Archean component. This conclusion is important because several studies have shown that sediments derived from Proterozoic crust display a distinctive uniform chemistry (Taylor and McLennan, 1985; McLennan, 1989; McLennan et al., 1990, 1993; McLennan and Taylor, 1991). It is this idea that led to the study by Cardenas et al. (1996), which showed that clastic sediments within a kilometer-scale block engulfed by the Sierra Nevada batholith have compositions that are like those determined for terrigenous materials in the lower part of the southwest Cordilleran miogeocline. Hence, Cardenas et al. (1996) argued that such a close similarity in composition supported the interpretation of Schweickert and Lahren (1990), that the kilometer-scaled block was structurally removed from its original position as part of the Cordilleran miogeocline.

## PROBLEM REPHRASED

Data and observations provided in the preceding section allow us to rephrase the question about the source of detritus in the Shoo Fly Complex into a more tractable form. Because the Shoo Fly Complex currently is west of the southern segment of the North American craton that is dominated by Proterozoic components, was the source of detritus in the Shoo Fly Complex restricted to this southern segment (Fig. 3)? A positive response to this question would imply that mudstones within the Shoo Fly Complex are similar in composition to mudstones composed of detritus derived from the interior of southwestern North America (Taylor and McLennan, 1985; McLennan, 1989; McLennan et al., 1990, 1993; McLennan and Taylor, 1991). In order to explore this possibility we collected a suite of samples from both the uppermost Precambrian and Cambrian sections of the Inyo and Death Valley facies of the Cordilleran miogeocline (Fig. 4), and from the Poison Canyon and Red Hill units for major, trace, and rare earth element (REE) analyses (Fig. 1). Stratigraphic trends and paleocurrent data consistently point to the source of clastic material in the Inyo and Death Valley facies as being located somewhere in the adjacent southwestern interior of North America (e.g., Stewart, 1980; Prave et al., 1991; Poole et al., 1992).

Latest Precambrian to Middle Cambrian samples of the Death Valley facies were collected in the Nopah Range, southeastern California, and were discussed by Cardenas et al. (1996). Samples of the Inyo facies were collected in the White and Inyo Mountains, eastern California, from the Andrews Mountain Member of the Lower Cambrian Campito Formation. Detailed sampling locations for all samples discussed in this paper can be found in Lawrence (1996). A tabulation of chemical data for samples from the Death Valley facies is provided in Cardenas et al. (1996). Tables 1 and 2 provide a summary of chemical data from the Andrews Mountain Member. Tables 3 to 6 summarize chemical data from the Shoo Fly Complex. Analytical methods and precisions are provided in the Appendix.

## DISCUSSION AND RESULTS

Prior to making statistical comparisons of the two sampling groups we must first make sure that we are comparing similar lithologies. For example, it would make little sense to compare the average composition of a suite of sandstones to the average composition of a suite of mudstones because these two rock types are known to have very different chemistries. As shown in Figure 5, all samples investigated during this study cluster within the shale (i.e., mudstone) field of the bivariate chemical classification scheme of Herron (1988). We are therefore reasonably confident that we are comparing similar rock types. Hence, our next step is to estimate some meas-

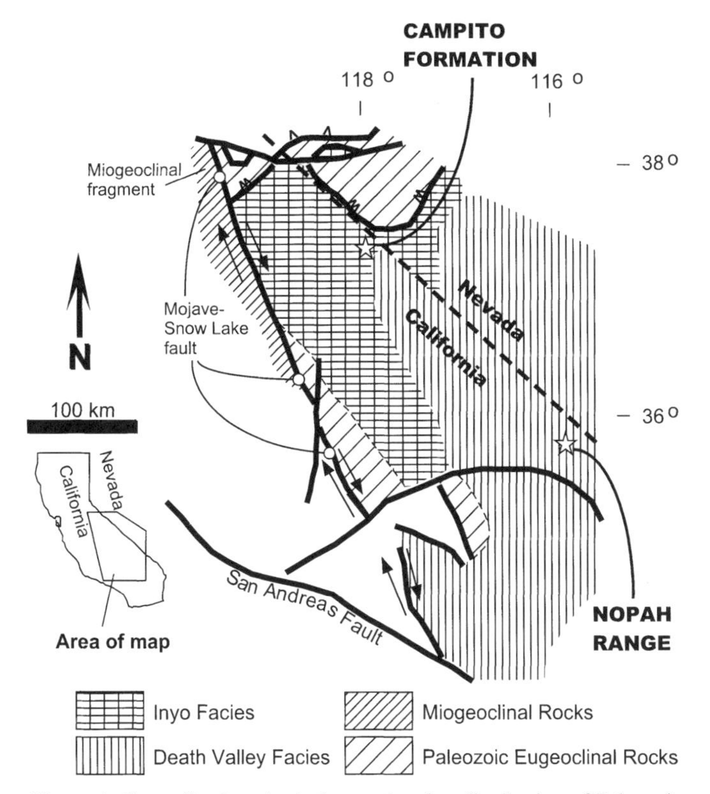


Figure 4. Generalized geological map showing distribution of Paleozoic miogeoclinal facies for southeastern California and southwestern Nevada. Map was modified from Schweickert and Lahren (1990). Stars indicate locations of sections sampled during this study.

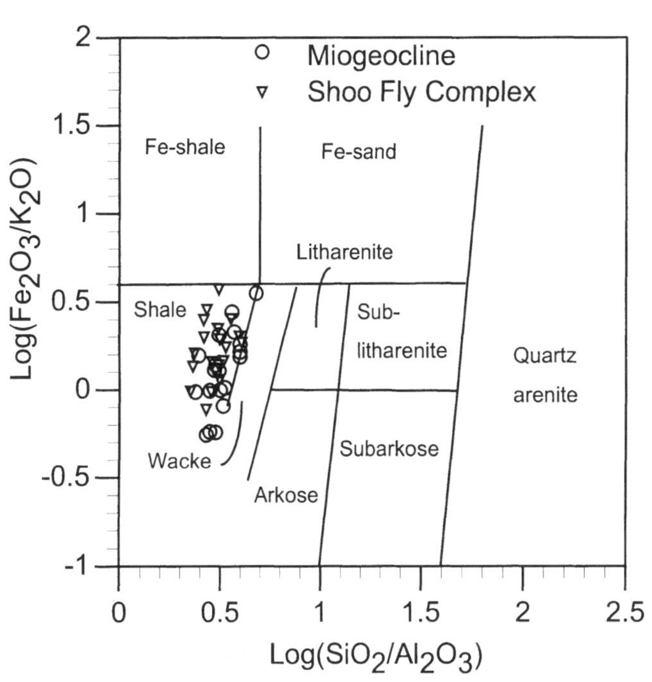


Figure 5. Chemical data derived from this study plotted on chemical discrimination diagram of Herron (1988).

ure of central tendency, e.g., the mean, for each group of analyzed mudstones.

Aitchison (1989) showed that calculating the mean directly from compositional data can lead to values that are sometimes not even included in the original composition. He suggested an alternative measure of central tendency that Ague (1994) and Ague and van Haren (1996) referred to as the Aitchison Measure of Location, or simply AML.

The AML is a compositional vector that is commonly expressed in the following form:

$$\left(\overline{C}_1,\overline{C},\cdots,\overline{C}_m\right)=\frac{(g_1,g_2,\ldots,g_m}{0.01\sum_{m=1}^{M}g_m}, \quad (1)$$

where $\overline{C}_m$ is the concentration of constituent $m$, the factor 0.01 converts to weight percent, and $M$ is the total number of constituents. The geometric mean concentration, $g_m$, of constituent $m$ is given by:

$$g_m=\exp\left[N^{-1}\sum_{n=1}^{N}\ln(c_{n,m})\right], \quad (2)$$

where $c_{n,m}$ is the concentration of $m$ in sample $n$, and $N$ is the total number of samples. The normalization to 100 wt% in equation 1 forms a composition by relating the geometric means of all of the constituents to each other through the process of closure (Ague and van Haren, 1996).

If mudstones from the Shoo Fly Complex are compositionally similar to mudstones from the miogeocline, then each constituent of the AML for each group of mudstones should have a similar value. Under this condition,

$$\overline{c}_{1mio}/\overline{c}_{1SFC}=\overline{c}_{2mio}/\overline{c}_{2SFC}=,\ldots,=\overline{c}_{nmio}/\overline{c}_{nSFC}=1, \quad (3)$$

where mio refers to a constituent of the miogeoclinal sampling group, subscript SFC refers to a constituent of the Shoo Fly Complex, and $\overline{c}_1,\overline{c}_2,\ldots,\overline{c}_m$ are as defined above. Unfortunately, compositional heterogeneities in both mudstone suites as well as variations in analytical uncertainties are likely to produce random variations in such values. We therefore are forced to address whether our estimates of the ratios $\overline{c}_{1mio}/\overline{c}_{1SFC}, \overline{c}_{2mio}/\overline{c}_{2SFC},\ldots, \overline{c}_{nmio}/\overline{c}_{nSFC}$ are statistically different from 1. Although parametric statistical procedures, e.g., the Student's $t$ test, are commonly used to estimate uncertainties associated with compositional data such procedures assume that samples approximately follow a normal distribution. If this assumption is significantly invalidated, then unwieldy results may follow. Because we do not know the nature of the underlying sampling distribution for our samples, what we need is a nonparametric procedure that allows us to assess whether the parameters $\overline{c}_{1mio}/\overline{c}_{1SFC}, \overline{c}_{2mio}/\overline{c}_{2SFC},\ldots, \overline{c}_{nmio}/\overline{c}_{nSFC}$ are statistically different from 1.

The bootstrap method, popularized by Efron and Tibshirani (1993), is a nonparametric Monte Carlo statistical technique that is ideally suited for estimating the uncertainties in complex parameters. For example, Ague (1994) and Ague and van Haren

**TABLE 1. MAJOR AND TRACE ELEMENT DATA FOR MUDSTONES IN THE ANDREWS MOUNTAIN MEMBER, CAMPITO FORMATION**

| Sample | JL52 | JL53 | JL54 | JL56 | JL57 | JL58 | JL59 | JL60 |
|---|---|---|---|---|---|---|---|---|
| Major Elements (wt%) | | | | | | | | |
| $SiO_2$ | 59.9 | 57.3 | 62.8 | 56.4 | 66.4 | 60.0 | 58.0 | 69.8 |
| $Al_2O_3$ | 19.2 | 19.5 | 16.0 | 18.2 | 14.0 | 16.3 | 16.1 | 11.9 |
| $TiO_2$ | 1.39 | 2.10 | 1.18 | 2.48 | 1.65 | 1.39 | 2.10 | 1.47 |
| FeO* | 6.87 | 8.53 | 8.93 | 11.0 | 8.76 | 10.51 | 11.64 | 4.53 |
| MnO | 0.07 | 0.10 | 0.07 | 0.09 | 0.06 | 0.06 | 0.10 | 0.08 |
| CaO | 0.71 | 0.68 | 0.64 | 0.68 | 0.61 | 0.54 | 0.57 | 2.81 |
| MgO | 2.08 | 2.34 | 2.57 | 2.41 | 2.06 | 2.93 | 3.27 | 1.18 |
| $K_2O$ | 7.11 | 6.75 | 5.07 | 5.53 | 2.54 | 5.07 | 4.37 | 3.36 |
| $Na_2O$ | 2.03 | 1.06 | 1.95 | 1.76 | 3.75 | 2.58 | 2.52 | 3.79 |
| $P_2O_5$ | 0.23 | 0.28 | 0.21 | 0.32 | 0.24 | 0.27 | 0.28 | 0.40 |
| Totals | 99.6 | 98.6 | 99.4 | 98.9 | 100.1 | 99.7 | 99.0 | 99.3 |
| Trace elements (ppm) | | | | | | | | |
| Ni | 26 | 35 | 33 | 42 | 33 | 36 | 34 | 10 |
| Cr | 62 | 93 | 42 | 80 | 46 | 49 | 59 | 52 |
| Sc† | 8 | 18 | 9 | 18 | 11 | 10 | 13 | 11 |
| V | 113 | 204 | 133 | 232 | 146 | 154 | 216 | 101 |
| Ba§ | 1068 | 924 | 631 | 797 | 312 | 588 | 561 | 509 |
| Rb§ | 258 | 222 | 173 | 198 | 90.6 | 174 | 152 | 76.9 |
| Sr§ | 95 | 56 | 69 | 59 | 78 | 68 | 56 | 89 |
| Zr | 422 | 346 | 184 | 331 | 217 | 211 | 421 | 417 |
| Y§ | 46.1 | 66.8 | 39.2 | 65.3 | 38.9 | 41.9 | 39.5 | 54.6 |
| Nb§ | 30.5 | 31.6 | 19.3 | 30.9 | 19.4 | 20.6 | 27.8 | 20.5 |
| Ga | 29 | 28 | 25 | 25 | 18 | 25 | 26 | 14 |
| Zn | 99 | 84 | 67 | 77 | 55 | 80 | 41 | 65 |
| Hf§ | 12.3 | 10.3 | 5.57 | 9.67 | 6.34 | 5.89 | 12.9 | 11.7 |
| Ta§ | 1.91 | 1.97 | 1.20 | 1.83 | 1.17 | 1.23 | 1.67 | 1.29 |
| Cs§ | 7.75 | 6.43 | 6.55 | 6.20 | 2.77 | 4.05 | 4.91 | 1.50 |
| U§ | 3.59 | 2.87 | 1.44 | 2.65 | 1.71 | 1.39 | 2.35 | 1.71 |
| Pb§ | 7.90 | 9.50 | 6.90 | 6.74 | 6.79 | 8.14 | 6.50 | 7.09 |
| Th§ | 12.5 | 10.1 | 4.87 | 8.67 | 5.64 | 4.61 | 7.25 | 5.10 |

*Note:*
* Total iron as FeO.
† Concentrations too low to be quantitatively detected by x-ray fluorescence (XRF). Listed values should be considered semi-quantitative.
§ Analyzed by inductively coupled plasma-source mass spectrometry (ICP-MS). All others analyzed by XRF.

**TABLE 2. RARE EARTH ELEMENT DATA FOR MUDSTONES IN THE ANDREWS MOUNTAIN MEMBER, CAMPITO FORMATION***

| Sample | JL52 | JL53 | JL54 | JL56 | JL57 | JL58 | JL59 | JL60 |
|---|---|---|---|---|---|---|---|---|
| La | 46.9 | 38.2 | 32.4 | 38.8 | 43.4 | 25.1 | 26.8 | 31.6 |
| Ce | 99.5 | 90.8 | 76.4 | 83.6 | 88.8 | 64.7 | 65.0 | 80.2 |
| Pr | 10.6 | 10.3 | 8.07 | 9.68 | 9.61 | 6.41 | 6.53 | 8.21 |
| Nd | 42.3 | 41.7 | 33.1 | 40.9 | 37.5 | 26.3 | 26.8 | 35.9 |
| Sm | 9.15 | 10.4 | 8.05 | 10.1 | 8.02 | 6.35 | 6.38 | 8.80 |
| Eu | 1.89 | 2.45 | 1.71 | 2.33 | 1.50 | 1.49 | 1.48 | 2.32 |
| Gd | 7.03 | 10.2 | 6.93 | 9.64 | 6.30 | 5.87 | 5.90 | 8.70 |
| Tb | 1.19 | 1.87 | 1.22 | 1.89 | 1.15 | 1.10 | 1.06 | 1.54 |
| Dy | 7.49 | 11.8 | 7.40 | 12.0 | 7.32 | 7.14 | 6.83 | 9.49 |
| Ho | 1.62 | 2.41 | 1.51 | 2.43 | 1.47 | 1.55 | 1.51 | 1.97 |
| Er | 5.08 | 6.89 | 4.28 | 6.66 | 4.20 | 4.68 | 4.75 | 5.71 |
| Tm | 0.80 | 0.96 | 0.58 | 0.91 | 0.57 | 0.66 | 0.73 | 0.82 |
| Yb | 5.26 | 5.97 | 3.56 | 5.70 | 3.50 | 4.03 | 4.80 | 5.23 |
| Lu | 0.89 | 0.93 | 0.54 | 0.88 | 0.54 | 0.62 | 0.81 | 0.84 |

*Note:*
*Analyzed by inductively coupled plasma-source mass spectrometry (ICP-MS).
All data are in ppm.

**TABLE 3. MAJOR AND TRACE ELEMENT DATA FOR MUDSTONES IN THE POISON CANYON UNIT**

| Sample | JL7 | JL 9 | JL10 | JL11 | GG8 | GG9 | GG11 | GG12 | GG13 | JL26 | JL27 | JL28 | JL29 | JL30 | JL32 |
|---|---|---|---|---|---|---|---|---|---|---|---|---|---|---|---|
| Major Elements (wt%) | | | | | | | | | | | | | | | |
| $SiO_2$ | 56.6 | 57.3 | 56.5 | 59.7 | 66.7 | 62.4 | 64 | 59.4 | 61.2 | 63.4 | 63.6 | 59.3 | 62.3 | 67.4 | 62.5 |
| $Al_2O_3$ | 24.5 | 21.8 | 21.6 | 19.9 | 17 | 17.5 | 19 | 19.1 | 20.0 | 19.3 | 20.3 | 19.2 | 20.1 | 16.7 | 22.0 |
| $TiO_2$ | 1.06 | 0.93 | 1.16 | 1.15 | 0.81 | 1.11 | 1.1 | 1.06 | 0.83 | 1.16 | 0.87 | 0.87 | 0.97 | 0.90 | 1.20 |
| FeO* | 7.38 | 9.65 | 8.97 | 7.96 | 5.82 | 8.22 | 6.09 | 10 | 7.8 | 6.72 | 5.48 | 10.59 | 6.4 | 6.43 | 5.02 |
| MnO | 0.09 | 0.03 | 0.08 | 0.08 | 0.05 | 0.08 | 0.04 | 0.07 | 0.06 | 0.067 | 0.09 | 0.05 | 0.03 | 0.07 | 0.08 |
| CaO | 0.15 | 0.8 | 1.24 | 1.09 | 0.63 | 1.05 | 0.23 | 0.56 | 0.57 | 0.07 | 0.17 | 0.16 | 0.33 | 0.13 | 0.19 |
| MgO | 2.31 | 2.13 | 2.91 | 2.85 | 2.16 | 2.33 | 2.96 | 3.64 | 2.27 | 2.79 | 2.13 | 2.28 | 2.47 | 2.7 | 2.24 |
| $K_2O$ | 6.08 | 5.44 | 4.01 | 3.94 | 3.24 | 3.63 | 3.9 | 3 | 3.9 | 5.15 | 5.36 | 5.31 | 5.18 | 3.89 | 5.77 |
| $Na_2O$ | 0.23 | 0.33 | 1.5 | 1.67 | 1.62 | 1.13 | 1.4 | 1.02 | 1.6 | 0.19 | 0.38 | 0.27 | 0.67 | 0.18 | 0.24 |
| $P_2O_5$ | 0.13 | 0.13 | 0.24 | 0.20 | 0.09 | 0.2 | 0.12 | 0.2 | 0.14 | 0.08 | 0.12 | 0.15 | 0.16 | 0.10 | 0.16 |
| Totals | 98.5 | 98.5 | 98.2 | 98.5 | 98.1 | 97.7 | 98.8 | 98.1 | 98.4 | 98.9 | 98.5 | 98.2 | 98.6 | 98.5 | 99.4 |
| Trace elements (ppm) | | | | | | | | | | | | | | | |
| Ni | 47 | 51 | 83 | 60 | 48 | 20 | 47 | 48 | 40 | 19 | 61 | 33 | 35 | 16 | 20 |
| Cr | 112 | 118 | 115 | 103 | 99 | 96 | 102 | 92 | 90 | 80 | 89 | 82 | 100 | 89 | 118 |
| Sc† | 12 | 12 | 12 | 9 | 11 | 16 | 17 | 25 | 15 | 12 | 9 | 12 | 11 | 8 | 12 |
| V | 154 | 146 | 138 | 128 | 78 | 106 | 118 | 118 | 93 | 111 | 108 | 106 | 106 | 105 | 145 |
| Ba§ | 1100 | 1382 | 1087 | 763 | 502. | 612 | 1518 | 703 | 861 | 645 | 681 | 902 | 893 | 648 | 713 |
| Rb§ | 129 | 180 | 129 | 139 | 113 | 118 | 131 | 93 | 141 | 155 | 189 | 195 | 169 | 113 | 166 |
| Sr§ | 33.2 | 51.3 | 92.4 | 86.4 | 69.0 | 70.0 | 52.0 | 64.0 | 107 | 18.5 | 36.3 | 35.8 | 33.0 | 14.0 | 25.8 |
| Zr | 113 | 119 | 167 | 156 | 208 | 152 | 193 | 144 | 142 | 161 | 121 | 121 | 130 | 134 | 223 |
| Y§ | 31.8 | 32.1 | 35.6 | 29.4 | 23.0 | 31.0 | 27.0 | 26.0 | 34.0 | 16.7 | 27.4 | 21.6 | 42.9 | 17.8 | 35.4 |
| Nb§ | 18.3 | 17.8 | 31.1 | 30.0 | 17.7 | 27.0 | 24.0 | 26.1 | 21.8 | 22.9 | 18.1 | 17.5 | 20.9 | 20.1 | 24.2 |
| Ga | 30 | 28 | 28 | 27 | 26 | 27 | 26 | 31 | 27 | 26 | 30 | 24 | 27 | 24 | 29 |
| Zn | 83 | 99 | 84 | 67 | 69 | 77 | 55 | 80 | 41 | 65 | 86 | 79 | 69 | 55 | 55 |
| Hf§ | 5.91 | 3.48 | 4.60 | 4.38 | 5.64 | 3.98 | 5.84 | 4.05 | 4.03 | 4.86 | 3.70 | 3.65 | 3.81 | 4.00 | 6.52 |
| Ta§ | 1.18 | 1.20 | 2.08 | 1.98 | 1.28 | 1.83 | 1.34 | 1.63 | 1.24 | 1.53 | 1.30 | 1.30 | 1.46 | 1.32 | 1.58 |
| Cs§ | 1.36 | 1.88 | 1.52 | 2.02 | 1.54 | 1.70 | 1.46 | 0.93 | 1.88 | 2.36 | 2.23 | 3.60 | 3.82 | 1.22 | 1.86 |
| U§ | 2.44 | 1.30 | 2.10 | 2.29 | 1.35 | 1.43 | 1.97 | 1.61 | 1.41 | 2.25 | 2.57 | 1.92 | 1.80 | 1.69 | 3.41 |
| Pb§ | 5.93 | 7.11 | 2.63 | 4.15 | 5.00 | 4.00 | 2.00 | 3.00 | 0.00 | 10.1 | 32.6 | 8.88 | 6.94 | 2.82 | 8.24 |
| Th§ | 14.1 | 14.7 | 19.7 | 16.8 | 14.9 | 15.2 | 14.7 | 16.7 | 18.9 | 13.0 | 18.9 | 19.1 | 17.5 | 12.6 | 18.0 |

*Note:*

* Total iron as FeO.

† Concentrations too low to be quantitatively detected by x-ray fluorescence (XRF). Listed values should be considered semi-quantitative.

§ Analyzed by inductively coupled plasma-source mass spectrometry (ICP-MS). All others analyzed by XRF.

TABLE 4. RARE EARTH ELEMENT DATA FOR MUDSTONES IN THE POISON CANYON UNIT*

| Sample | JL7 | JL9 | JL10 | JL11 | GG8 | GG9 | GG11 | GG12 | GG13 | JL26 | JL27 | JL28 | JL29 | JL30 | JL32 |
|---|---|---|---|---|---|---|---|---|---|---|---|---|---|---|---|
| La | 73.1 | 66.3 | 81.6 | 62.5 | 37.4 | 53.1 | 52.5 | 69.2 | 77.7 | 34.1 | 65.8 | 69.3 | 82.7 | 45.0 | 77.6 |
| Ce | 136 | 117 | 153 | 116 | 68.0 | 102 | 95.6 | 131 | 144 | 67.4 | 122 | 130 | 157 | 84.0 | 141 |
| Pr | 14.9 | 13.1 | 16.1 | 12.4 | 7.80 | 11.2 | 10.6 | 13.2 | 15.3 | 7.33 | 13.7 | 13.6 | 16.5 | 9.10 | 15.5 |
| Nd | 55.8 | 49.1 | 60.3 | 46.0 | 29.4 | 41.9 | 40.5 | 50.6 | 56.7 | 27.7 | 50.4 | 49.2 | 62.8 | 34.3 | 57.0 |
| Sm | 10.7 | 9.52 | 11.7 | 8.98 | 5.78 | 8.54 | 8.30 | 9.49 | 10.8 | 5.56 | 9.30 | 9.57 | 12.1 | 6.05 | 10.8 |
| Eu | 2.01 | 2.44 | 2.10 | 1.69 | 1.04 | 1.83 | 1.82 | 2.02 | 1.83 | 1.27 | 1.66 | 1.97 | 2.45 | 1.17 | 2.21 |
| Gd | 7.76 | 7.12 | 8.23 | 6.53 | 4.34 | 6.65 | 5.98 | 6.44 | 7.54 | 3.96 | 6.44 | 6.41 | 9.40 | 3.99 | 7.74 |
| Tb | 1.18 | 1.16 | 1.31 | 1.06 | 0.71 | 1.10 | 0.95 | 1.05 | 1.16 | 0.66 | 1.00 | 0.99 | 1.48 | 0.61 | 1.22 |
| Dy | 6.43 | 6.43 | 7.16 | 6.01 | 4.27 | 6.18 | 5.38 | 5.77 | 6.56 | 3.59 | 5.35 | 4.99 | 7.89 | 3.62 | 6.77 |
| Ho | 1.22 | 1.19 | 1.38 | 1.15 | 0.86 | 1.19 | 1.07 | 1.06 | 1.26 | 0.70 | 1.01 | 0.88 | 1.44 | 0.69 | 1.29 |
| Er | 3.33 | 3.30 | 3.82 | 3.13 | 2.52 | 3.20 | 3.12 | 2.89 | 3.36 | 1.94 | 2.81 | 2.46 | 3.83 | 1.92 | 3.57 |
| Tm | 0.46 | 0.44 | 0.51 | 0.44 | 0.35 | 0.43 | 0.43 | 0.39 | 0.44 | 0.28 | 0.40 | 0.36 | 0.51 | 0.28 | 0.51 |
| Yb | 2.75 | 2.63 | 2.98 | 2.59 | 2.27 | 2.57 | 2.70 | 2.44 | 2.73 | 1.79 | 2.46 | 2.28 | 3.00 | 1.86 | 3.10 |
| Lu | 0.43 | 0.40 | 0.46 | 0.40 | 0.36 | 0.38 | 0.43 | 0.39 | 0.42 | 0.30 | 0.36 | 0.35 | 0.42 | 0.28 | 0.46 |

* Analyzed by inductively coupled plasma-source mass spectrometry (ICP-MS). All data are in ppm.

(1996) successfully used the bootstrap technique to estimate the uncertainties in elemental ratios used to calculate changes in rock mass brought on by prograde metamorphism. In order to perform a bootstrap analysis of the ratios $\bar{c}_{1\text{mio}}/\bar{c}_{1\text{SFC}}$, $\bar{c}_{2\text{mio}}/\bar{c}_{2\text{SFC}}$, ... , $\bar{c}_{n\text{mio}}/\bar{c}_{n\text{SFC}}$ we treated each mudstone data set following the general procedures outlined in Ague and van Haren (1996). Thus, we first expressed our data in matrix format as follows:

$$D_{\text{miogecline}} = \begin{bmatrix} c^o_{1,1} & c^o_{1,2} & \cdots & c^o_{1,m^o} \\ c^o_{2,1} & c^o_{2,2} & \cdots & c^o_{2,m^o} \\ \vdots & \vdots & \cdots & \vdots \\ c^o_{n^o,1} & c^{oo}_{n^o,2} & \cdots & c^o_{n^m,m^o} \end{bmatrix} \quad (4)$$

and

$$D_{\text{ShooFlyComplex}} = \begin{bmatrix} c^i_{1,1} & c^i_{1,2} & \cdots & c^i_{1,m^i} \\ c^i_{2,1} & c^i_{2,2} & \cdots & c^i_{2,m^i} \\ \vdots & \vdots & \cdots & \vdots \\ c^i_{n^i,1} & c^i_{n^i,2} & \cdots & c^i_{n^i,m^i} \end{bmatrix} \quad (5)$$

where subscript $n^o$ and $n^i$ are the number of samples in the miogeoclinal and Shoo Fly Complex data sets, respectively. Subscripts $m^i$ and $m^o$ represent the number of chemical constituents in each composition for each data set. In practice the number of chemical constituents in each composition were the same. Thus, $m^i = m^o$.

In bootstrap analysis the parameter of interest, θ, is estimated using some function of the observed data, s(x), such that $\hat{\theta} = s(x)$, where $\hat{\theta}$ is an estimate and $x$ is a random sample from the unknown probability distribution (Efron and Tibshirnai, 1993; Ague and van Haren, 1996). For the case at hand, the parameters of interest are the constituents of the ratio of $\text{AML}_{\text{miogeocline}}/\text{AML}_{\text{ShooFlyComplex}}$. Thus, the following boot-

TABLE 5. MAJOR AND TRACE ELEMENT DATA FOR MUDSTONES IN THE RED HILL UNIT

| Sample | JL1 | JL2 | JL3 | JL6 | GG1 | GG3 | GG4 |
|---|---|---|---|---|---|---|---|
| Major Elements (wt%) | | | | | | | |
| $SiO_2$ | 56.9 | 56.3 | 61.7 | 55.6 | 60.7 | 64.9 | 59.8 |
| $Al_2O_3$ | 21.0 | 25.1 | 19.6 | 23.6 | 22.6 | 18.0 | 20.8 |
| $TiO_2$ | 1.01 | 1.08 | 0.87 | 0.91 | 1.02 | 0.88 | 1.26 |
| FeO* | 11.2 | 6.03 | 7.95 | 8.60 | 4.75 | 8.03 | 6.95 |
| MnO | 0.05 | 0.07 | 0.08 | 0.05 | 0.03 | 0.05 | 0.07 |
| CaO | 0.1 | 0.14 | 0.06 | 0.13 | 0.17 | 0.31 | 0.18 |
| MgO | 3.06 | 2.22 | 3 | 2.59 | 1.68 | 2.22 | 2.74 |
| $K_2O$ | 4.37 | 6.85 | 4.59 | 5.96 | 6.88 | 3.52 | 5.41 |
| $Na_2O$ | 0.14 | 0.16 | 0.08 | 0.2 | 0.19 | 2.01 | 0.46 |
| $P_2O_5$ | 0.12 | 0.15 | 0.07 | 0.15 | 0.11 | 0.07 | 0.13 |
| Totals | 98.0 | 98.1 | 98.0 | 97.8 | 98.1 | 100.0 | 97.8 |
| Trace elements (ppm) | | | | | | | |
| Ni | 61 | 50 | 40 | 31 | 34 | 42 | 51 |
| Cr | 100 | 120 | 96 | 106 | 97 | 76 | 109 |
| Sc[†] | 15 | 12 | 8 | 12 | 19 | 21 | 20 |
| V | 125 | 137 | 111 | 127 | 83 | 87 | 107 |
| Ba[§] | 811 | 1209 | 883 | 1708 | 1063 | 963 | 913 |
| Rb[§] | 139 | 218 | 154 | 183 | 214 | 131 | 187 |
| Sr[§] | 19 | 33 | 17 | 28 | 30 | 64 | 27 |
| Zr | 142 | 137 | 98 | 111 | 168 | 128 | 231 |
| Y** | 39.0 | 37.4 | 24.0 | 31.9 | 28.0 | 25.0 | 26.0 |
| Nb[§] | 18.9 | 20.8 | 17.1 | 17.2 | 24.0 | 18.0 | 30.0 |
| Ga | 30 | 36 | 25 | 32 | 32 | 24 | 29 |
| Zn | 132 | 79 | 99 | 81 | 58 | 214 | 76 |
| Hf[§] | 4.07 | 4.03 | 2.90 | 2.97 | 4.17 | 3.43 | 6.29 |
| Ta[§] | 1.27 | 1.43 | 1.16 | 1.19 | 1.39 | 1.12 | 1.61 |
| Cs[§] | 2.06 | 1.93 | 1.87 | 1.65 | 3.50 | 2.08 | 2.57 |
| U[§] | 2.65 | 3.63 | 1.74 | 1.76 | 2.29 | 1.03 | 1.83 |
| Pb[§] | 13.5 | 10.3 | 4.59 | 3.21 | 2.00 | 9.00 | 1.00 |
| Th[§] | 15.4 | 19.1 | 14.1 | 15.2 | 18.6 | 13.8 | 16.4 |

*Note:*

* Total iron as FeO.

[†] Concentrations too low to be quantitatively detected by x-ray fluorescence (XRF). Listed values should be considered semi-quantitative.

[§] Analyzed by inductively coupled plasma-source mass spectrometry (ICP-MS). All others analyzed by XRF.

**TABLE 6. RARE EARTH ELEMENT DATA FOR MUDSTONES IN THE RED HILL UNIT***

| Sample | JL1 | JL2 | JL3 | JL6 | GG1 | GG3 | GG4 |
|---|---|---|---|---|---|---|---|
| La | 62.4 | 81.3 | 60.0 | 78.7 | 74.0 | 41.9 | 57.7 |
| Ce | 123 | 150 | 112 | 143 | 141 | 92.4 | 108 |
| Pr | 12.8 | 16.6 | 11.9 | 15.8 | 14.6 | 8.49 | 11.4 |
| Nd | 49.7 | 62.3 | 45.2 | 59.2 | 54.7 | 32.0 | 41.9 |
| Sm | 10.8 | 12.1 | 8.74 | 11.4 | 10.4 | 6.47 | 7.83 |
| Eu | 2.52 | 2.20 | 2.05 | 2.40 | 2.01 | 1.24 | 1.54 |
| Gd | 9.13 | 8.48 | 6.09 | 8.09 | 7.12 | 4.77 | 5.54 |
| Tb | 1.46 | 1.33 | 0.94 | 1.22 | 1.12 | 0.82 | 0.86 |
| Dy | 7.94 | 7.43 | 5.05 | 6.64 | 6.03 | 4.73 | 5.08 |
| Ho | 1.54 | 1.41 | 0.94 | 1.25 | 1.12 | 0.95 | 1.01 |
| Er | 4.22 | 3.96 | 2.55 | 3.31 | 3.08 | 2.73 | 2.90 |
| Tm | 0.55 | 0.54 | 0.36 | 0.45 | 0.41 | 0.38 | 0.42 |
| Yb | 3.21 | 3.31 | 2.21 | 2.69 | 2.59 | 2.34 | 2.65 |
| Lu | 0.48 | 0.51 | 0.35 | 0.41 | 0.41 | 0.37 | 0.42 |

* Analyzed by inductively coupled plasma-source mass spectrometry (ICP-MS). All data are in ppm.

strap algorithm was developed and implemented within a Visual Basic 6 program with graphical user interface. Note that the superscript *o* identifies a bootstrap sample or parameter

1. Form a bootstrap sample, $D^o_{\text{miogeocline}}$, by random sampling with replacement the rows of $D_{\text{miogeocline}}$.

2. Form a bootstrap replicate of the AML from $D^o_{\text{miogeocline}}$, and call it $\text{AML}^o_{\text{miogeocline}}$.

3. Repeat steps (1) and (2) using $D_{\text{ShooFlyComplex}}$ to form $D^o_{\text{ShooFlyComplex}}$ and $\text{AML}^o_{\text{ShooFlyComplex}}$.

4. Form the ratios $\bar{c}^o_{1\text{mio}}/\bar{c}^o_{1\text{SFC}}, \bar{c}^o_{2\text{mio}}/\bar{c}^o_{2\text{SFC}}, \ldots, \bar{c}^o_{n\text{mio}}/\bar{c}^o_{n\text{SFC}}$ from the resulting $\text{AML}^o$ data, and store them.

5. Repeat steps (1) through (4) *B* times, where *B* is the total number of bootstrap estimates to form.

A confidence interval can then be constructed using the bootstrap distribution of $\hat{\theta}$ (i.e., the ratios $\bar{c}^o_{1\text{mio}}/\bar{c}^o_{1\text{SFC}}, \bar{c}^o_{2\text{mio}}/\bar{c}^o_{2\text{SFC}}, \ldots, \bar{c}^o_{n\text{mio}}/\bar{c}^o_{n\text{SFC}}$) (Efron and Tibshirani, 1993; Ague and van Haren, 1996). Specifically, the lower and upper limits of a given percentile interval for a given parameter $\theta$, denoted by $\hat{\theta}_{\%,\text{lo}}$ and $\hat{\theta}_{\%,\text{hi}}$ (lo = low, hi = high) are approximated by

$$\left[\hat{\theta}_{\%,\text{lo}}, \hat{\theta}_{\%,\text{hi}}\right] \approx \left[\hat{\theta}_B^{o(\alpha)}, \hat{\theta}_B^{o(1-\alpha)}\right] \quad (6)$$

where *B* signifies that the estimated percentile interval is based on *B* bootstrap replications, and $\alpha$ specifies the desired lower and upper limits (Efron and Tibshirani, 1993; Ague and van Haren, 1996). For a 95% confidence interval, $\alpha = 0.025$. Hence, in order to calculate a 95% confidence band, stored data derived from steps 1 through 4 above are sorted, and the 2.5 and 97.5 percentiles are found for each constituent ratio following conventional statistical procedures (e.g., Mendenhall and Beaver, 1994).

The results of 5000 bootstrap replicates of $\bar{c}^o_{1\text{mio}}/\bar{c}^o_{1\text{SFC}}, \bar{c}^o_{2\text{mio}}/\bar{c}^o_{2\text{SFC}}, \ldots, \bar{c}^o_{n\text{mio}}/\bar{c}^o_{n\text{SFC}}$ are listed in Table 7. In order to simplify presentation the 95% confidence intervals in the averages of $\ln(\bar{c}^o_{1\text{mio}}/\bar{c}^o_{1\text{SFC}}), \ln(\bar{c}^o_{2\text{mio}}/\bar{c}^o_{2\text{SFC}}), \ldots, \ln(\bar{c}^o_{n\text{mio}}/\bar{c}^o_{n\text{SFC}})$ are plotted in Figure 6. Recalling that ln(1) = 0, note that the 95% confidence bars for the elements $TiO_2$, CaO, $K_2O$, Cr, Sc, Ba, Rb, Zr, Y,

**TABLE 7. RESULTS OF 5000 BOOTSTRAP REPLICATES OF $\bar{c}^{\text{element}}_{\text{mio}} / \bar{c}^{\text{element}}_{\text{SFC}}$**

| $\bar{c}^{\text{element}}_{\text{mio}^*} / \bar{c}^{\text{element}}_{\text{SFC}^\#}$ | Average | 2.5 Percentile | 97.5 Percentile |
|---|---|---|---|
| $SiO_2$ | 1.01 | 0.98 | 1.04 |
| $Al_2O_3$ | 0.94 | 0.87 | 1.02 |
| $TiO_2$ | 1.29 | 1.10 | 1.50 |
| FeO | 0.88 | 0.75 | 1.03 |
| MnO | 0.78 | 0.57 | 1.06 |
| CaO | 1.85 | 1.14 | 2.79 |
| MgO | 1.03 | 0.89 | 1.17 |
| $K_2O$ | 1.22 | 1.04 | 1.42 |
| $Na_2O$ | 1.39 | 0.6 | 2.60 |
| $P_2O_5$ | 1.18 | 0.9 | 1.52 |
| Ni | 0.98 | 0.81 | 1.18 |
| Cr | 0.80 | 0.66 | 0.96 |
| Sc | 1.32 | 1.06 | 1.61 |
| V | 1.22 | 0.98 | 1.48 |
| Ba | 0.69 | 0.57 | 0.83 |
| Rb | 1.24 | 1.07 | 1.42 |
| Sr | 1.19 | 0.85 | 1.60 |
| Zr | 1.70 | 1.44 | 2.00 |
| Y | 1.52 | 1.31 | 1.76 |
| Nb | 1.12 | 1.00 | 1.25 |
| Ga | 0.99 | 0.89 | 1.08 |
| Zn | 0.99 | 0.79 | 1.21 |
| Hf | 1.62 | 1.37 | 1.92 |
| Ta | 1.12 | 1.00 | 1.24 |
| Cs | 2.46 | 1.81 | 3.18 |
| U | 1.07 | 0.89 | 1.28 |
| Pb | 1.76 | 1.05 | 2.92 |
| Th | 0.64 | 0.53 | 0.76 |
| La | 0.73 | 0.61 | 0.88 |
| Ce | 0.81 | 0.70 | 0.96 |
| Pr | 0.84 | 0.71 | 0.99 |
| Nd | 0.88 | 0.76 | 1.03 |
| Sm | 0.97 | 0.85 | 1.12 |
| Eu | 0.99 | 0.86 | 1.13 |
| Gd | 1.08 | 0.94 | 1.25 |
| Tb | 1.21 | 1.05 | 1.40 |
| Dy | 1.35 | 1.18 | 1.56 |
| Ho | 1.46 | 1.28 | 1.68 |
| Er | 1.55 | 1.37 | 1.76 |
| Tm | 1.62 | 1.44 | 1.82 |
| Yb | 1.67 | 1.49 | 1.86 |
| Lu | 1.71 | 1.53 | 1.90 |

*Note:*
* miogeocline
# Shoo Fly Complex.

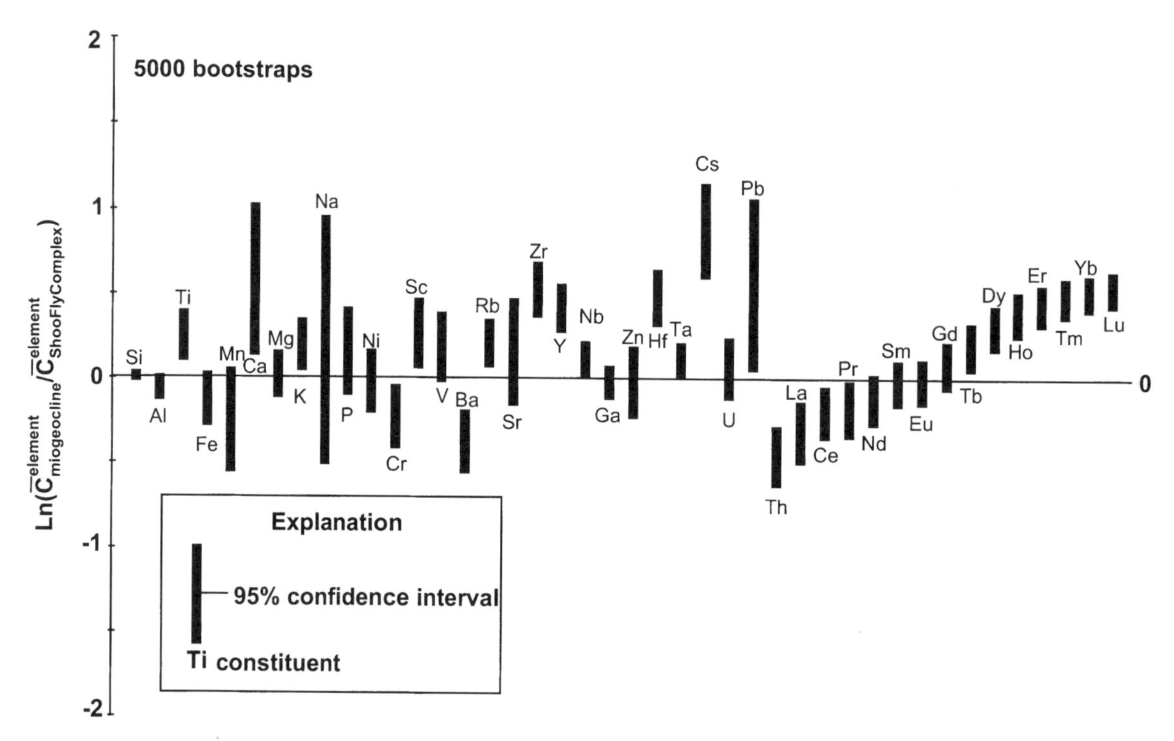


Figure 6. Plot of 95% confidence intervals for 42 elements analyzed during this study. Confidence intervals are based on 5000 bootstrap replicates of ratios $\bar{c}^o_{1mio}/\bar{c}^o_{1SFC}, \bar{c}^o_{2mio}/\bar{c}^o_{2SFC}, \ldots, \bar{c}^o_{nmio}/\bar{c}^o_{nSFC}$. See text for additional discussion.

Nb, Hf, Ta, Cs, Pb, Th, La, Ce, Pr, Tb, Dy, Ho, Er, Tm, Yb, and Lu plot either completely above or below zero. In contrast, the 95% confidence intervals for the remaining 17 elements include zero. These observations indicate that ~60% of the constituent ratios investigated during this study are statistically different from 1 at the 95% confidence level. Such a result implies that the Shoo Fly Complex was probably derived from a source that was compositionally unlike the source that supplied material to the southwest Cordilleran miogeocline. Hence, it is unlikely that the source of detritus in the Shoo Fly Complex was located in the southwestern interior of North America. Where was the source located?

Data presented by Harding et al. (this volume, Chapter 2) show that the detrital zircon population in a number of samples analyzed from the Shoo Fly Complex contains an Archean component. In addition, the cumulative probability diagram of single-crystal U-Pb detrital zircon ages from the Shoo Fly Complex is similar to that produced by detrital zircon ages derived from miogeoclinal strata, the source of which was located in northwestern North America (see Harding et al., this volume, Chapter 2). We therefore conclude that the source of detritus in samples analyzed from the Shoo Fly Complex was probably located in northwestern rather than southwestern North America, as Archean sources are not present anywhere in the latter area.

## CONCLUSIONS

Stratigraphic trends, paleocurrent data, and detrital U-Pb zircon geochronology unambiguously point to the source of material in the lower southwestern miogeocline as being located in the adjacent interior of North America (e.g., Stewart, 1980; Prave et al., 1991; Poole et al., 1992; Gehrels and Dickinson, 1995; Gehrels et al., 1996). In contrast, the source of clastic materials in the Shoo Fly Complex has been difficult to locate due to its position outboard of clear-cut aboriginal North American materials. However, data discussed and presented in this chapter point to a northwestern rather than southwestern North American source, and illustrate well the potential usefulness of the bootstrap technique in assessing the uncertainties in complex provenance-dependent parameters like the ratios of $\bar{c}^o_{1mio}/\bar{c}^o_{1SFC}, \bar{c}^o_{2mio}/\bar{c}^o_{2SFC}, \ldots, \bar{c}^o_{nmio}/\bar{c}^o_{nSFC}$.

## ACKNOWLEDGMENTS

This manuscript would not have been completed without the gracious and gentle nudging of George E. Gehrels, and was partially supported by National Science Foundation grant EAR-9505569. Mark Johnsson and Robert Cullers provided constructive and thorough reviews of an earlier draft of the manuscript. To request the Visual Basic 6 bootstrap program developed during this investigation contact G. Girty at ggirty@geology.sdsu.edu.

## APPENDIX. ANALYTICAL PROCEDURES

Diane Johnson and Charles Knaack completed all chemical analyses in the GeoAnalytical Laboratory of Washington State University, Pullman. Single 2:1 lithium tetraborate–rock-powder fused disks of each sample were analyzed for major and some trace elements following conventional X-ray fluorescence (XRF) techniques on an automatic Rigaku

3370 spectrometer (Hooper et al., 1993). Each elemental analysis was fully corrected for line interference and matrix effects. Loss on ignition was not determined, and therefore all chemical arguments presented in this chapter bear on differences in concentrations other than those contributed by changes in volatile components such as $H_2O$. Repeated analysis of well-characterized standards suggests that precisions for major element data, as measured by one standard deviation values, are <1 relative percent, whereas precisions for trace element data analyzed by XRF are less than ~5–20 relative percent, depending upon the analyzed trace element and its concentration (Hooper et al., 1993).

Following fusion of a 1:1 lithium tetraborate–rock-powder mixture at 1000 °C, sample digestion for rare earth element (REE) and trace element analyses followed conventional techniques as outlined in Knaack et al. (1994). REEs and some trace elements were analyzed on a Sciex Elan 250 inductively coupled plasma source mass spectrometer (ICP-MS), with Babington nebulizer, water-cooled spray chamber, and Brooks mass flow controllers. Oxide interference corrections were determined for each run using two solutions prepared from single element standards.

Sample intensities were corrected for isobaric oxide interference (Lichte et al., 1987), and instrumental drift using a mass-weighted average of the drift of In and Re (Doherty, 1989). Elemental intensities are calibrated against a curve constructed from three in-house standards: BCR-P, GMP-01, and MON-01 (Knaack et al., 1994). These standards are dissolved along with each batch of unknowns. The in-house standards have been characterized using a suite of international reference standards, and also have been analyzed by J.N. Walsh of Kings College and Allen Meier of the U.S. Geological Survey. Repeated analyses of BCR-P indicate that precisions for REE abundances are <5 relative percent, whereas precisions for trace element abundance values determined by ICP-MS are <10 relative percent (Knaack et al., 1994).

## REFERENCES CITED

Ague, J.J., 1994, Mass transfer during Barrovian metamorphism of pelites, south-central Connecticut. I: Evidence for changes in composition and volume: American Journal of Science, v. 294, p. 989–1057.

Ague, J.J., and van Haren, J.L.M., 1996, Assessing metasomatic mass and volume changes using the bootstrap, with application to deep crustal hydrothermal alteration of marble: Economic Geology, v. 91, p. 1169–1182.

Aitchison, J., 1989, Measures of location of compositional data sets: Mathematical Geology, v. 21, p. 787–790.

Burchfiel, B.C., and Davis, G.A., 1975, Nature and controls of Cordilleran orogenesis, western United States: Extensions of an earlier synthesis: American Journal of Science, v. 275-A, p. 363–396.

Burchfiel, B.C., and Royden, L.H., 1991, Antler orogeny: A Mediterranean-type orogeny: Geology, v. 19, p. 66–69.

Burchfiel, B.C., Cowan, D.S., and Davis, G.A., 1992, Tectonic overview of the Cordilleran orogen in the western United States, *in* Burchfiel, B.C., et al., eds., The Cordilleran orogen: Conterminous U.S.: Boulder, Colorado, Geological Society of America, Geology of North America, v. G-3, p. 407–479.

Cardenas, A., Girty, G.H., Hanson, A.D., Lahren, M.M., Knaack, C., and Johnson, D., 1996, Assessing differences in composition between low metamorphic grade mudstones and high-grade schists using logratio techniques: Journal of Geology, v. 104, p. 279–293.

Dickinson, W.R., Beard, L.S., Brakenridge, G.R., Erjavec, J.A., Ferguson, R.C., Inman, K.F., Knepp, R.A., Lindberg, F.A., and Ryber, P.T., 1983, Provenance of North American Phanerozoic sandstones in relation to tectonic setting: Geological Society of America Bulletin, v. 94, p. 222–235.

Doherty, W., 1989, An internal standardization procedure for the determination of yttrium and the rare earth elements in geological materials by inductively coupled plasma-mass spectrometry: Spectrochemica Acta, v. 44B, p. 263–280.

Efron, B., and Tibshirani, R.J., 1993, An introduction to the bootstrap: New York, Chapman and Hall, 436 p.

Gehrels, G.E., and Dickinson, W.R., 1995, Detrital zircon provenance of Cambrian to Triassic miogeoclinal and eugeoclinal strata in Nevada: American Journal of Science, v. 295, p. 722–734.

Gehrels, G.E., Butler, R.F., and Bazard, D.R., 1996, Detrital zircon geochronology of the Alexander terrane, southeastern Alaska: Geological Society of America Bulletin, v. 108, p. 722–734.

Girty, G.H., and Schweickert, R.A., 1984, The Culbertson Lake allochthon, a newly identified structure within the Shoo Fly Complex, California: Evidence for four phases of deformation and extension of the Antler orogeny to the northern Sierra Nevada: Modern Geology, v. 8, p. 181–198.

Girty, G.H., and Wardlaw, M.S., 1985, Petrology and provenance of pre-Late Devonian sandstones, Shoo Fly Complex, northern Sierra Nevada, California: Geological Society of America Bulletin, v. 96, p. 516–521.

Girty, G.H., Gester, K.C., and Turner, J.B., 1990, Pre-Late Devonian geochemical, stratigraphic, sedimentologic, and structural patterns, Shoo Fly Complex, northern Sierra Nevada, California, *in* Harwood, D.S., and Miller, M.M., eds., Paleozoic and early Mesozoic paleogeographic relations; Sierra Nevada, Klamath Mountains, and related terranes: Geological Society of America Special Paper 255, p. 43–56.

Girty, G.H., Gurrola, L.D., Taylor, G.W., Richards, M.J., and Wardlaw, M.S., 1991, The pre-Upper Devonian Lang and Black Oak Springs sequences, Shoo Fly Complex, northern Sierra Nevada, California: Trench deposits composed of continental detritus, *in* Cooper, J.D., and Stevens, C.H., eds., Paleozoic paleogeography of the Western United States—II, Volume 2: Pacific Section, Society for Sedimentary Geology (SEPM), p. 703–716.

Girty, G.H., Hanson, R.E., Girty, M.S., Schweickert, R.A., Harwood, D.S., Yoshinobu, A.S., Bryan, K.A., Skinner, J.E., and Hill, C.A., 1995, Timing of emplacement of the Haypress Creek and Emigrant Gap plutons: Implications for the timing and controls of Jurassic orogenesis, northern Sierra Nevada, California, *in* Miller, D., and Busby, C., eds., Jurassic Magmatism and Tectonics of the North American Cordillera: Geological Society of America Special Paper 299, p. 191–201.

Girty, G.H., Lawrence, J., Burke, T., Fortin, A., Gallarano, C.S., Wirths, T.A., Lewis, J.G., Peterson, M.M., Ridge, D.L., Knaack, C., and Johnson, D., 1996, The Shoo Fly Complex: Its origin and tectonic significance, *in* Girty, G. H., et al., eds., The northern Sierra terrane and associated Mesozoic magmatic units: Implications for the tectonic history of the western Cordillera: Pacific Section, Society for Sedimentary Geology (SEPM) book 81, p. 1–24.

Hannah, J.L., and Moores, E.M., 1986, Age relationships and depositional environments of Paleozoic strata, northern Sierra Nevada, California: Geological Society of America Bulletin, v. 97, p. 787–797.

Harwood, D.S., 1988, Tectonism and metamorphism in the northern Sierra terrane, northern California, *in* Ernst, W.D., ed., Metamorphism and crustal evolution of the western United States (Rubey Volume VII): Englewood Cliffs, New Jersey, Prentice-Hall, p. 765–788.

Harwood, D.S., 1992, Stratigraphy of Paleozoic and lower Mesozoic rocks in the northern Sierra terrane, California: U.S. Geological Survey Bulletin 1957, 78 p.

Herron, M.M., 1988, Geochemical classification of terrigeneous sands and shales from core or log data: Journal of Sedimentary Petrology, v. 58, p. 820–829.

Hoffman, P.F., 1989, Precambrian geology and tectonic history of North America, *in* Bally, A.W., and Palmer, A.R., eds., the geology of North America—An overview: Boulder, Colorado, Geological Society of America, Geology of North America, v. A, p. 447–512.

Hooper, P.R., Johnson, D., and Conrey, R.M., 1993, Major and trace element analyses of rocks and minerals by automated x-ray spectometry: Pullman, Washington State University, Department of Geology Open File Report, 36 p.

Knaack, C., Cornelius, S., and Hooper, P.R., 1994, Trace element analyses of rocks and minerals by ICP-MS: Pullman, Washington State University, Department of Geology Open-File Report, 18 p.

Lawrence, J.L., 1996, Assessing the source of terrigenous detritus in the Shoo Fly

Complex, northern Sierra Nevada, California: Implications for the Paleozoic development of western North America [M.S. thesis]: San Diego, California, San Diego State University, 119 p.
Lichte, F.E., Meier, A.L., and Crock, J.G., 1987, Determination of rare earth elements in geological materials by inductively coupled mass spectrometry: Analytical Chemistry, v. 59, p. 1150–1157.
McLennan, S.M., 1989, Rare earth elements in sedimentary rocks: Influences of provenance and sedimentary processes: Mineralogical Society of America Reviews in Mineralogy, v. 21, p. 169–200.
McLennan, S.M., and Taylor, S.R., 1991, Sedimentary rocks and crustal evolution: Tectonic setting and secular trends: Journal of Geology, v. 99, p. 1–21.
McLennan, S.M., Taylor, S.R., McCulloch, M.T., and Maynard, J.B., 1990, Geochemical and Nd-Sr isotopic composition of deep-sea turbidites: Crustal evolution and plate tectonic associations: Geochemica et Cosmochimica Acta, v. 54, p. 2015–2050.
McLennan, S.M., Hemming, S., McDaniel, D.K., and Hanson, G.N., 1993, Geochemical approaches to sedimentation, provenance, and tectonics, *in* Johnsson, M.J., and Basu, A., eds., Processes controlling the composition of clastic sediments: Geological Society of America Special Paper 284, p. 21–40.
Mendenhall, W., and Beaver, R.J., 1994, Introduction to probability and statistics: Belmont, California, Duxbury Press, 704 p.
Poole, F.G., Stewart, J.H., Palmer, A.R., Sandberg, C.A., Madrid, R.J., Ross, R.J., Jr., Hintze, L.F., Miller, M.M., and Wrucke, C.T., 1992, Latest Precambrian to latest Devonian time; development of a continental margin, *in* Burchfiel, B.C., et al., eds., The Cordilleran orogen: Conterminous U.S.: Boulder, Colorado, Geological Society of America, Geology of North America, v. G-3, p. 9–56.
Prave, A.R., Fedo, C.M., and Cooper, J.D., 1991, Lower Cambrian depositional and sequence stratigraphic framework of the Death Valley and eastern Mojave Desert region, *in* Walawender, M.J., and Hanan, B.B., eds., Geological excursions in southern California and Mexico; San Diego, California, San Diego State University, Department of Geological Sciences, p. 147–170.
Schweickert, R.A., and Lahren, M.M., 1990, Speculative reconstruction of the Mojave–Snow Lake fault: Implications for Paleozoic and Mesozoic orogenesis in the western United States: Tectonics, v. 9, p. 1609–1629.
Schweickert, R.A., and Snyder, W.S., 1981, Paleozoic plate tectonics of the Sierra Nevada and adjacent regions, *in* Ernst, W.D., ed., The geotectonic development of California (Rubey Volume I): Englewood Cliffs, New Jersey, Prentice-Hall, p. 182–202.
Stewart, J.H., 1980, Geology of Nevada, a discussion to accompany the geologic map of Nevada: Nevada Bureau of Mines and Geology Special Publication 4, 132 p.
Stewart, J.H., Poole, F.G., Ketner, K.B., Madrid, R.J., Roldan-Quintana, J., and Amaya-Martinez, R., 1990, Tectonics and stratigraphy of the Paleozoic and Triassic southern margin of North America, Sonora, Mexico, *in* Gehrels, G.E., and Spencer, J.E., eds., Geologic excursions through the Sonoran Desert region, Arizona and Sonora: Arizona Geological Survey Special Paper 7, p. 183–202.
Taylor, S.R., and McLennan, S.M., 1985, The continental crust: Its composition and evolution: Oxford, Blackwell Scientific Publications, 312 p.
Van Schmus, W.R., and 24 others, 1993, Transcontinental Proterozoic Provinces, *in* Reed, J.C., et al., eds., Precambrian: Conterminous U.S.: Boulder, Colorado, Geological Society of America, Geology of North America, v. C-2, p. 171–334.

Manuscript Accepted by the Society January 24, 2000

Geological Society of America
Special Paper 347
2000

# *Triassic evolution of the arc and backarc of northwestern Nevada, and evidence for extensional tectonism*

**Sandra J. Wyld**
*Department of Geology, University of Georgia, Athens, Georgia 30602, USA*

## ABSTRACT

**The early Mesozoic marine province of Nevada includes a shallow marine shelf terrane to the east, the Black Rock arc terrane to the west, and an intervening deep marine basinal terrane. A new integrated analysis of the Triassic record across the northern marine province indicates the following history. From the Early to late Middle Triassic, the province was affected by differential uplift and subsidence, subaerial to basinal sedimentation, and intermittent volcanism. Regional subsidence then occurred in the late Middle to early Late Triassic, leading to marine deposition on top of previously exposed areas. During this time, carbonate deposition occurred in all areas, coarse clastics from nearby basement uplifts were shed into the margins of the province, and volcanism occurred in the basinal and shelf terranes. Regional subsidence then continued in the latest Triassic, as manifested by the accumulation of very thick (up to 6 km) successions of marine strata: but in the Black Rock terrane, these strata consist of volcanogenic arc deposits; whereas, in the shelf and basinal terranes, these strata consist exclusively of continentally-derived clastics. Based on these data, the Black Rock, basinal and shelf terranes are interpreted to have evolved together in the Triassic in an extensional tectonic regime that culminated in the opening of a wide basin between the shelf and a volcanic arc. Extensional tectonism led to differential uplift and subsidence in the early history of the province, regional subsidence and accumulation of thick marine successions across the province in the late Middle to Late Triassic, and progressive isolation of the Black Rock terrane from the basinal and shelf terranes as the extensional basin widened.**

## INTRODUCTION

Triassic strata in western Nevada can be divided into three basic lithologic successions (Fig. 1; Speed, 1978a): a shallow-marine succession to the east that contains abundant carbonate and was deposited in a shelf environment (shelf terrane); a volcanic-rich marine succession to the west; and an intervening deep-marine basinal succession dominated by fine-grained siliciclastic rocks (basinal terrane). The volcanic-rich sequences to the west form part of a group of early Mesozoic magmatic arc assemblages that extend into California and Arizona (Fig. 1). Collectively, the volcanic arc, basinal, and shelf terranes of western Nevada form elements of an early Mesozoic marine province that evolved in the region following the latest Permian to earliest Triassic Sonoma orogeny (Speed, 1978a; Saleeby and Busby-Spera, 1992). This orogenic event involved deformation and thrusting of Paleozoic basinal assemblages onto the continental margin (Golconda allochthon) and accretion of a Paleozoic oceanic arc to the margin outboard of the allochthon (Miller et al., 1992). The Paleozoic arc forms the basement for early Mesozoic arc assemblages in northwest Nevada and northern California (Miller and Harwood, 1990; Wyld, 1990), and the Golconda allochthon forms the basement for the early Mesozoic shelf terrane (Nichols and Silberling, 1977). Early

Wyld, S.J., 2000, Triassic evolution of the arc and backarc of northwestern Nevada, and evidence for extensional tectonism, *in* Soreghan, M.J., and Gehrels, G.E., eds., Paleozoic and Triassic paleogeography and tectonics of western Nevada and northern California: Boulder, Colorado, Geological Society of America Special Paper 347, p. 185–207.

Mesozoic subsidence above the Sonoma orogenic belt and development of the marine province was terminated in the Jurassic by shortening deformation within the arc and backarc region. The principal manifestation of this deformation is the Luning-Fencemaker fold-and-thrust belt, which developed within the basinal terrane and involved thrusting of the basinal terrane over the shelf terrane (Fig. 1; Oldow, 1984).

Although the basic features of the western Nevada marine province are understood, two key problems remain unresolved in the evolution of this province. First, it is unclear how or why a marine province developed on top of the Sonoma orogenic belt in the early Mesozoic. Of particular interest is the basinal terrane in which at least 6 km of strata accumulated in latest Triassic (Norian) time (Speed, 1978a), a thickness that requires either substantial syndepositional subsidence or development of a very deep basin prior to the Norian. Several contrasting models have been suggested to explain these relations, including thermal subsidence of the underlying Paleozoic arc (Speed, 1978a, 1979), regional extension (Wyld, 1990, 1992; Burchfiel et al., 1992), inherited Paleozoic bathymetry (Saleeby and Busby-Spera, 1992), and dynamic subsidence associated with initiation of subduction beneath the continental margin (Lawton, 1994), but no consensus has been reached.

Second, it remains unclear how evolution of the early Mesozoic magmatic arc assemblages relates to the evolution of the backarc basinal and shelf terranes. Numerous studies have shown that, despite the fact that the basinal and shelf terranes are everywhere separated by thrust faults, there are many stratigraphic and facies links between the two terranes (Speed, 1978a, 1978b; Lupe and Silberling, 1985; Oldow et al., 1990). The same cannot be said, however, for the volcanic arc assemblages, the relation of which to the adjacent backarc basin is less clear. One reason for this uncertainty is that the only detailed studies within the basinal terrane have been in its eastern part, near the contact with the shelf terrane; little is known about the character of the basinal terrane near its contact with the arc assemblages. Another reason is that a detailed stratigraphy of arc assemblages west of the basinal terrane was lacking until recently, thus making comparisons with the basinal or shelf terranes difficult. An additional complication is the recent recognition that arc assemblages of northern California and west-central Nevada may be substantially displaced from their site of origin by Jurassic-Cretaceous strike-slip faults (see MSL in Fig. 1; Oldow, 1984; Schweickert and Lahren, 1990; Wyld et al., 1996). There are therefore only a few areas in which the original paleogeographic relation between the arc and backarc basin can be evaluated.

In this chapter, I address these two problems by summarizing and integrating existing and new data from the shelf, basinal, and volcanic arc terranes of northern Nevada. For the arc terrane, I focus on rocks in northwest Nevada, which constitute the Black Rock terrane of Silberling et al. (1987) (Fig. 1). This region, the comprehensive Triassic record of which has not previously been documented, contains one of the few early Mesozoic arc assemblages that is not potentially out of place with respect to the basinal terrane (Fig. 1). I also present new data from the westernmost exposed rocks of the basinal terrane immediately adjacent to the Black Rock arc province. These rocks are of particular interest because they include the oldest rocks so far recognized in the basinal terrane and because they provide fundamental new information on how the arc, basinal, and shelf terranes can be related.

The focus in this chapter is on an integrated, up-to-date analysis of the Triassic stratigraphic record of the entire marine province in northwestern Nevada, with the goal of providing a framework for reinterpreting the paleogeographic and tectonic evolution of this province in the early Mesozoic. This approach builds on the pioneering work of Speed (1978a), and leads to the conclusion that the volcanic arc, basinal, and shelf terranes of northwest Nevada evolved together during the Triassic within an extensional tectonic

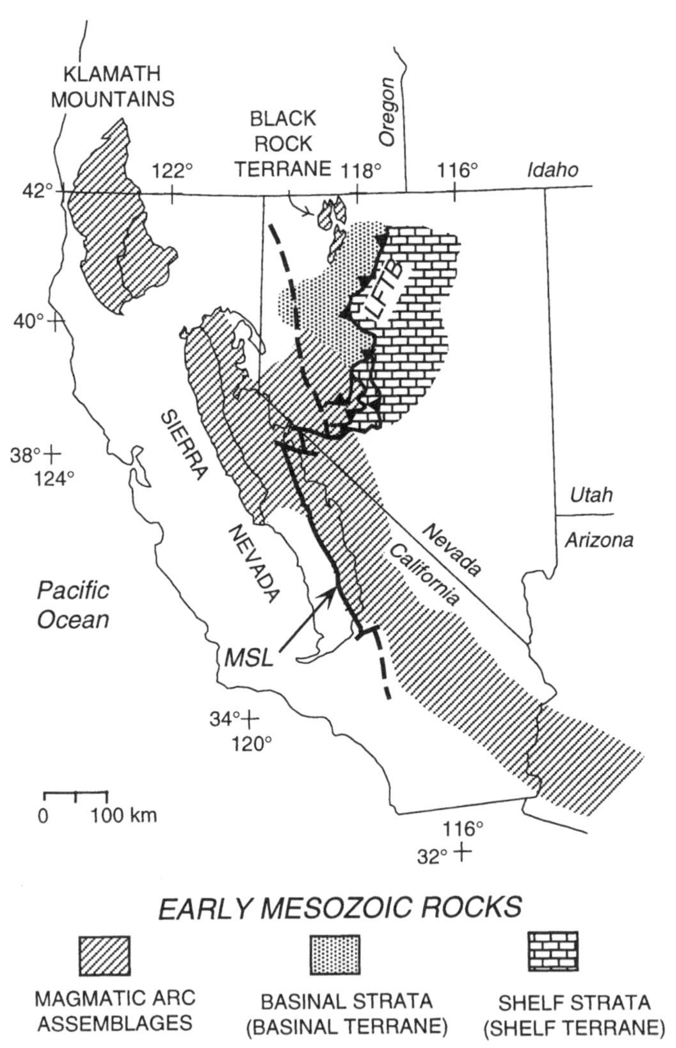


Figure 1. Early Mesozoic (Triassic and Early Jurassic) tectonostratigraphic assemblages of western U.S. Cordillera. LFTB is Luning-Fencemaker fold-and-thrust belt; only frontal thrusts are shown (Oldow, 1984). MSL is Mojave–Snow Lake fault of Schweickert and Lahren (1990), which accommodated ~450 km of dextral Early Cretaceous strike-slip displacement; dashed where location is speculative. Portion of MSL in west-central Nevada corresponds approximately to location of Jurassic-Cretaceous Pine Nut strike-slip fault of Oldow (1984).

setting that ultimately resulted in development of a deep marine backarc basin behind an active Late Triassic arc.

## TERRANE TERMINOLOGY

Groups of rocks within the marine province of northwestern Nevada have been assigned a variety of different names by different workers (Speed, 1978a; Oldow, 1984; Lupe and Silberling, 1985; Silberling et al., 1987). In this paper, I follow Speed (1978a) in calling the shallow-marine succession to the east the shelf terrane and the deep-marine sedimentary succession to the west of this the basinal terrane (Fig. 1). These terranes include, respectively, the Humboldt and Lovelock assemblages discussed in other chapters of this volume. Speed (1978a) originally included early Mesozoic rocks of the Black Rock terrane within his basinal terrane; however, subsequent studies indicate that these rocks are better considered as part of a volcanic arc terrane (Russell, 1984; Lupe and Silberling, 1985; Silberling et al., 1987; Wyld, 1990; Quinn et al., 1997).

In this chapter, I focus primarily on the northern part of the marine province. This facilitates stratigraphic comparisons and circumvents the complication that the Triassic stratigraphy of the arc, basinal, and shelf terranes is somewhat different to the south than it is to the north. Figure 2 shows the mountain ranges in northern Nevada that contain the principal exposures of the shelf, basinal, and arc terranes.

In the following sections, I first summarize the Triassic stratigraphic record in the northern shelf and basinal terranes to provide a basis for comparison with the Triassic record of rocks exposed farther west. I then discuss the Triassic evolution of the Black Rock terrane and relations in the adjacent westernmost basinal terrane. The final part of the chapter integrates data from all these areas into a comprehensive stratigraphic, paleogeographic, and tectonic model.

## STRATIGRAPHY OF THE NORTHERN SHELF TERRANE

Shelf terrane strata can be divided into two groups (Fig. 3): the Lower Triassic to lower Upper Triassic Star Peak Group, which contains abundant carbonate (Nichols and Silberling, 1977); and the overlying Upper Triassic Auld Lang Syne Group, which consists mostly of clastic rocks (Burke and Silberling, 1973). The Star Peak Group was deposited unconformably across older Mesozoic and Paleozoic rocks, including the Lower Triassic Koipato Group and deformed Paleozoic rocks of the Golconda allochthon. The Koipato Group consists of a thick (>3 km) sequence of rhyolitic and less common mafic volcanic strata that were deposited in a shallow-marine to subaerial environment during an episode of block faulting that was probably related to regional extensional tectonism (Silberling and Roberts, 1962; Burke, 1973; Speed, 1978a). Subsequently, erosion of part, and locally all, of the Koipato Group occurred in the latest Early Triassic prior to deposition of the Star Peak Group (Fig. 3). The following description of the Star Peak Group is based on the detailed studies of Silberling and Wallace (1969) and Nichols and Silberling (1977).

Uppermost Spathian to lower Anisian units of the Star Peak Group consist of a mixture of carbonate and locally derived clastic rocks deposited in a shallow-marine to subaerial environment; clastic rocks were derived from either the Koipato Group or internal parts of the Star Peak system. Apparently parts of the shelf terrane in the Spathian to early Anisian were undergoing differential uplift and erosion at the same time that nearby areas were accumulating sediment, a pattern that implies syndepositional tectonism. A substantial unconformity in the central part of the Star Peak outcrop area reflects differential uplift and erosion in the early Anisian (Fig. 3). Scarce mafic volcanic rocks were deposited during this time frame in the northwestern part of the succession (Fig. 3).

Subsequently, in the late Anisian, the entire shelf terrane underwent relative subsidence, resulting in deposition of the

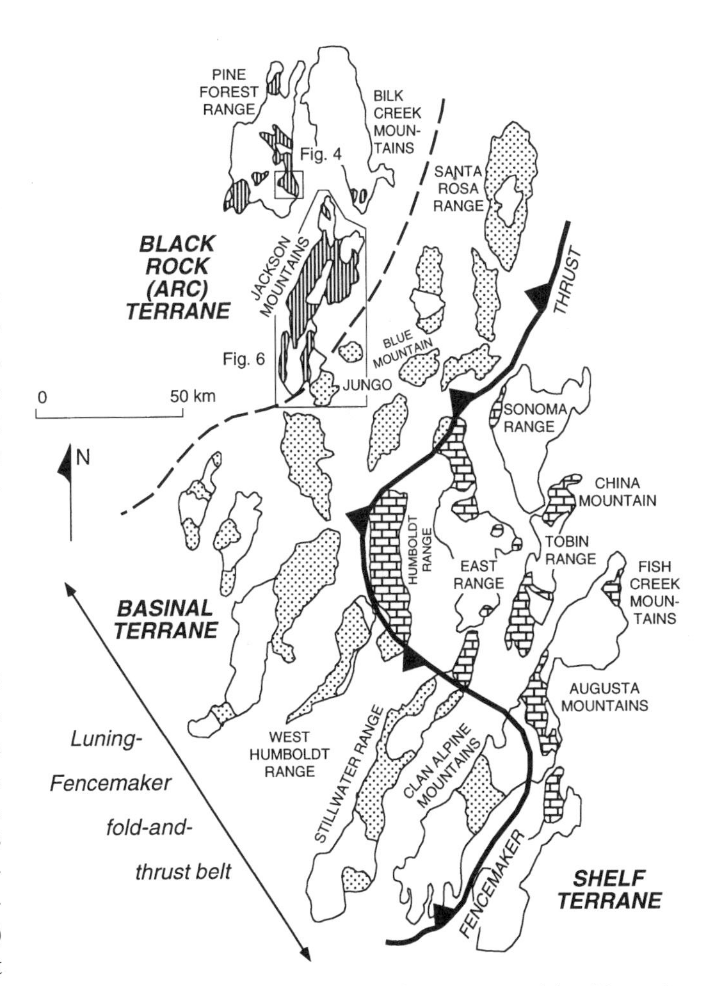


Figure 2. Location map showing mountain ranges containing Mesozoic rocks in Black Rock, basinal, and shelf terranes of northwest Nevada. See Figure 1 for location of terranes. Vertical lines, lower Mesozoic and Paleozoic rocks in Black Rock terrane; dot pattern, lower Mesozoic strata of basinal terrane; limestone pattern, Triassic rocks of shelf terrane. Fold-and-thrust belt is essentially confined to basinal terrane (Oldow, 1984; Wyld, 1998; Rogers, 1999; Folsom, 2000).

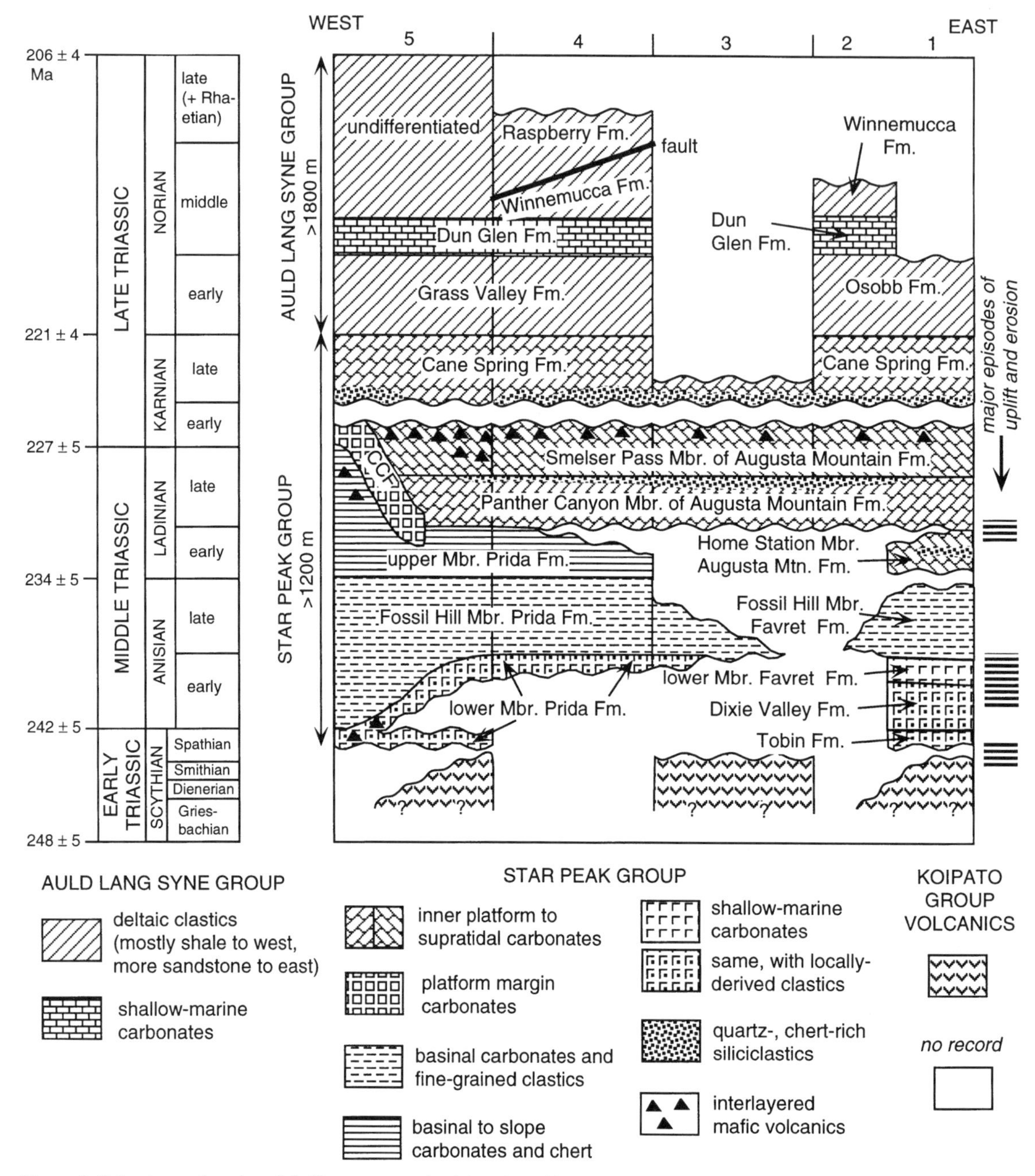


Figure 3. Triassic stratigraphy of shelf terrane, emphasizing depositional environments, episodes of erosion, and stratigraphic variations from west to east (based on Fig. 5 of Nichols and Silberling, 1977). CCF is Congress Canyon Formation. Locations of columns (see Fig. 2): 1 is Augusta Mountains, 2 is southern Tobin Range, 3 is northern Tobin Range, 4 is northern East Range, 5 is Humboldt Range. Stratigraphic data from Silberling and Wallace (1969), Burke and Silberling (1973), Nichols and Silberling (1977), and Lupe and Silberling (1985). Data sources for time scale: absolute ages of stage boundaries are from Gradstein et al. (1994); relative durations of substages are from Silberling and Wallace (1969), Nichols and Silberling (1977), and Lupe and Silberling (1985). Mbr., member, Fm., formation.

basinal Fossil Hill Member of the Prida and Favret Formations across the terrane (Fig. 3). This unit consists of calcareous shale intercalated with thin- to medium-bedded, fine-grained limestone bearing a pelagic fauna. In the early Ladinian, basinal conditions continued in the western part of the shelf, as manifested by the upper member of the Prida Formation, which consists mostly of laminated dark limestone and intercalated chert. Farther east, however, the Home Station member of the Augusta Mountains Formation consists mostly of shallow-marine to supratidal carbonates that record construction of a carbonate platform. Some siliciclastic strata were deposited on this platform; these strata are rich in chert and quartzose detritus and were apparently derived from basement rocks in the central shelf terrane that were differentially uplifted and eroded during a period of mid-Ladinian tectonism (Fig. 3).

Mid-Ladinian uplift and erosion were followed by a period of regional subsidence during which carbonate platform development resumed, and the platform margin migrated westward to the western part of the shelf terrane. The platform succession is represented by shallow-marine to supratidal carbonates of the upper Ladinian to lower Karnian Panther Canyon and Smelser Pass members of the Augusta Mountains Formation and the middle to upper Karnian Cane Spring Formation (Fig. 3). Carbonate platform development appears to have occurred during an episode of syndepositional subsidence, as the shallow-marine carbonate sequence is on average about 500 m thick. Carbonate deposition was interrupted briefly in the mid-Karnian by a period of relative uplift that produced a regional disconformity, and by episodic influx of coarse siliciclastic material (silt to cobble size) (Fig. 3). These clastic strata are rich in chert and quartzose detritus, and were derived from basement sources to the north and east, based on spatial variations in thickness and grain size of the clastics. Volcanism also affected the shelf region during the late Ladinian to earliest Karnian, as manifested by the widespread presence of mafic lavas, breccias, and tuffs in the upper Smelser Pass member and uppermost Prida Formation (Fig. 3).

The record of sedimentation and erosion within the Star Peak Group provides evidence for syndepositional tectonism. Key lines of evidence are complex facies patterns within the Star Peak Group, both spatial and temporal; the presence of intraformational unconformities and evidence for differential uplift and/or erosion and subsidence and/or deposition across the shelf; episodic influx of coarse-grained clastic detritus from local or nearby source terranes; and episodes of regional subsidence that are required by the substantial thickness of shallow marine strata contained within the Star Peak Group (Fig. 3; Nichols and Silberling, 1977; Lupe and Silberling, 1985). These features appear most consistent with extensional tectonism and block faulting during Star Peak deposition, as first suggested by Nichols and Silberling (1977). Thus the extensional tectonism that was ongoing during deposition of the Lower Triassic Koipato Group appears to have continued during deposition of the upper Lower Triassic to lower Upper Triassic Star Peak Group. Intermittent bimodal and basaltic volcanism within the shelf terrane during this time frame may also be related to extensional tectonism (cf. Speed, 1978a), and is suggestive of magmatism associated with continental rifting (e.g., Barberi et al., 1982).

Processes affecting the shelf region changed substantially in the Norian, when siliciclastic material of the Auld Lang Syne Group overwhelmed carbonate deposition, and regional relative subsidence resulted in uniform patterns of sedimentation across the region (Fig. 3). Auld Lang Syne deposition began in the earliest Norian and continued into the Early Jurassic. Most of this sequence is composed of Norian clastic strata, found in the Grass Valley, Osobb, Winnemucca, and Raspberry Formations, and local undifferentiated units (Fig. 3; Silberling and Wallace, 1969; Burke and Silberling, 1973; Lupe and Silberling, 1985). These units consist largely of shale with subordinate quartz sandstone, although the abundance of sandstone increases to the east. Facies analysis, compositional data, and regional relations indicate that Auld Lang Syne Group clastic strata were derived from the continental interior and transported via fluvial systems to the shelf, where they were deposited in a shallow-marine deltaic environment (Burke and Silberling, 1973; Lupe and Silberling, 1985). The abrupt influx of clastic material to the shelf in the earliest Norian, followed by near cessation of carbonate deposition, is interpreted to reflect a reorganization of drainage patterns within the western part of the continent such that the shelf was flooded by clastic detritus and conditions there became generally unfavorable for carbonate buildup (Lupe and Silberling, 1985). Regional subsidence during deposition of the Auld Lang Syne Group is implied by the substantial thickness (~2 km) of the Norian deltaic strata.

## STRATIGRAPHY OF THE NORTHERN BASINAL TERRANE

The basinal terrane, as defined by Speed (1978a, 1978b), consists of early Mesozoic siliciclastic and minor carbonate strata deposited in a deep-marine basin west of the shelf terrane. These strata were also referred to as part of the Auld Lang Syne Group by Silberling and Wallace (1969), Burke and Silberling (1973), and Lupe and Silberling (1985). Correlation is based on the observation that the basinal terrane strata are the same age as and composed of rock types similar to those found in the Auld Lang Syne Group of the shelf terrane, while recognizing that different facies and depositional environments are represented by the two groups of rocks. As pointed out by Speed (1978a) and Oldow et al. (1990), it is not necessarily useful or valid to correlate these two different packages of rocks at the formational level, but no other group name has been proposed. For purposes of clarity, I refer to the western sequence as the basinal ALS group (short for basinal Auld Lang Syne Group equivalents).

Unlike the shelf terrane, the stratigraphy of the basinal terrane has been worked out in detail in only a few areas, but reconnaissance studies throughout the terrane have confirmed its monotonously uniform character, which has led to the common field term "mudpile" for the basinal ALS group (Compton, 1960; Willden, 1964; Burke and Silberling, 1973; Speed, 1978a, 1978b; Lupe and Silberling, 1985; Oldow et al., 1990; my own mapping). The dominant rock type in the basinal ALS group is shale (now slate or phyllite). Much less common, but locally abundant, is quartz sandstone. Carbonate and calcareous rocks are in general rare, although they are locally common near the contact with the shelf terrane. Biostratigraphic data, coupled with correlations with Auld Lang Syne Group units of the shelf terrane, indicate that the basinal ALS group primarily spans the early through latest Norian; locally, however, Early Jurassic strata have been recognized at the top of the Norian section (Compton, 1960; Burke and Silberling, 1973; Speed, 1978a, 1978b; Lupe and Silberling,

1985; Oldow et al., 1990; Thole and Prihar, 1998). Facies data indicate deposition in a deep-marine environment throughout most of the Norian, with shoaling of the basin to shallower marine depths in the latest Norian to Early Jurassic (Speed, 1978a; Lupe and Silberling, 1985; Oldow et al., 1990). Basement rocks are not exposed in most places; the only known exception is described in a later section.

Speed (1978a, 1978b), Lupe and Silberling (1985), and Oldow et al. (1990) have provided compelling evidence that Norian rocks of the basinal ALS group represent the deeper marine equivalents of the Norian deltaic clastics of the shelf terrane; this conclusion is based on facies patterns, and on similarities in composition and stratigraphic architecture between the two groups of strata. One of the most distinctive features of the Norian basinal ASL group, however, is the stratigraphic thickness of this succession, which is at least 6 km (Compton, 1960; Burke and Silberling, 1973; Speed, 1978b), in comparison with the stratigraphic thickness of less than 2 km for correlative Auld Lang Syne Group strata on the shelf (Fig. 3). This thickness indicates that either the basinal ALS group strata were deposited in a deep basin that developed prior to the early Norian or that they were deposited in an actively subsiding basin.

## TRIASSIC EVOLUTION OF THE BLACK ROCK TERRANE

Triassic strata in the Black Rock terrane are found in the southeastern Pine Forest Range, the southern Bilk Creek Mountains, and the western, eastern, and southeast Jackson Mountains (Fig. 2). Although potential correlations between the strata in these various areas have been mentioned in some prior publications (Silberling et al., 1987; Wyld, 1990; Quinn et al., 1997), they have not previously been linked in any detail or discussed in terms of a regional stratigraphic record. It is the purpose of this section to describe and compare the Triassic stratigraphy of and processes affecting the different areas, and to use these data to interpret the Triassic evolution of the terrane. It is important to stress that the regional record emerging from this analysis provides a crucial new framework for evaluating the role of the Black Rock terrane within the northern Nevada marine province; all previous attempts to interpret the Triassic history of the Black Rock terrane or to integrate its history with that of the larger marine province have been hindered by incomplete, erroneous, or conflicting data from the terrane.

Three key aspects of the Triassic geology of the Black Rock terrane are necessary to mention prior to describing the Triassic stratigraphy. First, Triassic strata of the southeastern Pine Forest Range form a coherent stratigraphic succession that is well dated from base to top and that has been divided into a number of distinct units (Wyld, 1990, 1992). This stratigraphy serves as an important template for interpreting scattered exposures of less well dated Triassic rocks in the Jackson Mountains. Second, the Mesozoic geology of the Jackson Mountains has recently been substantially reinterpreted based on the recognition that a large unit previously thought to be volcanic is mostly intrusive, and based on new correlations with the Triassic rocks of the Pine Forest Range (Quinn, 1996; Quinn et al., 1997). The new unit divisions of Quinn et al. (1997) are used in this chapter; see this reference for a complete discussion of older unit divisions (Willden, 1964; Russell, 1981, 1984; Maher, 1989). Third, all areas in the Black Rock terrane were regionally deformed and metamorphosed in the Jurassic (Quinn, 1996; Wyld, 1996). These younger processes did not signficantly obscure protolith textures or compositions, and the rocks are therefore all described in terms of protoliths. The Triassic record across the region can be divided conveniently into three distinct periods—Early to late Middle Triassic, late Middle to early Late Triassic, and latest Triassic—and the next discussion follows this division.

### *Early to Middle Triassic*

Early(?) to Middle Triassic strata in the Black Rock terrane are exposed only in the southern Bilk Creek Mountains (Fig. 2), where they have been interpreted as either conformable or disconformable on Upper Permian rocks (Ketner and Wardlaw, 1981; Jones, 1990; Blome and Reed, 1995). These strata (the Quinn River Formation) were deposited in a relatively deep marine environment, and consist of ~130 m of shale and siltstone, both variably siliceous to cherty, with less common tuffaceous sandstone. Fossil ages range from Early(?) Triassic to Anisian in the lower part of the unit to Ladinian and early Karnian at the top. Volcaniclastic input appears to have been primarily in the Early(?) Triassic and Ladinian (Jones, 1990).

In striking contrast, the latest Permian to late Middle Triassic is missing in the Pine Forest Range across a slightly angular unconformity (Fig. 4; Wyld, 1990). In the Jackson Mountains, Permian strata are everywhere separated from Triassic strata by faults or intrusions (Quinn, 1996; Quinn et al., 1997), but no Triassic rocks older than Late Triassic are known (Russell, 1981, 1984; Fuller, 1986; Maher, 1989), suggesting that an Early to Middle Triassic unconformity is also present in the Jackson Mountains. In light of the large area of exposure of Triassic strata in the Pine Forest Range and Jackson Mountains, relative to that in the Bilk Creek Mountains (Fig. 2), these relations collectively suggest that most of the Black Rock terrane either was emergent in the Early to Middle Triassic or underwent substantial uplift and erosion at the end of the Middle Triassic. Local areas (Bilk Creek Mountains), however, were obviously submerged basins that were never uplifted and eroded away. These relations imply a complex paleogeography and history of uplift and subsidence for the Early and Middle Triassic, although detailed relations between the different areas are now obscured by Cenozoic cover (Fig. 2).

The only other data concerning the evolution of the Black Rock terrane in the Early to Middle Triassic come from the central to northern Pine Forest Range, where dioritic stocks and plutons dated as ca. 230–235 Ma (preliminary U-Pb zircon) locally intrude Paleozoic strata (Wyld, 1990; Wyld et al., 1996; Wyld and Wright, 1997). These intrusions and the volcaniclastic interlayers in the Bilk Creek Mountains indicate magmatic activity in the northern Black Rock terrane in the Ladinian and Early Triassic.

### *Late Middle to early Late Triassic*

Only a few tens of meters of post-Anisian strata (mostly siltstone and shale) are exposed in the Bilk Creek Mountains (Ketner and Wardlaw, 1981; Jones, 1990; Blome and Reed, 1995), but extensive exposure of younger Triassic rocks is found in the Pine Forest Range and Jackson Mountains. In the latter areas, these strata can be divided into two distinct lithologic successions; the older consists of carbonates and siliciclastic rocks yielding Ladinian or Karnian to early Norian fossils, and the younger consists mostly of Norian volcanogenic rocks. The carbonate-siliciclastic succession is the focus of this section. Stratigraphic relations are best defined in the southeastern Pine Forest Range, described first.

***Pine Forest Range.*** Middle(?) to Late Triassic carbonates and siliciclastic rocks in the Pine Forest Range unconformably overlie Paleozoic rocks, and were originally described by Smith (1966, 1973), and then by Wyld (1990), who referred to them as the Triassic limestone and clastic unit. These rocks were renamed the Bishop Canyon formation (informal) by Wyld (1992; Fig. 4). Three members are defined in the Bishop Canyon formation (Fig. 5): a lower limestone member (TrB1), a middle siliciclastic member (TrB2), and an upper limestone member (TrB3). Fossil data indicate that the formation ranges from Ladinian or Karnian at the base to early Norian at the top (Fig. 5; Wyld, 1990). Because the unit is quite thin (200–400 m; Fig. 5), it is considered most likely that the base is no older than Karnian; this is consistent with age data from correlative strata in the Jackson Mountains. The following description is based on Wyld (1992).

Both limestone members contain an abundance of massive to thinly laminated limestone, which is generally lacking in any

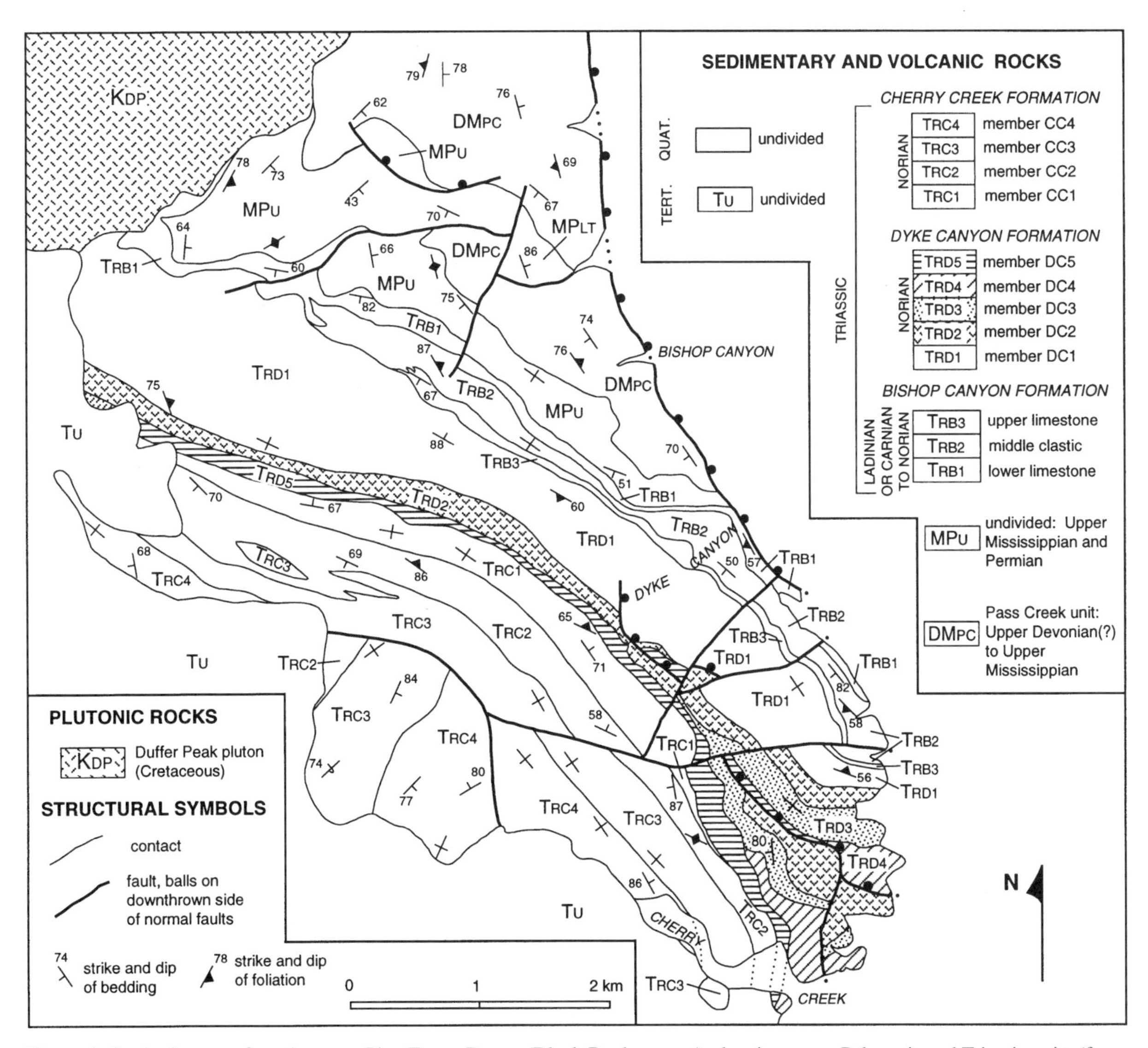


Figure 4. Geologic map of southeastern Pine Forest Range (Black Rock terrane), showing upper Paleozoic and Triassic units (from Wyld, 1992). For location, see Figure 2.

macrofossils. Other rock types in members TrB1 and TrB3 include thinly interbedded limestone and shale, and massive to normally graded limestone conglomerate and sandstone. These strata represent a combination of pelagic to hemipelagic and sediment gravity-flow deposits, and are consistent with deposition at the base of the slope, offshore from a carbonate bank. The middle siliciclastic member consists of a poorly organized association of massive and inversely to normally graded conglomerate (pebble to cobble), massive to graded sandstone, siltstone, and limestone turbidites. These rocks are best interpreted as reflecting deposition in the proximal part of a debris apron at the base of the slope. The clastic debris includes a mixture of sedimentary lithic fragments (argillite, siliceous argillite, and less common quartzo-feldspathic clastics and carbonate) and recycled volcanic material (plagioclase crystal fragments and intermediate to silicic volcanic lithic fragments); quartz and chert clasts are minor constituents. The possible source of this clastic detritus is discussed in the following. The prevalence of carbonates in the Bishop Canyon formation and restriction of siliciclastics to the middle of the unit suggest that influxes of siliciclastic debris episodically interrupted what was otherwise a carbonate-dominated system.

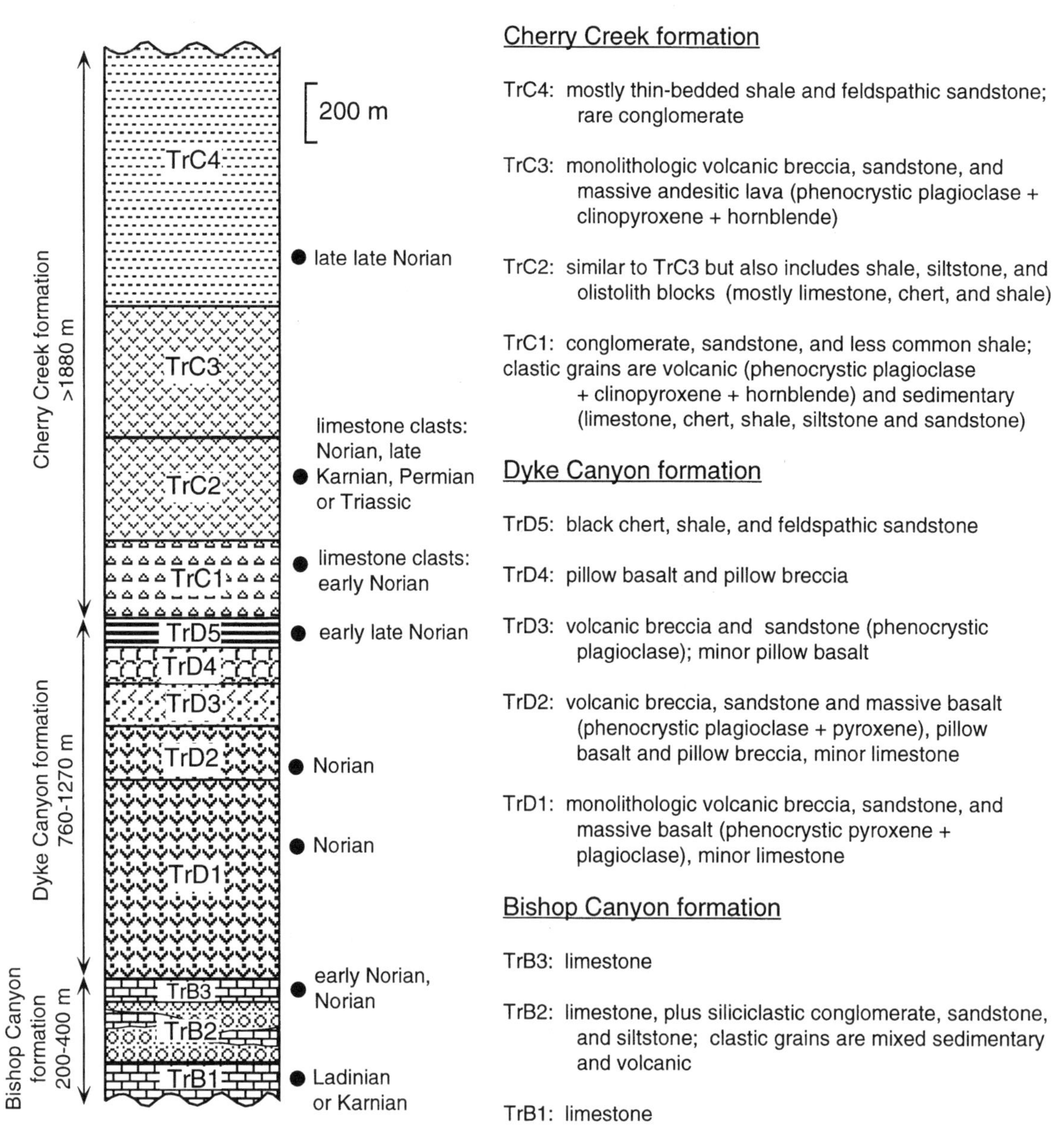


Figure 5. Stratigraphy of Triassic rocks in Pine Forest Range (Black Rock terrane). Data are from Wyld (1990, 1992). Solid circles indicate stratigraphic level of dated fossil localities.

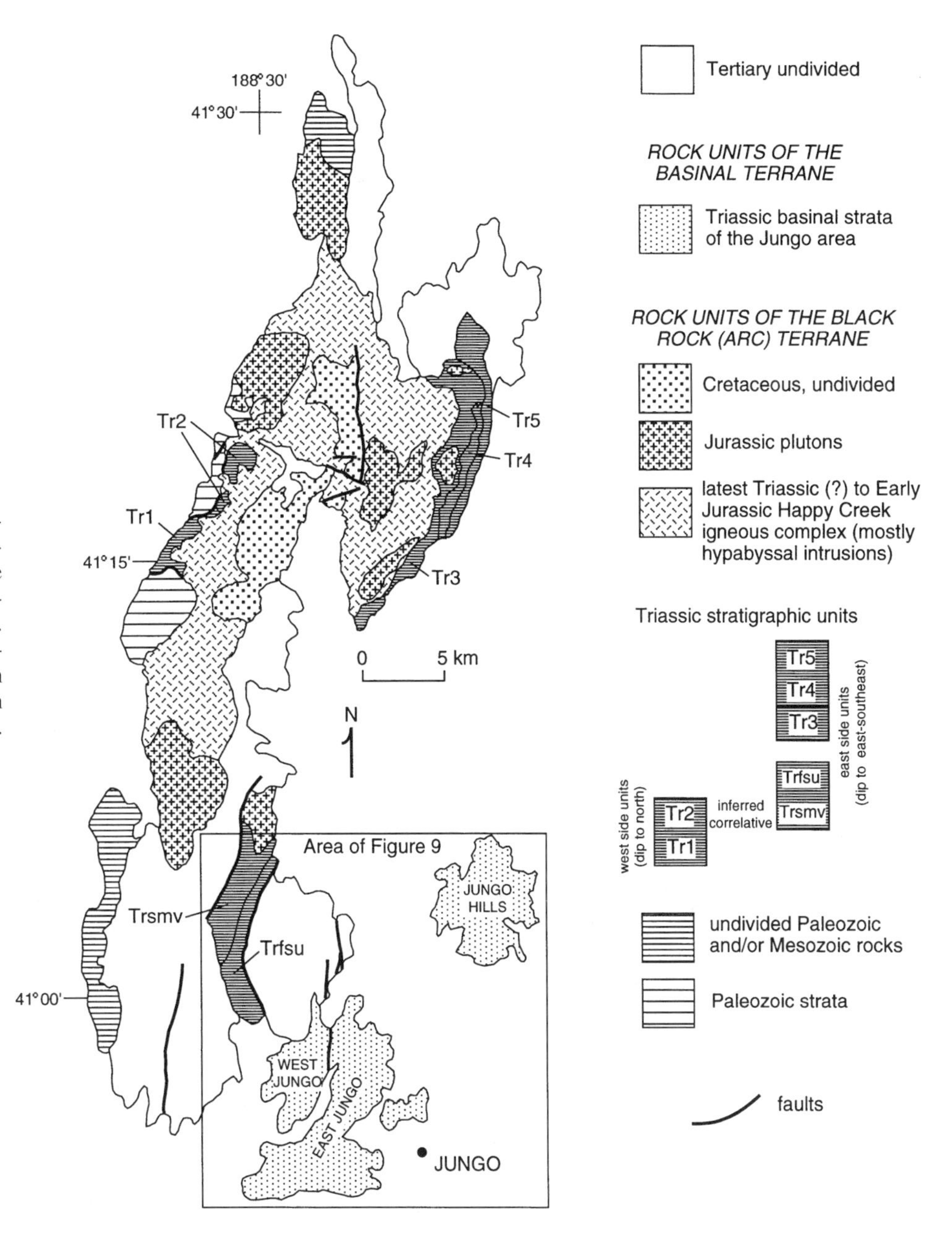


Figure 6. Geologic map of Jackson Mountains and Jungo area (Black Rock and western basinal terranes). For location, see Figure 2. Geology of central Jackson Mountains from Quinn (1996) and Quinn et al. (1997). Geology of northern and southwestern Jackson Mountains from Willden (1964). Geology of southeast Jackson Mountains from Wyld (my own mapping).

***Jackson Mountains.*** Strata similar to the Bishop Canyon formation are exposed on the west side of the Jackson Mountains (Fig. 6) and were referred to as unit Tr1 by Quinn (1996) and Quinn et al. (1997). Unit Tr1 corresponds to part of the Triassic Boulder Creek beds defined by Russell (1981, 1984). The following description of unit Tr1 is taken from Quinn (1996) and Quinn et al. (1997), except where noted.

Unit Tr1 consists of siliciclastic sedimentary rocks and limestone from which in situ fossils indicate deposition in the Karnian and early (to middle?) Norian (Fig. 7; Russell, 1981; Fuller, 1986; Maher, 1989; Quinn, 1996). Siliciclastic rocks include a poorly organized assemblage of argillite, massive pebbly debris flows, and graded sandstone and pebble conglomerate interbedded with siltstone. Much of the limestone in this unit occurs as large olistoliths (to 700 m long) enclosed by coarse siliciclastic strata; these limestone blocks yield Late Triassic fossils and therefore were apparently derived from a coeval carbonate terrane. Some thin- to medium-bedded limestone is also interlayered locally with the fine-grained clastic strata. Quinn (1996) interpreted these strata to have been deposited in a debris-apron setting, near the base of the slope and close to a carbonate bank; episodic influx of coarse clastic detritus was separated by quieter periods of pelagic to hemipelagic sedimentation, and carbonate olistolith blocks represent gravity-driven slide blocks derived from the carbonate bank. Clasts in the coarser siliciclastic rocks include plagioclase and felsic to intermediate volcanic lithic fragments, with less common clasts of limestone, quartzo-feldspathic clastics, argillite, chert, and quartz (Quinn, 1996).

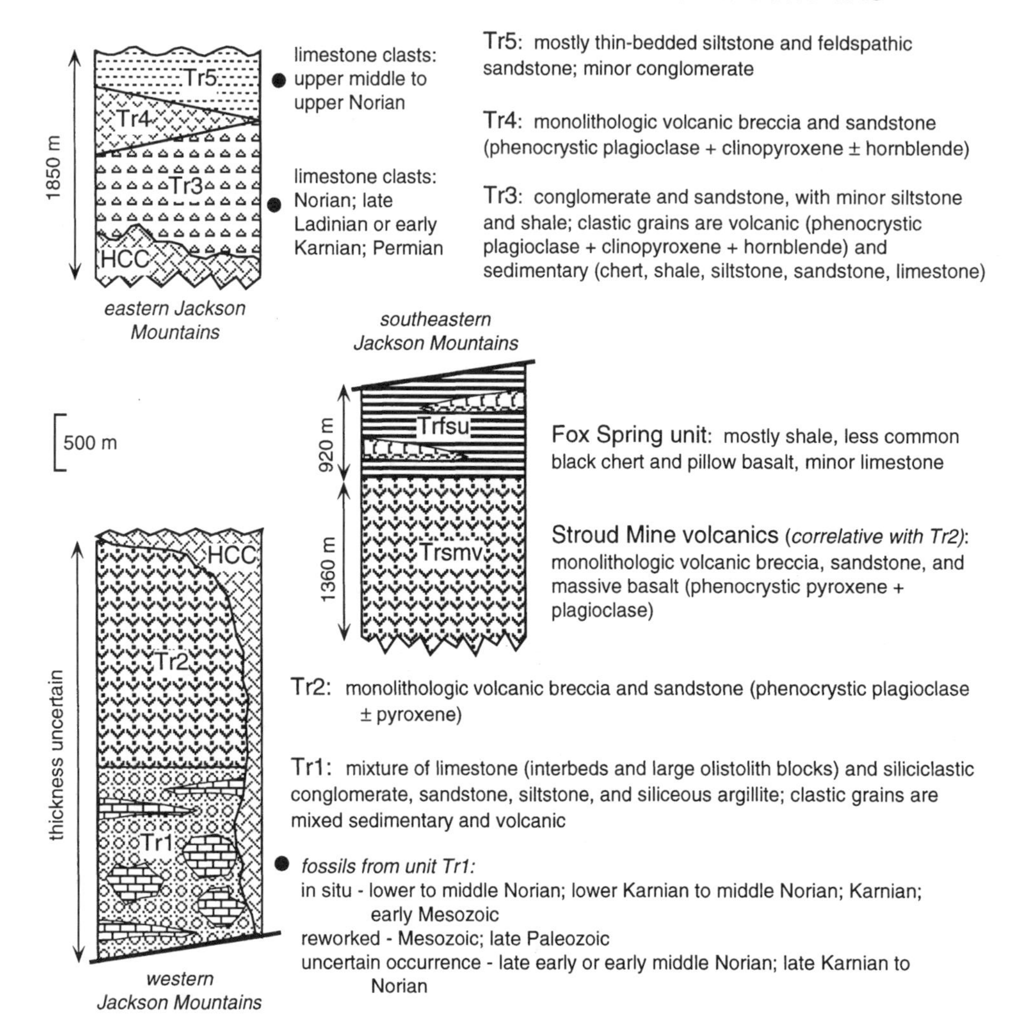


Figure 7. Stratigraphy of Triassic rocks in Jackson Mountains (Black Rock terrane). Eastern and western Jackson Mountains data are from Quinn (1996). Southeastern Jackson Mountains data are from Wyld (my own mapping). Solid circles indicate biostratigraphic data. Biostratigraphic data sources (summarized in Quinn, 1996) are Willden (1964), Russell (1981), Fuller (1986), and Maher (1989); uncertain occurrence indicates that it is unclear whether fossils are in situ or reworked. HCC, hypabyssal intrusive rocks of latest Triassic(?) to Early Jurassic Happy Creek complex (Quinn et al., 1997). Stratigraphic thicknesses of units Tr1 and Tr2 are uncertain due to polyphase deformation, faulting, and widespread intrusion by HCC (Quinn, 1996).

Unit Tr1 is similar in age, overall composition, and general depositional environment to the Bishop Canyon formation, leading to the conclusion that these units are correlative, as argued by Quinn (1996). Clastic material in both units was all evidently derived from the same source terrane, which was composed mostly of sedimentary plus felsic to intermediate volcanic rocks. The coarse grain size of much of the debris further indicates that the source terrane was relatively close to the site of deposition. These requirements suggest that the most likely source terrane consisted of Paleozoic basement rocks either within the Black Rock terrane or in nearby arc assemblages of the Cordillera, all of which consist of a mixture of appropriate source rock types. This conclusion is supported by detrital zircon analyses of the Bishop Canyon formation clastics (Darby et al., this volume, Chapter 5) and by the presence of reworked late Paleozoic fossils in clastic rocks of unit Tr1 (Fig. 7; Russell, 1981).

### *Latest Triassic*

A succession of dominantly volcanogenic Norian strata conformably overlies the Karnian to lower Norian carbonate-siliciclastic assemblage in the Pine Forest Range, and very similar strata are also found in the Jackson Mountains. These volcanogenic rocks have magmatic arc chemistry and reflect the onset of voluminous arc volcanism in the Black Rock terrane (Russell, 1981; Wyld, 1990, 1992; Quinn, 1996; Quinn et al.,

1997). Key features of the volcanogenic succession in the Pine Forest Range are described first, followed by descriptions of correlative rocks in the Jackson Mountains.

***Pine Forest Range.*** Norian volcanogenic rocks in the Pine Forest Range were originally described by Smith (1966, 1973), who erroneously interpreted them to be Paleozoic. Later studies (Silberling et al., 1987; Wyld, 1990) demonstrated the Norian age and led to division of the succession into five units. This stratigraphy was then revised by Wyld (1992), who divided the volcanogenic succession into the (informal) Dyke Canyon formation and overlying Cherry Creek formation, each of which contains a number of distinct members (Fig. 4). The description that follows is based on Wyld (1992).

The Dyke Canyon formation consists of a thick succession of mafic volcanogenic and minor sedimentary rocks that can be divided into five members (Fig. 5). Biostratigraphic data indicate that the formation extends from the early or middle Norian to the early late Norian (Fig. 5; Wyld, 1990). The lower four members are volcanogenic, consisting almost entirely of basaltic lava and generally coarse volcaniclastic detritus. Member TrD1 comprises more than half the formation and consists of a monolithologic suite of massive basalt lava and volcanic breccia, with less common volcanic sandstone and scattered limestone lenses. Phenocrystic or fragmental plagioclase and pyroxene are abundant in these rocks, and the large size (to 1 cm) of the pyroxene leads to the field term "augite porphyry" for this unit. Member TrD2 is similar to TrD1 but also contains pillow basalt and pillow breccia, and many of the volcaniclastic rocks have a calcareous or chlorite-rich (after mafic glass) matrix. Breccias and sandstones in these two units are generally very thick bedded and internally structureless. Member TrD3 contains an abundance of well-bedded volcanic breccia and sandstone, deposited by sediment gravity flows, with minor pillow basalt and pillow breccia. Member TrD4 consists of pillow basalt (variably vesicular and generally aphyric) and pillow breccia with interpillow and/or interclast carbonate.

Collectively, the lower four members of the Dyke Canyon formation are manifestations of a cycle of mafic volcanism, during which lavas were interlayered with compositionally similar, coarse-grained, primary volcaniclastic debris. Deposition was proximal to volcanic vents and, on the basis of facies analysis, occurred in a relatively deep marine environment at the base of the slope (Wyld, 1992). In contrast with the lower four members, the upper member of the Dyke Canyon formation (TrD5) consists of thinly interbedded black chert and shale, massive to laminated shale, and minor thin-bedded limestone and feldspathic sandstone. This member is interpreted to reflect cessation of the Dyke Canyon volcanic cycle and a return to quiescent pelagic to hemipelagic sedimentation, interrupted only by episodic deposition of distal detritus eroded from the volcanic pile (Wyld, 1992).

The Cherry Creek formation depositionally overlies the Dyke Canyon formation, and consists of a thick succession of clastic rocks with minor interlayered lavas that can be divided into four members (Fig. 4). Biostratigraphic data indicate deposition from the early late Norian to the latest Norian (Wyld, 1990). Member TrC1 consists mostly of conglomerate and sandstone containing volcanic and sedimentary clasts, with less common shale. A distinctive feature of this member is an abundance of debris flows and normally graded conglomerate and sandstone beds. Another distinctive feature is the presence of numerous anomalously large limestone clasts, one of which has yielded early Norian fossils (Fig. 5). Member TrC2 is dominated by massive volcanic breccia, but also includes andesite lava flows, and interbedded volcanic breccia, sandstone, and shale. A distinctive feature of this unit is the presence, in coarser clastic beds, of olistolith blocks to 16 m in diameter, composed of limestone and thinly bedded chert, argillite, or sandstone, which have yielded Norian and late Karnian fossils. Member TrC3 consists of andesite lava flows plus monolithologic andesitic volcanic breccias and sandstones; the breccia and sandstone beds are commonly graded.

Considerable compositional similarity exists between the lower three members of the Cherry Creek formation. Lavas contain the phenocryst assemblage plagioclase + hornblende + clinopyroxene, and most clastic material consists of angular lithic and crystal fragments identical in composition to the lavas and their phenocrysts (Wyld, 1992). This leads to the conclusion that the lower three members were deposited during a cycle of compositionally homogeneous volcanism, with clastic deposits representing volcanic material remobilized downslope during active eruptions. Facies analysis indicates that deposition occurred primarily from sediment gravity flows in a proximal debris-apron environment at the base of a volcanic slope (Wyld, 1992). In contrast to the lower three members of the Cherry Creek formation, the upper member (TrC4) consists largely of thin- to medium-bedded shale and sandstone with rare pebble conglomerate, and contains clastic detritus dominated by plagioclase and plagioclase-phyric volcanic lithic fragments. Facies within this unit, including an abundance of graded beds and partial Bouma sequences, indicate deposition in a distal, deep-marine debris-apron or submarine-fan setting. Member TrC4 is therefore interpreted to reflect cessation of the Cherry Creek volcanic cycle followed by erosion of the volcanic pile (Wyld, 1992).

***Jackson Mountains.*** Rocks that can be correlated with the Dyke Canyon formation occur in two places in the Jackson Mountains. Quinn (1996) and Quinn et al. (1997) described a mafic volcaniclastic unit on the west side of the Jackson Mountains, called unit Tr2, which depositionally overlies unit Tr1 (Fig. 6). This unit corresponds to part of the Triassic Boulder Creek beds defined by Russell (1981, 1984). According to the studies by Quinn, unit Tr2 consists of monolithologic volcanic breccia and sandstone, composed almost exclusively of mafic volcanic detritus with phenocrystic plagioclase and pyroxene (Fig. 7). These rocks are very thick bedded and internally structureless, and are interpreted to have been deposited by mass-flow processes in a slope apron adjacent to an active volcanic center (Quinn, 1996). Although not directly dated, unit Tr2 is constrained by crosscutting relations to be younger than early Norian and older than Early Jurassic (Fig. 7; Quinn et al., 1997). Unit Tr2 is very similar to member DC1 of the Dyke Canyon formation, with which it is correlated.

In the southeastern Jackson Mountains, two units have been mapped that can also be correlated with the Dyke Canyon formation (my own mapping). These units are here named the Stroud Mine volcanics (Trsmv) and the Fox Spring unit (Trfsu) (Fig. 6). The Stroud Mine volcanics consist of monolithologic, mafic, augite porphyry lava, volcanic breccia, and less common volcanic-lithic sandstone, which are interlayered as very thick, internally structureless beds (Fig. 7). This unit is indistinguishable in composition and facies from the lower member of the Dyke Canyon formation (TrD1), and appears quite similar to Quinn's unit Tr2, with which it is correlated. The depositionally overlying Fox Spring unit consists mostly of shale and thinly interbedded black chert and shale, but also contains thick horizons of variably vesicular and generally aphyric pillow basalt with interpillow carbonate, plus some thin limestone interbeds (Fig. 7). This unit shares obvious similarities in composition and facies with the upper two members of the Dyke Canyon formation (TrD4 and TrD5), with which it is correlated.

Rocks that can be correlated with the Cherry Creek formation occur on the east side of the central Jackson Mountains (Fig. 6). These strata were first described in detail by Russell (1981, 1984), who included them in his Triassic Boulder Creek beds, and were correlated with Triassic rocks in the Pine Forest Range by Silberling et al. (1987) on the basis of similarities in age and lithology. Quinn (1996) and Quinn et al. (1997) subsequently divided the strata into three conformable units; in ascending order, these are Tr3, Tr4, and Tr5 (Fig. 6). The following description of these units is based on Quinn (1996) and Quinn et al. (1997), except where noted.

Unit Tr3 consists mostly of debris flows, massive conglomerate, and massive to graded sandstone (Fig. 7), and is interpreted to have been deposited in a proximal debris-apron setting adjacent to an active volcanic system. Clastic material includes abundant crystal fragments of plagioclase, clinopyroxene, and hornblende, and volcanic lithic fragments with the same phenocryst assemblage, in addition to less common chert, shale, limestone, siliciclastic lithic fragments, and quartz. A distinctive feature of this unit is the presence of large carbonate clasts and blocks (to 350 m long) that have yielded Norian, Ladinian or Karnian, and Permian fossils (Fig. 7; Willden, 1964; Russell, 1981; Maher, 1989; Quinn, 1996). Unit Tr3 is probably entirely Norian in age, based on these fossil data and the fact that the unit is intruded on its west side by the latest Triassic(?) to Early Jurassic Happy Creek hypabyssal complex. Unit Tr4 consists of monolithologic, internally structureless, and coarse-grained volcanic breccia and sandstone (Fig. 7), and is interpreted to reflect an influx of more proximal volcanic debris into the basin (Quinn, 1996). The volcaniclastic material in Tr4 includes crystal clasts of plagioclase, clinoyproxene, and hornblende, and lava clasts with the same phenocryst assemblage; virtually no nonvolcanic detritus is present. Unit Tr5, constrained by fossils to be middle Norian or younger (Maher, 1989; Quinn, 1996), consists mostly of thin-bedded, fine-grained sandstone and siltstone, with less common conglomeratic horizons (Fig. 7), and is interpreted to have been deposited in either a distal debris apron or basinal setting. Sandstones are rich in feldspar and volcanic lithic fragments. Collectively, units Tr3–Tr5 in the eastern Jackson Mountains are remarkably similar in composition, facies, and age to the Cherry Creek formation in the Pine Forest Range, and correlations between these units are obvious: unit Tr3 is strikingly similar to Cherry Creek formation member TrC1 and parts of TrC2; unit Tr4 is very similar to member TrC3; and unit Tr5 is a clear match for member TrC4.

***Summary***

With the new detailed data base on Triassic rocks throughout the Black Rock terrane, as summarized herein, it is now possible for the first time to present a region-wide synthesis of the Triassic stratigraphy and evolution of the terrane (Fig. 8). This synthesis is a critical tool in evaluating how Triassic processes in the arc region of Nevada may relate to processes in the backarc basin and shelf.

The terrane was evidently largely emergent during the Early and/or Middle Triassic, because rocks of this age are absent in the Pine Forest Range and Jackson Mountains (Fig. 8). Local areas (Bilk Creek Mountains), however, underwent basinal sedimentation during this time frame. These apparently conflicting relations argue for a history of differential uplift and/or erosion and subsidence and/or deposition in the Black Rock terrane during the Early and Middle Triassic; tectonism and development of structurally controlled basins seems the most likely explanations for this record. Some magmatic activity occurred during this time period, but was restricted to the Early Triassic and the Ladinian (Fig. 8).

An important episode of regional subsidence then affected the Black Rock terrane in the Ladinian to Karnian, resulting in widespread deposition of deep-marine, slope to base-of-slope Karnian strata (Fig. 8). From the Karnian to the early Norian, the Black Rock terrane was magmatically inactive, and was receiving two forms of sediment input, as represented by the Bishop Canyon formation in the Pine Forest Range and unit Tr1 in the Jackson Mountains. (1) Carbonate input, which, on the basis of relations in the Pine Forest Range, appears to have been the background nature of sedimentation in the region but was interrupted during periods of siliciclastic input. (2) Siliciclastic input, which was derived from a mixed volcanic and sedimentary source terrane, most likely from erosion of uplifted Paleozoic basement rocks within the Black Rock terrane and nearby arc assemblages of the Cordillera. The sediment transport direction is uncertain, but a finer average grain size for clastic rocks in the Jackson Mountains versus the Pine Forest Range (Russell, 1981; Wyld, 1992; Quinn, 1996) suggests sediment transport from the west and/or north; this is consistent with the location of potential arc basement source terranes (Fig. 1). Conversely, sediment transport from the east can be ruled out by the fact that coeval siliciclastics deposited to the east on the shelf are of a very different composition (quartz and chert rich) from those deposited in the Black Rock terrane.

Active arc volcanism then began abruptly in the early or

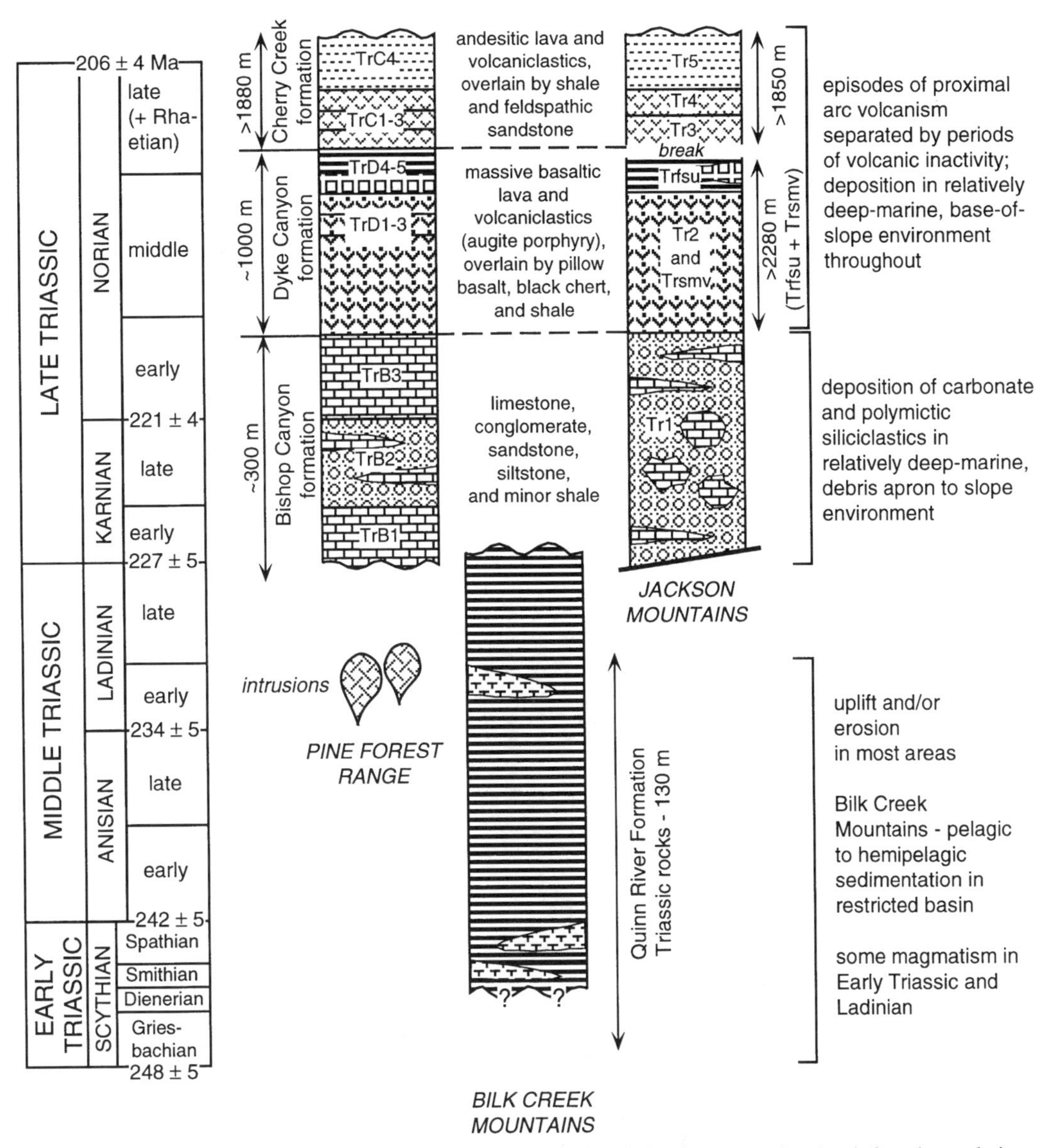


Figure 8. Summary of Triassic stratigraphy and evolution of Black Rock terrane, showing inferred correlations (dashed lines) between units in Jackson Mountains and Pine Forest Range.

middle Norian and continued to be the dominant process in the Black Rock terrane throughout the remainder of the Triassic. Two broad episodes of volcanism can be distinguished throughout the terrane. The first is exemplified by the Dyke Canyon formation in the Pine Forest Range, and units Tr2, Trsmv, and Trfsu in the Jackson Mountains (Fig. 8). This episode was characterized by eruption of massive and pillowed basaltic lavas, and by voluminous deposition of compositionally similar volcanic breccias and sandstones. The end of this eruptive cycle was associated with deposition of black chert and argillite. The second eruptive cycle is exemplified by the Cherry Creek formation in the Pine Forest Range and units Tr3, Tr4, and Tr5 in the Jackson Mountains (Fig. 8). This episode was characterized by deposition of coarse volcaniclastic material rich in andesitic volcanic lithics and fragmental plagioclase, clinopyroxene, and hornblende, as well as some compositionally similar lavas. The end of this eruptive cycle is marked by deposition of finer grained clastic strata that are interpreted to reflect erosion of the volcanic pile.

As emphasized by Wyld (1990, 1992) and further supported by Quinn (1996), there is no evidence of any progradation of facies during deposition of the Norian volcanogenic rocks in the Black Rock terrane. All of these strata were deposited in a relatively deep marine environment, either within base-of-slope debris aprons, or in distal submarine fans and/or hemipelagic basins (Wyld, 1992; Quinn, 1996). In light of the fact that ~3–4 km of strata were deposited during this volcanogenic period (Fig. 8), this lack of progradation implies syndepositional subsidence. Thus the episode of regional subsidence in the Black Rock terrane that began in the Ladinian to Karnian apparently continued through the remainder of the Triassic.

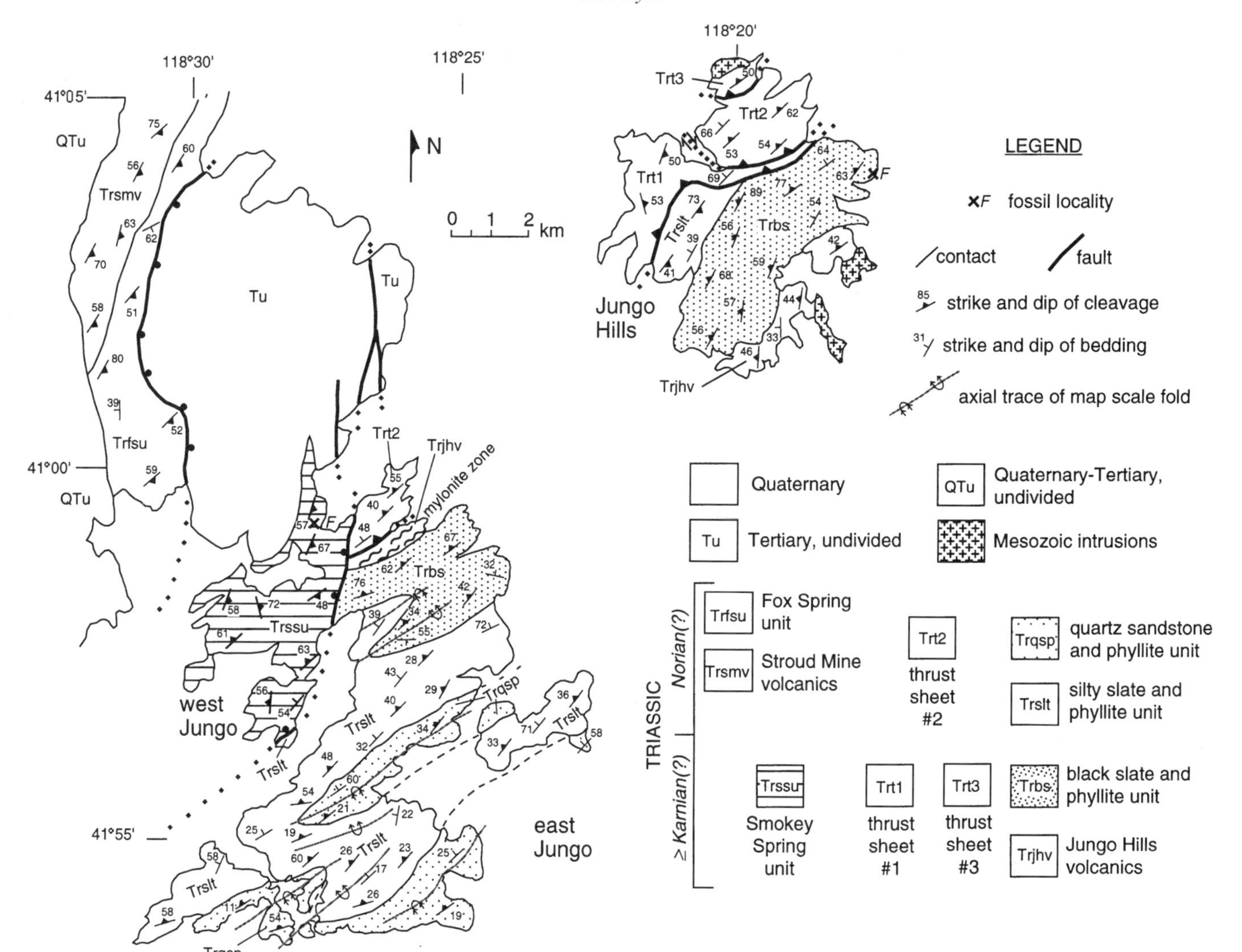


Figure 9. Geologic map of Jungo area and southeast Jackson Mountains. For location, see Figure 6.

## WESTERN BASINAL TERRANE—JUNGO AREA

In the southeasternmost part of the Jackson Mountains, near the ghost town site of Jungo (Figs. 2, 6, and 9), a suite of Mesozoic sedimentary and volcanic rocks sheds new light on the Triassic paleogeographic evolution of the backarc region. This area, which has previously been studied only in reconnaissance (Willden, 1964), is herein called the Jungo area (Fig. 9). It is convenient to refer to three different parts of the Jungo area (Figs. 6 and 9). The Jungo Hills is an isolated area of exposure of Mesozoic rocks to the northeast. The large area of Mesozoic rocks to the southwest of this can be divided into two parts by a north-south–trending Cenozoic fault. The area west of this fault is called west Jungo, and the area east of the fault is called east Jungo.

To the northwest of the Jungo area are igneous and sedimentary rocks of the Black Rock terrane, including the Triassic Stroud Mine volcanics and Fox Spring unit (Figs. 6 and 9). To the east and south of the Jungo area are sedimentary rocks of the basinal terrane (Fig. 2). The Jungo area thus is near or contains the boundary between the Black Rock terrane and the backarc basinal terrane. Drawing a distinct boundary between the two terranes is difficult on stratigraphic grounds, however, and any Mesozoic structural contact between the two terranes is concealed by Cenozoic faulting.

The structural evolution of the Jungo area is complex and beyond the scope of this chapter. Structural features are thus only briefly described to provide a framework for stratigraphic data; detailed structural analysis of the Jungo area will be presented elsewhere. Except where noted, all data and relations described in the following are based on new mapping and structural analysis.

### *Stratigraphic and structural framework*

West Jungo contains only one unit, herein named the Smokey Spring unit (Trssu; Fig. 9). This unit occurs within a downdropped fault block, between two normal faults of known and inferred Cenozoic age that separate Triassic rocks of the Black Rock terrane (Stroud Mine volcanics and Fox Spring unit) from the Triassic strata of east Jungo and Jungo Hills (Fig. 9).

Two phases of deformation have affected the Smokey Spring unit. D1 produced a regional cleavage that is generally north-northeast striking and east-southeast dipping, and broadly parallel to bedding (Fig. 9). Open D2 folds refold the older cleavage and have northeast-southwest–trending axes.

A more complex stratigraphy and structural framework is found in east Jungo and Jungo Hills. Detailed mapping in these areas has resulted in the recognition of seven distinct units (Fig. 9) and a structural evolution involving one principal phase of deformation. This deformation produced tight folds from outcrop to map scale, a penetrative axial-planar cleavage that strikes northeast and dips moderately northwest, and ductile thrust faults that are broadly parallel to the cleavage and reflect transport to the southeast (Fig. 9). The thrusts carry in their upper plates relatively thin thrust sheets that are each composed of one distinct unit. Thrust sheets are numbered 1 to 3, from base (southeast) to top (northwest) in the Jungo Hills, and the units contained within these thrust sheets are accordingly called Trt1, Trt2, and Trt3. Only one thrust sheet is exposed in east Jungo, but it is composed of unit Trt2 and is therefore interpreted to be part of thrust sheet 2 as defined in the Jungo Hills; thrust sheet 2 is interpreted to have overriden thrust sheet 1 in the Jungo area. In the lower plate of the thrust pile are four conformable units that are folded by several map-scale folds (Fig. 9); these units are called the Jungo Hills volcanics (Trjhv), black slate and phyllite unit (Trbs), silty slate and phyllite unit (Trslt), and quartz sandstone and phyllite unit (Trqsp).

Structures within east Jungo and Jungo Hills are identical in style and orientation to Jurassic Luning-Fencemaker structures found throughout the northern basinal terrane (Oldow, 1984; Elison and Speed, 1989; Rogers and Wyld, 1998; Wyld, 1998), leading to the conclusion that all of these rocks were deformed together during development of the backarc Luning-Fencemaker fold-and-thrust belt. The structural evolution of the west Jungo area is very different, however. Work in progress indicates that D1 deformation in west Jungo is kinematically related to deformation in the Black Rock terrane that apparently predates development of the Luning-Fencemaker thrust belt, whereas D2 structures in west Jungo likely reflect thrusting of the west Jungo rocks over the east Jungo–Jungo Hills rocks during development of the Luning-Fencemaker belt. Unfortunately any original Jurassic structural contacts between these groups of rocks are concealed by Cenozoic normal faulting.

All rocks in the Jungo area are regionally deformed and metamorphosed to such a degree that shale protolith rocks have been transformed into slates or phyllites, and these terms are therefore used to describe these rocks. Other rock types are simply described in terms of protolith lithology.

### *Description of Triassic units and age relations*

Figure 10 describes all the units in the Jungo area. Only two units have been dated, but a number of inferences can be made about the ages of all the units. Key relations relevant to evaluating ages of units are as follows.

1. Three of the units in east Jungo and the Jungo Hills (units Trqsp, Trslt, and Trt2) consist of abundant slate and phyllite, variably abundant quartz sandstone, and minor limy rocks (Fig. 10), that were most likely deposited in a relatively deep marine environment by turbidity currents and hemipelagic processes. These units are very similar in facies and composition to the Norian basinal ALS group in the basinal terrane. Insofar as the basinal ALS group is rather unique in the pre-Cenozoic geology of Nevada, this similarity provides a compelling argument that units Trqsp, Trslt, and Trt2 are all of Norian age and are part of the basinal ALS group.

2. Other units in east Jungo and Jungo Hills are quite different and are dominated by some combination of limestone, limy slate, and slate (units Trbs, Trt3, Trt1), or volcanogenic rocks (unit Trjhv). These units differ from the Norian basinal ALS group in their abundance of limy strata, lack of quartz sandstone, and/or presence of volcanogenic rocks. Because deposition of the basinal ALS group (and their shelf equivalents, the Auld Lang Syne Group) spanned the entire Norian, it therefore seems reasonable to conclude that the lithologically different east Jungo–Jungo Hills units noted here are not of Norian age. Evidence that these units are pre-Norian is presented in the following.

3. A conformable stratigraphic sequence, comprising the units Trjhv, Trbs, Trslt, and Trqsp, is present in the folded lower plate of the thrusts in east Jungo and the Jungo Hills (Fig. 9). Because folds in this sequence of rocks are overturned, structurally higher rocks are not necessarily stratigraphically higher. However, unit Trqsp is exposed only in the cores of several map-scale folds, two of which are sufficiently well exposed to determine that they are overturned synforms (Fig. 9), which implies that Trqsp is the youngest unit of the sequence. As noted in point 1, unit Trqsp and Trslt are both inferred to be Norian. Unit Trbs has yielded late Karnian to early Norian fossils (Fig. 10; N.J. Silberling, 1997, personal commun.); the older (Karnian) part of the age range is considered more likely for this unit because of the arguments in point 2. This is consistent with the structural argument that Trbs is older than Trslt and Trqsp. Unit Trjhv is therefore interpreted to be older than unit Trbs. Because the dated fossils in unit Trbs occur almost 1 km above the top of unit Trjhv, the latter unit is most likely pre-late Karnian, probably early Karnian and/or Ladinian.

4. Four units in the Jungo area are dominated by some combination of limestone, limy slate, and slate or phyllite—these include the Smokey Spring unit of west Jungo, and units Trt1, Trt3, and Trbs of east Jungo and Jungo Hills (Fig. 10). The Smokey Spring unit, Trt1, and Trt3 also contain minor amounts of felsic tuff, and Trbs depositionally overlies felsic volcanic strata (Fig. 10). Thus, there are marked lithologic similarities between all of these units, suggesting that they may be facies equivalents. This concept is supported by the apparent similarity in ages of two of these units: the Smokey Spring unit, which has yielded late Karnian fossils (Fig. 10; N.J. Silberling, 1997, personal commun.), and unit Trbs, which is interpreted to be most likely late Karnian, as explained in point 3. I therefore suggest that units Trt1 and Trt3 may also be of Karnian age.

### *Discussion and interpretation*

Columns summarizing the stratigraphy of the Jungo area are shown in Figure 11. Mesozoic shortening deformation and Cenozoic faulting within this region have shuffled stratigraphic units from their original positions relative to one another during deposition. Some interpretation about original relative positions of different units can be made, however, and this is done in Figure 11 based on the following rationale. First, west Jungo rocks are currently west of east Jungo and Jungo Hills rocks, and are assumed to have always been located west of the latter rocks. Second, the various thrust sheets in east Jungo and the Jungo Hills record tectonic transport to the southeast, and rocks contained within these thrust sheets are therefore interpreted to have originally been deposited west of the rocks contained within the lower plate of the thrust pile.

The relations and arguments summarized in the preceding section provide evidence that strata of both Norian and Karnian (and possibly Ladinian) age are present in the Jungo area. Strata of inferred Norian age (units Trslt, Trbs, and Trt2) are found only in east Jungo and Jungo Hills, and can be correlated on the basis of composition and facies with Norian rocks of the basinal ALS group. The east Jungo–Jungo Hills area is therefore considered to be part of the basinal terrane. It is unique in the basinal terrane, however, because strata of known and inferred Karnian (and Ladinian?) age are also exposed; these rocks provide the only known information on the pre-Norian history of the basinal terrane.

Strata of known and inferred pre-Norian age in east Jungo and the Jungo Hills include unit Trbs, most likely of late Karnian age, and the inferred lower Karnian (and Ladinian?) unit Trjhv. Unit Trjhv consists of mafic to felsic volcanogenic rocks

Quartz sandstone and phyllite unit (Trqsp) - exposed in east Jungo and Jungo Hills
Mostly gray phyllite and well-bedded quartz sandstone; minor black slate, limy slate, gray limestone, silty phyllite and siltstone. Average phyllite:sandstone ratio ~65:35. Sandstone mostly thin to medium-bedded; beds to 2.5 m thick; amalgamated sandstones to 6 m thick. Plane laminations and cross-bedding common in thinner sandstone beds; thicker beds typically massive; some graded beds. Phyllite varies from thin partings to thicker horizons (up to 7 m). Age - undated but inferred Norian, based on compositional and facies similarity to Norian basinal terrane ALS group.

Silty slate and phyllite unit (Trslt) - exposed in east Jungo and Jungo Hills
Mostly gray slate or phyllite that is commonly silty and locally limy; generally minor siltstone, limy siltstone, fine-grained quartz sandstone, dark gray impure limestone. Strata mostly thin bedded, but some limestone to 1.5 m thick. Gradational contact with overlying unit Trqsp. Age - same as Trqsp.

Black slate and phyllite unit (Trbs) - exposed in east Jungo and Jungo Hills
Mostly black or dark gray, graphitic rocks, including slate, phyllite, limy slate, and minor impure limestone. Thin to medium bedded and laminated where stratification visible. Gradational contact with overlying unit Trslt. Age - late Karnian to early Norian fossils ~950 m above base; inferred late Karnian (see text).

Jungo Hills volcanics (Trjhv) - exposed in east Jungo and Jungo Hills
Lower mafic member: plagioclase-phyric andesitic lava and compositionally similar volcanic breccia and sandstone; less common vesicular basalt lava, and basaltic breccia and sandstone with carbonate matrix; minor tuffaceous argillite and black graphitic argillite. Upper felsic member: quartz and feldspar-phyric lava. Sharp to gradational contact between felsic member and overlying unit Trbs. Age - pre-late Karnian, based on age of overlying unit Trbs; inferred early Karnian or possibly Ladinian (see text).

Thrust sheet #3 (unit Trt3) - exposed in Jungo Hills
Mostly dark gray and locally limy slate, with less common gray limestone and quartz-phyric felsic tuff. Limestone beds and tuff horizons to 3 m thick; stratification obscure in slate. In thrust contact with unit Trt2. Age - same as unit Trt1.

Thrust sheet #2 (unit Trt2) - exposed in east Jungo and Jungo Hills
Approximately 70% gray phyllite, 25% fine- to medium-grained quartz sandstone, 5% siltstone and limy graphitic slate. Sandstone beds thin- to thick-bedded, often lens-shaped and discontinuous, in amalgamated sequences to 20 m thick; some plane lamination and cross-bedding. In thrust contact with units Trt3, Trt1 and Trjhv. Age - same as unit Trqsp.

Thrust sheet #1 (unit Trt1) - exposed in Jungo Hills
Mostly limestone, slate, and phyllite. Limestone varies from black and graphitic with abundant black fossil fragments to medium or dark gray and argillaceous. Slate and phyllite vary from dark to medium gray. Other rock types include dark gray to black limy slate, quartz-phyric felsic tuff, and rare quartz sandstone. Medium- to thick-bedded where stratification visible. In thrust contact with units Trt2, Trslt and Trbs. Age - undated but inferred Karnian based on similarity to units Trbs, upper Trjhv, and Trssu.

Smokey Spring unit (Trssu) - exposed in west Jungo
Mostly limestone and slate; rare siltstone and tuffaceous(?) volcaniclastic rocks. Limestone is medium- to thick-bedded , massive, light to dark gray, commonly contains small fossil fragments. Slate is variably silty or limy, mostly tan to gray, but also brown, red or purple. Age - late Karnian fossils present in middle(?) of unit.

Figure 10. Key features of Triassic units in the Jungo area. ALS group defined in text. Biostratigraphic data from N.J. Silberling and B. Wardlaw (N.J. Silberling, 1997, personal commun.).

(Fig. 10). A marine environment of deposition is evident for this unit, based on the presence of carbonate matrix and graded beds in some clastic rocks. This unit is more than 650 m thick (Fig. 11) and records a major episode of volcanism in the basinal terrane. The overlying unit Trbs is a distinctive unit composed primarily of black graphitic rocks including slate, limy slate, and less common limestone (Fig. 10). The bedding and facies characteristics of this unit argue for deposition in a restricted, anoxic basin, certainly below wave base and most likely in a relatively deep marine environment, based on the ~1850 m thickness of the unit (Fig. 11). Other units in east Jungo and the Jungo Hills that are inferred to be of probable Karnian age include units Trt1 and Trt3. Trt1 contains abundant limestone and slate and/or phyllite, Trt3 contains abundant slate and minor limestone, and both units contain quartz-phyric felsic tuffs. Both were deposited in a marine environment near a felsic volcanic center; relatively deep marine depths are suggested by the abundance of fine-grained clastic rocks, the absence of any macrofauna, and the lack of any evidence of stratification by currents. Dark gray to black graphitic rocks are also present in these two units, although they are not as prominent as in unit Trbs, and likewise suggest that the depositional basin for these units was in part anoxic.

The history that emerges from the pre-Norian strata in the basinal terrane of east Jungo and the Jungo Hills is one involving mafic to felsic volcanism in the Karnian and possibly Ladinian, followed by deposition of carbonate, mud, and local felsic tuffs in a restricted basinal environment in the Karnian. This history is reminiscent of the evolution of the shelf terrane during the late Ladinian and Karnian (Fig. 3), except that shallower marine conditions are represented by the shelf strata and the shelf volcanic rocks are exclusively mafic. The implications are that Karnian (and Ladinian) strata of the western basinal terrane in east Jungo–Jungo Hills may form the offshore, basinal equivalent of coeval Star Peak group units on the shelf, that a relatively deep marine environment existed in the basinal terrane by at least the Karnian, and that both the shelf and basinal terranes were magmatically active in the Karnian and Ladinian.

The late Karnian Smokey Spring unit of west Jungo is somewhat more difficult to interpret. This unit, which consists of abundant limestone and slate, with minor tuffaceous rocks (Fig. 10), was deposited in a marine environment near a volcanic center. The abundance of shaly (protolith) rocks, general lack of macrofauna, and lack of any sedimentary structures indicative of current action argue against deposition in a shallow-marine environment, and suggests deposition below wave base. It is not immediately obvious, however, whether this unit should be considered part of the Black Rock terrane or part of the basinal terrane. It is similar to Karnian rocks of the Black Rock terrane in that both groups of rocks were deposited in a marine setting and both consist of carbonate and clastic rocks (e.g., Figs. 8 and 10). It also appears to share a common structural history with the Black Rock terrane, as noted earlier, which suggests that it was in proximity to the latter terrane during its D1 phase of deformation. It differs from the Karnian section in the Black Rock terrane in that Smokey Spring clastic rocks are much finer grained and sparse volcaniclastic horizons are present. These differences could, however, reflect facies variations, a conclusion that is consistent with decreasing average grain size of Karnian clastics to

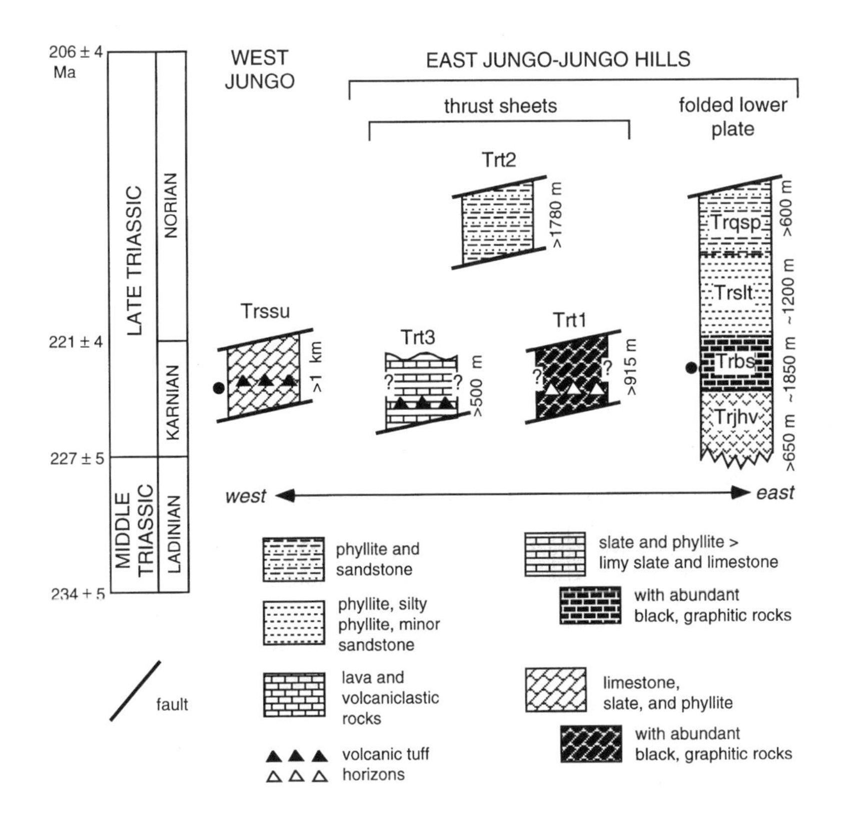


Figure 11. Stratigraphy of Jungo area. Unit abbreviations defined in Figure 9. Solid circles denote fossil data. See text for explanation of inferred ages of undated units.

the south and east in the Black Rock terrane and decreasing abundance of Karnian volcanogenic rocks to the west in the Jungo area (Fig. 11).

Conversely, the Smokey Spring unit is similar to Karnian and inferred Karnian strata in east Jungo–Jungo Hills in the sense that both groups of rocks were deposited in a marine setting and both contain a mixture of carbonate, fine-grained clastic rocks and volcanogenic deposits. It differs in containing slightly coarser clastic rocks and no evidence for deposition in an anoxic basin. In addition, the Smokey Spring unit was affected by an early phase of deformation that is not represented in the east Jungo–Jungo Hills rocks, which suggests that west Jungo and east Jungo–Jungo Hills rocks were not immediately adjacent to one another during this early phase of deformation.

It may be best to view the Smokey Spring unit as a transitional unit between the Black Rock arc terrane and the backarc basinal terrane. This point is important in evaluating relations between the arc and backarc in the Triassic, as explored further in the following section.

## SUMMARY OF THE TRIASSIC RECORD IN THE NORTHERN MARINE PROVINCE

Data discussed in the preceding sections allow a new regional synthesis of the Triassic evolution of the marine province in northern Nevada. This synthesis goes beyond that presented in previous regional analyses of this area (Speed, 1978a; Lupe and Silberling, 1985; Saleeby and Busby-Spera, 1992) in that it includes new data, from the Black Rock terrane and adjacent basinal terrane strata in the Jungo area, that allow a more comprehensive understanding of the Triassic evolution of the province. In this section I summarize key elements of the Triassic evolution of the shelf, basinal, and Black Rock terranes, emphasizing how processes affecting individual areas can be linked together in terms of a coherent paleogeographic reconstruction through the Triassic time period. This reconstruction serves as the basis for a tectonic model of the evolution of the arc-basin-shelf system during the Triassic that is presented in the following section. Figure 12 summarizes the Triassic stratigraphy of the different areas, in addition to noting other important features or processes of each region. It is convenient to divide the following discussion into three periods of time: the Early Triassic to mid-Ladinian, the late Ladinian to late Karnian, and the Norian.

### *Early Triassic to mid-Ladinian*

Information about the Early Triassic to mid-Ladinian time period is found only in the Black Rock and shelf terranes; no strata of this age or older are known to be exposed in the basinal terrane. During this time frame, the shelf underwent a complex evolution (Fig. 12). Volcanism was widespread in the Early Triassic; this volcanism is represented by the Koipato Group volcanics which were deposited in a subaerial to shallow-marine setting during an episode of extensional tectonism. Subsequent differential uplift and erosion in the late Early Triassic removed part to all of the Koipato Group. During the remainder of this time period, differential uplift and subsidence resulted in a complex pattern of sedimentation and erosion across the shelf; depositional environments varied spatially and temporally from subaerial to basinal, sediment input varied from carbonate-dominated to clastics derived from uplifted areas of the shelf, and unconformities developed that span part to all of this time frame. Because there is no evidence for shortening deformation on the shelf in the Triassic, these features appear most consistent with a history of block faulting and differential uplift and subsidence associated with extensional tectonism, as was implied by Nichols and Silberling (1977).

Differential uplift and subsidence are also recorded in the Black Rock terrane during this time frame, as indicated by the fact that Early(?) Triassic to Ladinian basinal strata are preserved in the Bilk Creek Mountains, whereas strata of this age range are missing elsewhere in the terrane (Fig. 12). Because the Early(?) to Middle Triassic section of strata in the Bilk Creek Mountains constitutes a tiny area within the larger terrane (Fig. 2), these data suggest that either the Black Rock terrane was largely emergent in the Early to Middle Triassic or that it underwent more substantial episodes of uplift and erosion during this time frame than did the shelf. As on the shelf, there is evidence for Early Triassic volcanism in the Black Rock terrane, although only distal volcaniclastics are found in the latter area (Fig. 12). There is also some early to mid-Ladinian magmatism recorded in the Black Rock terrane; volcanic rocks in the upper Prida Formation on the shelf (Fig. 3) may be coeval with these Ladinian magmatic rocks in the Black Rock terrane, but the exact age of the shelf volcanics is not certain.

Collectively, these relations suggest a broadly similar pattern of differential uplift and erosion, with episodic magmatism, in the shelf and Black Rock terranes during the Early Triassic to mid-Ladinian and provide a link between these areas during this time frame.

### *Late Ladinian to late Karnian*

This time period is represented by strata of the shelf terrane, western basinal terrane (Jungo area), and Black Rock terrane (Fig. 12). Regional subsidence of the shelf terrane occurred at the beginning of this time period, resulting in widespread deposition of platform carbonates; this environment then persisted across the shelf for the remainder of the Karnian, except for a brief episode of uplift in the mid-Karnian (Fig. 12). Deposition of ~500 m of shallow-marine carbonates during this time frame attests to syndepositional subsidence. Episodes of coarse siliciclastic input interrupted carbonate deposition twice and reflect erosion of uplifted basement sources to the north and east. Mafic volcanism was also widespread on the shelf during the latest Ladinian to early Karnian (Fig. 12).

Further west, in the western basinal terrane, strata of known and inferred Karnian (and Ladinian?) age include basinal carbon-

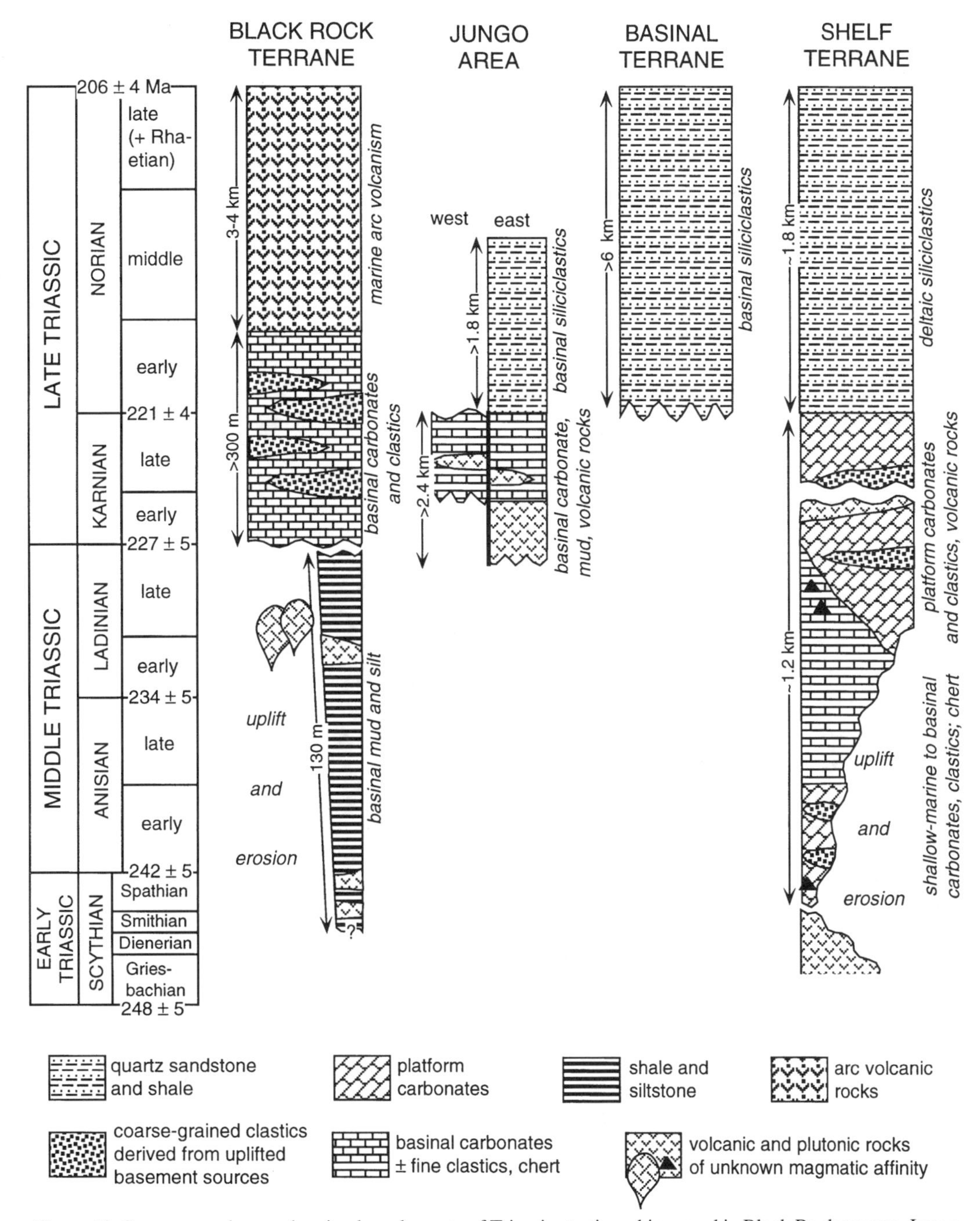


Figure 12. Summary columns showing key elements of Triassic stratigraphic record in Black Rock terrane, Jungo area, basinal terrane, and shelf terrane.

ates and shales, as well as mafic to felsic volcanogenic rocks (Fig. 12). The greatest thickness of volcanic strata is found in sequences that are inferred, based on structural relations, to have been in a more eastern location, whereas only thin tuffs and volcaniclastic horizons were deposited farther west. Deposition in the east also tended to occur in restricted, anoxic basins, whereas an anoxic environment is not represented in strata that were deposited to the west. A greater amount of silt-size clastic material is also found in western sequences, implying that the source of clastic material was from the west. The combined thickness of known and inferred Karnian (and Ladinian?) strata in this part of the basinal terrane is at least 1–2 km (Fig. 12), implying deposition in a deep or actively subsiding basin.

Farther west, in the Black Rock terrane, regional subsidence in the late Ladinian or early Karnian led to deposition of relatively deep marine slope to base-of-slope strata across the region through the Karnian period (Fig. 12). These strata include abundant carbonates as well as generally coarse grained clastic strata that were derived from erosion of basement sources within the terrane and/or from more outboard arc assemblages (Darby et al., this volume, Chapter 5). Based on spatial variations in average grain size of clastics, sediment transport was from the west or north.

Collectively, these relations indicate a number of important similarities in the history of the Black Rock, basinal, and shelf terranes in the late Ladinian to late Karnian, including: (1) regional subsidence in at least the shelf and Black Rock terranes, and possibly in the basinal terrane; (2) widespread carbonate deposition in all areas; (3) coarse siliciclastic influx in the shelf and Black Rock terranes derived from nearby uplifted basement rocks; (4) a systematic spatial pattern for siliciclastic sedimentation, varying from proximal (coarse grained) near the margins of the marine province (Black Rock and shelf terranes) to distal (fine grained) in the center of the province (basinal terrane); and (5) volcanism in the basinal terrane and on the shelf. These similarities provide a link between the three terranes during this time frame. Significant differences are that the Black Rock and western basinal terranes subsided to greater depths than did the shelf, and that the abundance of volcanogenic rocks dies out to the west.

### *Norian*

The Norian period marked a substantial change in processes across the northern marine province. Starting in the earliest Norian and continuing through the latest Norian, the shelf was flooded by muds and quartz-rich sands that were derived from continental sources to the east and deposited in a deltaic environment (Fig. 12). Farther west, in the basinal terrane, similar Norian clastic sequences were deposited in a deeper marine environment that is interpreted to represent the offshore facies equivalent of the deltaic complex. These mud and quartz sand-dominated sequences are recognized as far west as the Jungo area.

A very different history characterized the Black Rock terrane during the Norian, however. Here, carbonate deposition continued during the early Norian and was then replaced by voluminous and generally proximal arc volcanism during the remainder of the Norian (Fig. 12). There is no evidence in the Norian record of the Black Rock terrane of the mud-quartz sand influx that is so prominent in Norian sequences farther east.

Thus, by the early Norian, something had happened within the marine province to isolate the Black Rock terrane from sedimentation processes affecting the basinal and shelf terranes. The only similarity between all three terranes in the Norian is that very thick sequences of marine strata were deposited in all areas: ~3–4 km of deep-marine strata in the Black Rock terrane, more than 6 km of deep marine strata in the basinal terrane, and >1.8 km of deltaic strata on the shelf (Fig. 12). These relations require syndepositional subsidence on the shelf and either syndepositional subsidence or deposition in a preexisting deep basin in the basinal and arc terranes; if a preexisting basin characterized the Black Rock arc terrane, however, it could only have developed in the late Ladinian to Karnian.

## TECTONIC MODELS

The model that appears most consistent with the relations summarized herein is one involving prolonged regional extension across the northern marine province during the Triassic, coupled with opening of a deep-marine extensional basin in the Late Triassic (Fig. 13). In this model, there is no fundamental distinction between the Black Rock, basinal, and shelf terranes following the Sonoma orogeny in the Early to Middle Triassic; all areas were in proximity and subjected to differential uplift and subsidence related to incipient extension during this time frame (Fig. 13A). Continued extension from the late Ladinian to the Norian then led to regional subsidence and the progressive opening of a basin between the shelf terrane and an evolving magmatic arc in the Black Rock terrane (Fig. 13, B–D).

This extensional model is attractive for a number of reasons. First, it provides an explanation for the history of differential uplift and subsidence in the shelf and Black Rock terranes in the Early Triassic to mid-Ladinian; this history can be interpreted in terms of block faulting associated with incipient extension (Fig. 13A). Second, it can explain the history of regional subsidence across the province during the late Ladinian to early Karnian, as continued extension would eventually lead to crustal thinning and, in the absence of thermal anomalies, regional subsidence (Fig. 13B). Similarly, the accumulation of very thick, nonprogradational marine sequences across the province throughout the remainder of the Triassic is consistent with continued regional subsidence related to ongoing extension (Fig. 13, C and D). Third, it can explain the record of coarse clastic influx from uplifted basement sources in the Black Rock and shelf terranes during the late Ladinian through Karnian; the coarse clastic sediment may have been shed from uplifted fault blocks at the margins of the extensional province (Fig. 13B). Fourth, it may provide an explanation for local magmatism that occurred in the northern marine province during the Early Triassic and mid-Ladinian to early Karnian (Fig. 13, A and B); this magmatism, which is not obviously associated with construction of a volcanic arc, could reflect processes associated with continental rifting. Fifth, it offers the only obvious explanation for the shift over time from broadly similar processes affecting the Black Rock, basinal and shelf terranes in the Early Triassic through Karnian to very different processes affecting the Black Rock versus the basinal, and shelf terranes in the Norian (Fig. 13, A–D). This shift can be interpreted in terms of progressive widening of a rift basin due to continued extension, eventually leading to isolation of the Black Rock terrane from continental margin sedimentation (Fig. 13D).

This model is also consistent with widespread evidence elsewhere in the western U.S. Cordillera for extensional tectonism during the early Mesozoic. In particular, there is abundant evidence that arc assemblages of the western Cordillera evolved in an extensional tectonic regime during the Triassic and, at least locally, the Early Jurassic (Busby-Spera, 1988; Saleeby and Busby-Spera, 1992). Because extension in magmatic arcs is com-

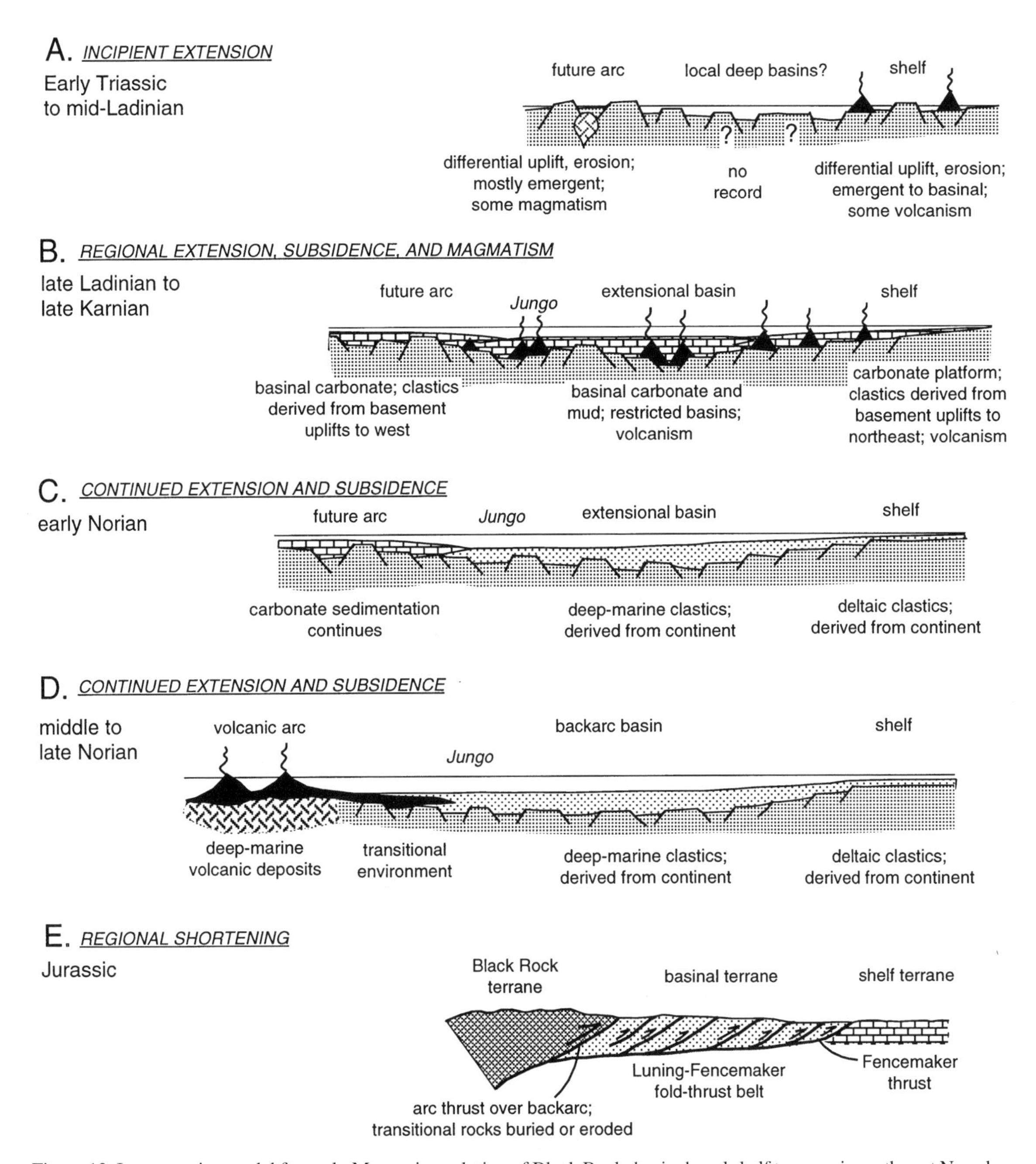


Figure 13. Interpretative model for early Mesozoic evolution of Black Rock, basinal, and shelf terranes in northwest Nevada.

monly associated with extension in the backarc region (e.g., Dewey, 1980), it is reasonable to assume that extensional tectonism affected the backarc region in Nevada during the early Mesozoic (e.g., Fig. 13D).

An intriguing feature within Triassic successions of northern Nevada is the absence of any transitional facies between the carbonate and volcanic-dominated Norian succession of the Black Rock terrane and the fine-grained siliciclastic-dominated succession of the basinal terrane. Because there are clear links between the two terranes prior to the Norian, it is logical to suppose that there was originally a transitional environment between the two terranes in the Norian. The lack of any facies consistent with such an environment implies that the two terranes were juxtaposed structurally in post-Norian time in such a way as to conceal or remove the transitional environment. One obvious possibility is that the arc terrane was thrust over the backarc basinal terrane during development of the Jurassic backarc Luning-Fencemaker fold-and-thrust belt, thereby concealing the transitional environment by tectonic burial or removing it by overthrusting and erosion (Fig. 13E). This interpretation is consistent with ongoing structural studies in the Black Rock terrane and Jungo area (my own work in progress; Rogers, 1999; Folsom, 2000), and with interpretations originally presented by Oldow (1984).

Other models have been proposed to explain the Triassic evolution of this part of the Cordillera, and, in particular, the development of a deep-marine backarc basinal terrane in central Nevada. Some of these models can now be discounted.

Speed (1978a, 1979) proposed that the marine province, and in particular the deep-marine basinal terrane, developed due to thermal subsidence of an underlying Paleozoic arc terrane that collided with the continent in the latest Permian–earliest Triassic Sonoma orogeny and then ceased to be magmatically active, resulting in thermal contraction and development of a successor deep-marine basin. This model relied on an incomplete database on Paleozoic rocks in northwest Nevada, and is clearly flawed because the main locus of Paleozoic arc magmatism is now known to have been to the west in the Klamath Mountains and Sierra Nevada, not in northwest Nevada (Miller and Harwood, 1990; Wyld, 1991, 1992). Therefore, any thermal subsidence of the Paleozoic arc should have been to the west, not in northwest Nevada. In addition, it is evident from the data presented in this chapter that magmatism occurred in the Black Rock, basinal, and shelf terranes at least intermittently during the Triassic; it is therefore difficult to conclude that this part of the Cordillera was subsiding in the Triassic due to cooling and thermal contraction. Thermal subsidence is therefore an implausible explanation for the western Nevada marine province or the basinal terrane.

Saleeby and Busby-Spera (1992) speculated that the deep-marine Triassic paleogeography of western Nevada may have been inherited from a preexisting Paleozoic paleotopography. This model seems unlikely in view of the data discussed here, indicating that regional susidence to relatively deep marine depths in northwest Nevada did not begin until the late Ladinian to Karnian; the available data indicate that prior to this time much of northwest Nevada was emergent to shallow marine.

Lawton (1994) proposed that Triassic subsidence and development of marine sedimentary basins in western Nevada may have been related to the initiation of east-dipping subduction beneath the continental margin following the Sonoma orogeny, and to the process of dynamic subsidence (Gurnis, 1992) associated with a nascent subducting oceanic plate. This model hinges on the question of whether east-dipping subduction beneath this part of the Cordillera was already ongoing in the late Paleozoic, a question still unresolved (e.g., Gehrels et al., this volume, Chapter 10). If east-dipping subduction initiated in the Triassic, dynamic subsidence potentially played a role in the evolution of the Nevada marine province. This model by itself, however, cannot easily explain the evidence supporting Early to Middle Triassic tectonism, differential subsidence and uplift, and localized magmatism in the northern marine province, or the shift from broadly similar processes affecting the future arc-basin-shelf system in the Early Triassic through Karnian to the contrasting processes affecting the arc and backarc in the Norian. I therefore conclude that while dynamic subsidence may have played a role in the evolution of the early Mesozoic marine province, extensional tectonism was the dominant process.

In conclusion, the data discussed in this chapter provide a compelling argument that this part of the western U.S. Cordillera was strongly influenced by extensional tectonism in the Triassic. In this regard, the Triassic evolution of this part of the Cordillera reflects a continuation of the extensional processes that dominated the Cordillera in the Paleozoic (Burchfiel et al., 1992; Miller et al., 1992), and represents the last major episode of extension in this region prior to widespread shortening deformation in the Jurassic and Cretaceous (Saleeby and Busby-Spera, 1992; Smith et al., 1993).

## ACKNOWLEDGMENTS

This project was supported by National Science Foundation grant EAR-9796174, and benefited from discussions with J.E. Wright, N.J. Silberling, M.J. Quinn, J.W. Rogers, and H.K. Folsom. Reviews by T.F. Lawton and C.J. Busby improved the early version of this manuscript.

## REFERENCES CITED

Barberi, F., Santacroce, R., and Varet, J., 1982, Chemical aspects of rift magmatism/ Continental and oceanic rifts, in G., Palmason, ed., Continental and oceanic rifts: American Geophysical Union Geodynamics Series, v. 8, p. 223–258.

Blome, C.D., and Reed, K.M., 1995, Radiolarian biostratigraphy of the Quinn River Formation, Black Rock terrane, north-central Nevada: Correlations with eastern Klamath terrane geology: Micropaleontology, v. 41, p. 49–68.

Burchfiel, B.C., Cowan, D.S., and Davis, G.A., 1992, Tectonic overview of the Cordilleran orogen in the western United States, *in* Burchfiel, B.C., et al., eds., The Cordilleran orogen: Conterminous U.S.: Boulder, Colorado, Geological Society of America, Geology of North America, v. G-3, p. 407–479.

Burke, D. B., 1973, Reinterpretion of the "Tobin thrust": Pre-Tertiary geology of the southern Tobin Range, Pershing County, Nevada [Ph.D. thesis]: Stanford, California, Stanford University, 82 p.

Burke, D.B., and Silberling, N.J., 1973, The Auld Lang Syne Group, of Late Triassic and Jurassic (?) age, north-central Nevada: U.S. Geological Survey Bulletin 1394-E, 14 p.

Busby-Spera, C.J., 1988, Speculative tectonic model for the early Mesozoic arc of the southwest Cordilleran United States: Geology, v. 16, p. 1121–1125.

Compton, R.R., 1960, Contact metamorphism in Santa Rosa Range, Nevada: Geological Society of America Bulletin, v. 71, p. 1383–1416.

Dewey, J.F., 1980, Episodicity, sequence, and style at convergent plate boundaries: Geological Association of Canada Special Paper, v. 20, p. 553–573.

Elison, M.W., and Speed, R.C., 1989, Structural development during flysch basin collapse: The Fencemaker allochthon, East Range, Nevada: Journal of Structural Geology, v. 11, p. 523–538.

Folsom, H.K., 2000, Analysis of Mesozoic deformation in a backarc basin, northwestern Nevada [Senior honor's thesis]: Athens, University of Georgia, 92 p.

Fuller, L. R., 1986, General geology of Triassic rocks at Alaska Canyon in the Jackson Mountains, Humboldt County, Nevada [M.S. thesis]: Reno, University of Nevada, 110 p.

Gradstein, F.M., Agterberg, F.P., Ogg, J.G., Hardenbol, J., van Veen, P., Thierry, J., and Huang, Z., 1994, A Mesozoic time scale: Journal of Geophysical Research, v. 99, p. 24,051–24,074.

Gurnis, M., 1992, Rapid continental subsidence following the initiation and evolution of subduction: Science, v. 255, p. 1556–1558.

Jones, A.E., 1990, Geology and tectonic significance of terranes near Quinn River Crossing, Humboldt County, Nevada, *in* Harwood, D.S., and Miller, M.M., eds., Paleozoic and early Mesozoic paleogeographic relations; Sierra Nevada, Klamath Mountains, and related terranes: Geological Society of America Special Paper 255, p. 239–254.

Ketner, K.B., and Wardlaw, B.R., 1981, Permian and Triassic rocks near Quinn River Crossing, Humboldt county, Nevada: Geology, v. 9, p. 123–126.

Lawton, T.F., 1994, Tectonic setting of Mesozoic sedimentary basins, Rocky Mountain region, United States; *in* Caputo, M.V., et al., eds., Mesozoic systems of the Rocky Mountain region, USA: Rocky Mountain Section, SEPM (Society for Sedimentary Geology), p. 1–26.

Lupe, R., and Silberling, N.J., 1985, Genetic relationship between lower Mesozoic continental strata of the Colorada Plateau and marine strata of the western Great Basin: Significance for accretionary history of Cordilleran lithotectonic terranes; *in* Howell, D.G., ed., Tectonostratigraphic terranes of the Circum-Pacific region: Houston, Texas, Circum-Pacific Council for Energy and Mineral Resources, Earth Science Series, no. 1, p. 263–271.

Maher, K.A., 1989, Geology of the Jackson Mountains, northwest Nevada [Ph.D. thesis]: Pasadena, California Institute of Technology, 491 p.

Miller, E.L., Miller, M.M., Stevens, C.H., Wright, J.E., and Madrid, R., 1992, Late Paleozoic paleogeographic and tectonic evolution of the western U.S. Cordillera, *in* Burchfiel, B.C., et al., eds., The Cordilleran orogen: Conterminous U.S.: Boulder, Colorado, Geological Society of America, Geology of North America, v. G-3, p. 57–106.

Miller, M.M., and Harwood, D.S., 1990, Stratigraphic variation and common provenance of upper Paleozoic rocks in the northern Sierra and eastern Klamath terranes, northern California, *in* Harwood, D.S., and Miller, M.M., eds., Paleozoic and early Mesozoic paleogeographic relations; Sierra Nevada, Klamath Mountains, and related terranes: Geological Society of America Special Paper 255, p. 175–192.

Nichols, K.M, and Silberling, N.J., 1977, Stratigraphy and depositional history of the Star Peak Group (Triassic), northwestern Nevada: Geological Society of America Special Paper 178, 73 p.

Oldow, J.S., 1984, Evolution of a late Mesozoic back-arc fold and thrust belt, northwestern Great Basin, U.S.A.: Tectonophysics, v. 102, p. 245–274.

Oldow, J.S., Bartel, R.L., and Gelber, A.W., 1990, Depositional setting and regional relationships of basinal assemblages: Pershing Ridge Group and Fencemaker Canyon sequence in northwestern Nevada: Geological Society of America Bulletin, v. 102, p. 193–222.

Quinn, M.J., 1996, Pre-Tertiary stratigraphy, magmatism, and structural history of the central Jackson Mountains, Humboldt County, Nevada [Ph.D. thesis]: Houston, Texas, Rice University, 243 p.

Quinn, M.J., Wright, J.E., and Wyld, S.J., 1997, Happy Creek igneous complex and tectonic evolution of the early Mesozoic arc in the Jackson Mountains, northwest Nevada: Geological Society of America Bulletin, v. 109, p. 461–482.

Rogers, J.W., 1999, Jurassic-Cretaceous deformation in the Santa Rosa Range, Nevada: Implications for the development of the northern Luning-Fencemaker fold-and-thrust belt [M.S. thesis]: Athens, University of Georgia, 195 p.

Rogers, J.W., and Wyld, S.J., 1998, Jurassic-Cretaceous deformation in the early Mesozoic marine province of central Nevada: New data from the Santa Rosa Range: Geological Society of America Abstracts with Programs, v. 30, no. 5, p. 62.

Russell, B.J., 1981, Pre-Tertiary paleogeography and tectonic history of the Jackson Mountains, northwestern Nevada [Ph.D. thesis]: Evanston, Illinois, Northwestern University, 206 p.

Russell, B.J., 1984, Mesozoic geology of the Jackson Mountains, northwestern Nevada: Geological Society of America Bulletin, v. 95, p. 313–323.

Saleeby, J.B., and Busby-Spera, C., 1992, Early Mesozoic tectonic evolution of the western U.S. Cordillera, *in* Burchfiel, B.C., et al., eds., The Cordilleran orogen: Conterminous U.S.: Boulder, Colorado, Geological Society of America, Geology of North America, v. G-3, p. 107–168.

Schweickert, R.A., and Lahren, M.M., 1990, Speculative reconstruction of the Mojave–Snow Late fault: Implications for Paleozoic and Mesozoic orogenesis in the western United States: Tectonics, v. 9, p. 1609–1629.

Silberling, N.J., Jones, D.L., Blake, M.C., Jr., and Howell, D.G. 1987, Lithotectonic terrane map of the western conterminous United States: U.S. Geological Survey Miscellaneous Field Studies Map MF-1874-C, scale 1:2 500 000.

Silberling, N.J., and Roberts, R.J., 1962, Pre-Tertiary stratigraphy and structure of northwestern Nevada: Geological Society of America Special Paper 72, p. 1–58.

Silberling, N.J., and Wallace, R.E., 1969, Stratigraphy of the Star Peak Group (Triassic) and overlying lower Mesozoic rocks, Humboldt Range, Nevada: U.S. Geological Survey Professional Paper 592, 50 p.

Smith, D.L., Wyld, S.J., Wright, J.E., and Miller, E.L., 1993, Progession and timing of Mesozoic crustal shortening in the northern Great Basin, western U.S.A., *in* Dunne, G., and McDougall, K., eds., Mesozoic paleogeography of the western United States—II: Pacific Section, SEPM, book 71, p. 389–406.

Smith, J.G., 1966, Petrology of the southern Pine Forest Range, Humboldt County, Nevada [Ph.D. thesis]: Stanford, California, Stanford University, 136 p.

Smith, J.G., 1973, Geologic map of the Duffer Peak quadrangle, Humboldt County, Nevada: U.S. Geological Survey Miscellaneous Field Investigations Map I-606, scale 1:48 000.

Speed, R.C., 1978a, Paleogeographic and plate tectonic evolution of the early Mesozoic marine province of the western Great Basin, *in* Howell, D.G., and McDougall, K.A., eds., Mesozoic paleogeography of the western U.S.: Pacific Section, Society of Economic Paleontologists and Mineralogists, Pacific Coast Paleogeography Symposium 2, p. 253–270.

Speed, R.C., 1978b, Basinal terrane of the early Mesozoic marine province of the western Great Basin, *in* Howell, D.G., and McDougall, K.A., eds., Mesozoic paleogeography of the western U.S.: Pacific Section, Society of Economic Paleontologists and Mineralogists, Pacific Coast Paleogeography Symposium 2, p. 237–252.

Speed, R.C., 1979, Collided Paleozoic microplate in the western U.S.: Journal of Geology, v. 87, p. 279–292.

Thole, R.H., and Prihar, D.W., 1998, Geologic map of the Eugene Mountains, northwestern Nevada: Nevada Bureau of Mines and Geology map 115, 12 p., scale 1:24 000.

Willden, R., 1964, Geology and mineral deposits of Humboldt County, Nevada: Nevada Bureau of Mines Bulletin 59, 154 p.

Wyld, S.J., 1990, Paleozoic and Mesozoic rocks of the Pine Forest Range, northwest Nevada, and their relation to volcanic arc assemblages of the western U.S. Cordillera, *in* Harwood, D.S., and Miller, M.M., eds., Late Paleozoic and early Mesozoic paleogeographic relations; Klamath Mountains, Sierra Nevada, and related rocks: Geological Society of America Special Paper 255, p. 219–237.

Wyld, S.J., 1991, Tectonic implications of newly documented early (?) through late Paleozoic age strata, Black Rock Desert region, northwest Nevada: American Association of Petroleum Geologists Bulletin, v. 75, p. 386–387.

Wyld, S.J., 1992, Geology and geochronology of the Pine Forest Range, northwest Nevada: Stratigraphic, structural and magmatic history, and regional implications [Ph.D. thesis]: Stanford, California, Stanford University, 429 p.

Wyld, S.J., 1996, Early Jurassic deformation in the Pine Forest Range, northwest Nevada, and implications for Cordilleran tectonics: Tectonics, v. 15, p. 566–583.

Wyld, S.J., 1998, Structural development of the Luning-Fencemaker fold-and-thrust belt: New constraints from northern Nevada: Geological Society of America Abstracts with Programs, v. 30, no. 5, p. 71.

Wyld, S.J., and Wright, J.E., 1997, Triassic-Jurassic tectonism and magmatism in the Mesozoic continental arc of Nevada: Classic relations and new developments, *in* Link, P.K., and Kowallis, B.J., eds., Proterozoic to recent stratigraphy, tectonics, and volcanology, Utah, Nevada, southern Idaho and central Mexico: Geological Society of America field trip guide book: Brigham Young University Geology Studies, v. 42, part 1, p. 197–224.

Wyld, S.J., Quinn, M.J., and Wright, J.E., 1996, Anomalous(?) Early Jurassic deformation in the western U.S. Cordillera: Geology, v. 24, p. 1037–1040.

Manuscript Accepted by the Society January 24, 2000

Printed in U.S.A

Geological Society of America
Special Paper 347
2000

# Geodynamic interpretation of Paleozoic tectonic trends oriented oblique to the Mesozoic Klamath-Sierran continental margin in California

**William R. Dickinson**
*Department of Geosciences, Box 210077, University of Arizona, Tucson, Arizona 85721, USA*

## ABSTRACT

**Devonian-Mississippian Antler and Permian-Triassic Sonoma orogenic trends are oriented northeast-southwest across Nevada and California at high angles to the Mesozoic-Cenozoic Cordilleran margin. The Roberts Mountains (Antler) and Golconda (Sonoma) allochthons of central Nevada were thrust over the miogeoclinal continental margin as accretionary prisms assembled and emplaced by episodic slab rollback toward the southeast during subduction downward to the northwest beneath an evolving offshore system of frontal and remnant magmatic arcs facing toward the continent. Diachronous Devonian and Permian arc assemblages forming superposed stratigraphic successions in the eastern Klamath Mountains and northern Sierra Nevada record alternating episodes of eruptive activity and dormancy along parallel but separate arc structures within the offshore island arc complex. Deformed lower Paleozoic rocks forming the substratum of the Klamath-Sierran arc assemblages display imbricated thrust panels shingled in a structural pattern indicative of southeast vergence within a regional accretionary prism that included the Roberts Mountains allochthon at its leading edge. Thrust emplacement of the Roberts Mountains allochthon over the miogeoclinal belt temporarily arrested subduction, to allow deposition of the Havallah sequence between the deformed continental margin and an offshore system of remnant island arcs. Renewed subduction and slab rollback emplaced the accretionary prism of the Golconda allochthon composed of the deformed Havallah sequence. Antler-Sonoma tectonic elements were truncated on the southwest by a sinistral late Paleozoic to earliest Mesozoic transform fault, which established the northwest-southeast trend of the California continental margin and displaced the miogeoclinal Caborca block southward into Mexico.**

## INTRODUCTION

Interpreting the geodynamic context of the Antler and Sonoma orogenies has been a continuing challenge for the past 30 years. The two events involved the thrusting of intricately deformed oceanic sedimentary facies, together with subordinate submarine volcanic rocks, over the western miogeoclinal flank of the North American craton in Late Devonian to Early Mississippian (Antler) and Late Permian to earliest Triassic (Sonoma) times. The preserved overthrust assemblages in central Nevada form the Roberts Mountains and Golconda allochthons, respectively, with the latter emplaced structurally above a sedimentary overlap assemblage that depositionally overlies the former.

Two contrasting tectonic models have been proposed for Antler-Sonoma events: (1) episodic backarc thrusting behind a long-lived west-facing island arc beneath which subduction was downward to the east (Burchfiel and Davis, 1972, 1975, 1981a; E.L. Miller et al., 1984, 1992; M.M. Miller, 1987; Burchfiel et al.,

Dickinson, W.R., 2000, Geodynamic interpretation of Paleozoic tectonic trends in oriented oblique to the Mesozoic Klamath-Sierran continental margin in California, *in* Soreghan, M.J., and Gehrels, G.E., eds., Paleozoic and Triassic paleogeography and tectonics of western Nevada and northern California: Boulder, Colorado, Geological Society of America Special Paper 347, p. 209–245.

1992) and (2) two successive arc-continent collisions involving east-facing island arcs beneath which subduction was downward to the west (Moores, 1970; Speed, 1977, 1979, 1984; Schweickert, 1978; Dickinson, 1981a, 1981b; Snyder and Brueckner, 1983; Lapierre et al., 1986; Speed et al., 1988). Evidence for thrusting of offshore tectonic elements toward the continental block is compatible with either model (Wyld, 1991). Proponents of each model have typically allowed for the other as a possible, though disfavored, alternative, and hybrid models involving sequential reversals of arc polarity have also been suggested (Churkin, 1974a, 1974b; Churkin and Eberlein, 1977; Dickinson, 1977; Hamilton, 1978; Schweickert and Snyder, 1981; M.M. Miller and Harwood, 1990; Girty et al., 1996). Backarc thrusting is probably the most favored current model, but suffers from a lack of clearcut actualistic analogy with modern tectonic features of active island arcs, and from the difficulty of establishing arc polarity without ambiguity (Oldow et al., 1989).

Resolution of any longstanding interpretive dichotomy in geoscience commonly requires improved root concepts, as well as additional data to sharpen constraints, and typically involves development of a third alternative model combining aspects of both preceding controversial models. For interpretation of Antler-Sonoma events, the most revealing data sets developed over the past 25 years, apart from incremental advances in knowledge of key field relationships and the petrochemistry of relevant igneous suites, stem from two facets of geochronology: (1) isotopic ages, especially as based on U/Pb systematics, and (2) radiolarian-conodont biostratigraphy derived from microfossils extracted from otherwise unfossiliferous cherts and argillites by HF dissolution. Freshly acquired detrital zircon ages from associated clastic sedimentary strata provided the intellectual springboard for my reappraisal of Antler-Sonoma events, but represent only an incremental addition to the body of data reconsidered here.

The most significant zircon data for analysis of Antler-Sonoma tectonism are geochronological results indicating that lower Paleozoic quartzose sandstones of the Yreka terrane in the eastern Klamath Mountains, the Shoo Fly Complex in the northern Sierra Nevada, and the bulk of the Roberts Mountains allochthon in central Nevada contain essentially indistinguishable detrital zircon populations of Precambrian ages incompatible with derivation from the immediately adjacent craton (Dickinson and Gehrels, this volume, Chapter 11). The conclusions presented in this chapter arose from devising a tectonic model that allows for derivation of all three stratal assemblages from a common but unexpected provenance.

## APENNINE ANALOGY

The perception that Mediterranean-style subduction (Malinverno and Ryan, 1986) played a central role for Antler events (Burchfiel and Royden, 1991; Burchfiel et al., 1992), and by analogy for Sonoma events (Burchfiel et al., 1992), offers the means to develop a third alternative to backarc thrusting or arc-continent collision as a model for the Antler and Sonoma orogenies. Intermittent slab rollback of lithosphere descending to the west beneath a nearby offshore island arc system, facing toward the continent, allowed migration of successive subduction zones toward the continental margin. This style of tectonic behavior, driven by retrograde slab descent (Garfunkel et al., 1986; Doglioni, 1991), requires no large shift in the positions of successive active and remnant island arc structures stranded within the oceanic region beyond the migratory subduction zone approaching the continent. Resulting tectonism in Nevada emplaced two separate accretionary prisms, first the Roberts Mountains allochthon and then the Golconda allochthon, across the edge of the continental block by sequential episodes of slab rollback.

In the western Mediterranean arena, the classic Apennine overthrusting of an accretionary prism over the Apulian and Sicilian platforms during Neogene time has been coeval with opening of the oceanic Tyrrhenian Sea behind the thrust complex, and with associated subduction-related magmatism along various nascent and remnant arc segments within or beyond the newly formed oceanic domain. The locus of subduction-related arc volcanism migrated in prograde fashion across the incrementally widening oceanic domain of the Tyrrhenian Sea to build successive volcanogenic structures of modest bulk within the expanse of generally oceanic crust (Savelli, 1988; Gueguen et al., 1997, 1998; Argnani and Savelli, 1999). Sedimentary cover was stripped from the underthrusting plate by the migrating subduction zone, to be assembled into the Apennine accretionary prism that has been overthrust above the continental platforms that form the lower plate of the Apennine thrust system in Italy. Apennine overthrusting and Tyrrhenian extension have been contemporaneous in parallel geotectonic domains for much of Neogene time (Keller et al., 1994; Cavinato and De Celles, 1999).

Adoption of a third, Mediterranean-style, option for the interpretation of Antler-Sonoma events allows for east-facing subduction, downward to the west, to emplace overthrust allochthons without requiring either arc-continent collisions or reversals in arc polarity. With a Mediterranean-style model, the long-continued presence of island arc structures near the continental margin does not require west-facing subduction to pin them in place. In the tectonic scheme developed here, preceding or simultaneous west-facing subduction, downward to the east beneath the offshore island arcs involved in Mediterranean-style behavior (Burchfiel and Royden, 1991; Burchfiel et al., 1992), is discounted for two reasons. Intermittent polarity reversal is an unnecessary hypothesis and simultaneous subduction on both sides of the same island arc is unlikely (Hamilton, 1988).

Figure 1 shows the comparable scales of Paleozoic Antler-Sonoma and Cenozoic Apennine-Tyrrhenian tectonic systems. Although their configurations vary in detail, the relative positions and sizes of key geotectonic elements are similar. For example, the extent of the offshore Sardinia-Corsica remnant arc (Savelli et al., 1979) and the net expanse of segments of the compound Paleozoic island arc system preserved in the eastern Klamath

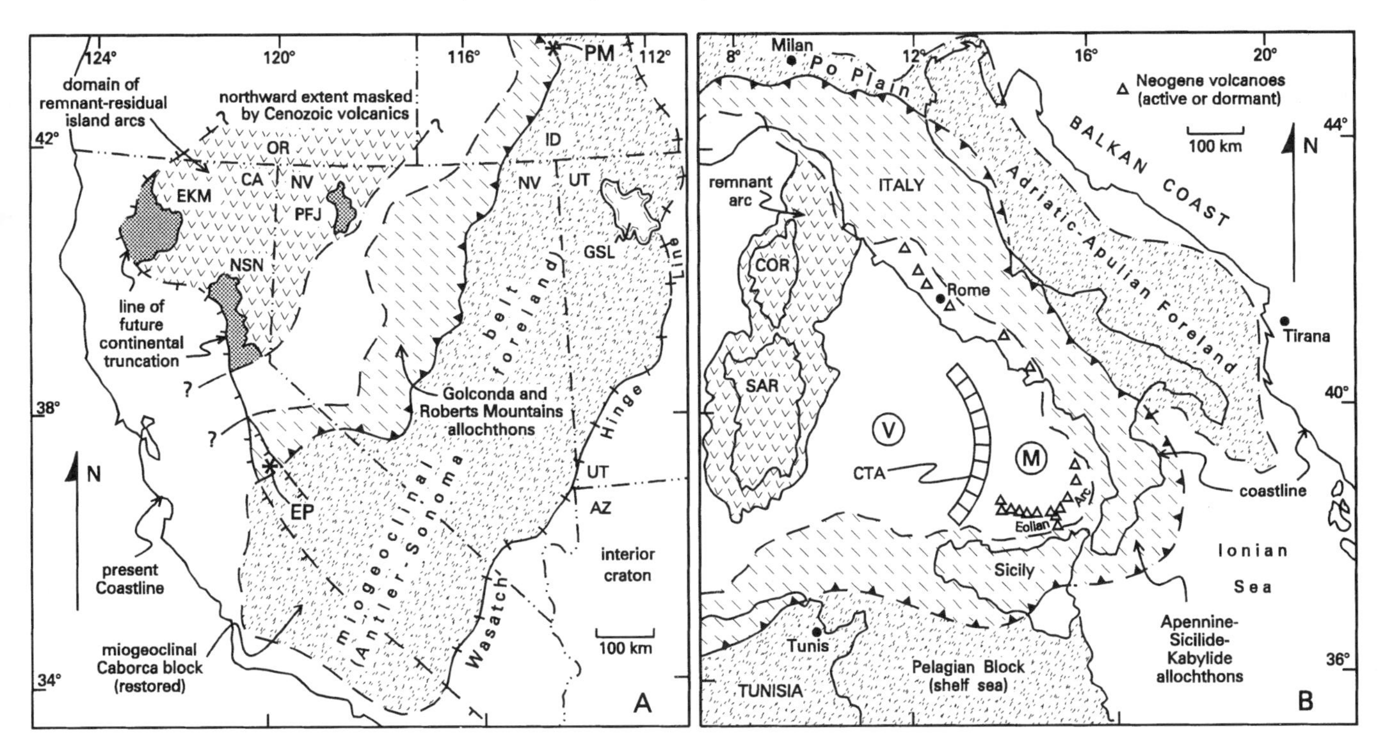


Figure 1. Geotectonic elements (at same scale) of Paleozoic Antler-Sonoma (A) and Cenozoic Apennine-Tyrrhenian (B) orogenic systems. In A (regional relationships adapted from Figs. 2 and 7), bedrock exposures of remnant-residual arc domain include eastern Klamath Mountains (EKM), northern Sierra Nevada (NSN), and Pine Forest–Jackson Ranges (PFJ); asterisks denote El Paso (EP) and Pioneer (PM) Mountains, and GSL is Great Salt Lake. In B (modified after Boccaletti and Dainelli, 1982), exposed segments of remnant arc include Corsica (COR) and Sardinia (SAR), with Miocene Vavilov (V) and Quaternary Marsili (M) oceanic subbasins of the Tyrrhenian Sea separated by intervening Pliocene Central Tyrrhenian arc (CTA) after Argnani and Savelli (1999); multiple intraplate seamounts (Savelli and Wezel, 1980) occur within Vavilov and Masili subbasins.

Mountains, the northern Sierra Nevada, and the Pine Forest–Jackson Ranges are roughly coordinate. In the Antler-Sonoma system, however, the preserved width of the once oceanic domain between the overthrust allochthons and remnant-residual island arcs, which stood offshore in Paleozoic time, was reduced by Mesozoic crustal shortening during subsequent evolution of the Cordilleran orogen. In the Tyrrhenian-Apennine system, by contrast, subduction of the slab rolling back under the active Eolian arc is still underway (Anderson and Jackson, 1987), and the oceanic domain west of the overthrust Apennine allochthons remains at full width.

A direct analogy can be drawn between the modern Apennine-Tyrrhenian system and tectonic events of the Antler orogeny (Burchfiel and Royden, 1991), but not for the Sonoma orogeny. A Mediterranean-style model for the Sonoma orogeny requires the postulate of renewed post-Antler subduction beneath offshore remnant arcs to consume syn-Antler sea floor. In Apennine-Tyrrhenian terms, this behavior would require future subduction of Tyrrhenian sea floor beneath the Sardinia-Corsica remnant arc (Fig. 1B), followed by a second episode of slab rollback and migratory subduction to emplace a second-generation allochthon above the existing Apennine thrust sheets. Such postulated future destruction of Tyrrhenian oceanic crust could also dispose of the fragmentary arc structures that have been built within the oceanic domain by progradational arc volcanism.

## OVERVIEW

Seminal for interpretation of Paleozoic orogenic events in California and Nevada is the observation that pre-Mesozoic tectonic trends are oriented in a northeast-southwest direction, nearly at right angles to the Mesozoic continental margin (Fig. 2), and imply oblique tectonic truncation of Paleozoic assemblages to create anew the present trend of the Cordilleran margin near the Paleozoic-Mesozoic time boundary (Hamilton and Myers, 1966). Consequently, Paleozoic subduction and arc magmatism cannot be treated successfully as just precursors of the well known Mesozoic-Cenozoic arc-trench system of the continental margin. Paleozoic tectonic elements had a northeast, rather than northwest strike, and flanked a continental margin from which some segment of uncertain dimensions, toward the southwest, was removed by continental truncation.

As a corollary, the tectonic vergence of Paleozoic orogenic assemblages in California and Nevada was either southeastward or northwestward, rather than southwestward or northeastward, and there is no strike continuity, northwest to southeast, between Paleozoic tectonic elements of the Klamath Mountains and the Sierra Nevada. In this light, for example, the similarities in the pre-Jurassic geology of the Pine Forest Range and Quinn River Crossing in northwestern Nevada (Black Rock terrane of Silberling et al., 1987, 1992) to the eastern Klamath Mountains (Wyld,

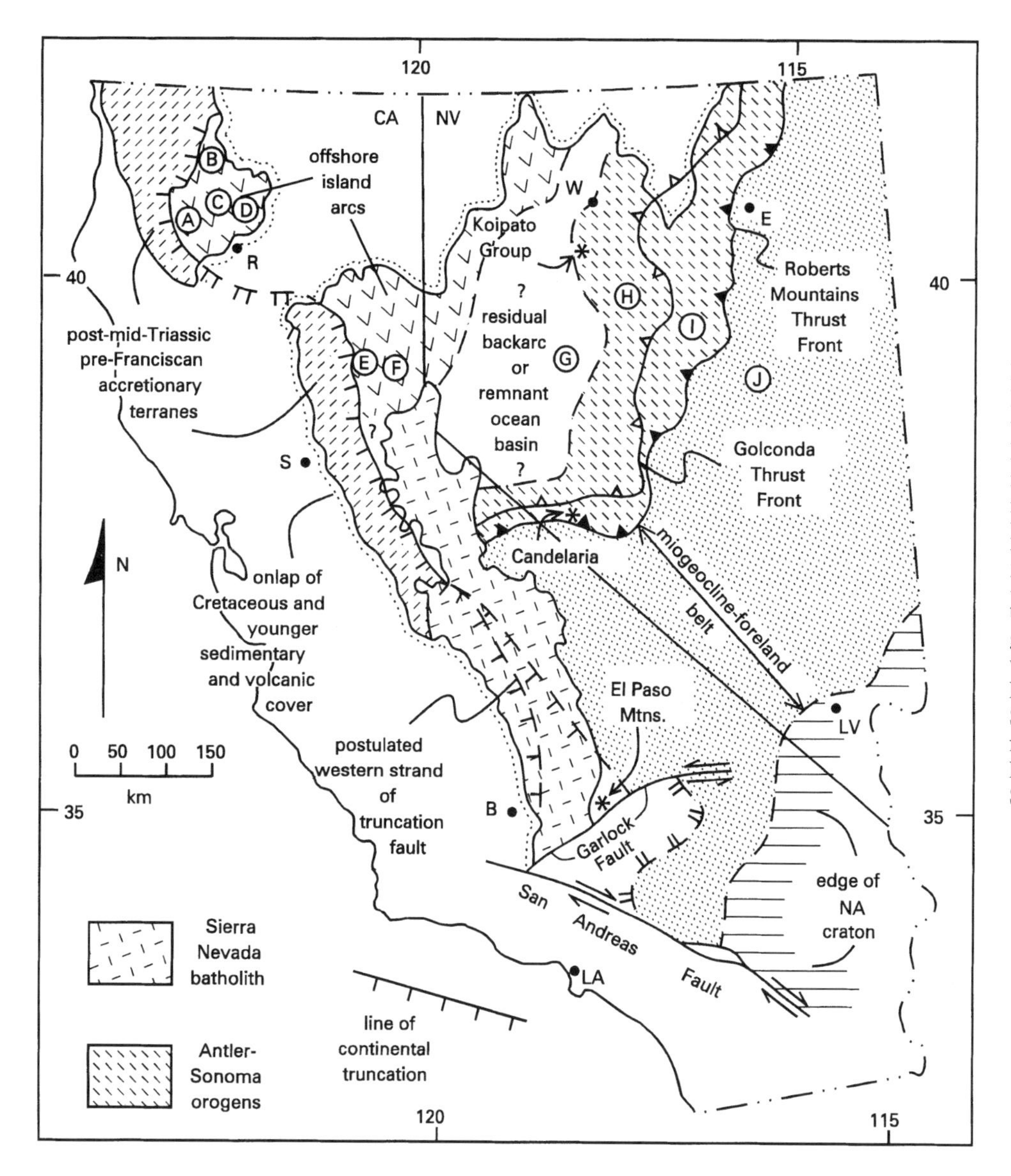


Figure 2. Geotectonic relations of late Paleozoic to earliest Mesozoic truncation of edge of continental block to form Mesozoic continental margin trending northwest-southeast, as opposed to northeast-southwest trend of preceding Paleozoic continental margin. NA is North America. Asterisks denote key localities discussed in text. Circled letters denote locations of tectonostratigraphic columns of Figure 3 (A–D within eastern Klamath Mountains and E–F within northern Sierra Nevada). Selected cities for location: B—Bakersfield; E—Elko; LA—Los Angeles; LV—Las Vegas; R—Redding; S—Sacramento; W—Winnemucca.

1990; Jones, 1990; Blome and Reed, 1995) is expected, because Paleozoic assemblages of the Klamath Mountains and northwestern Nevada lay nearly along tectonic strike prior to basin-range extension.

### *Regional geochronology*

Figure 3 correlates available isotopic and biostratigraphic age data for Paleozoic and lower Mesozoic rock assemblages of northeastern California and northwestern Nevada. Because correlation of isotopic and biostratigraphic ages is important for understanding tectonostratigraphic and tectonomagmatic relations, the time scale incorporates significant revisions, derived from recent geochronological studies, to the conventional Decade of North American Geology (DNAG; Palmer, 1983) and Geological Society of London (GSL; Harland et al., 1990) time scales. The time scale used here was devised as follows.

1. The base of the Cambrian at 545 Ma (vs. 570 Ma of both DNAG and GSL) is taken from Tucker and McKerrow (1995) and Landing et al. (1998), with intra-Cambrian divisions after Bowring and Erwin (1998).

2. The base of the Ordovician at 490 Ma (vs. 505–510 Ma of DNAG and GSL) is taken from Davidek et al. (1998), with intra-Ordovician divisions adapted from Tucker and McKerrow (1995).

3. The base of the Silurian at 440 Ma is midway between the DNAG and GSL figures (438–439 Ma) and the suggestion (442 Ma) of Tucker and McKerrow (1995), with intra-Silurian divisions adapted from Tucker and McKerrow (1995) and Tucker et al. (1998).

4. The base of the Devonian at 417 Ma (vs. 408–409 Ma of DNAG and GSL) is taken from Tucker and McKerrow (1995), with intra-Devonian divisions adapted from Tucker et al. (1998).

5. The base of the Carboniferous at 362 Ma is the average of DNAG (360 Ma) and GSL (363 Ma) scales, with intra-Carboniferous divisions interpolated as averages from those two scales.

6. The base of the Permian at 288 Ma is the average of DNAG (286 Ma) and GSL (290 Ma) scales, with intra-Permian divisions interpolated as averages from those two scales, but adjusted proportionally for 251 Ma as the Paleozoic-Mesozoic boundary.

7. The base of the Triassic at 251 Ma is taken from Bowring et al. (1998), with intra-Triassic divisions adjusted proportionally from Gradstein et al. (1994), who inferred 248 Ma for the base of the Triassic.

8. The base of the Jurassic and all intra-Jurassic divisions are taken directly from Gradstein et al. (1994), which supersedes the Mesozoic time scale (Kent and Gradstein, 1985) used for the DNAG scale.

### *Regional correlations*

Inspection of Figure 3 reveals the following key points.

1. The complexly deformed Yreka (or Yreka-Callahan) terrane of the Klamath Mountains, Shoo Fly Complex of the northern Sierra Nevada, and Roberts Mountains allochthon of central Nevada are composed dominantly of similar lower Paleozoic strata; all three assemblages contain comparable detrital zircon populations (Dickinson and Gehrels, this volume, Chapter 11) that are compatible with the diachronous assembly of the three assemblages by the subduction of sea-floor strata representing parts of the same regional depositional system.

2. The mid-Paleozoic and latest Paleozoic arc assemblages of the Klamath Mountains and Sierra Nevada form integral parts of depositionally stacked stratigraphic successions (Burchfiel and Davis, 1972; M.M. Miller, 1987; Burchfiel et al., 1992), and cannot represent unrelated arc structures of contrasting age that accreted to the continent at markedly different times.

3. The ages of the Paleozoic arc assemblages in the Klamath Mountains and Sierra Nevada are systematically diachronous (M.M. Miller and Harwood, 1990; Blein et al., 1994, 1996), and cannot represent contemporaneously active arc segments, although the Sierran and Klamath assemblages could well be remnants of tectonically related arc structures analogous to modern frontal and remnant arcs of the western Pacific region (Rouer and Lapierre, 1989).

4. Eruption of the mid-Paleozoic arc assemblage of the northern Sierra Nevada coincided with emplacement of the Roberts Mountains allochthon in central Nevada, suggesting that both events were integral facets of the Antler orogeny (Harwood and Murchey, 1990; Harwood et al., 1991; Hanson et al., 1996).

5. Because the oldest components of the upper Paleozoic Havallah sequence, which forms the Golconda allochthon, are also coeval with emplacement of the Roberts Mountains allochthon, inception of the Havallah basin preceded or accompanied the Antler orogeny (E.L. Miller et al., 1981, 1984; Snyder and Brueckner, 1983; Whiteford, 1990; Burchfiel et al., 1992); the Havallah basin could not have been initiated by post-Antler rifting, whether such rifting is postulated after either backarc thrusting or continental collision (Churkin, 1974a, 1974b; Dickinson, 1977, 1981b; Dickinson et al., 1983; E.L. Miller et al., 1992).

6. Deposition of a thin sequence of ribbon chert in the northern Sierra Nevada spans at least 50 m.y. of Carboniferous time (Harwood, 1983), and was coeval with much of the chert-rich Havallah sequence in northwestern Nevada, suggesting a lateral depositional tie between the Havallah basin and a dormant arc structure farther away from the continental block during the interval between the Antler and Sonoma orogenies (Murchey, 1990; Harwood and Murchey, 1990).

7. Eruption of Permian arc assemblages in the Klamath Mountains and Sierra Nevada preceded Early Triassic emplacement of the Golconda allochthon in central Nevada, but an unconformity between Permian and Triassic strata in the northern Sierra Nevada can be regarded as a local expression of the Sonoma orogeny (D'Allura et al., 1977; Harwood, 1983, 1992; Wyld, 1991; Hannah and Moores, 1986; Hanson et al., 1996).

8. Lower Mesozoic arc volcanics began to accumulate widely along the Klamath-Sierran trend in the Middle Triassic to earliest Jurassic time frame (Schweickert, 1978; Busby-Spera, 1988; Dilles and Wright, 1988; Saleeby and Busby-Spera, 1992).

## EARLY-MIDDLE PALEOZOIC (ANTLER) TECTONISM

Southeasterly tectonic vergence of structural elements in northern California and Nevada prior to Carboniferous time is indicated by the direction of overthrusting of the Roberts Mountains allochthon (Fig. 3I) in Nevada and by the structural relationship of the Yreka terrane (Fig. 3B) to the polygenetic Trinity ophiolitic complex (Fig. 3C) in the Klamath Mountains (Fig. 4A). It is also compatible with the internal imbrication of the partly correlative and tectonically analogous Shoo Fly Complex (Fig. 3F) in the northern Sierra Nevada (Fig. 4B), and with the structural juxtaposition of the Central Metamorphic belt (Fig. 3A) immediately below both the Yreka and Trinity assemblages (Fig. 4A).

With a Mediterranean-style model for subduction associated with the mid-Paleozoic Antler orogeny, the development of lower Paleozoic Klamath-Sierran rock assemblages can be viewed as the record of migratory subduction drawn progressively to the southeast by slab rollback, coupled with arc volcanism that advanced progressively across successively assembled subduction complexes to leave dormant remnant arcs in the wake of the progradational magmatism. The evolving subduction system was arrested by emplacement of the Roberts Mountains allochthon, as a frontal accretionary prism, over the miogeoclinal flank of the continental block (Burchfiel and Royden, 1991; Burchfiel et al., 1992).

### *Roberts Mountains allochthon*

The Roberts Mountains allochthon, composed of internally disrupted and imbricated thrust panels of lower Paleozoic chert-argillite, quartzose turbidites, and subordinate greenstone (Dick-

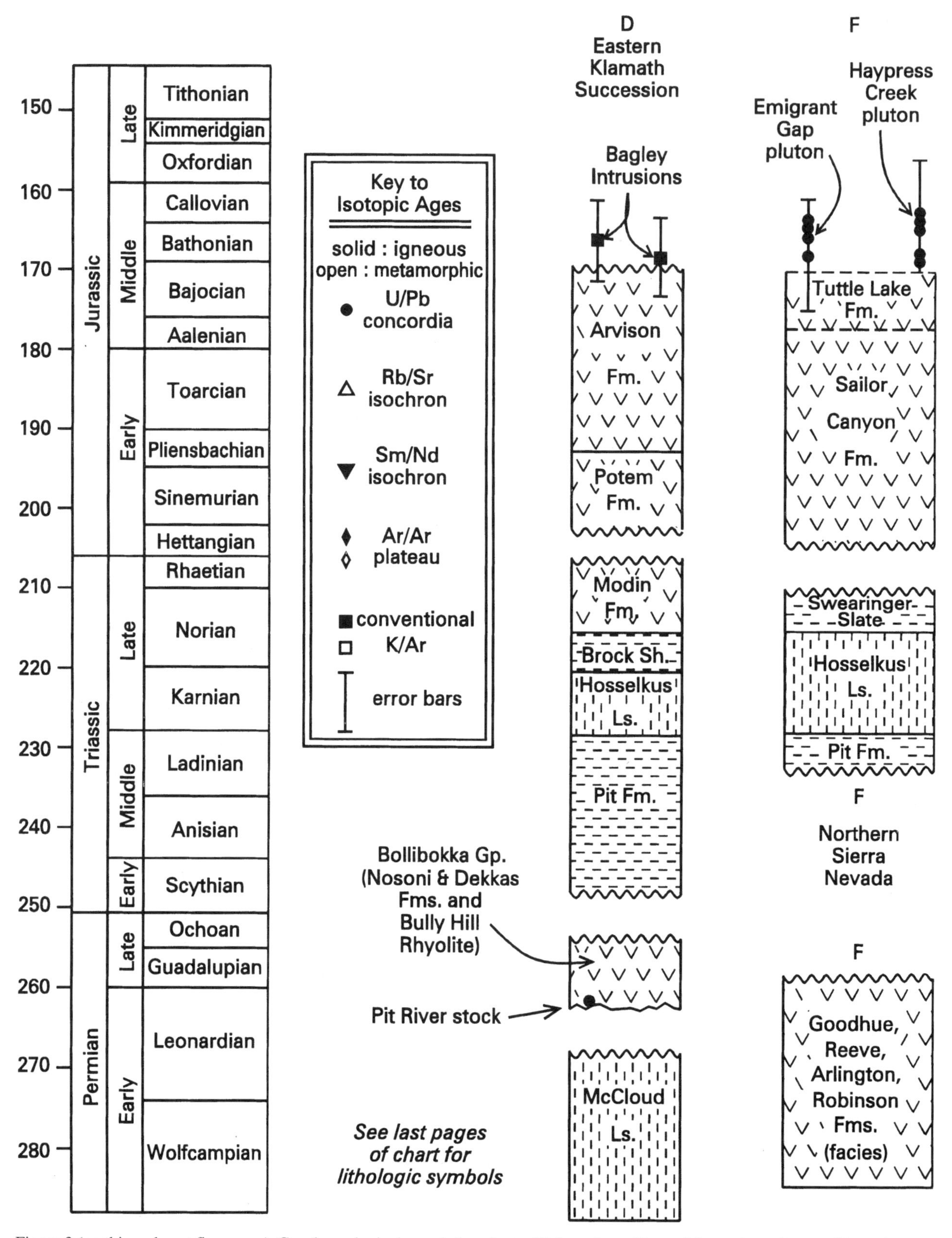


Figure 3 (on this and next five pages). Geochronological correlation chart of Paleozoic and lower Mesozoic rock assemblages in eastern Klamath Mountains, northern Sierra Nevada, and west-central Nevada (see Fig. 2 for locations of columns). Spatial relations of Klamath (A–D) and Sierra Nevada (E–F) assemblages are shown in Figure 4. Fm—Formation. Gp—Group. Ls—limestone. Sources of isotopic data: A, Hacker and Peacock (1990); B, Cashman (1980), Cotkin (1987, 1992), Cotkin et al. (1992), Wallin et al. (1988);

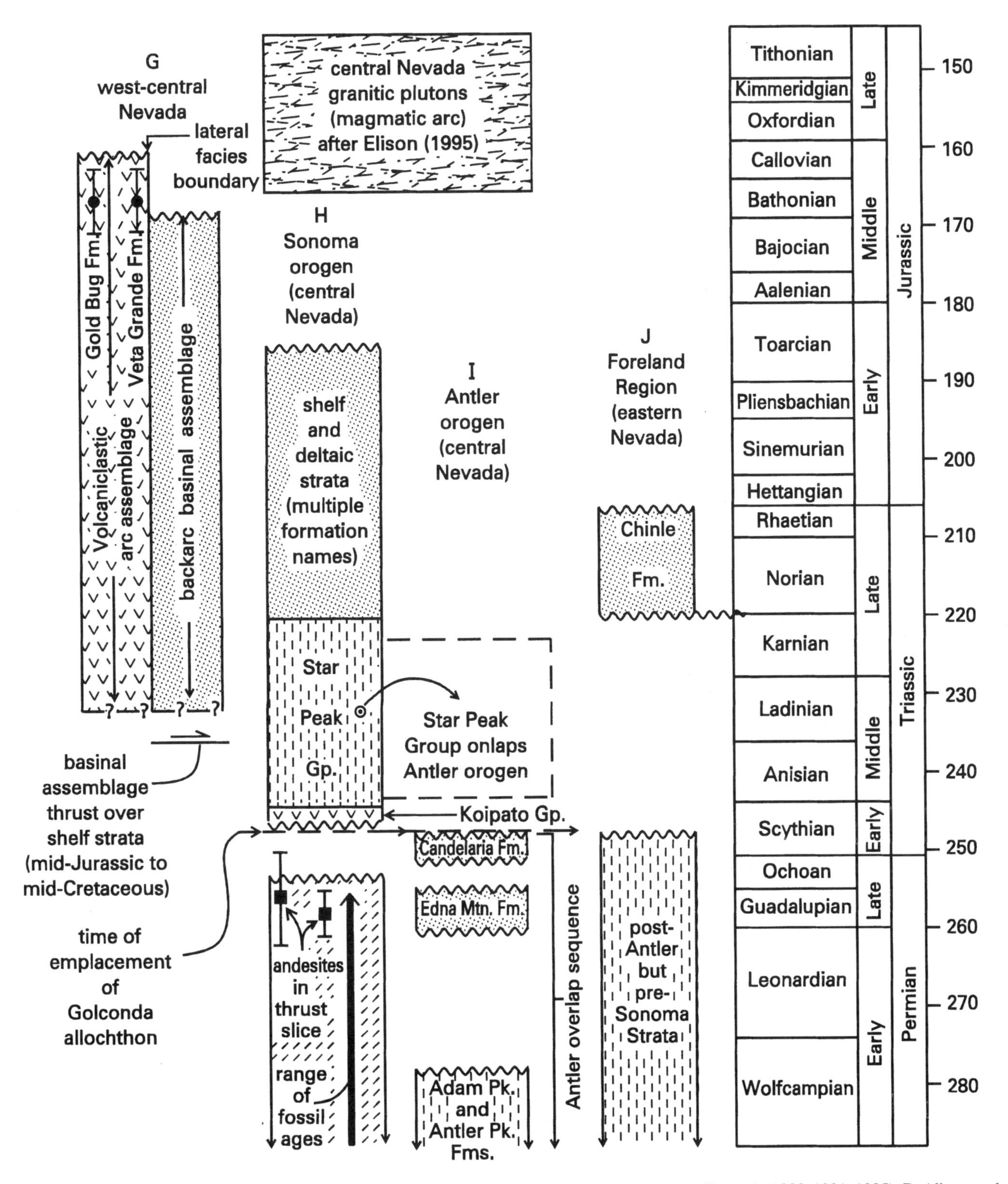


(Figure 3, continued) C, Jacobsen et al. (1984), Lanphere et al. (1968), Wallin and Metcalf (1998), Wallin et al. (1988, 1991, 1995); D, Albers et al. (1981), Fraticelli et al. (1985), Renne and Scott (1988); E, Edelman and Sharp (1989), Edelman et al. (1989), Hacker and Peacock (1990); F, Girty et al. (1991b, 1993, 1995), Hanson et al. (1988, 1996), Harwood et al. (1991), M.M. Miller and Harwood (1990), Saleeby et al. (1987); G, Wyld and Wright (1993). H, E.L. Miller et al. (1992). Sources of biostratigraphic data: B, Hotz (1977), Lindsley-Griffin (1977a, 1977b), Lindsley-Griffin and Griffin (1983), Potter et al. (1977, 1990), Savage (1977); D, Alibert et al. (1991), Charvet et al. (1990), Lapierre et al. (1985a, 1987), E.L. Miller et al. (1992), M.M. Miller (1989), M.M. Miller and Cui (1989), M.M. Miller and Harwood (1990), Sanborn (1960), Watkins (1973, 1985, 1986, 1990, 1993a, 1993b); F, Durrell and D'Allura (1977), Hannah and Moores (1986), Hanson and Schweickert (1986), Hanson et al. (1996),

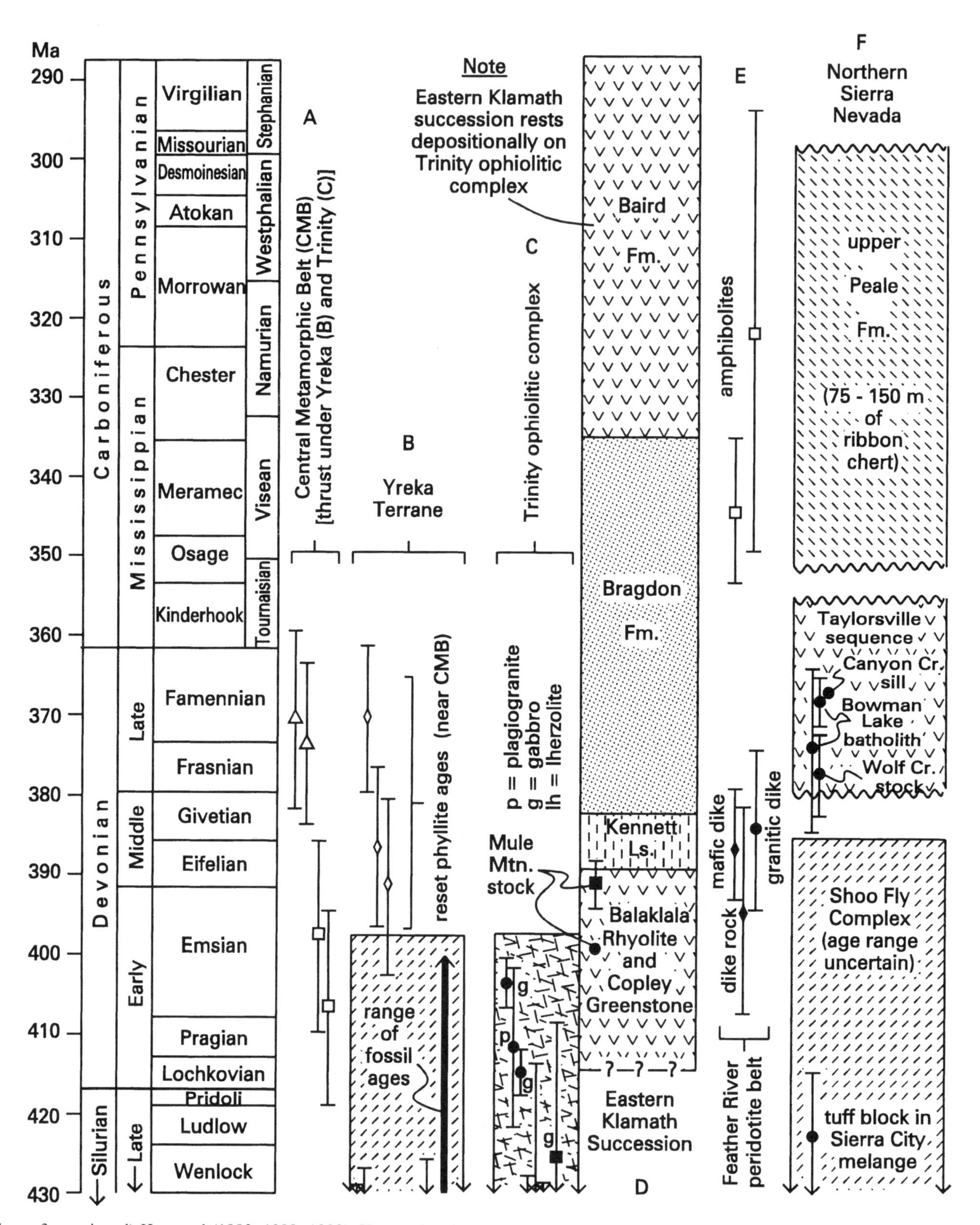


(Figure 3, continued) Harwood (1983, 1988, 1992), Harwood and Murchey (1990), Harwood et al. (1991, 1995), E.L. Miller et al. (1992), M.M. Miller and Harwood (1990), Poole et al. (1992); G, Burke and Silberling (1973), Elison and Speed (1988), Oldow et al. (1990), Speed (1978a, 1978b), Speed and Jones (1969), Wyld and Wright (1993); H, Dickinson et al. (1983), Harwood and Murchey (1990), Jones (1991), E.L. Miller et al. (1982, 1984, 1992), Murchey (1990), Nichols and Silberling (1977), Silberling (1973), Silberling and Wallace (1969), Speed (1978b, 1984), Stewart et al. (1986), Taylor et al. (1983); I, Dickinson et al. (1983), Giles and Dickinson (1995), Goebel (1991), E.L. Miller et al. (1981, 1992), Jones and Jones (1991), Poole et al. (1992); J, Dickinson et al. (1983), Giles and Dickinson (1995), Poole and Sandburg (1991), Poole et al. (1992), Stewart (1980), Stewart and Poole (1974).

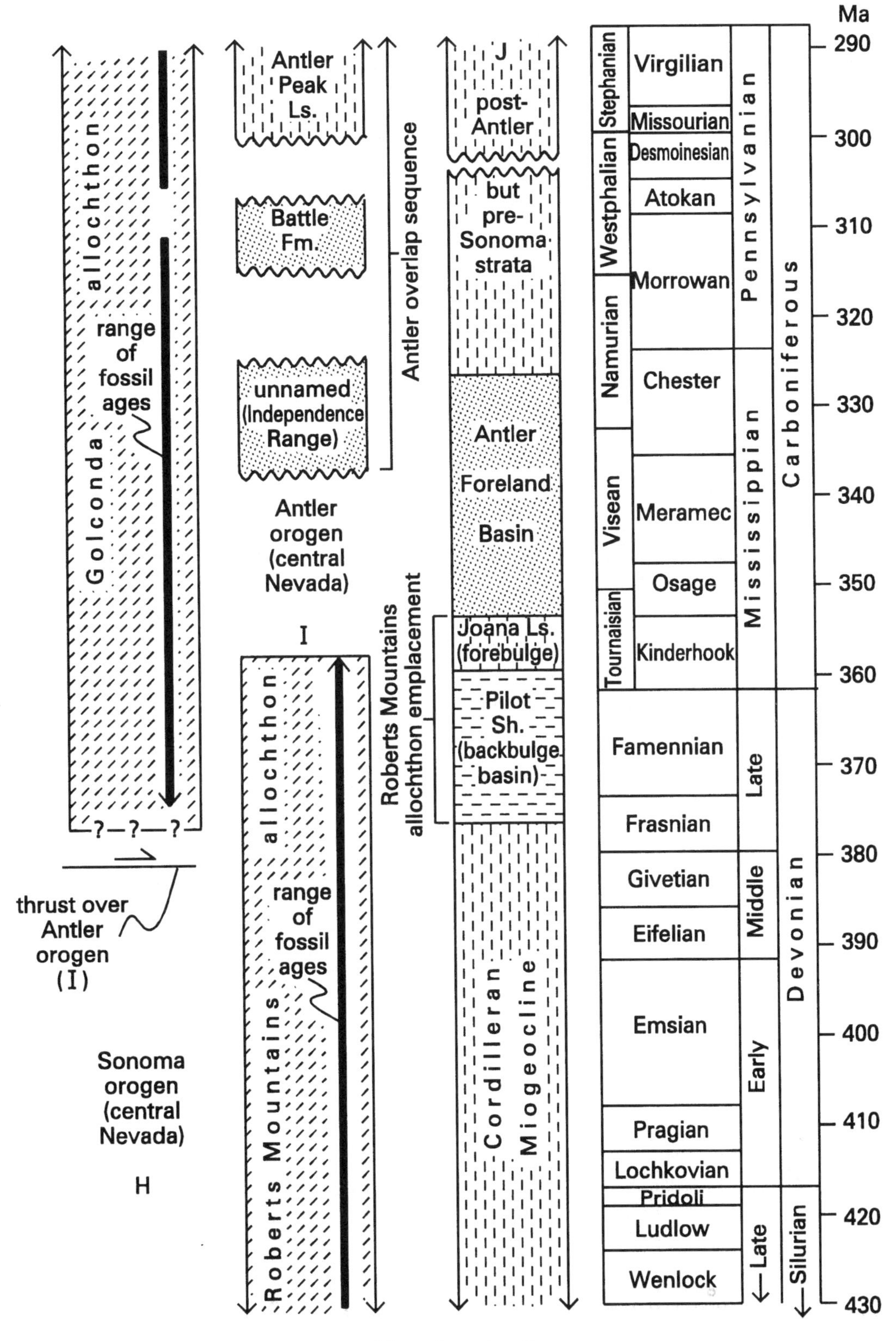


(Figure 3, continued)

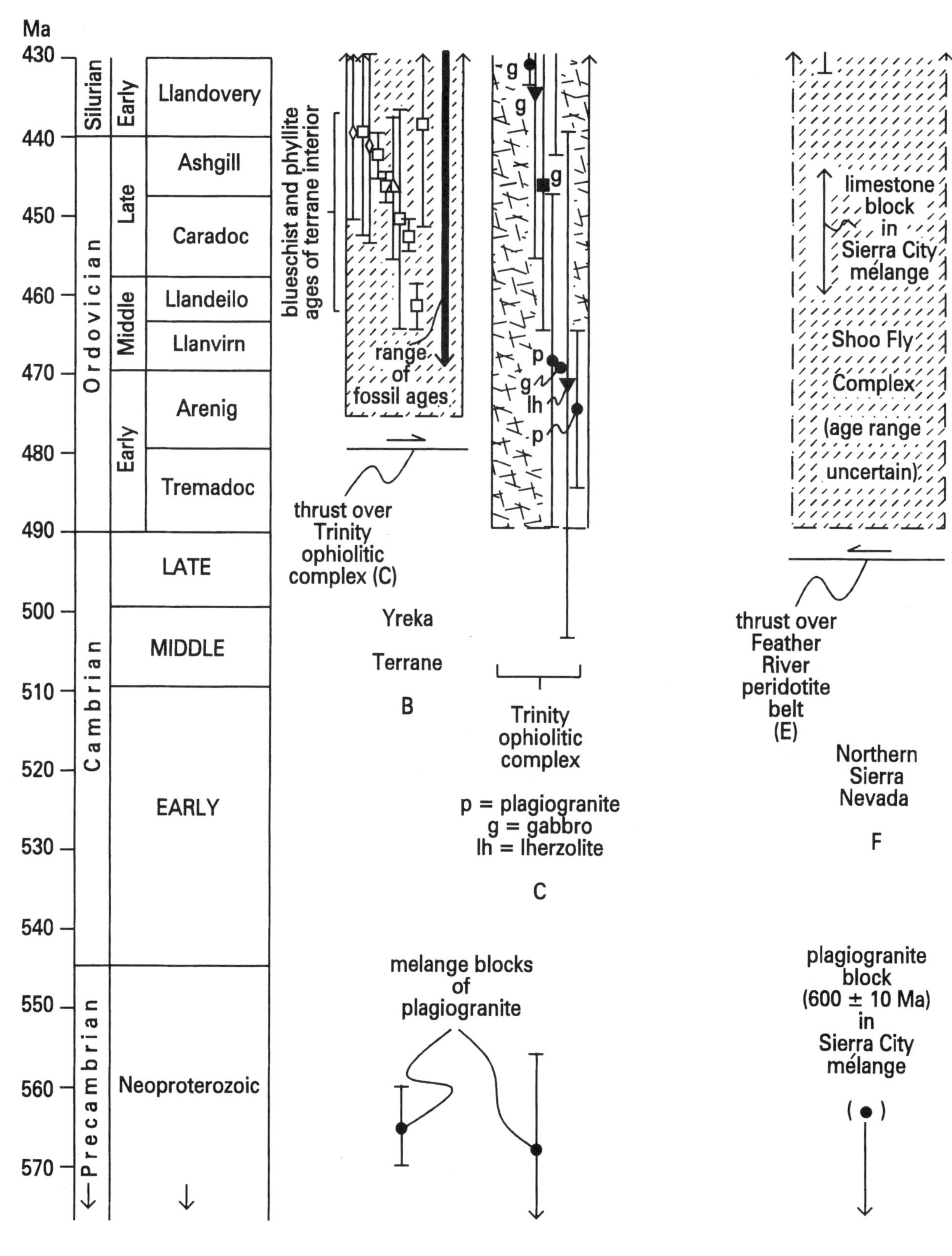


(Figure 3, continued)

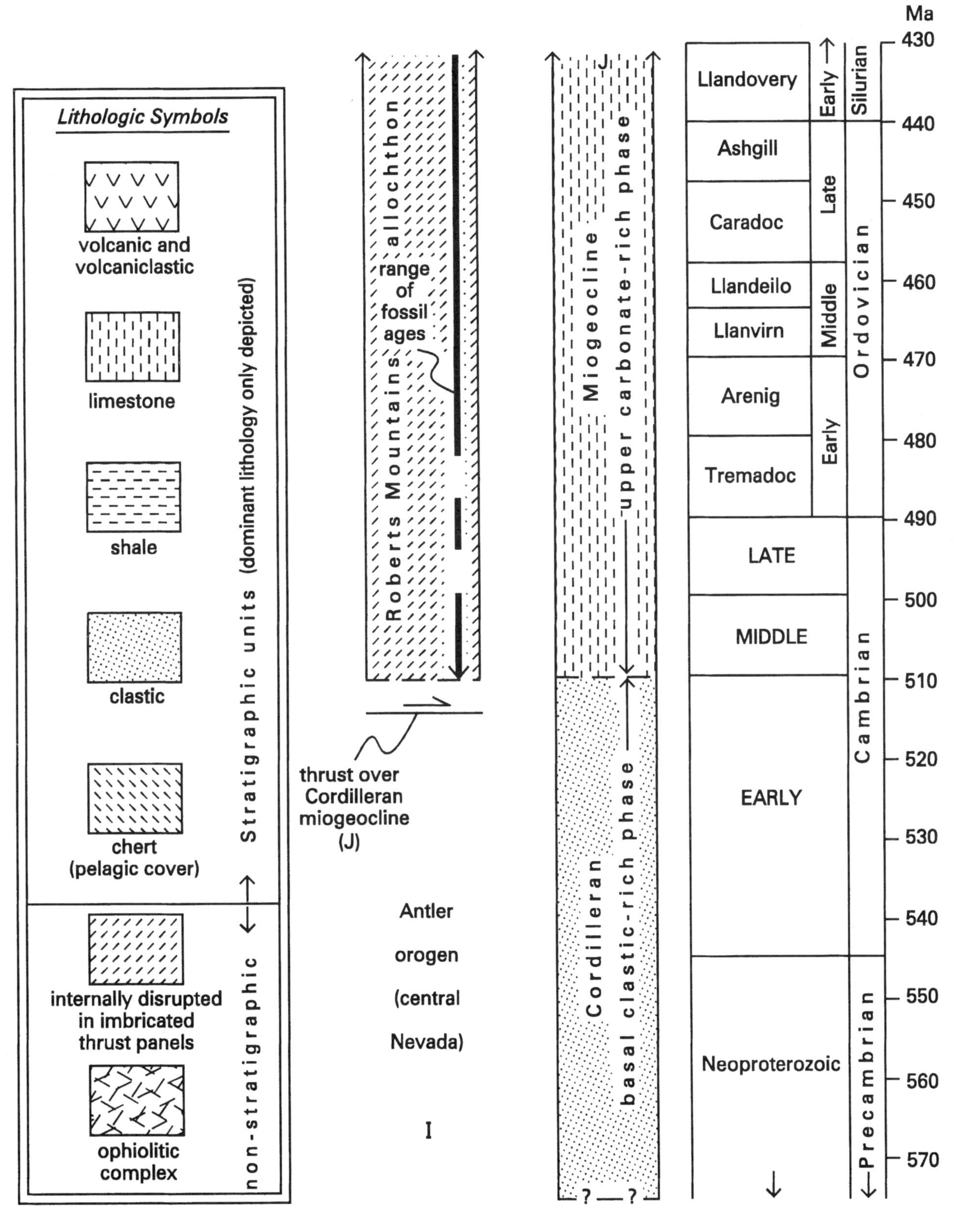


(Figure 3, continued)

inson et al., 1983; E.L. Miller et al., 1992), was emplaced across the outer flank of the Cordilleran miogeocline, from an oceanic region to the west, during latest Devonian to earliest Mississippian time (Johnson and Pendergast, 1981; Speed and Sleep, 1982; Poole et al., 1992). The alternate interpretation that most displaced strata of the Roberts Mountains allochthon were deposited close to the Cordilleran miogeocline above a rifted belt of continental crust (Turner et al., 1989; Poole et al., 1992, p. 28–32; E.L. Miller et al., 1992, p. 62–65) is called into question by information on detrital zircons from the Roberts Mountains allochthon (Dickinson and Gehrels, this volume, Chapter 11). Sandstones forming the bulk of the Roberts Mountains allochthon contain zircon age assemblages that could not have been derived from the adjacent segment of the continental block, as would be expected for strata in rift basins subsiding along the continental margin. Deformed strata of the Vinini Formation, which contains detrital zircons of ages compatible with nearby derivation, form the structurally lowest component of the Roberts Mountains allochthon (Schweickert and Snyder, 1981). Vinini strata are inferred to have been incorporated into basal horizons of the Roberts Mountains accretionary prism from continental slope and rise deposits as a subduction zone neared the continental margin. From their detrital zircon content, other turbidites of the Roberts Mountains allochthon could have been deposited at any arbitrary distance from the edge of the continental block.

Stratigraphic analysis of backbulge and forebulge deposits (Goebel, 1991) depositionally overlying strata of the Cordilleran miogeocline (Fig. 3J) indicates that the Roberts Mountains allochthon, including slope-basin deposits capping the imbricated thrust panels, was approaching the continental margin by early Late Devonian time and was in place, with a nearly stationary forebulge to the east, by early Early Mississippian time (Goebel, 1991; Giles and Dickinson, 1995). Clastic strata derived from the Roberts Mountains allochthon subsequently filled a downflexed foreland basin in eastern Nevada and western Utah through most of the remainder of Mississippian time (Poole, 1974; Dickinson et al., 1983; Giles and Dickinson, 1995). The uplifted bulk of the allochthon evidently shielded the foreland basin from any arc-derived detritus (Dickinson et al., 1983). A depositional overlap sequence of upper Mississippian to Upper Permian, and locally lowermost Triassic, strata was deposited unconformably but discontinuously upon the Roberts Mountains allochthon (Fig. 3I).

The synchroneity, within inherent uncertainties of combined biostratigraphic and isotopic age control, of arc volcanism in the northern Sierra Nevada with emplacement of the Roberts Mountains allochthon is striking (Fig. 3). The coincidence in timing suggests a genetic relationship between descent, downward to the northwest, of a subducted oceanic slab, to stimulate arc magmatism within an offshore island arc, and the slab rollback responsible for Roberts Mountains thrusting toward the southeast. The interruption of arc magmatism can be attributed to the arresting of further subduction by arrival of the migratory subduction zone at the edge of the buoyant continental block. Thrusting of the Roberts Mountains allochthon above the Cordilleran miogeocline was then achieved by partial subduction of the attenuated edge of the continental block beneath the accretionary prism of the migratory subduction zone.

### *Shoo Fly Complex*

The Shoo Fly Complex of the northern Sierra Nevada extends along tectonic trend from the Roberts Mountains allochthon, which strikes into the central Sierra Nevada at nearly right angles to the Sierra Nevada batholith (Schweickert and Lahren, 1987, 1993b; Greene et al., 1997). Shoo Fly stratal components include alkalic metabasalts interpreted as fragments of oceanic seamounts, chert intervals, and quartzose turbidites interpreted as either trench-fill or subsea-fan deposits (Bond and Devay, 1980; Girty and Schweickert, 1984; Schweickert et al., 1984; Girty and Wardlaw, 1985; Girty et al., 1990, 1991b, 1996; Harwood, 1988; M.M. Miller and Harwood, 1990; Harwood et al., 1991). The complex is composed of several "allochthons" (thrust panels) of partially intact strata overlain structurally by the Sierra City mélange, and can be interpreted as an accretionary prism laterally equivalent to parts of the Roberts Mountains allochthon (Schweickert, 1981; Schweickert et al., 1984; Schweickert and Snyder, 1981; Harwood et al., 1991).

The present steep easterly dip of bedding and thrust faults within the Shoo Fly Complex is the result of Mesozoic deformation, for overlying Paleozoic and Mesozoic volcanic and volcaniclastic strata (Fig. 3F) also display steep easterly dips, and the main internal structures of the Shoo Fly Complex were subhorizontal in Paleozoic time. The present exposures of the Shoo Fly Complex can thus be viewed as the intersection of the northwesterly trending Mesozoic continental margin with a northeasterly trending Paleozoic subduction complex, now tipped to the east by Mesozoic deformation. The shingling of thrust panels within the Shoo Fly Complex records originally southeast vergence (D'Allura et al., 1977; Schweickert et al., 1984; Girty and Pardini, 1987; Harwood, 1988, 1992; Harwood and Murchey, 1990; Harwood et al., 1991), and the unconformity with the overlying mid-Paleozoic arc assemblage (Fig. 3F) oversteps a structural thickness of at least 10 km of imbricated thrust panels over a northwest-southeast distance of little more than 100 km.

By inference, migration of the subduction zone that assembled the combined Roberts Mountains and Shoo Fly accretionary prism toward the southeast culminated in the Antler orogeny when the forward edge of the accretionary prism was thrust over the Paleozoic margin of the continental block (D'Allura et al., 1977; Schweickert and Snyder, 1981; Varga and Moores, 1981; Schweickert et al., 1984; Hannah and Moores, 1986; Harwood et al., 1991). As now preserved, the Shoo Fly Complex apparently represents a somewhat older evolutionary stage of the lower Paleozoic accretionary prism than the Roberts Mountains allochthon, as finally assembled. Arc volcanics, coupled in time with Roberts Mountains thrusting, unconformably overlie the Shoo Fly Complex (Fig. 3F). The Shoo Fly Complex was apparently stranded well west of the continental block, by slab rollback, to form the underpinnings of an island arc that was active while migratory subduction, which emplaced the Roberts Mountains allochthon, continued within an oceanic region to the southeast.

### *Yreka terrane*

All observers agree that the Yreka terrane (Fig. 3B) of the Klamath Mountains is composed of an imbricate stack of internally disrupted mélange and broken formation, internally coherent thrust slices, and isolated tectonic blocks thrust structurally above the western flank of the Trinity ophiolitic complex (Lindsley-Griffin, 1977a, 1977b, 1991; Potter et al., 1977; Lindsley-Griffin and Griffin, 1983; Lapierre et al., 1986, 1987; Charvet et al., 1989, 1990; Saleeby, 1990; Cotkin, 1992; Wallin and Metcalf, 1998). A structurally complex zone of mélanges containing opholitic blocks and slabs is along the contact between the two assemblages, which are generally coeval, although ages remain uncertain for several individual map units within the Yreka terrane (Hotz, 1977; Potter et al., 1990). Many fossil collections from the Yreka terrane derive from limestone blocks in mélange (Lindsley-Griffin et al., 1991), and no systematic distinctions in age can be drawn between fossil collections from the basal ophiolitic mélange zone as opposed to structurally higher mélange belts. Combined fossil collections from the Yreka terrane and the mélange belt document the presence, in one place or another, of every stage from Middle Ordovician to Lower Devonian in the standard biostratigraphic time scale (Fig. 3B). The Yreka terrane has all the earmarks of an accretionary prism assembled over time by long-continued subduction (Lindsley-Griffin, 1977b, 1991; Lindsley-Griffin et al., 1991), and blueschist mineral assemblages of middle Ordovician to earliest Silurian age (Fig. 3B) are compatible with evolution of the Yreka terrane within a lower Paleozoic subduction zone (Cotkin, 1987, 1992; Cotkin et al., 1992).

The disrupted strata of the Yreka terrane are capped, with local structural modification of an originally depositional contact (Lindsley-Griffin, 1977a), by an undisrupted Lower Devonian sequence (restricted Gazelle Formation of Lindsley-Griffin et al., 1991) interpreted as the transient record of a trench slope basin perched atop the Yreka subduction complex (Lindsley-Griffin et al., 1991; Wallin and Metcalf, 1998). Sandstones within the slope-basin deposits include (1) a dominant volcaniclastic petrofacies indicative of derivation from an undissected magmatic arc, (2) a petrofacies rich in chert and argillite grains present low in the succession and inferred to represent sand reworked into the slope basin from the subjacent accretionary prism, and (3) a mixed petrofacies somewhat higher in the section inferred to record mixing of detritus from both those provenances at a time when the trench slope break was temporarily emergent (Wallin and Trabert, 1994). Volcanic rocks that locally overlie the slope-basin deposits provide a provisional link to widespread Devonian volcanics overlying the Trinity ophiolitic complex to the southeast (Lindsley-Griffin, 1977b, 1991; Wallin et al., 1991).

Past interpretations that the Yreka subduction complex was within the forearc of a west-facing arc-trench system, with subduction downward to the east (Saleeby, 1990; Potter et al., 1990; Hacker and Peacock, 1990; Lindsley-Griffin, 1977b, 1991; Lindsley-Griffin et al., 1991; Cotkin et al., 1992; Wallin and Metcalf, 1998), encounter the difficulty that emplacement of the Yreka terrane over the Trinity ophiolitic complex requires the hypothesis of massive forearc backthrusting (Potter et al., 1990). Past mapping and recent structural analysis indicate, however, that all known thrust faults and folds stacked within the interior of the Yreka terrane are east vergent (Eschelbacher and Wallin, 1998), including structures that involve blueschist metamorphic assemblages. This pervasive structural fabric seems compatible with the straightforward interpretation that the Trinity ophiolitic complex was thrust beneath the Yreka terrane by subduction downward to the northwest.

### *Trinity ophiolitic complex*

The Trinity ultramafic mass, with its associated mafic intrusions, is a polygenetic and partly dismembered ophiolite (Lindsley-Griffin, 1991; Wallin et al., 1988, 1991), and cannot represent a simple slab of oceanic crust. The main ultramafic sheet, dominantly harzburgite and lherzolite (Quick, 1981), is apparently Ordovician in age (Lindsley-Griffin, 1991; Wallin and Metcalf, 1998), but intrusions of gabbro and plagiogranite range in age from Early Ordovician to Early Devonian (Fig. 3C), a time span of at least 60 m.y. Moreover, blocks of Neoproterozoic plagiogranite occur within mélange near the contact of the Trinity ophiolitic complex (Fig. 3C) with the Yreka terrane, as well as within mélange associated with blueschists in the interior of the Yreka terrane (Fig. 3B). Some of the older Phanerozoic intrusions that have been dated may also be parts of ophiolitic blocks or slabs within the mélange zone bounding exposures of the main ultramafic sheet (Wallin et al., 1995). The abundance of plagioclase-bearing lherzolite within the Trinity sheet is atypical of ophiolites generally, and may reflect initial emplacement at crustal levels by mantle diapirism within an arc-trench system (Quick, 1981; Rubin et al., 1990).

Ordovician gabbro and plagiogranite of the Trinity ophiolitic complex may represent magmatism allied to peridotite emplacement, but Silurian and Devonian counterparts reflect much later magmatism, most probably of suprasubduction zone (SSZ) character (Wallin and Metcalf, 1998). Pegmatitic Lower Devonian gabbro and trondhjemite may be cogenetic with Devonian volcanic rocks that overlie and overlap the Trinity ophiolitic complex, (discussed further in a later section), but Silurian gabbro and tonalite or plagiogranite reflect earlier magmatism, either of precursor suprasubduction zone affinity or perhaps associated with slab rollback within an oceanic region (Lindsley-Griffin et al., 1991). The petrologic affinity of ophiolitic blocks and slabs within the mélange zone that structurally overlies the Trinity ultramafic sheet on the northwest remains unclear, although available information implies both tholeiitic and calc-alkalic components (Brouxel and Lapierre, 1988; Brouxel et al., 1988;

Lapierre et al., 1986, 1987). The isolated blocks of Neoproterozoic plagiogranite within mélanges west of the Trinity ophiolitic complex could conceivably represent scraps of oceanic lithosphere formed west of the Cordilleran miogeocline by latest Precambrian rifting.

### *Central Metamorphic belt*

Metavolcanic and subordinate metasedimentary rocks of the Central Metamorphic belt, which displays an inverted metamorphic gradient ranging from greenschist to amphibolite (Davis, 1968; Peacock and Norris, 1989), were thrust beneath the Trinity ophiolitic complex during Devonian time, as inferred from isotopic ages for metamorphic mineral assemblages (Fig. 3A). Protoliths are compatible with the interpretation that the Central Metamorphic belt represents an underthrust slab of metamorphosed oceanic crust with a thin sediment cover (Hamilton, 1969; Peacock and Norris, 1989).

Rocks of the Central Metamorphic belt also occur structurally beneath the western flank of the Yreka terrane (Hotz, 1973, 1977; Lindsley-Griffin, 1991), where exposures of the Central Metamorphic belt are underlain structurally in turn by an ultramafic belt (Fig. 4A) composed largely of serpentinite (Davis et al., 1965; Davis, 1969; Hotz, 1973, 1977; Wallin and Metcalf, 1998). This serpentinite body is commonly interpreted as a fault sliver of the Trinity ophiolitic complex, but is not in normal structural order above the Central Metamorphic belt (Irwin, 1977; Hacker and Peacock, 1990). The ultramafic body west of the Yreka terrane is also dissimilar in both texture and mineralogy to the Trinity ophiolitic complex, and apparently represents a separate ultramafic sheet thrust beneath the Central Metamorphic belt (Lindsley-Griffin, 1977b, 1991; Cashman, 1980). The serpentinite body may represent part of the oceanic lithosphere that underlay the crustal protoliths of the Central Metamorphic belt, or an unrelated tectonic slice emplaced later during Mesozoic time. Thrusting of the Central Metamorphic belt directly beneath the Yreka terrane apparently reset originally Ordovician (Silurian) isotopic systems, in phyllites of the Yreka terrane lying immediately adjacent to the fault contact, to Devonian ages that are indistinguishable from isotopic ages for the Central Metamorphic belt (Fig. 3B).

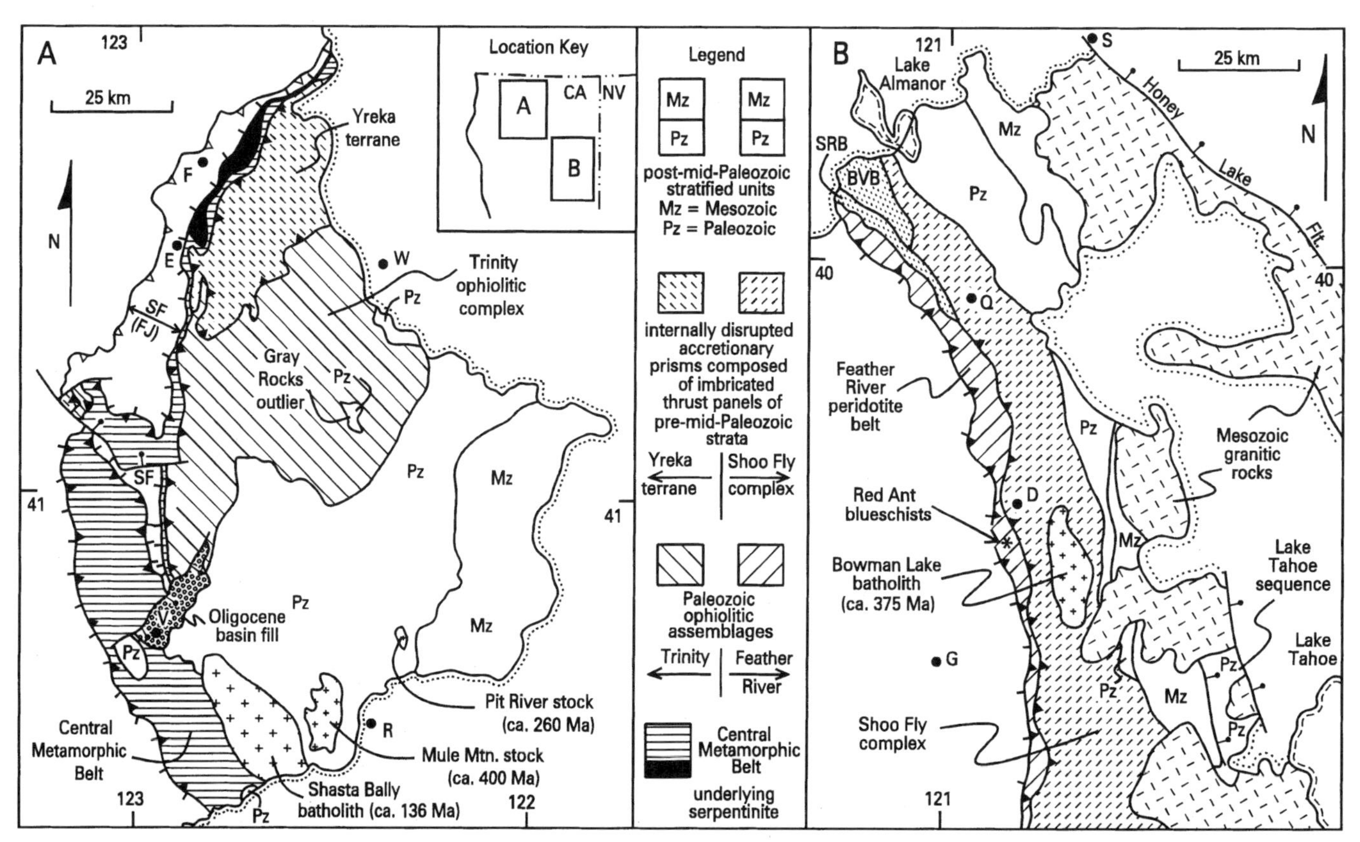


Figure 4. Distribution of Paleozoic and lower Mesozoic rock assemblages in eastern Klamath Mountains (A; modified after Irwin, 1994) and northern Sierra Nevada (B; modified after Harwood, 1992). Dotted line parallels depositional base of Cretaceous and younger sedimentary and volcanic cover. For clarity of wall-rock relationships, only selected plutons are shown. Hachured line is eastern edge of Mesozoic accretionary terranes (see text for discussion); SF (FJ) = blueschist-bearing (Late Triassic isotopic ages) Stuart Fork terrane (Fort Jones terrane of Silberling et al., 1987, 1992). See Figure 3 for Paleozoic (Pz) and Mesozoic (Mz) stratigraphy of each domain (Fig. 3, A–D, eastern Klamath Mountains; Fig. 3, E–F, northern Sierra Nevada). Stippled pre-Cretaceous blocks in northwestern Sierra Nevada (B): BVB—Butt Valley block (Pz-Mz stratigraphy similar to northern Sierra Nevada); SRB—Soda Ravine block (Pz-Mz stratigraphy similar to eastern Klamath Mountains). Selected towns for location: D—Downieville; E—Etna; F—Fort Jones; G—Grass Valley; Q—Quincy; S—Susanville; R—Redding; V—Weaverville; W—Weed.

Appreciation that the Central Metamorphic belt is juxtaposed structurally beneath both the Trinity ophiolitic complex and the structurally overlying Yreka terrane suggests that the Trinity ophiolitic complex is a huge megaduplex panel within a vast southeast-vergent subduction complex (Fig. 5). The structural base of the imbricated megaduplex of ophiolitic rock is obscured by post-Paleozoic cover below and around the northeastern end of the Great Valley (Fig. 4). The common assumption, based on the hypothesis of subduction downward to the east, that the Central Metamorphic belt roots eastward beneath the Trinity ophiolitic complex is not confirmed by seismic reflection profiling. In east-west profile, the exposed Trinity ophiolitic complex has a subhorizontal base, and is underplated by 6–7 km of subducted materials overlying ophiolitic mantle at a depth of 14–15 km below the surface (Zucca et al., 1986). Beneath the Trinity ophiolitic complex, subhorizontal reflectors inferred to represent the Central Metamorphic belt (Fuis et al., 1987) thicken from the north and west (~2 km) toward the south and east (4–6 km), forming an areal pattern more compatible with subduction downward to the northwest than to the east. Original subduction of the Central Metamorphic belt downward to the northwest is also compatible with the observation (Hacker and Peacock, 1990, p. 78) that the grade of its constituent metamorphic mineral assemblages is slightly higher to the north, west of the Yreka terrane, than to the south, in contact with the Trinity ophiolitic complex. The present easterly dip of the Central Metamorphic belt along the western flank of the composite Yreka-Trinity mass evidently reflects Mesozoic deformation, in common with the easterly regional dip of Mesozoic strata along the eastern flank of the Klamath Mountains, rather than the initial orientation of the Devonian subduction zone into which the Central Metamorphic belt was drawn.

### *Island arc successions*

Devonian volcanic sequences of the Klamath Mountains and Sierra Nevada are composed of facies characteristic of largely submerged island arcs, but are of entirely different ages in the two domains (Fig. 3, D and F). Eruptions spanned an estimated interval of 15–25 m.y. in each case, but the two successions contrast as much in age as, e.g., Eocene and Miocene volcanic sequences of the Tertiary Cordillera.

In the eastern Klamath Mountains, Lower to Middle Devonian volcanics that reach a thickness of 2500–3000 m (Albers and Bain, 1985) overlie the Trinity ophiolitic complex. The volcanic succession includes partly interfingering components of bimodal character, including a mafic lower interval (Copley Greenstone) of commonly pillowed lava and breccia, and an upper felsic interval (Balaklala Rhyolite) of pyroclastics and associated domes. Laterally equivalent volcanic strata apparently overstep the extent of the Trinity ophiolitic complex to form a cap over the youngest slope-basin deposits of the Yreka terrane to the west. The coeval Mule Mountain stock (Figs. 3D and 4A) is inferred from isotopic systematics to be cogenetic with the volcanic pile (Kistler et al., 1985), and Devonian intrusions within the Trinity ophiolitic complex are probably also cogenetic, at least within the same supra-subduction zone (SSZ) environment (Metcalf et al., 1998). Kuroko-type massive sulfide deposits, together with the general bimodality of the volcanic suite, reflect extensional tectonism during arc volcanism (Lapierre et al., 1985b, 1986; Miller and Harwood, 1990). This extensional signature of the volcanic suite is compatible with intra-arc extension during slab rollback at a subduction zone to the southeast. The volcanic pile is overlain depositionally by Middle Devonian Ken-

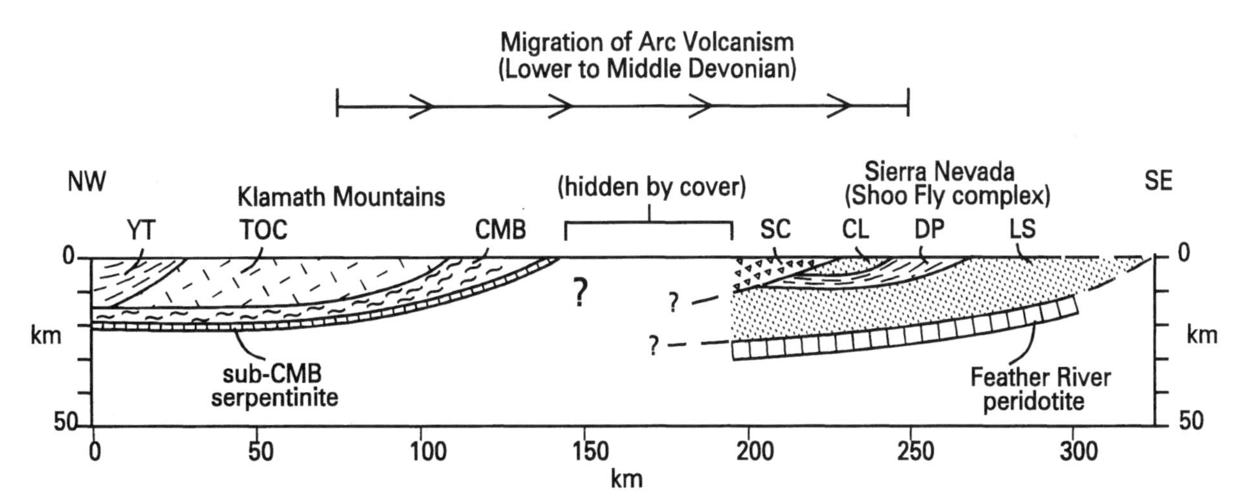


Figure 5. Schematic northwest-southeast profile (vertical exaggeration 2X) depicting inferred subsurface relationships within southeast-vergent Klamath-Sierran accretionary prism of imbricated lower Paleozoic tectonic elements. Ground surface is time-transgressive (younger to southeast) infra-Devonian unconformity beneath volcanic arc assemblages prograding to southeast. Eastern Klamath Mountains: YT, Yreka terrane; TOC, Trinity ophiolitic complex; CMB, Central Metamorphic belt. Northern Sierra Nevada (Shoo Fly components adapted from Schweickert et al., 1984, Fig. 9, and Harwood, 1992, Fig. 4): SC—Sierra City mélange; CL—Culbertson Lake allochthon; DP—Duncan Peak allochthon; LS—Lang Halstead sequence.

nett Limestone, which formed shallow platform caps and reworked slope or basinal deposits covering and infilling a dormant arc structure (Watkins and Flory, 1986).

In the Sierra Nevada, Upper Devonian to Lower Mississippian volcanics of the Taylorsville sequence, which reaches a thickness of 4750–5000 m (Durrell and D'Allura, 1977; Harwood, 1983; Hanson et al., 1996), unconformably overlie the previously deformed Shoo Fly Complex. The volcanic succession is composed of partly interfingering components, both largely pyroclastic, including a lower rhyolitic interval (Sierra Buttes Formation) and an upper andesitic interval (Taylor Formation), with more felsic rocks (lower Peale or "Keddie Ridge" Formation) present again at the top of the volcanic pile. Coeval and cogenetic intrusions include the Bowman Lake batholith (Fig. 4B) and the Wolf Creek stock in the northern Sierra Nevada (Fig. 3F). The gneissoid Confluence pluton of similar age (U/Pb 370 Ma) in the central Sierra Nevada probably represents a more southerly counterpart of the arc roots (Merguerian and Schweickert, 1987). The basal unconformity beneath the volcanic assemblage is overlain locally by lensoid turbidite bodies of clastic sedimentary strata occupying submarine channels incised into the subjacent Shoo Fly Complex (Hanson and Schweickert, 1986; Miller and Harwood, 1990; Girty et al., 1991a; Hanson et al., 1996). The channels were perhaps cut into a trench slope just prior to progradational migration of arc volcanism across a growing subduction complex from the eastern Klamath Mountains in Early Devonian time to the northern Sierra Nevada in Late Devonian time. The Sierran arc assemblage is overlain unconformably by a thin (~100 m) but widespread interval of Carboniferous ribbon chert (upper Peale Formation) that apparently represents pelagic cover draped over an extinct or dormant arc structure (D'Allura et al., 1977).

The Devonian volcanic assemblages of the Klamath Mountains and Sierra Nevada differ petrochemically as well as in age. The Early to Middle Devonian Klamath suite reflects eruption in an immature, low-K, tholeiitic intraoceanic arc (Lapierre, 1983; Lapierre and Cabanis, 1985; Lapierre et al., 1985a, 1986, 1987; Rouer and Lapierre, 1989), with positive $\varepsilon_{Nd}$ values dominantly in the range of +6 to +8 (Brouxel et al., 1987). By contrast, the Upper Devonian to Lower Mississippian Sierran suite represents eruption in a more mature intraoceanic arc, transitional from tholeiitic (Brooks and Coles, 1980; Brooks et al., 1982) to calc-alkalic (Rouer et al., 1989) in character, with negative $\varepsilon_{Nd}$ values dominantly in the range of –3 to –9 (Rouer and Lapierre, 1989; Rouer et al., 1989). Rocks of shoshonitic affinity are present near the top of the Sierran volcanic pile (Rouer and Lapierre, 1989).

### *Island arc polarity*

Several lines of evidence indicate that the evolving mid-Paleozoic island arc system of the Klamath Mountains and northern Sierra Nevada faced southeast, with subduction downward to the northwest. The most general argument is based on the migration of arc volcanism from northwest to southeast during Devonian time. Progradation of arc activity across an incrementally assembled subduction complex is readily understood (Varga and Moores, 1981), especially during Mediterranean-style subduction with slab rollback, but retrograde movement of arc activity is not expected. In Early to Middle Devonian time, the active frontal arc of the subduction system was in the eastern Klamath Mountains, and the Shoo Fly Complex of the northern Sierra Nevada evolved as an accretionary prism in the forearc region to the southeast. By Late Devonian time, the axis of active volcanism had prograded across the Shoo Fly subduction complex to reach the northern Sierra Nevada, and a remnant arc in the eastern Klamath Mountains had accumulated the limestone cover of the Middle Devonian Kennett Formation. As discussed in a later section, the distance between the Klamath remnant arc and the Sierran frontal arc during Late Devonian time is uncertain because of later tectonic deformation, but analogies with modern systems of frontal and remnant arcs suggest that an oceanic interarc basin may have intervened between the two arc structures. As interpreted here, the migratory arc assemblage and the associated subduction complex were both truncated on the southwest by transform faulting that established the trend of the younger Mesozoic continental margin.

Other, more specific, arguments regarding arc polarity stem from the following depositional facies and petrochemical variations within the arc assemblages.

1. The location of the forearc basin of the Devonian-Mississippian frontal arc in the northern Sierra Nevada can be inferred from stratigraphic relations along a transect at nearly a right angle to the northeast-southwest trend of the Paleozoic arc system. In the northern Sierra Nevada, the Taylorsville sequence of arc volcanic and volcaniclastic rocks thins to the southeast from almost 5000 m of vent-proximal facies near Sierra Buttes to <1000 m of more distal arc-apron volcaniclastic turbidites near Cisco Grove <50 km away (Harwood, 1983, 1992; Miller and Harwood, 1990; Harwood et al., 1991, 1995). Although analogous stratigraphic variations might occur along strike within an island arc having volcanic centers spaced widely apart (Hannah and Moores, 1986), the distal volcaniclastic strata interfinger within another 25 km to the southeast with nonvolcanogenic quartzite-pelite units of the Tahoe sequence (Harwood, 1988; Harwood and Murchey, 1990; Harwood et al., 1991). The siliciclastic detritus was apparently derived from the the subjacent Shoo Fly Complex, or possibly from the analogous Roberts Mountains allochthon (Harwood, 1988; Miller and Harwood, 1990; Harwood et al., 1991; Hanson et al., 1996), interpreted here as forming an accretionary prism in the forearc region. Such a source is compatible with the detrital zircon content of the Tahoe sequence (Dickinson and Gehrels, this volume, Chapter 11), and the provenance of the nonvolcanic detritus in the Tahoe sequence is inferred here to have formed an emergent trench slope break within the forearc region. The deformed Devonian-Mississippian basin between Sierra Buttes and Lake Tahoe, exposed now in profile with its floor tipped longitudinally on edge by Mesozoic deformation, is interpreted as a forearc basin between the arc axis and the trench slope break, the former being intersected by the trend of the younger Mesozoic continental margin near Sierra

Buttes, and the latter near Lake Tahoe. Equivalents of the pelagic drape of postvolcanic ribbon chert forming the upper Peale Formation are present within the Tahoe sequence (Harwood et al., 1991; Harwood, 1992), and serve as a stratigraphic link between the proximal and distal flanks of the forearc basin.

2. Lateral variations in the petrochemistry of the andesitic Taylor Formation within the Taylorsville sequence of the northern Sierra Nevada also support the inference of southeasterly arc vergence. Within a span of 60–75 km along present bedding strike, $\varepsilon_{Nd}$ values for Taylor samples decline from −1 to −4 on the southeast to −7 to −9 on the northwest, indicating increasing involvement of crustal materials in magma genesis in a northwesterly direction. As there is no reason to suspect the presence of pre-Mesozoic continental basement beneath the northern Sierra Nevada, the crustal materials with which magmas interacted were presumably within the subjacent Shoo Fly Complex (Hanson et al., 1996). As thickening of an accretionary prism toward the controlling arc is anticipated in the general case, the petrochemistry of the Taylor andesitic rocks is compatible with an arc-trench system facing to the southeast.

3. In the Klamath Mountains, pillow basalts of the Gray Rocks outlier (Fig. 4A), overlying the Trinity ophiolitic complex to the northwest of the main exposures of Devonian volcanics in the Klamath Mountains, have been interpreted as the product of backarc eruptions behind an island arc (Brouxel and Lapierre, 1984, 1988; Charvet et al., 1990). Thin Devonian volcanics that overlie the Yreka terrane are presumably analogous distal products of suprasubduction-zone magmatism, and indicate tectonic assembly of the Yreka terrane and the Trinity ophiolitic complex into a composite terrane before the end of Early Devonian time (M.M. Miller and Cui, 1989; M.M. Miller and Harwood, 1990; Rubin et al., 1990). Underthrusting of the Central Metamorphic belt as a subducted slab beneath the composite terrane was evidently accompanied by supra-subduction (SSZ) Devonian magmatism affecting both its tectonic components. Before the end of Devonian time, clastic strata of the Bragdon Formation (Fig. 3D) had covered the Copley-Balaklala remnant arc, and overlapped unconformably across the Trinity ophiolitic complex, with a basal conglomerate containing clasts of peridotite, gabbro, basalt, and plagiogranite (Brouxel and Lapierre, 1988; Charvet et al., 1989, 1990). In the southernmost Klamath Mountains, the Bragdon Formation may also have onlapped exposures of the Central Metamorphic Belt (Fig. 4B), although contact relations south of Weaverville are equivocal owing to structural disturbance of unknown age (Albers and Bain, 1985).

Quartzolithic sandstones typical of the largely Mississippian Bragdon Formation (3000–5000 m), and most abundant near its base, were probably derived from the Yreka terrane, whereas subordinate volcaniclastic sandstones higher in the section may have been derived from the Devonian-Mississippian arc structure of the northern Sierra Nevada (M.M. Miller and Cui, 1989; M.M. Miller, 1989; M.M. Miller and Harwood, 1990). The mixed population of detrital zircons in a nonvolcanogenic sample of quartzolithic sandstone rich in chert grains is compatible with contributions from both the Yreka terrane and the Trinity ophiolitic complex (Dickinson and Gehrels, this volume, Chapter 11). The Bragdon Formation is a generally progradational subsea-fan complex that effectively buried the Devonian arc assemblage of the Klamath Mountains (Watkins, 1986; M.M. Miller and Cui, 1989), suggesting subsidence of a rifted remnant arc to leave higher standing sources of sedimentary detritus to either side. Minor pyroclastic beds present near the top of the Bragdon Formation may be precursors of the overlying volcaniclastic Baird Formation (3000–5000 m) of largely Pennsylvanian age.

### *Summary*

Figure 5 is a schematic summary of the tectonic relations of lower Paleozoic rocks, as displayed in northwest-southeast transverse profile oriented subparallel to the trend of the younger Mesozoic Klamath-Sierran continental margin, and Figure 6 (A–D) depicts the inferred sequence of Antler orogenic events leading to emplacement of the Roberts Mountains allochthon above the Cordilleran miogeocline. Structural relations internal to the Klamath Mountains and the northern Sierra Nevada are depicted with comparative confidence, whereas relations between the two domains are hidden by younger cover (Fig. 5), and relations to the southeast toward the central Sierra Nevada are largely obscured by the Cretaceous Sierra Nevada batholith. Conceivably, as successive frontal and remnant arcs evolved during slab rollback, an oceanic expanse of indeterminate width may have intervened between Klamath and Sierran domains in mid-Paleozoic time (Fig. 6, C–D).

Near the northwestern end of the Sierra Nevada (Fig. 4B), a fault of uncertain overall character juxtaposes a Klamath-like lithic assemblage of Permian through Triassic strata in the Soda Ravine block on the west against a Sierran-like lithic assemblage of Devonian volcanics overlain unconformably by Triassic-Jurassic strata in the Butt Valley block on the east (Jayko, 1990). Both fault blocks are east of the Feather River peridotite belt, and are unrelated to the Mesozoic subduction complex exposed farther west, but with present information their geology can be interpreted as indicative either of a gradation between pre-Mesozoic Klamath-Sierran geology or of later interdomain contraction.

The Feather River peridotite belt (Ehrenberg, 1975) of the northern Sierra Nevada (Fig. 4B) has been correlated in a general way with the Trinity ophiolitic complex (Davis, 1969; Saleeby et al., 1989; Hacker and Peacock, 1990), but the former lacks the plagioclase-bearing lherzolite so common in the latter (Hacker and Peacock, 1990), and available isotopic ages suggest that the Feather River body (Fig. 3E) may be distinctly younger than the Trinity body (Fig. 3C). Each ultramafic body structurally underlies an accretionary prism, the Yreka terrane in the case of the Trinity mass and the Shoo Fly Complex in the case of the Feather River mass, but the underthrusting events are inferred provisionally here to have been diachronous, in sequence from northwest to southeast (Fig. 5). The failure of the Central Metamorphic belt to cross the northern end of the Great Valley (Davis, 1969), at least

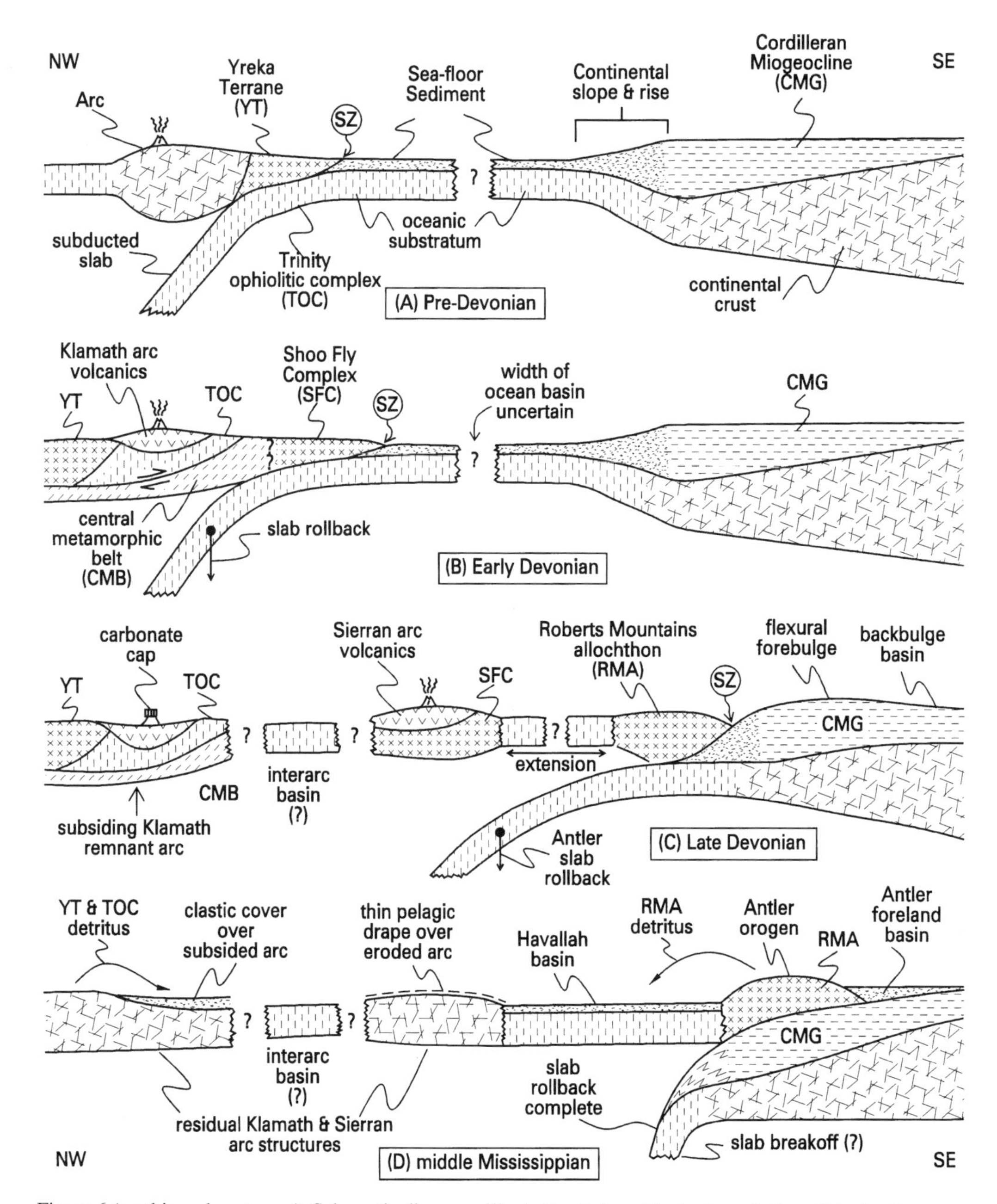


Figure 6 (on this and next page). Schematic diagrams illustrating inferred tectonic evolution of Antler-Sonoma orogens including thrust emplacement of Roberts Mountains and Golconda allochthons by episodic slab rollback associated with subduction downward to northwest beneath southeast-facing island arc system (active and remnant arcs) located offshore from Cordilleran miogeoclinal margin (possible transient migratory arc volcanism within oceanic domains between offshore remnant arcs and continental margin not depicted). Queried gaps denote oceanic domains of uncertain width, and other queries denote various unclear relationships: in A, proximity of Alexander arc is uncertain (see text); in B, relation of Central Metamorphic belt to Shoo Fly Complex is inferential; in C–D, possible existence of oceanic interarc basin is hypothetical; in E, alternate positions are shown for subduction zone; in F–G, distance of separation between Klamath and Sierran domains is unknown; in H, subduction zone postdates time frame shown. Possible intraarc diapirism to emplace Trinity ophiolitic complex at crustal levels is not illustrated.

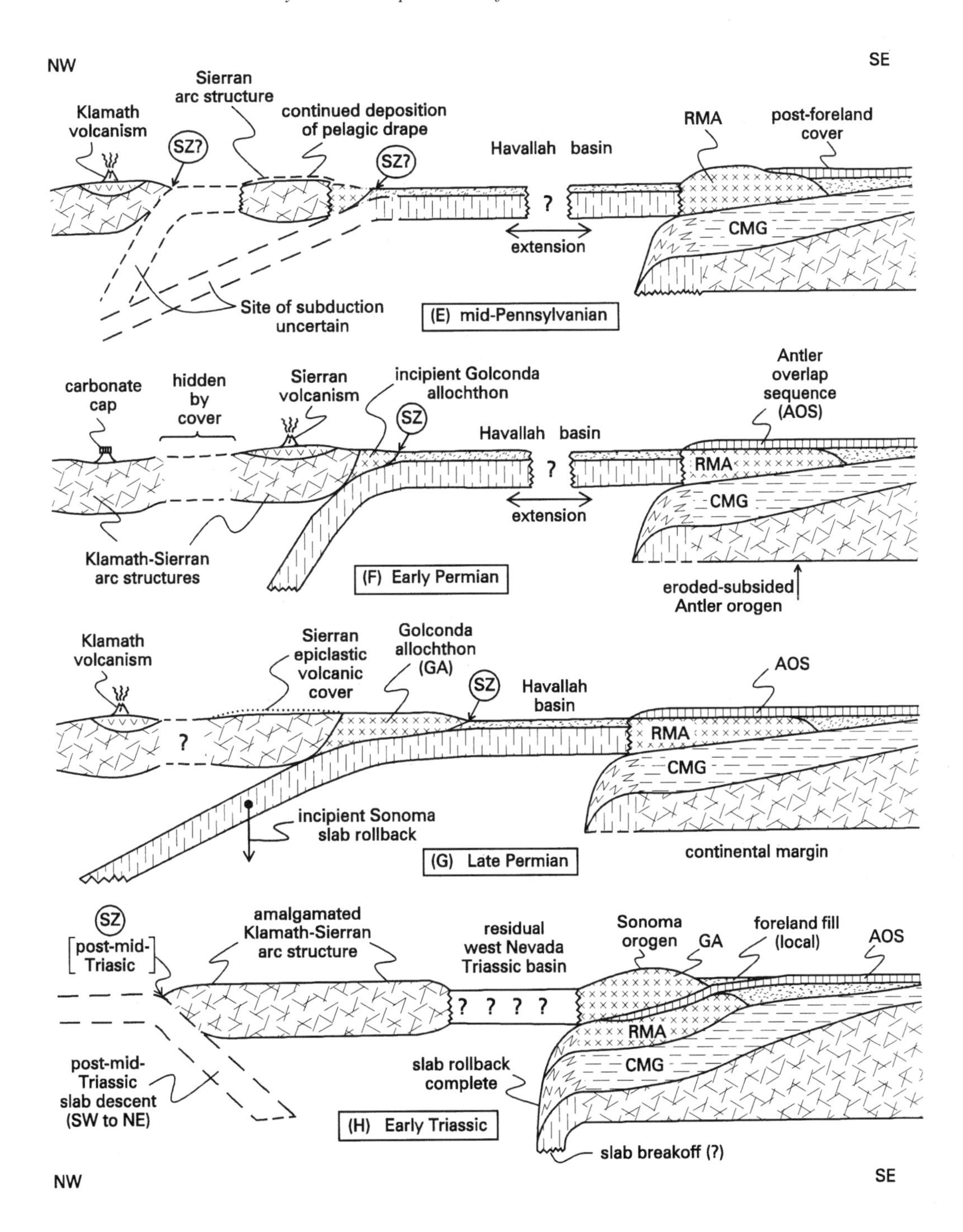


on outcrop, makes any firm interpretation difficult (Fig. 6B). The genesis of the oldest ultramafic rocks in either body remains uncertain; mid-ocean ridge spreading or intra-arc extension are both conceivable as end-member options, and hybrid settings related to slab rollback within a remnant ocean basin are also possible.

The sequential arc assemblages of the Klamath Mountains and northern Sierra Nevada specify the sites of Devonian magmatism that can be related to migratory, southeast-vergent subduction, downward to the northwest (Fig. 6, B and C). The original distance of the evolving active-remnant arc system from the continental margin is indeterminate with present information. Any ancestral island arc having the same polarity, however, and paired with the blueschist-bearing components of the Yreka terrane metamorphosed in Late Ordovician to earliest Silurian time (Fig. 3B), must have been farther to the northwest (Metcalf et al., 1998). No Paleozoic assemblages are exposed in that direction, perhaps due to tectonic removal during the plate motions that truncated the continental margin in latest Paleozoic to earliest Mesozoic time.

Several have suggested that the missing arc may still be present much farther north in the Alexander Archipelago (Jones et al., 1972; Schweickert and Snyder, 1981), from which selected blocks in the Sierra City mélange of the Shoo Fly Complex may have been derived (Girty and Wardlaw, 1984; Girty and Pardini, 1987). The age range (438–472 Ma) of Middle Ordovician to earliest Silurian arc plutons in the Alexander terrane closely matches the metamorphic ages of blueschists within the Yreka terrane, and the Middle Silurian to earliest Devonian Klakas orogeny in the Alexander terrane immediately preceded the inception of Devonian arc magmatism in the Klamath Mountains (Gehrels and Saleeby, 1987a, 1987b). Moreover, local Silurian-Devonian carbonate platforms built on Ordovician-Silurian arc volcanic piles within the Alexander terrane are suggestive of deposition on the volcanic edifices of a waning and remnant island arc (Soja, 1988, 1990, 1993). However, paleomagnetic studies suggest a Paleozoic position for the Alexander terrane near Baltica, rather than southern Laurentia (Bazard et al., 1995; Butler et al., 1997), and the detrital zircon geochronology of Lower Devonian sandstone in the Alexander terrane makes its proximity to the Cordilleran margin unlikely in mid-Paleozoic time (Gehrels et al., 1996). Full evaluation of the putative connection between Alexander Archipelago and Klamath geology is beyond the scope of this paper (Fig. 6A).

## LATE PALEOZOIC—EARLIEST MESOZOIC (SONOMA) TECTONISM

The polarity of Permian arc-trench systems in California and Nevada is more difficult to discern with confidence than the polarity of mid-Paleozoic tectonic systems. Continued southeastward tectonic vergence, with subduction downward to the northwest (Fig. 6, E–G), is suggested in the absence of post-Devonian, pre-Middle Triassic subduction complexes west of the Central Metamorphic belt of the Klamath Mountains or the Feather River peridotite belt of the northern Sierra Nevada. Subsequent west-facing Mesozoic subduction, downward to the east beneath a newly formed continental margin oriented northwest-southeast (Fig. 6H), is discussed in a later section. With an Apennine-Tyrrhenian analogy in mind (Fig. 1), the tectonic behavior assumed here for Paleozoic Sonoma events could be duplicated in the Cenozoic Mediterranean arena by future subduction of Tyrrhenian sea floor beneath the Sardinia-Corsica remnant arc to rejuvenate magmatism there.

The distances by which lower Paleozoic remnant arcs, stranded within the eastern Klamath Mountains and northern Sierra Nevada by termination of slab rollback at the time of the Antler orogeny, may have further approached each other during late Paleozoic time is uncertain. Resolution of the question depends upon understanding Klamath-Sierran relations across the covered area between the two domains (Figs. 4), and upon better understanding of the Paleozoic geologic history of the central Sierra Nevada, where Mesozoic batholith emplacement has obscured relations among pre-batholithic rock assemblages. The width of the late Paleozoic Havallah basin (Fig. 6, D–G), to the southeast of offshore Klamath-Sierran arc assemblages and from which the Golconda allochthon was derived, is also difficult to reconstruct palinspastically with present information. Final emplacement of the Golconda allochthon may have been associated, however, with some finite reduction in the distance between the offshore asemblage of active and remnant arcs in the Klamath-Sierran region and the pre-Sonoma continental margin in central Nevada (Fig. 6, F–H).

### *Golconda allochthon*

The Golconda allochthon of central Nevada (Fig. 3H) is composed of complexly deformed and imbricated thrust panels of upper Paleozoic strata, termed the Havallah sequence, with an intricately disrupted internal structure reminiscent of the older Roberts Mountains allochthon in the same region (Burchfiel et al., 1992). The Schoonover sequence of northeastern Nevada is treated here as part of the Havallah sequence (following E.L. Miller et al., 1992). Stratal components include chert-argillite intervals, pillow basalts of sea-floor and seamount affinity, and varied turbidites of quartzolithic, calcarenitic, quartzose, and volcaniclastic character (Dickinson et al., 1983; Snyder and Brueckner, 1983; Brueckner and Snyder, 1985; Whiteford, 1990; Murchey, 1990; Jones, 1991; Jones and Jones, 1991). The previously assembled allochthon was emplaced by thrusting over the Antler overlap sequence (Silberling, 1973; E.L. Miller et al., 1981; Wyld, 1991; Daly et al., 1991) in mid-Early Triassic time (Fig. 3H). The alternatives of backarc thrusting and arc-continent collision for emplacement of the allochthon have been discussed at length (Tomlinson et al., 1987) without achieving a definitive consensus. By analogy with earlier emplacement of the Roberts Mountains allochthon during the Antler orogeny, the Golconda allochthon is interpreted here as an accretionary prism emplaced during the Sonoma orogeny. Partial subduction of the continental margin in central Nevada is again envisioned as the result of slab rollback beneath a migratory subduction zone associated with subduction downward to the northwest beneath a southeast-facing island arc system (Fig. 6, E–H).

The presence locally within the Havallah sequence of Upper Devonian to Lower Mississippian chert, depositionally overlying greenstone, indicates that inception of the Havallah basin occurred either before or during emplacement of the Roberts Mountains allochthon (E.L. Miller et al., 1984, 1992; Jones, 1991), with initial deposition on sea floor inferred to have been produced by slab rollback during the Antler orogeny (Burchfiel et al., 1992). The occurrence within the Havallah sequence of Mississippian, Middle Pennsylvanian, and Lower Permian greenstones (Snyder and Brueckner, 1983; Murchey, 1990; Whiteford, 1990; E.L. Miller et al., 1992) implies, however, that the floor of the Havallah basin was continuously or intermittently rejuvenated or broadened by sea-floor magmatism during Havallah sedimentation (Burchfiel et al., 1992). The intrabasinal extension may have occurred during westward subduction of sea floor without slab rollback (Fig. 6, E and F).

Derivation of most Havallah detritus from the Roberts Mountains allochthon (Dickinson et al., 1983; Whiteford, 1990; Jones, 1991) in any case reflects proximity of the Havallah basin to the latter source. Available biostratigraphic control for the ages of multiple thrust slices from place to place within the Golconda allochthon (E. L. Miller et al., 1982, 1984, 1992; Stewart et al., 1986; Murchey, 1990; Jones, 1991) implies that turbidites increase in relative abundance upward stratigraphically within the Havallah sequence. This general style of sedimentary evolution would be expected from progradation of subsea fans fed either from exposures of the Roberts Mountains allochthon or from overlying terrigenous and carbonate shelf sediments. In detail, however, depositional patterns were complex, and clastic-free chert-argillite successions are common at nearly all stratigraphic levels.

The Golconda thrust strikes into the central Sierra Nevada at a high angle to the trend of the Sierra Nevada batholith (Schweickert and Lahren, 1987, 1993b), and has never been located on the western side of the batholith. Its possible presence there is suggested, however, by the observation that Permian arc volcanics are in depositional sequence above the Shoo Fly Complex of the northern Sierra Nevada, but appear in thrust slices above the Havallah strata of the Golconda allochthon in Nevada not far east of the batholith (Speed, 1977; Blein et al., 1994, 1996). Near Candelaria (Fig. 2), the Lower Triassic Candelaria Formation (Fig. 3H) is composed of a lower member (~100 m) of quartzose and calcareous shelf and slope strata deposited on top of the Roberts Mountains allochthon before the approach of the Golconda allochthon, and an upper member (~1000 m) of volcaniclastic strata composed of detritus shed from arc volcanics atop the Golconda allochthon (Speed, 1977). A slice of serpentinite (Speed, 1984) occurs between the volcaniclastic beds and the overriding allochthon, and analogous ultramafic bodies are exposed near the base of the Golconda allochthon in exposures farther west near the flank of the Sierra Nevada batholith (Schweickert and Lahren, 1993b). The upward-coarsening volcaniclastic strata of the Candelaria Formation evidently represent a locally preserved stratigraphic record of a Sonoma foreland basin.

Farther north in Nevada, fossiliferous pyroclastic strata, of felsic composition and mid-Early Triassic age (Silberling and Wallace, 1969), occur near the top of the volcanic Koipato Group (~3000 m) and unconformably overlie deformed strata of the Golconda allochthon (Fig. 3H). Although the entire Koipato Group (Fig. 3H) may represent a precursor phase of post-Sonoma arc volcanism established along the Mesozoic Klamath-Sierran continental margin after Golconda allochthon emplacement, an internal unconformity separates the dated uppermost part of the group from the underlying Limerick Greenstone (~2000 m), of contrasting andesitic character. The latter may represent Permian volcanic rocks erupted before final emplacement of the Golconda allochthon (Speed, 1977, 1978b; Stewart, 1980), and analogous to the Permian arc assemblage preserved in the northern Sierra Nevada. In that case, the overlying felsic volcanic rocks of the uppermost Koipato Group may represent transient volcanism of uncertain character associated with emplacement of the Golconda allochthon. Before the end of Early Triassic time, shelf limestones of the Star Peak Group (~1000 m) had covered the Koipato Group, and by Middle Triassic time had overlapped the residual Antler orogen to the east (Fig. 3, H and I).

### *Island arc assemblages*

Permian volcanic sequences of the Klamath Mountains and Sierra Nevada are composed of facies characteristic of partly emergent island arcs, and are diachronous, although there was apparently some overlap in the timing of eruptions within the two domains (Fig. 3, D and F). As in the case of older mid-Paleozoic arc assemblages in the two areas, there are also contrasts in the petrochemistry of the two suites. In the eastern Klamath Mountains, the volcaniclastic uppermost Mississippian through Pennsylvanian Baird Formation (Fig. 3D) is poorly understood, and has no counterpart in the northern Sierra Nevada, where the same time interval is represented partly by continued deposition of pelagic chert covering a Devonian-Mississippian remnant arc, and partly by the succeeding hiatus preceding Permian volcanism (Fig. 3F).

The Permian arc assemblage of the northern Sierra Nevada includes complexly interfingering volcanic and volcaniclastic facies forming a succession 2750–3250 m thick (Harwood, 1983; Hanson et al., 1996) erupted within a calc-alkalic island arc (Rouer et al., 1988). Eruptions began early in Early Permian time and continued into the earliest part of Late Permian time (Fig. 3F). Through most of that time interval, the McCloud Limestone, composed of carbonate platforms and associated carbonate ramps and slopes which prograded over volcaniclastic beds of the underlying Baird Formation (Watkins, 1985; Miller, 1989), was accumulating without contemporary volcanism in the eastern Klamath Mountains (Fig. 3D). Fusulinids of McCloud affinity within the volcanogenic succession of the northern Sierra Nevada indicate, however, that a biostratigraphic link existed between the two domains (Harwood, 1992; Hanson et al., 1996).

In the Klamath Mountains, the Permian arc assemblage of the Bollibokka Group (Fig. 3D), which is 2500–5000 m thick (Lapierre et al., 1986; M.M. Miller and Harwood, 1990), unconformably overlies the McCloud Limestone and was erupted from latest Early Permian through much of Late Permian time. The Pit River stock is inferred to be cogenetic with the Bully Hill Rhyolite of the Bollibokka Group, although the time correlation between isotopic and biostratigraphic ages is not perfect (Fig. 3D). Volcanogenic massive sulfide deposits probably related to local extensional deformation occur at some of the rhyolitic centers (Lapierre et al., 1987). The petrochemistry of the Bollibokka Group reflects eruption in a mature andesitic island arc, of low-K tholeiitic character, with subordinate associated basalts and rhyolites (Martin et al., 1984; Lapierre et al., 1986, 1987; Charvet et al., 1990; Alibert et al., 1991).

Dominant $\varepsilon_{Nd}$ values for the Bollibokka Group are in the range of +4 to +8, similar to values for the underlying Devonian arc suite of the eastern Klamath Mountains, whereas dominant $\varepsilon_{Nd}$ values for the Permian volcanic suite of the northern Sierra Nevada are in the lower range of 0 to +4 (Alibert et al., 1991), although not so low as for the Sierran Devonian arc suite, which displays negative $\varepsilon_{Nd}$ values. In westernmost Nevada, $\varepsilon_{Nd}$ values of –1.4 to +5.3 for allochthonous Permian volcanic rocks (265–275 Ma, $Ar^{40}/Ar^{39}$) of the Black Dyke (or Black Dike) Formation, exposed as a klippe within or structurally overlying the Luning allochthon or thrust complex of Mesozoic age, implies affinity with the Sierran Permian arc suite (Blein et al., 1994, 1996).

### *Island arc polarity*

Late Paleozoic arc magmatism apparently migrated from the Klamath domain (Pennsylvanian Baird) to the Sierran domain in Early Permian time. Early Permian McCloud Limestone accumulated on and near volcanic edifices of a remnant arc in the Klamath domain. In Late Permian time, magmatism migrated back to the Klamath domain, with volcanism evidently contemporaneous in both domains for some interval of mid-Permian time. The unsystematic shift of arc activity from the northwest to the southeast, and back again, is inferred here to reflect stages in the subduction, down to the northwest, of the floor of the Havallah basin, perhaps as the dip of a subducting slab varied (Fig. 6, E–G). As a result, the locus of active arc magmatism alternated between the twin Klamath and Sierran remnant arc structures inherited from mid-Paleozoic time as magmatism was reactivated diachronously during late Paleozoic time. Alternatively, Pennsylvanian volcanism restricted to the Klamath domain might be attributed to closure of a cryptic interarc basin between the Klamath and Sierran domains (Fig. 6E). As discussed in the next section, truncation of the continental margin farther south was underway during the time frame of late Paleozoic volcanism in the Klamath Mountains and northern Sierra Nevada.

Pennsylvanian volcanism, confirmed by exclusively Mississippian-Pennsylvanian (320–325 Ma) detrital zircons in volcaniclastic strata of the Baird Formation (Dickinson and Gehrels, this volume, Chapter 11), may have been accompanied by sea-floor spreading without slab rollback within the Havallah basin (Fig. 6E), with only Permian volcanism accompanied by slab rollback (Fig. 6, F and G). The inferred slab rollback initially pulled volcanism closer to the subduction zone in the northern Sierran domain, and then renewed arc magmatism in the Klamath domain as the northern Sierran domain neared the contintental margin just prior to emplacement of the Golconda allochthon during the Sonoma orogeny. Paleozoic magmatism was finally terminated by emplacement of the Golconda allochthon, and the Upper Triassic Hosselkus Limestone (Fig. 3, D and F), common to both the Sierra Nevada and the Klamath Mountains (Jayko, 1990; Miller and Harwood, 1990), reflects passive shelf sedimentation capping the Paleozoic assemblages of both domains.

Sonoma deformation of the northern Sierra Nevada is reflected by an angular unconformity above which Triassic beds overlap across the entire Permian arc assemblage to directly overlie rocks of the Shoo Fly Complex to the southeast (Harwood and Murchey, 1990). In the eastern Klamath Mountains, a hiatus between Permian and Triassic volcanogenic strata can also be regarded as the record of Sonoma tectonism (Wyld, 1991). Farther east, in the Pine Forest–Jackson domain (Fig. 1), a generally correlative hiatus is present between mid-Upper Permian and Middle Triassic strata, but records only subdued deformation (Wyld, 1991). Relatively mild Sonoma deformation of Klamath-Sierran Permian volcanics, in comparison to intricately deformed coeval strata within the Golconda allochthon, can be attributed to the minimal crustal shortening, or even crustal extension, undergone by the interiors of accreting island arcs, as opposed to basinal strata caught tectonically between arc structures and continental margins (Wyld, 1991). Mild extensional deformation of Klamath-Sierran Permian arc assemblages thus could have accompanied slab rollback that produced intense contractional deformation of the Havallah basin to form the Golconda allochthon.

## PALEOZOIC-MESOZOIC CONTINENTAL TRUNCATION

Many have noted previously that the northeast-southwest trend of Paleozoic tectonic elements across Nevada is truncated abruptly at the northwest-southeast trend of the Mesozoic Cordilleran margin in California (Hamilton, 1969, 1978; Burchfiel and Davis, 1972; Schweickert, 1976, 1978; Davis et al., 1978; Schweickert and Snyder, 1981; Saleeby and Busby-Spera, 1992; Burchfiel et al., 1992), and have inferred a tectonic truncation of the edge of the continental block by rifting or strike slip near the Paleozoic-Mesozoic time boundary. The Paleozoic tectonic boundaries that were truncated along the newly formed Mesozoic continental margin include contacts or transition zones between craton, miogeocline, overthrust Antler-Sonoma allochthons (Roberts Mountains and Golconda), and volcanic arc assemblages exposed in the northern Sierra Nevada and Klamath Mountains (Fig. 2). The distance that the continental block extended to the southwest in Paleozoic time remains uncertain, but the Antler-Sonoma Cordilleran edge of the continent eventually curved around through present-day Mexico to join the Ouachita-Marathon mesoamerican continental margin (Peiffer-Rangin, 1977; Stewart, 1988).

The alignment of the continental truncation, although doubtless overprinted by younger structural deformation, can be specified closely along a currently curvilinear trend marking the western edges of the Central Metamorphic belt in the Klamath Mountains, the Feather River peridotite body in the northern Sierra Nevada, and the Shoo Fly Complex in the central Sierra Nevada, where the Feather River peridotite belt is absent (Davis et al., 1978; Schweickert, 1976, 1978; Burchfiel and Davis, 1981a; Saleeby, 1981, 1982; Varga and Moores, 1981; Saleeby and Busby-Spera, 1992; Dilek and Moores, 1993). From the lack of rift facies along the line of truncation, the structure is inferred to have been a transform fault, with sinistral motion to displace the missing fragment of the pre-Mesozoic continental margin southward into Mexico (Davis et al., 1978; Burchfiel and Davis, 1981a; Saleeby and Busby-Spera, 1992; Burchfiel et al., 1992; Saleeby and Busby, 1993; Sedlock et al., 1993), where miogeoclinal strata similar to those of the Great Basin occur as the Caborca block of northwestern Sonora (Stewart et al., 1984, 1990, 1997). Pre-Mesozoic ophiolitic basement exposed along the western flank of the southern Sierra Nevada may have been placed adjacent to the continental block by transform slip during continental truncation (Saleeby, 1977, 1990, 1992; Saleeby and Busby-Spera, 1992).

### *California-Coahuila transform*

Displacement of the Caborca block southward was originally attributed to Jurassic strike slip along the so-called Mojave-Sonora megashear (Silver and Anderson, 1974; Anderson and Silver, 1979; Anderson and Schmidt, 1983), but subsequent work has shown that associated continental truncation in the Mojave region clearly predated Jurassic time (Walker, 1987, 1988). The configuration and depositional facies of Middle Pennsylvanian to mid-Permian sedimentary basins in the Death Valley region of southeastern California document a reorientation of local tectonic trends from northeasterly to northwesterly, and are inferred to reflect transform deformation of a borderland belt during continental truncation (Stone and Stevens, 1988). Conglomeratic Lower Triassic deposits within the Mojave region (Walker, 1988) are taken here to imply continuation of transform tectonism until mid-Early Triassic time (Stevens et al., 1997).

Recent geologic mapping in southwestern Arizona and northwestern Sonora (Calmus and Sosson, 1995; Stewart et al., 1997) has shown that the locus of strike-slip displacement responsible for shifting the position of the Caborca block relative to the interior of the continent followed a more westerly course than the inferred projection of the Mojave-Sonora megashear to the northwest across the Mojave block. The line of continental truncation evidently passed entirely west of the Mojave block, beyond the exposures of miogeoclinal strata in the San Bernardino Mountains (Dickinson, 1981b; Stevens et al., 1997). To avoid possible confusion with the inferred location and timing of the Mojave-Sonora megashear of supposed Jurassic age, the Permian-Triassic structure responsible both for truncation of the California continental margin and for southward displacement of the Caborca block is here termed the California-Coahuila transform (Fig. 7). My reinterpretation of the geodynamic relationships and timing of transform offset in no way detracts, however, from the fundamental contribution of Silver and Anderson (1974) in calling attention to the displaced nature of the Caborca block.

Perhaps the strongest argument for Jurassic displacement of the Caborca block has been the similarity of Triassic and Lower Jurassic strata in the Antimonio Formation of Sonora to supposedly offset counterparts in western Nevada (Stanley and González-León, 1995; Lucas, 1996; González-León et al., 1996; González-León, 1997; Lucas et al., 1997). In detail, however, biostratigraphic and structural dissimilarities between the broadly comparable successions in Nevada and Sonora do not support the hypothesis that they are offset counterparts (Gómez-Luna and Martínez-Cortés, 1997). Moreover, the Antimonio strata occupy a forearc position with respect to the Jurassic Cordilleran magmatic arc, which extended across southern Arizona to the northeast, whereas the strata in Nevada to which comparisons have been drawn occupy a backarc position with respect to the same Jurassic arc terrane where it passes northward up the eastern flank of the Sierra Nevada and along the adjacent fringe of Nevada. The trend of the Jurassic arc is unbroken, however, from Arizona across the Mojave region toward the California-Nevada border (Busby-Spera, 1988, Fig. 1; Saleeby and Busby-Spera, 1992, Plate 5). There seems little likelihood that strike slip sufficient to displace the Caborca block into Sonora from the miogeoclinal belt in Nevada could have occurred as late as Jurassic time without disrupting and offsetting the Jurassic arc trend across the Mojave region.

### *Transform geodynamics-kinematics*

Geochronological studies in eastern Mexico (Torres et al., 2000) shed light on the regional geotectonic context of Califor-

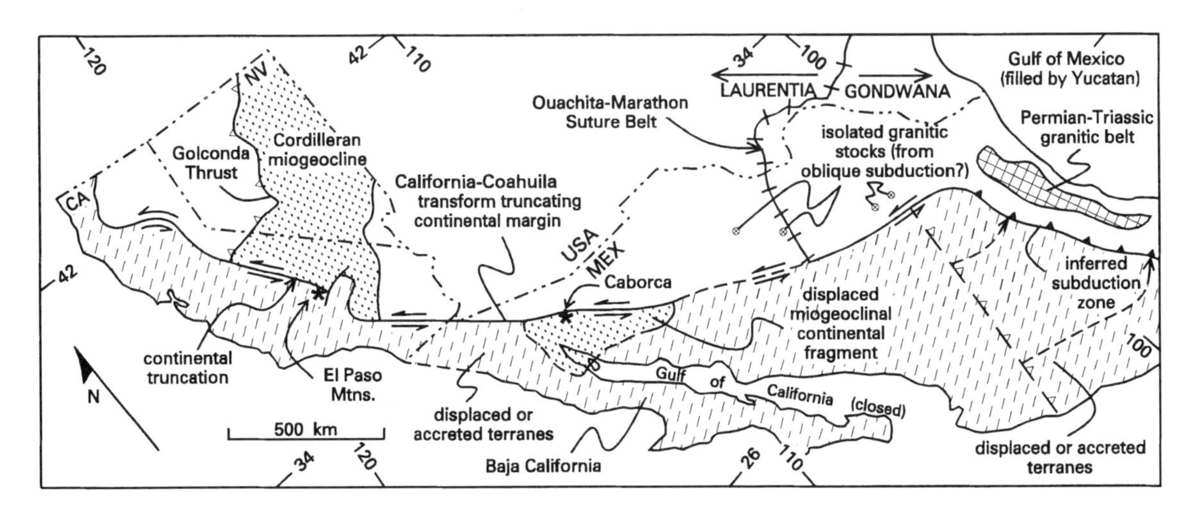


Figure 7. Speculative regional tectonic relations of California-Coahuila transform fault connecting Permian-Triassic subduction zones (with plate consumption) in northwestern Nevada and eastern Mexico. Dashed arrows indicate probable eastward displacement of Mexican subduction zone from its inferred Permian-Triassic position (dashed line with open barbs) to its present position (solid line with closed barbs) during breakup of Pangaea and Jurassic opening of Gulf of Mexico. Asterisks denote key localities discussed in text. Approximate extent of offset miogeoclinal rocks after Stewart et al. (1990) and Gastil et al. (1991), with Baja California restored to its pre-Neogene position.

nia-Coahuila transform displacement. Inland from the east coast of Mexico, a north-south belt of granitic plutons represents a magmatic arc of Early Permian to Middle Triassic age (284–232 Ma; n = 44). The arc was apparently built along the western edge of Gondwana shortly after the amalgamation of Gondwana to Laurentia along the Ouachita-Marathon suture belt in Pennsylvanian to earliest Permian time. Postsuture development of the arc is indicated by extension of northern outlying stocks across the Ouachita-Marathon suture belt in Chihuahua (Fig. 7). Lithosphere carrying the Caborca block and other crustal fragments torn off the truncated Cordilleran margin is inferred here to have been drawn into a trench west of the Permian-Triassic magmatic arc of eastern Mexico (Fig. 7). Crustal components of Mexico west of the magmatic arc and its associated trench were accreted to the Mexican prong of the continental block after Middle Triassic time (Lapierre et al., 1992; Tardy et al., 1992, 1994; Centeno-Garcia and Silva-Romo, 1997; Moores, 1998; Torres et al., 2000). In this context, the California-Coahuila transform was a kinematic connector between subduction associated with the Sonoma orogeny and the coeval subduction zone in eastern Mexico. Others have previously suggested a similar kinematic linkage through northern Mexico between an active Cordilleran continental margin and a trench associated with the Permian-Triassic granite belt of eastern Mexico (Sedlock et al., 1993, Fig. 31; Ortega-Gutiérrez et al., 1994, Fig. 10).

Displacement of the Caborca miogeoclinal block from the projection of the Cordilleran miogeoclinal belt where it crosses the Mojave block of California would require 950 ± 25 km of net strike slip. That estimate for inferred slip assumes that metamorphosed miogeoclinal rocks of northeastern Baja California were once contiguous with the Caborca block prior to the separation of the Baja California peninsula from the mainland by Neogene rifting within the Gulf of California (Fig. 7). Assuming that transform slip ended by mid-Early Triassic time (ca. 247.5 Ma), initiation of transform displacement in Middle Pennsylvanian (Desmoinesian) time (ca. 302.5 Ma), when deep borderland basins first began to develop in the Death Valley region (Stone and Stevens, 1988), would retrodict a slip rate of 17 mm/yr for 55 m.y. If actual offset was delayed until transpressional deformation became pronounced in the Death Valley region near the beginning of middle Wolfcampian time (284 Ma), with earlier configurations of local basins attributed to preslip wrench deformation, then the inferred slip rate would be 26 mm/yr for only 36.5 m.y. Either figure seems compatible with our current general knowledge of relative plate motions along transform faults, although the latter interpretation is more coordinate with the oldest isotopic age (284 Ma) reported for the Permian-Triassic granitic belt of eastern Mexico. Slip at a rate comparable to modern San Andreas transform motion (~47.5 mm/yr) would, however, allow only 20 m.y. for full California-Coahuila transform displacement; such a fast rate seems incompatible with available timing constraints.

### *Mojave region*

The course of the truncational transform through the Mojave region is difficult to trace, probably in part because of complex posttruncational deformation of the structure (Martin and Walker, 1992; Stevens et al., 1992). One strand of the transform apparently passed to the east of the El Paso Mountains, where an offset segment of the Roberts Mountains allochthon is exposed near the southern end of the Sierra Nevada (Burchfiel and Davis, 1981b; Carr and Christiansen, 1984; Carr et al., 1997), with counterparts forming roof pendants within the Mesozoic Sierra Nevada batholith on the Kern Plateau to the northwest (Dunne and Suczek, 1991). As the total displacement of the Caborca block from the Cordilleran miogeocline is roughly twice the displacement of the El Paso Mountains domain from the western extension of the Roberts Mountains allochthon into the central Sierra Nevada, one principal strand of the transform system probably passed farther west (Davis et al., 1978; Burchfiel and Davis, 1981a, 1981b), leaving the El Paso Mountains domain within a subregional fault sliver (Burchfiel et al., 1992; Stevens et al., 1992).

Termination of transform motion within the Mojave region before the end of Early Triassic time is confirmed by widespread granitic plutons, formed within a magmatic arc, that have yielded U/Pb ages (247–208 Ma), based on concordia intercepts (Miller et al., 1995; Barth et al., 1997), indicative of late Early Triassic to late Late Triassic emplacement. A chain of analogous Middle to Upper Triassic plutons (232–210 Ma) extending up the eastern flank of the Sierra Nevada batholith into Nevada, and representing the northern continuation of the posttruncation magmatic arc (Dilles and Wright, 1988; Barth et al., 1990), was emplaced obliquely across the trend of truncated Paleozoic tectonic elements (Busby-Spera et al., 1990; Saleeby and Busby-Spera, 1992; Schweickert, 1996). A single gneissic pluton within the El Paso Mountains domain has yielded an older Late Permian U/Pb age of 260 Ma (Miller et al., 1995), although the intercept with concordia is at a very low angle and the affinity of the pluton is uncertain from tectonostratigraphic relations within the El Paso Mountains.

Exposures of the Roberts Mountains allochthon in the El Paso Mountains are overlain unconformably by a conglomeratic overlap assemblage of Mississippian age (Carr and Christiansen, 1984; Carr et al., 1997). Younger Pennsylvanian and Lower Permian strata also unconformably overlie the Roberts Mountains allochthon of the El Paso Mountains, and display complex facies relationships suggesting affinity with lithologically similar coeval strata of the transform borderland assemblage exposed farther east in the Death Valley region (Walker, 1987). This local borderland sedimentary succession, which was probably deposited on a substratum disrupted by transform tectonism, grades upward through a mid-Permian volcaniclastic interval into andesitic Upper Permian volcanic rocks (Carr and Christiansen, 1984; Carr et al., 1997).

Although the Permian volcanic cap is missing, the Paleozoic

assemblage of central Sonora is broadly analogous to that of the El Paso Mountains, with intensely deformed Ordovician-Devonian chert-argillite and varied turbidites overlain unconformably by a Carboniferous overlap sequence succeeded depositionally by a Permian succession grading upward into turbidites (Stewart et al., 1997). The stratigraphic analogies are supportive of fault offset along the California-Coahuila transform. Sonora exposures also include presumably overthrust Carboniferous-Permian eugeoclinal strata analogous to the Golconda allochthon of Nevada (Stewart et al., 1990), suggesting that Permian-Triassic subduction analogous to Sonoma deformation affected the oceanward flank of the Caborca block during its transform displacement southward. Farther north, to the east of the El Paso Mountains, emplacement of the Last Chance allochthon of southern Nevada and the Death Valley region of southeastern California in earliest Permian (Wolfcampian) time apparently reflected intense contractional deformation in the region east of the California-Coahuila transform as well (Snow, 1992). If transform displacement was delayed until mid-Wolfcampian time, however, Last Chance thrusting may have preceded significant strike slip.

The presence of a Permian volcanic sequence in the El Paso Mountains pointedly raises the question of the regional tectonic relationship between the Sonoma orogeny and the continental truncation of generally comparable age. As perceived here (Fig. 7), the transform that truncated the edge of the continental block linked subduction responsible for arc assemblages related to northwest-southeast convergence farther north, along the obliquely oriented Paleozoic continental margin, with the subduction zone flanking the Permian-Triassic magmatic arc in eastern Mexico (Sedlock et al., 1993). Such a regional tectonic framework implies some degree of kinematic coordination between northwest-southeast transform slip and northwest-southeast emplacement of the Golconda allochthon to the east of the line of continental truncation (Speed, 1979; Oldow et al., 1989; Burchfiel et al., 1992; Saleeby, 1992). In that context, the Permian volcanic rocks of the El Paso Mountains may have been displaced southward, together with the subjacent segment of the Roberts Mountains allochthon, from counterparts farther north.

## POST-TRUNCATION SUBDUCTION

Although Permian-Triassic truncation of the continental block is widely accepted, its implication for the history of subduction along the continental margin has not always been fully appreciated. Several recent syntheses imply, for example, that Paleozoic arc assemblages of the northern Sierra Nevada and Klamath Mountains, with flanking subduction complexes farther west, were aligned along strike, parallel to the younger Mesozoic continental margin, rather than to the obliquely oriented pre-Mesozoic continental margin (Oldow et al., 1989, p. 159; Poole et al., 1992, Fig. 1; E. L. Miller et al., 1992, Plate 4; Burchfiel et al., 1992, Figs. 5–8). A common tendency to depict Antler-Sonoma tectonic elements in cross-sectional view (Dickinson, 1981a; Ingersoll, 1997) also tends to obscure the spatial analysis of tectonic relations in three dimensions, relative to changing Paleozoic and Mesozoic configurations of the continental margin.

Geochronological data from both the Klamath Mountains and the Sierra Nevada indicate, however, that no pre-Late Triassic subduction complexes, which might be indicative of west-facing Paleozoic arc-trench systems, are present west of the line of Permian-Triassic truncation of the continental block. In the Klamath Mountains, blueschists of the Stuart Fork terrane (Fort Jones terrane of Silberling et al., 1987, 1992), structurally below and immediately to the west of the band of serpentinite bounding a northern portion of the Central Metamorphic belt on the west (Fig. 4B), have yielded Late Triassic isotopic ages (white mica K-Ar) of 214–222 Ma (Hotz et al., 1977). In the Sierra Nevada, Red Ant blueschists (Fig. 4B) in an analogous structural position, associated with the Feather River peridotite body, have yielded even younger (Jurassic) isotopic ages (white mica and whole rock K-Ar) of 173–176 Ma (Schweickert et al., 1980). Suspected argon loss implies, however, that the Red Ant blueschists may actually be comparable in age to the tectonically analogous Stuart Fork blueschists (Ernst, 1983). In both cases, the blueschist mineral assemblages reflect the highest pressures recorded by the metamorphic mineral facies present in the accretionary terranes of the western Klamath Mountains and Sierra Nevada (Hacker and Goodge, 1990).

In subduction complexes the world over, the oldest and most intensely developed blueschist metamorphism occurs systematically along the arcward flanks of accretionary prisms (Ernst, 1975), suggesting that the Stuart Fork and Red Ant blueschists record the inception of subduction along the truncated Mesozoic continental margin (Ernst, 1984; Schweickert, 1978; Goodge, 1990; Burchfiel et al., 1992). There is no room for older accretionary components between the blueschists and the truncated Mesozoic edge of the continental block. Speculation that portions of the accreted terranes west of the blueschists were assembled during pre-Late Triassic subduction (M.M. Miller, 1987) requires the unlikely assumption that the locus of most intense blueschist metamorphism stepped arcward during accretion, in a pattern of metamorphic evolution for which there is no analogy elsewhere. Speculation that a pre-Late Triassic subduction complex once existed along the western flank of the Klamath Mountains, but was either translated away during continental truncation or later masked beneath Mesozoic thrusts (Davis et al., 1978), is more difficult to reject. Neither alternative for wholesale removal of such an undetected subduction complex seems as attractive, however, as the straightforward interpretation that the Stuart Fork and Red Ant blueschists record the inception of subduction along a nascent continental margin.

### *Sierran-Klamath accretion*

Fossils recovered from the various accretionary belts or terranes west of the blueschist-bearing units in both the Klamath Mountains and the Sierra Nevada indicate assembly of a composite accretionary prism exclusively during Mesozoic time, with no record of preceding infra-Paleozoic subduction. Numerous Paleozoic limestone blocks are present in rock assemblages described variously by different workers as mélanges, broken formations, and olistostromes, but structurally interleaved and intermingled cherts and argillites have yielded widespread radiolarian and conodont faunas commonly as young as Early, Middle, or Late Triassic age, and locally as young as Early Jurassic age (Irwin, 1981; Irwin et al., 1977, 1978, 1982, 1983; Blome and Irwin, 1983; Bateman et al., 1985).

The age range of the fossils within a disrupted terrane cannot be taken as a measure of the time span of subduction and accretion (cf. M.M. Miller and Saleeby, 1991), because it is the youngest fossils present within a mélange or olistostrome that provide a maximum measure of its age of formation (Hsu, 1968). Disrupted lower Mesozoic chert-argillite strata are present, along with Permian limestone blocks, within the Calaveras terrane (Merced River terrane of Silberling et al., 1987, 1992) of the central Sierra Nevada, the so-called Central belt (Bucks Lake terrane of Silberling et al., 1987, 1992) of the northern Sierra Nevada, and the North Fork, Hayfork, and Rattlesnake Creek terranes of the western Klamath Mountains. The more easterly of these terranes abut the inferred line of Permian-Triassic continental truncation or the bordering blueschist belt in both the Klamath Mountains and the Sierra Nevada, and all are therefore regarded here as part of a composite Mesozoic accretionary prism. Various other pre-Cretaceous terranes farther west in the Sierra Nevada foothills and the far-western Klamath Mountains are of undoubted Jurassic age. Older Paleozoic limestone blocks that were incorporated into the growing accretionary prism can be interpreted as part of the sedimentary cover of sea floor or seamounts that were subsequently buried beneath Mesozoic chert and argillite before subduction of oceanic lithosphere along the continental margin (Dickinson, 1977). In some cases, olistostromal Permian limestone blocks may have been shed from reef-flanked seamounts prior to subduction (M.M. Miller and Wright, 1987), but provide no evidence for Permian accretion.

In northern California, the absence of pre-Mesozoic subduction complexes along strike from Paleozoic orogenic assemblages aligned regionally from northeast to southwest (Fig. 2) is not surprising. Any related subduction zones would be expected to be either to the southeast in Nevada, as inferred here, or else to the northwest in Oregon. In central Oregon, however, known blueschists have yielded a Late Triassic K-Ar age of 223 Ma (Hotz et al., 1977), and mélange within the John Day inlier (Dickinson and Thayer, 1978) of the Blue Mountains contains Devonian, Mississippian, and Permian limestone blocks associated with much greater volumes of disrupted Permian through Early Triassic chert and argillite (Blome and Nestell, 1991). Although bedrock over much of southern Oregon is masked by Tertiary volcanics, the central Oregon exposures suggest a continuation of Mesozoic Sierran-Klamath accretionary assemblages, which strike northeasterly toward central Oregon in the northernmost Klamath Mountains (Hietanen, 1981; Renne and Scott, 1988).

### *Sierran-Klamath arc*

Lower Mesozoic volcanic and volcaniclastic assemblages of the Sierra Nevada (Fig. 3F), westernmost Nevada (Fig. 3G), and the eastern flank of the Klamath Mountains (Fig. 3D) represent the earliest components of the magmatic arc built on the truncated continental margin and related to Mesozoic subduction complexes farther west. Basal units of the new arc assemblage were deposited locally as early as Middle Triassic time. Lower to Middle Jurassic volcanogenic strata reach a thickness of 7500 m in the northern Sierra Nevada (Harwood, 1993; Hanson et al., 1996), and analogous Jurassic volcanic successions extend southward along the eastern flank of the Sierra Nevada (Dunne and Walker, 1993). Jurassic quartzose sandstones interbedded with Jurassic volcaniclastic strata near Lake Tahoe (Harwood, 1988; Fisher, 1990; Miller and Harwood, 1990; Harwood et al., 1991) serve to tie the Sierran Mesozoic arc assemblage to counterparts in the nearer ranges of the Great Basin to the east (Speed and Jones, 1969; Wyld and Wright, 1993). At the northernmost end of the Sierra Nevada, the Jurassic volcaniclastic section at Mount Jura (Harwood, 1992, 1993) is both lithostratigraphically and chronostratigraphically similar to the analogous Jurassic section exposed in the John Day inlier of central Oregon (Dickinson and Thayer, 1978).

### *Nevada basinal assemblage*

In western Nevada east of the Sierran-Klamath arc assemblage, a thick sequence (5500–6000 m) of Middle to Upper Triassic and Lower to Middle Jurassic turbidites and associated strata (Fig. 3G) were deposited within a backarc basin behind the magmatic arc newly established to the west along the truncated continental margin (Fig. 3G). This basinal sequence grades westward into the arc assemblage (Speed, 1978a), with a belt of interfingering volcanic and sedimentary strata especially well displayed farther south (Dunne et al., 1998), and was bounded on the east by westward-prograding shelf and deltaic strata (Elison and Speed, 1988; Oldow et al., 1990). Basin inversion from Middle Jurassic to mid-Cretaceous time, perhaps related to sinistral transpression (Oldow et al., 1993), thrust the basinal assemblage eastward over coeval shelf equivalents (Speed, 1978b; Oldow, 1983, 1984). Although the substratum of the basin is nowhere exposed, it was probably formed by a comparatively thin crustal profile composed of remnant arcs, segmented subduction complexes, and perhaps remnant sea floor remaining after slab rollback to emplace the Golconda allochthon (Fig. 6H). Post-Middle Triassic crustal attenuation, stemming from backarc extension behind the Sierran-Klamath magmatic arc, may also have contributed to syndepositional subsidence (Oldow and Bartel, 1987; Wyld, this volume, Chapter 13). Mesozoic arc plutons which in time intruded the deformed basinal assemblage doubtless later contributed to crustal thickening along the trend of the initially backarc basin. Eastern elements of the magmatic arc were incorporated locally into thrust sheets that rode eastward over the flank of the basin (Dilek and Moores, 1995).

## TROUBLING INTERPRETIVE PROBLEMS

Three topics that provide evidence potentially refuting one or more of the tectonic interpretations outlined here deserve special discussion: (1) structural data for thrust vergence within the Central Metamorphic belt of the Klamath Mountains and the Shoo Fly Complex of the northern Sierra Nevada; (2) outcrops within roof pendants of the Sierra Nevada batholith suggesting dextral displacement along a cryptic Mesozoic structure termed the Mojave–Snow Lake fault; and (3) paleomagnetic data that argue for significant pre-Cretaceous clockwise rotation of the Klamath Mountains block. Each of the negative lines of evidence is, however, either equivocal to some extent or susceptible to alternate interpretations that allow the viewpoints developed here to stand intact.

### *Klamath-Sierran thrust vergence*

The geometry of mesoscopic structures within the Central Metamorphic belt (Fig. 4A) has been interpreted as compatible with generally west-directed overthrusting of the Trinity ophiolitic complex (Peacock and Norris, 1989), but overprinting by Mesozoic deformation is widespread (Davis et al., 1965). Moreover, the dominant trend of hornblende lineation in metabasalts of the Central Metamorphic belt is actually north-northwest to south-southeast (Davis, 1968, Fig. 4). Previous structural analysis has not entertained the possibility of tectonic transport toward the southeast, but the available structural data seem as broadly compatible with that option as with any other.

Folds within the Shoo Fly Complex (Fig. 4B) have been interpreted as compatible with tectonic transport generally up the present dip of associated internal thrusts, which are inclined to the east-northeast (Girty et al., 1990, 1996), but any structural analysis is complicated by the necessity for restoring post-Devonian tilt of the complex as a whole. Present steeply dipping internal thrusts were gently dipping (10°–25°) prior to tilt (Girty et al., 1990, 1996), and have been overprinted to varying degrees by Mesozoic deformation that accompanied or followed tilting. Stereonet analysis of the dextral and sinistral vergence of asymmetric drag folds, in relation to associated thrust faults, suggests tectonic transport toward a more southerly azimuth than a direction at right angles to the present outcrop belt. Given the complexity of deformation within accretionary prisms assembled by subduction, extant results of structural analysis do not seem incompatible with southerly tectonic transport, as envisioned in this chapter.

### *Mojave–Snow Lake fault*

Exposures of miogeoclinal strata within a roof pendant of the central Sierra Nevada argue for dextral Cretaceous strike slip of ~400 km along a cryptic structure (Mojave–Snow Lake fault) trending generally along the longitudinal axis of the Sierra Nevada batholith, which has largely obliterated the trace of the fault (Lahren, 1989; Lahren and Schweickert, 1994; Lahren et al., 1990; Schweickert and Lahren, 1990, 1991, 1993a; Schweickert, 1996). Northward, the trend of the structure is inferred to swing eastward into Nevada, roughly along the western flank of the lower Mesozoic basinal assemblage discussed herein as a backarc feature behind the Sierran-Klamath continental-margin arc. Southward, the trace of the fault is inferred to pass to the west of remnants of the Roberts Mountains allochthon in the El Paso Mountains (Fig. 2) and in roof pendants of the Sierra Nevada batholith on the nearby Kern Plateau.

Restoration of the indicated strike slip along the inferred Mojave–Snow Lake fault would place the Paleozoic arc assemblages of the northern Sierra Nevada and the Klamath Mountains adjacent to the Cordilleran miogeoclinal belt prior to offset, and directly athwart the trend of the latter. This initial position for Klamath-Sierran arc assemblages seems an unlikely hypothesis, at least prior to Permian-Triassic continental truncation that offset the miogeoclinal belt, but the line of truncation passes west of the Paleozoic arc assemblages (Fig. 2). Other explanations for the occurrence of miogeoclinal rocks in the Snow Lake pendant involving lesser Cretaceous strike slip may prove preferable to Mojave–Snow Lake displacement of the magnitude originally suggested (Saleeby and Busby, 1993).

Hindcasting the past positions of blocks possibly offset by the Mojave–Snow Lake fault is complicated by older late Paleozoic to earliest Mesozoic displacements associated with continental truncation along the northwestern extension of the California-Coahuila transform (Fig. 7). The amount of inferred dextral offset across the Mojave–Snow Lake fault is virtually the same as the inferred sinistral offset, across the easternmost strand of the truncational structure, which displaced the El Paso Mountains segment of the Roberts Mountains allochthon from counterparts farther north. To some extent, therefore, Cretaceous Mojave–Snow Lake dextral displacement may have served merely to recover Permian-Triassic California-Coahuila sinistral displacement in the Sierra Nevada region. However, the inferred position, between the two structures, of both the Klamath Mountains and the northern Sierra Nevada, and of the El Paso Mountains as well, seemingly precludes simple reversal of motion along the same trend.

The regionally criss-crossing pattern formed by the inferred trends of the Permian-Triassic truncational fault and the Cretaceous Mojave–Snow Lake fault makes any time-motion plan incorporating both structures difficult to envision. For example, reversing Cretaceous motion along the Mojave–Snow Lake fault would spoil the apparent continuity of the older truncational fault by shifting the northern trace of the latter out of alignment with its southern continuation. Consequently, large strike slip along the Mojave–Snow Lake fault seems incompatible with the interpretations preferred here for Paleozoic rock assemblages of northern California and Nevada. Resolution of the paradox must await further data or improved analysis.

### *Klamath tectonic rotation*

Paleomagnetic data from Permian, Triassic, and Jurassic strata of the eastern Klamath Mountains northeast of Redding (Figs. 2 and 4A) show no significant discordance in paleolatitude (Fagin and Gose, 1983; Mankinen, 1984; Renne and Scott, 1988; Mankinen et al., 1989; Mankinen and Irwin, 1990), in harmony with the tectonic interpretations of this chapter, but declination anomalies raise questions about the orientation of the Klamath domain through time. The available data from Paleozoic (Devonian-Permian) and Triassic rocks imply essentially uniform pre-Jurassic clockwise rotation of ~100°, but results from Jurassic strata are inconsistent. Some of the youngest sites reflect apparent clockwise rotation of as much as 60°, but other sites, including the oldest in Jurassic strata, record no paleomagnetically discernible rotation. Jurassic rocks of comparable age in the northern Sierra Nevada are also apparently unrotated (Mankinen and Irwin, 1990).

The correct interpretation of the paleomagnetic data is unclear. Bulk rotation of Triassic and older rock assemblages in the eastern Klamath Mountains by the indicated amount would, if restored, swing the line of inferred continental truncation from its present generally north-south orientation (Figs. 2 and 7) to a generally east-west original orientation, which is difficult to reconcile with any extant tectonic hypotheses for the geologic history of the continental margin. Those who infer west-facing Paleozoic arcs in the Klamath Mountains, with subduction downward to the east in terms of the present orientation of the domain, would then be confronted by south-facing arcs, with subduction downward to the north. Although this implied azimuth of subduction is not at large variance with the direction of subduction preferred in this chapter, the seeming concordance is spurious, because restoration of the same bulk rotation of the Klamath domain would convert the inferred southeasterly vergence of structures within the Yreka terrane to a northeasterly vergence, essentially at right angles to the southeasterly vergence inferred for the Shoo Fly Complex in the northern Sierra Nevada The paleomagnetic rotations observed locally for rocks of the eastern Klamath Mountains can perhaps be reconciled better with internal deformation involving areally restricted oroclinal bending and associated local structures (Renne and Scott, 1988).

## DISCUSSION

The fresh perspective developed here for Paleozoic Antler-Sonoma orogenesis in Nevada and California is encapsulated by the schematic diagrams of Figure 6, and rests on the following premises implicit in past work, but not previously combined into the same conceptual synthesis.

1. Pre-Middle Triassic tectonic trends were oriented northeast-southwest, at nearly a right angle to the present continental margin, hence Paleozoic tectonic elements were not parallel to the younger Cordilleran continental margin of Mesozoic age.

2. Klamath-Sierran arc assemblages of Paleozoic age include Devonian and Permian components that are stacked stratigraphically, to form parts of the same composite arc structures, but are diachronous between Klamath and Sierran domains, which were across rather than along tectonic strike in Paleozoic time.

3. Mediterranean-style subduction allows for the evolution of extensional island arc systems, involving multiple remnant arcs, at the same time that accretionary prisms develop along migratory subduction zones associated with slab rollback controlled by retrograde slab descent.

4. The Yreka terrane, Trinity ophiolitic complex, Shoo Fly Complex, and Roberts Mountains allochthon can be interpreted as parts of the same southeast-vergent early Paleozoic subduction complex, which continued to develop diachronously as Devonian magmatism prograded across its eroded surface from northwest to southeast.

5. Successive episodes of slab rollback, teminated by emplacement of accretionary prisms as overthrust allochthons across the miogeoclinal flank of the continental block, can account for emplacement of the Roberts Mountains and Golconda allochthons of the Antler and Sonoma orogens in central Nevada.

6. Episodes of arc magmatism in the Klamath-Sierran region were linked in time with emplacement of the Roberts Mountains and Golconda allochthons during the Antler and Sonoma orogenies, but separated by intervals of sedimentation that buried remnant arc structures beneath carbonate platforms, pelagic drapes, or turbidite fill.

7. Mediterranean-style subduction, with slab rollback, allowed a Klamath-Sierran island arc system that faced the continental block to remain close to the continental margin, without either arc-continent collision or arc polarity facing away from the continent required to pin remnant arcs in place.

8. The Antler and Sonoma orogenic systems included both the accretionary prisms of overthrust allochthons in central Nevada and generically related island arc structures of the Klamath-Sierran region, with southeastward tectonic vergence characteristic of both.

9. The composite Antler-Sonoma orogenic belt was truncated obliquely, but at a high angle, by a transform fault that offset a continuation of the Cordilleran miogeoclinal belt from California into Sonora, and linked Sonoma subduction kinematically with a subduction zone paralleling the Permian-Triassic granitic belt of eastern Mexico.

10. Accretionary belts of mélanges and associated deformed strata in the western Klamath Mountains and Sierra Nevada foothills were added to the continental margin exclusively during post-Middle Triassic time, as subduction complexes associated with the Mesozoic magmatic arc along the continental margin, and contain no record of west-facing (down to the east) Paleozoic subduction.

## ACKNOWLEDGMENTS

I thank S.J. Finney, G.E.Gehrels, G.S. Girty, R.E. Hanson, M.M. Miller, J.I. Satterfield, R.A. Schweickert, W.S. Snyder, S.J. Wyld, and especially D.S. Harwood for stimulating campfire, carryall, and motel discussions about eastern Klamath, northern Sierra Nevada, and Nevada geology. Valuable reviews by R.V. Ingersoll, T.F. Lawton, R.A. Schweickert, S.J. Wyld, and an anonymous reader, and an especially detailed and careful review by G.A. Davis, markedly improved the manuscript; however, the ideas expressed are entirely my own. Figures were prepared by Jim Abbott of SciGraphics. The stimulus for rethinking Antler-Sonoma orogenesis was the detrital zircon project of George Gehrels, who was in some sense the godfather of my effort.

## REFERENCES CITED

Albers, J.P., and Bain, J.H.C., 1985, Regional setting and new information on some critical geologic features of the West Shasta district, California: Economic Geology, v. 80, p. 2072–2091.

Albers, J.P., Kistler, R.W., and Kwak, L., 1981, The Mule Mountain stock, an early Middle Devonian pluton in northern California: Isochron/West, no. 31, p. 17.

Alibert, C., Martin, P., and Lapierre, H., 1991, The origin of geochemical variations in a Late Permian volcanic arc, eastern Klamath Mountains, California: Journal of Volcanology and Geothermal Research, v. 46, p. 299–322.

Anderson, H., and Jackson, J., 1987, The deep seismicity of the Tyrrhenian Sea: Royal Astronomical Society Geophysical Journal, v. 91, p. 613–637.

Anderson, T.H., and Schmidt, V.A., 1983, The evolution of Middle America and the Gulf of Mexico–Caribbean Sea region during Mesozoic time: Geological Society of America Bulletin, v. 94, p. 941–966.

Anderson, T.H., and Silver, L.T., 1979, The role of the Mojave-Sonora megashear in the tectonic evolution of northern Sonora, *in* Anderson, T.H., and Roldán-Quintana, J., eds., Geology of northern Sonora: Hermosillo, Sonora Instituto de Geología, Universidad Nacional Autónoma de México, p. 59–68.

Argnani, A., and Savelli, C., 1999, Cenozoic volcanism and tectonics in the southern Tyrrhenian Sea: Space-time distribution and geodynamic significance: Geodynamics, v. 27, p. 409–432.

Barth, A.P., Tosdal, R.M., and Wooden, J.L., 1990, A petrologic comparison of Triassic plutonism in the San Gabriel and Mule Mountains, southern California: Journal of Geophysical Research, v. 95, p. 20075–20096.

Barth, A.P., Tosdal, R.M., Wooden, J.L., and Howard, K.A., 1997, Triassic plutonism in southern California: Southward younging of arc initiation along a truncated continental margin: Tectonics, v. 16, p. 290–304.

Bateman, P.C., Harris, A.G., Kistler, R.W., and Krauskopf, K.B., 1985, Calaveras reversed: Westward younging is indicated: Geology, v. 13, p. 338–341.

Bazard, D.R., Butler, R.F., Gehrels, G., and Soja, C.M., 1995, Early Devonian paleomagnetic data from the Lower Devonian Karheen Formation suggest Laurentia-Baltica connection for the Alexander terrane: Geology, v. 23, p. 707–710.

Blein, O., Lapierre, H., Schweickert, R.A., Monie, P., and Pecher, A., 1994, La formation permienne de Black Dyke (Nevada centro-occidental) témoin le plus oriental de l'arc paléozoique de Sierra Nevada: Paris, Académie des Sciences Comptes Rendus (Sér. II), v. 319, p. 201–208.

Blein, O., Lapierre, H., Schweickert, R.A., Monie, P., Maluski, H., and Pecher, A., 1996, Remnants of the northern Sierra Nevada Paleozoic island arc in western Nevada: Journal of Geology, v. 104, p. 485–492.

Blome, C.D., and Irwin, W.P., 1983, Tectonic significance of late Paleozoic to Jurassic radiolarians from the North Fork terrane, Klamath Mountains, California, *in* Stevens, C.H., ed., Pre-Jurassic rocks in North American suspect terranes: Los Angeles, Pacific Section, Society of Economic Paleontologists and Mineralogists, p. 77–89.

Blome, C.D., and Nestell, M.K., 1991, Evolution of a Permo-Triassic sedimentary mélange, Grindstone terrane, east-central Oregon: Geological Society of America Bulletin, v. 103, p. 1280–1296.

Blome, C.D., and Reed, K.M., 1995, Radiolarian biostratigraphy of the Quinn River Formation, Black Rock terrane, north-central Nevada: Correlations with eastern Klamath terrane geology: Micropaleontology, v. 41, p. 49–68.

Boccalletti, M., and Dainelli, P., 1982, Schema tettonico dell'area Mediterranea con i principali elementi strutturalli Neogenico-Quaternari: Rome, Consigilio Nazionale della Richerche (Italia), scale 1:5 000 000.

Bond, G.C., and Devay, J.C., 1980, Pre-Upper Devonian quartzose sandstones in the Shoo Fly Formation, northern California—Petrology, provenance and implications for regional tectonics: Journal of Geology, v. 88, p. 285–308.

Bowring, S.A., and Erwin, D.H., 1998, A new look at evolutionary rates in deep time: Uniting paleontology and high-precision geochronology: GSA Today, v. 8, no. 9, p. 2–8.

Bowring, S.A., Erwin, D.H., Jin, Y.G., Martin, M.W., Davidek, K., and Wang, W., 1998, U/Pb geochronology and tempo of the end-Permian mass extinction: Science, v. 280, p. 1039–1045.

Brooks, E.R., and Coles, D.G., 1980, Use of immobile trace elements to determine original tectonic setting of eruption of metabasalts, northern Sierra Nevada, California: Geological Society of America Bulletin, v. 91, p. 665–671.

Brooks, E.R., Wood, M.M., and Garbutt, P.L., 1982, Origin and metamorphism of peperite and associated rocks in the Devonian Elwell Formation, northern Sierra Nevada, California: Geological Society of America Bulletin, v. 93, p. 1208–1231.

Brouxel, M., and Lapierre, H., 1984, La série basaltique de Trinity (Klamaths orientales, Nord Californie): Témoin de l'existence d'un bassin marginal au Dévonien moyen: Paris, Académie des Sciences Comptes Rendus, (Sér. II), v. 299, p. 457–462.

Brouxel, M., and Lapierre, H., 1988, Geochemical study of an early Paleozoic island-arc–back-arc basin system; Part 1: The Trinity ophiolite (northern California): Geological Society of America Bulletin, v. 100, p. 1111–1119.

Brouxel, M., Lapierre, H., Michard, A., and Albarede, F., 1987, The deep layers of a Paleozoic arc: Geochemistry of the Copley-Balaklala series, northern California: Earth and Planetary Science Letters, v. 85, p. 386–400.

Brouxel, M., Lapierre, H., Michard, A., and Albarede, F., 1988, Geochemical study of an early Paleozoic island-arc–back-arc basin system; Part 2: Eastern Klamath, early to middle Paleozoic island-arc volcanic rocks (northern California): Geological Society of America Bulletin, v. 100, p. 1120–1130.

Brueckner, H.K., and Snyder, W.S., 1985, Structure of the Havallah sequence, Golconda allochthon, Nevada: Evidence for prolonged evolution in an accretionary prism: Geological Society of America Bulletin, v. 96, p. 1113–1130.

Burchfiel, B.C., and Davis, G.A., 1972, Structural framework and evolution of the southern part of the Cordilleran orogen, western United States: American Journal of Science, v. 272, p. 97–118.

Burchfiel, B.C., and Davis, G.A., 1975, Nature and controls of Cordilleran orogenesis, western United States: Extensions of an earlier synthesis: American Journal of Science, v. 275-A, p. 363–396.

Burchfiel, B.C., and Davis, G.A., 1981a, Triassic and Jurassic tectonic evolution of the Klamath Mountains—Sierra Nevada geologic terrane, *in* Ernst, W.G., ed., The geotectonic development of California (Rubey Volume I): Englewood Cliffs, New Jersey, Prentice-Hall, p. 50–66.

Burchfiel, B.C., and Davis, G.A., 1981b, Mojave Desert and environs, *in* Ernst, W.G., ed., The geotectonic development of California (Rubey Volume I): Englewood Cliffs, New Jersey, Prentice-Hall, p. 217–252.

Burchfiel, B.C., and Royden, L.H., 1991, Antler orogeny: A Mediterranean-type orogeny: Geology, v. 19, p. 66–69.

Burchfiel, B.C., Cowan, D.S., and Davis, G.A., 1992, Tectonic overview of the Cordilleran orogen in the western United States, *in* Burchfiel, B.C., et al., eds., The Cordilleran orogen: Conterminous U.S.: Boulder, Colorado, Geological Society of America, Geology of North America, v. G-3, p. 407–479.

Burke, D.B., and Silberling, N.J., 1973, The Auld Lang Syne Group of Late Triassic and Jurassic(?) age, north-central Nevada: U.S. Geological Survey Bulletin 1394-E, p. E1-E14.

Busby-Spera, C.J., 1988, Speculative tectonic model for the early Mesozoic arc of the southwest Cordilleran United States: Geology, v. 16, p. 1121–1125.

Busby-Spera, C.J., Mattinson, J.M., Riggs, N.R., and Schermer, E.R., 1990, The Triassic-Jurassic magmatic arc in the Mojave-Sonoran deserts and the Sierran-Klamath region: Similarities and differences in paleogeographic evolution, *in* Harwood, D.S., and Miller, M.M., eds., Paleozoic and early Mesozoic paleogeographic relations; Sierra Nevada, Klamath Mountains, and related terranes: Geological Society of America Special Paper 255, p. 325–337.

Butler, R.F., Gehrels, G.E., and Bazard, D.R., 1997, Paleomagnetism of Paleozoic strata of the Alexander terrane, southeastern Alaska: Geological Society of America Bulletin, v. 109, p. 1372–1388.

Calmus, T., and Sosson, M., 1995, Southwestern extension of the Papago terrane into the Altar Desert region, northwestern Sonora, and its implications, *in* Jacques-Ayala, C., et al., eds., Studies on the Mesozoic of Sonora and adjacent areas: Geological Society of America Special Paper 301, p. 99–109.

Carr, M.D., and Christiansen, R.L., 1984, Pre-Cenozoic geology of the El Paso Mountains, southwestern Great Basin, California—A summary, *in* Lintz, J., Jr., ed., Western geological excursions, Volume 4: Reno, Department of Geological Sciences, Mackay School of Mines, p. 84–93.

Carr, M.D., Christiansen, R.L., Poole, F.G., and Goodge, J.W., 1997, Bedrock geologic map of the El Paso Mountains in the Garlock and El Paso Peaks 7-1/2′ quadrangles, Kern County, California: U.S. Geological Survey Miscellaneous Investigations Series Map I-2389, 9 p., scale 1:24 000.

Cashman, S.M., 1980, Devonian metamorphic event in the northeastern Klamath Mountaina, California: Geological Society of America Bulletin, v. 91, Part I, p. 453–459.

Cavinato, G.P., and DeCelles, P.G., 1999, Extensional basins in the tectonically bimodal central Apennines fold-thrust belt, Italy: Response to corner flow above a subducting slab in retrograde motion: Geology, v. 27, p. 955–958.

Centeno-Garcia, E., and Silva-Romo, G., 1997, Petrogenesis and tectonic evolution of central Mexico during Triassic-Jurassic time: Revista Mexicana de Ciencias Geológicas, v. 14, p. 244–260.

Charvet, J., Lapierre, H., and Campos, C., 1989, Les effets de la phase Antler (Dévonien superieur–Carbonifére inferieur) dans les Klamath orientales (Californie, U.S.A.): Implications géodynamiques: Paris, Académie des Sciences Comptes Rendus, (Sér. II), v. 308, p. 1629–1635.

Charvet, J., Lapierre, H., Rouer, O., Coulon, C., Campos, C., Martin, P., and Lecuyer, C., 1990, Tectono-magmatic evolution of Paleozoic and early Mesozoic rocks in the eastern Klamath Mountains, California, and the Blue Mountains, eastern Oregon–western Idaho, *in* Harwood, D.S., and Miller, M.M., eds., Paleozoic and early Mesozoic paleogeographic relations; Sierra Nevada, Klamath Mountains, and related terranes: Geological Society of America Special Paper 255, p. 255–276.

Churkin, M., Jr., 1974a, Paleozoic marginal ocean basin-volcanic arc systems in the Cordilleran foldbelt, *in* Dott, R.H., Jr., and Shaver, R.H., eds., Modern and ancient geosynclinal sedimentation: Society of Economic Paleontologists and Mineralogists Special Publication 19, p. 174–192.

Churkin, M., Jr., 1974b, Deep-sea drilling for landlubber geologists—The southwest Pacific, an accordion plate tectonics analogue for the Cordilleran geosyncline: Geology, v. 2, p. 339–342.

Churkin, M., Jr., and Eberlein, G.D., 1977, Ancient borderland terranes of the North American Cordillera: Correlation and microplate events: Geological Society of America Bulletin, v. 88, p. 769–786.

Cotkin, S.J., 1987, Conditions of metamorphism in an early Paleozoic blueschist, schist of Skookum Gulch, northern California: Contributions to Mineralogy and Petrology, v. 96, p. 192–200.

Cotkin, S.J., 1992, Ordovician-Silurian tectonism in northern California: The Callahan event: Geology, v. 20, p. 821–824.

Cotkin, S.J., Cotkin, M.L., and Armstrong, R.L., 1992, Early Paleozoic blueschist from the schist of Skookum Gulch, eastern Klamath Mountains, northern California: Journal of Geology, v. 100, p. 323–338.

D'Allura, J.A., Moores, E.H., and Robinson, L., 1977, Paleozoic rocks of the northern Sierra Nevada: Their structural and paleogeographic implications, *in* Stewart, J.H., et al., eds., Paleozoic paleogeography of the western United States: Pacific Section, Society of Economic Paleontologists and Mineralogists Pacific Coast Paleogeography Symposium 1, p. 395–408.

Daly, W.E., Doe, T.C., and Loranger, R.J., 1991, Geology of the northern Independence Mountains, Elko County, Nevada, *in* Raines, G.L., et al., eds., Geology and ore deposits of the Great Basin: Reno, Geological Society of Nevada, p. 583–602.

Davidek, K., Landing, E., Bowring, S.A., Westrop, S.R., Rushton, A.W.A., Fortey, R.A., and Adrain, J.M., 1998, New uppermost Cambrian U-Pb date from Avalonian Wales and age of the Cambrian-Ordovician boundary: Geological Magazine, v. 135, p. 305–309.

Davis, G.A., 1968, Westward thrust faulting in the south-central Klamath Mountains, California: Geological Society of America Bulletin, v. 79, p. 911–934.

Davis, G.A., 1969, Tectonic correlations, Klamath Mountains and western Sierra Nevada, California: Geological Society of America Bulletin, v. 80, p. 1095–1108.

Davis, G.A., Holdaway, M.J., Lipman, P.W., and Romey, W.D., 1965, Structure, metamorphism, and plutonism in the south-central Klamath Mountains, California: Geological Society of America Bulletin, v. 76, p. 933–966.

Davis, G.A., Monger, J.W.H., and Burchfiel, B.C., 1978, Mesozoic construction of the Cordilleran "collage", central British Columbia to central California, *in* Howell, D.G., and McDougall, K.A., eds., Mesozoic paleogeography of the western United States: Pacific Section, Society of Economic Paleontologists and Mineralogists Pacific Coast Paleogeography Symposium 2, p. 1–32.

Dickinson, W.R., 1977, Paleozoic plate tectonics and the evolution of the Cordilleran continental margin, *in* Stewart, J.H., et al., eds., Paleozoic paleogeography of the western United States: Pacific Section, Society of Economic Paleontologists and Mineralogists Pacific Coast Paleogeography Symposium 1, p. 137–155.

Dickinson, W.R., 1981a, Plate tectonics and the continental margin of California, *in* Ernst, W.G., ed., The geotectonic development of California (Rubey Volume I): Englewood Cliffs, New Jersey, Prentice-Hall, p. 1–28.

Dickinson, W.R., 1981b, Plate tectonic evolution of the southern Cordillera, *in* Dickinson, W.R., and Payne, W.D., eds., Relations of tectonics to ore deposits in the southern Cordillera: Arizona Geological Society Digest, v. 14, p. 113–135.

Dickinson, W.R., and Thayer, T.P., 1978, Paleogeographic and paleotectonic implications of Mesozoic stratigraphy and structure in the John Day inlier of central Oregon, *in* Howell, D.G., and McDougall, K.A., eds., Mesozoic paleogeography of the western United States: Pacific Section, Society of Economic Paleontologists and Mineralogists Pacific Coast Paleogeography Symposium 2, p. 147–161.

Dickinson, W.R., Harbaugh, D.W., Saller, A.H., Heller, P.L., and Snyder, W.S., 1983, Detrital modes of upper Paleozoic sandstones derived from Antler orogen in Nevada: Implications for nature of Antler orogeny: American Journal of Science, v. 283, p. 481–509.

Dilek, Y., and Moores, E.M., 1993, Across-strike anatomy of the Cordilleran orogen at 40°N latitude: Implications for the Mesozoic paleogeography of the western United States, *in* Dunne, G.C., and McDougall, K.A., eds., Mesozoic paleogeography of the western United States—II: Pacific Section, SEPM (Society for Sedimentary Geology) book 71, p. 333–346.

Dilek, Y., and Moores, E.M., 1995, Geology of the Humboldt igneous complex, Nevada, and tectonic implications for the Jurassic magmatism in the Cordilleran orogen, *in* Miller, D.M., and Busby, C., eds., Jurassic magmatism and tectonics of the North American Cordillera: Geological Society of America Special Paper 299, p. 229–248.

Dilles, J., and Wright, J.E., 1988, The chronology of early Mesozoic arc magmatism in the Yerington district of western Nevada and its regional implications: Geological Society of America Bulletin, v. 100, p. 644–652.

Doglioni, C., 1991, A proposal for the kinematic modelling of W-dipping subductions—possible applications to the Tyrrhenian-Apennines system: Terra Nova, v. 3, p. 423–434.

Dunne, G.C., and Suczek, C.A., 1991, Early Paleozoic eugeoclinal strata in the Kern Plateau pendants, southern Sierra Nevada, California, *in* Cooper, J.D., and Stevens, C.H., eds., Paleozoic paleogeography of the western

United States—II: Pacific Section, SEPM (Society for Sedimentary Geology) book 67, p. 677–692.
Dunne, G.C., and Walker, J.D., 1993, Age of Jurassic volcanism and tectonism, southern Owens Valley region, east-central California: Geological Society of America Bulletin, v. 105, p. 1223–1230.
Dunne, G.C., Garvey, T.P., Osborne, M., Schneidereit, D., and Fritsche, A.E., 1998, Geology of the Inyo Mountains volcanic complex: Implications for Jurassic paleogeography of the Sierran magmatic arc in eastern California: Geological Society of America Bulletin, v. 110, p. 1376–1397.
Durrell, C., and D'Allura, J., 1977, Upper Paleozoic section in eastern Plumas and Sierra counties, northern Sierra Nevada, California: Geological Society of America Bulletin, v. 88, p. 844–852.
Edelman, S.H., and Sharp, W.D., 1989, Terranes, early faults, and pre-Late Jurassic amalgamation of the western Sierra Nevada metamorphic belt, California: Geological Society of America Bulletin, v. 101, p. 1420–1433.
Edelman, S.H., Day, H.W., Moores, E.M., Zigan, S.M., Murphy, T.P., and Hacker, B.R., 1989, Structure across a Mesozoic ocean-continent collision zone in the northern Sierra Nevada, California: Geological Society of America Special Paper 224, 56 p.
Ehrenberg, S.N., 1975, Feather River ultramafic body, northern Sierra Nevada, California: Geological Society of America Bulletin, v. 86, p. 1235–1243.
Elison, M.W., 1995, Causes and consequences of Jurassic magmatism in the northern Great Basin: Implications for tectonic development, *in* Miller, D.M., and Busby, C., eds., Jurassic magmatism and tectonics of the North American Cordillera: Geological Society of America Special Paper 299, p. 249–265.
Elison, M.W., and Speed, R.C., 1988, Triassic flysch of the Fencemaker allochthon, East Range, Nevada: Fan facies and provenance: Geological Society of America Bulletin, v. 100, p. 185–199.
Ernst, W.G., 1975, Systematics of large-scale tectonics and age progressions in Alpine and circum-Pacific blueschist belts: Tectonophysics, v. 26, p. 229–246.
Ernst, W.G., 1983, Phanerozoic continental accretion and the metamorphic evolution of northern and central California: Tectonophysics, v. 100, p. 287–320.
Ernst, W.G., 1984, Californian blueschists, subduction, and the significance of tectonostratigraphic terranes: Geology, v. 12, p. 436–440.
Eschelbacher, J.W., and Wallin, E.T., 1998, Assembly of a subduction complex during the Early to Middle Devonian, Yreka terrane, California: Geological Society of America Abstracts with Programs, v. 30, no. 5, p. 13.
Fagin, S.W., and Gose, W.A., 1983, Paleomagnetic data from the Redding section of the eastern Klamath belt, northern California: Geology, v. 11, p. 505–508.
Fisher, G.R., 1990, Middle Jurassic syntectonic conglomerate in the Mt. Tallac roof pendant, northern Sierra Nevada, California, *in* Harwood, D.S., and Miller, M.M., eds., Paleozoic and early Mesozoic paleogeographic relations; Sierra Nevada, Klamath Mountains, and related terranes: Geological Society of America Special Paper 255, p. 339–350.
Fraticelli, L.A., Albers, J.P., and Zartman, R.E., 1985, The Permian Pit River stock of the McCloud plutonic belt, Eastern Klamath terrane, northern California: Isochron/West, no. 44, p. 6–8.
Fuis, G.S., Zucca, J.J., Mooney, W.D., and Milkereit, B., 1987, A geologic interpretation of seismic-refraction results in northeastern California: Geological Society of America Bulletin, v. 98, p. 53–65.
Garfunkel, Z., Anderson, C.A., and Schubert, G., 1986, Mantle circulation and the lateral motion of subducted slabs: Journal of Geophysical Research, v. 91, p. 7205–7223.
Gastil, G., Miller, R., Anderson, P., Crocker, J., Campbell, M., Buch, P., Lothringer, C., Leier-Engelhardt, P., DeLattre, M., Hoobs, J., and Roldán-Quintana, J., 1991, The relation between the Paleozoic strata on opposite sides of the Gulf of California, *in* Pérez-Segura, E., and Jacques-Ayala, C., eds., Studies of Sonoran geology: Geological Society of America Special Paper 254, p. 7–17.
Gehrels, G.E., and Saleeby, J.B., 1987a, Geology of southern Prince of Wales Island, southeastern Alaska: Geological Society of America Bulletin, v. 98, p. 123–137.
Gehrels, G.E., and Saleeby, J.B., 1987b, Geologic framework, tectonic evolution, and displacement history of the Alexander terrane: Tectonics, v. 6, p. 151–173.
Gehrels, G.E., Butler, R.F., and Bazard, D.R., 1996, Detrital zircon geochronology of the Alexander terrane, southeastern Alaska: Geological Society of America Bulletin, v. 108, p. 722–734.
Giles, K.A., and Dickinson, W.R., 1995, The interplay of eustasy and lithospheric flexure in forming stratigraphic sequences in foreland settings: An example from the Antler foreland, Nevada and Utah, *in* Dorobek, S.L., and Ross, G.M., eds., Stratigraphic evolution of foreland basins: SEPM (Society for Sedimentary Geology) Special Publication 52, p. 187–211.
Girty, G.H., and Pardini, C.H., 1987, Provenance of sandstone inclusions in the Paleozoic Sierra City mélange, Sierra Nevada, California: Geological Society of America Bulletin, v. 98, p. 176–181.
Girty, G.H., and Schweickert, R.A., 1984, The Culbertson Lake allochthon, a newly defined structure within the Shoo Fly complex, California: Evidence for four phases of deformation and extension of the Antler orogeny to the northern Sierra Nevada: Modern Geology, v. 8, p. 181–198.
Girty, G.H., and Wardlaw, M.S., 1984, Was the Alexander terrane a source of feldspathic sandstones in the Shoo Fly complex, Sierra Nevada, California?: Geology, v. 12, p. 339–342.
Girty, G.H., and Wardlaw, M.S., 1985, Petrology and provenance of pre-Late Devonian sandstones, Shoo Fly complex, northern Sierra Nevada, California: Geological Society of America Bulletin, v. 96, p. 516–521.
Girty, G.H., Gester, K.A., and Turner, J.B., 1990, Pre-Late Devonian geochemical, stratigraphic, sedimentologic, and structural patterns, Shoo Fly complex, northern Sierra Nevada, California, *in* Harwood, D.S., and Miller, M.M., eds., Paleozoic and early Mesozoic paleogeographic relations; Sierra Nevada, Klamath Mountains, and related terranes: Geological Society of America Special Paper 255, p. 43–56.
Girty, G.H., Keller, A.M., Franklin, K.R., and Stroh, R.C., 1991a, The southernmost lens of the Upper Devonian Grizzly Formation, northern Sierra Nevada, California: Evidence for a trench-slope depositional setting, *in* Cooper, J.D., and Stevens, C.H., eds., Paleozoic paleogeography of the western United States—II: Pacific Section, SEPM (Society for Sedimentary Geology) book 67, p. 693–701.
Girty, G.H., Gurrola, L.D., Taylor, G.W., Richards, M.J., and Girty, M.S., 1991b, The pre-Upper Devonian Lang and Black Oak sequences, Shoofly complex, northern Sierra Nevada, California: Trench deposits composed of continental detritus, *in* Cooper, J.D., and Stevens, C.H., eds., Paleozoic paleogeography of the western United States—II: Pacific Section, SEPM (Society for Sedimentary Geology) book 67, p. 703–716.
Girty, G.H., Yoshinobu, A.S., Wracher, M.D., Girty, M.S., Bryan, K.A., Skinner, J.E., McNulty, B.A., and Bracchi, K.A., 1993, U-Pb geochronology of the Emigrant Gap composite pluton, northern Sierra Nevada, California: Implications for the Nevadan orogeny, *in* Dunne, G.C., and McDougall, K.A., ed., Mesozoic paleogeography of the western United States—II: Pacific Section SEPM (Society for Sedimentary Geology) book 71, p. 323–332.
Girty, G.H., Hanson, R.E., Girty, M.S., Schweickert, R.A., Harwood, D.S., Yoshinobu, A.S., Bryan, K.A., Skinner, J.E., and Hill, C.A., 1995, Timing of emplacement of the Haypress Creek and Emigrant Gap plutons: Implications for the timing and controls of Jurassic orogenesis, northern Sierra Nevada, California, *in* Miller, D.M., and Busby, C., eds., Jurassic magmatism and tectonics of the North American Cordillera: Geological Society of America Special Paper 299, p. 191–201.
Girty, G. H., Lawrence, J., Burke, T., Fortin, A., Gallarano, C.S., Wirths, T.A., Lewis, J.G., Peterson, M.M., Ridge, D.L., Knaack, C., and Johnson, D., 1996, The Shoo Fly complex: Its origin and tectonic significance, *in* Girty, G.H., et al., eds., The Northern Sierra terrane and associated Mesozoic magmatic units: Implications for the tectonic history of the western Cordillera: Pacific Section, SEPM (Society for Sedimentary Geology) book 81, p. 1–23.
Goebel, K.A., 1991, Paleogeographic setting of Late Devonian to Early Mississippian transition from passive to collisional margin, Antler foreland, eastern Nevada and western Utah, *in* Cooper, J.D., and Stevens, C.H., eds.,

Paleozoic paleogeography of the western United States—II: Pacific Section, SEPM (Society for Sedimentary Geology) book 67, p. 401–418.
Gómez-Luna, M.E., and Martínez-Cortés, A., 1997, Relationships and differences between the Triassic ammonoid successions of northwestern Sonora, Mexico, and west-central Nevada, U.S.A.: Revista Mexicana de Ciencias Geológicas, v. 14, p. 208–218.
González-León, C.M., 1997, Sequence stratigraphy and paleogeographic setting of the Antimonio Formation (Late Permian–Early Jurassic), Sonora, Mexico: Revista Mexicana de Ciencias Geológicas, v. 14, p. 136–148.
González-León, C.M., Taylor, D.W., and Stanley, G.D., Jr., 1996, The Antimonio Formation in Sonora, Mexico, and the Triassic-Jurassic boundary: Canadian Journal of Earth Sciences, v. 33, p. 418–428.
Goodge, J.W., 1990, Tectonic evolution of a coherent Late Triassic subduction complex, Stuart Fork terrane, Klamath Mountains, northern California: Geological Society of America Bulletin, v. 102, p. 86–101.
Gradstein, F.M., Agterberg, F.P., Ogg, J.G., Hardenbol, J., van Veer, P., Thierry, J., and Zehui Huang, 1994, A Mesozoic time scale: Journal of Geophysical Research, v. 99, p. 24051–24074.
Greene, D.C., Schweickert, R.A., and Stevens, C.H., 1997, Roberts Mountains allochthon and the western margin of the Cordilleran miogeocline in the northern Ritter Range pendant, eastern Sierra Nevada, California: Geological Society of America Bulletin, v. 109, p. 1294–1305.
Gueguen, E., Doglioni, C., and Fernandez, M., 1997, Lithospheric boudinage in the western Mediterranean back-arc basin: Terra Nova, v. 9, p. 184–187.
Gueguen, E., Doglioni, C., and Fernandez, M., 1998, On the post-25 Ma geodynamic evolution of the western Mediterranean: Tectonophysics, v. 298, p. 259–269.
Hacker, B.R., and Goodge, J.W., 1990, Comparison of early Mesozoic high-pressure rocks in the Klamath Mountains and Sierra Nevada, *in* Harwood, D.S., and Miller, M.M., eds., Paleozoic and early Mesozoic paleogeographic relations; Sierra Nevada, Klamath Mountains, and related terranes: Geological Society of America Special Paper 255, p. 277–295.
Hacker, B.R., and Peacock, S.M., 1990, Comparison of the Central Metamorphic belt and Trinity terrane of the Klamath Mountains with the Feather River terrane of the Sierra Nevada, *in* Harwood, D.S., and Miller, M.M., eds., Paleozoic and early Mesozoic paleogeographic relations; Sierra Nevada, Klamath Mountains, and related terranes: Geological Society of America Special Paper 255, p. 75–92.
Hamilton, W., 1969, Mesozoic California and the underflow of Pacific mantle: Geological Society of America Bulletin, v. 80, p. 2409–2430.
Hamilton, W., 1978, Mesozoic tectonics of the western United States, *in* Howell, D.G., and McDougall, K.A., eds., Mesozoic paleogeography of the western United States: Pacific Section, Society of Economic Paleontologists and Mineralogists Pacific Coast Paleogeography Symposium 2, p. 33–61.
Hamilton, W. B., 1988, Plate tectonics and island arcs: Geological Society of America Bulletin, v. 100, p. 1503–1527.
Hamilton, W., and Myers, W.B., 1966, Cenozoic tectonics of the western United States: Reviews of Geophysics, v. 4, p. 509–549.
Hannah, J.L., and Moores, E.M., 1986, Age relationships and depositional environments of Paleozoic strata, northern Sierra Nevada, California: Geological Society of America Bulletin, v. 97, p. 787–797.
Hanson, R.E., and Schweickert, R.A., 1986, Stratigraphy of mid-Paleozoic island-arc rocks in part of the northern Sierra Nevada, Sierra and Nevada counties, California: Geological Society of America Bulletin, v. 97, p. 986–998.
Hanson, R.E., Saleeby, J.B., and Schweickert, R.A., 1988, Composite Devonian island-arc batholith in the northern Sierra Nevada, California: Geological Society of America Bulletin, v. 100, p. 446–457.
Hanson, R.E., Girty, G.H., Girty, M.S., Hargrove, U.S., Harwood, D.S., Kulow, M.J., Mielke, K.L., Phillipson, S.E., Schweickert, R.A., and Templeton, J.H., 1996, Paleozoic and Mesozoic arc rocks in the Northern Sierra terrane, *in* Girty, G.H., et al., eds., The Northern Sierra terrane and associated Mesozoic magmatic units: Implications for the tectonic history of the western Cordillera: Pacific Section, SEPM (Society for Sedimentary Geology) book 81, p. 25–55.
Harland, W.B., Armstrong, R.L., Cox, A.V., Craig, L.E., Smith, A.G., and Smith, D.G., 1990, A geologic time scale 1989: New York, Cambridge University Press, 265 p.
Harwood, D.S., 1983, Stratigraphy of upper Paleozoic volcanic rocks and regional unconformities in part of the northern Sierra terrane, California: Geological Society of America Bulletin, v. 94, p. 413–422.
Harwood, D.S., 1988, Tectonism and metamorphism of the northern Sierra terrane, northern California, *in* Ernst, W.G., ed., Metamorphism and crustal evolution of the western United States (Rubey Volume VII): Englewood Cliffs, New Jersey, Prentice-Hall, p. 764–788.
Harwood, D.S., 1992, Stratigraphy of Paleozoic and lower Mesozoic rocks in the northern Sierra terrane, California: U.S. Geological Survey Bulletin 1957, 78 p.
Harwood, D.S., 1993, Mesozoic geology of Mt. Jura, northern Sierra Nevada: A progress report, *in* Dunne, G.C., and McDougall, K.A., eds., Mesozoic paleogeography of the western United States—II: Pacific Section, SEPM (Society for Sedimentary Geology) book 71, p. 263–274.
Harwood, D.S., and Murchey, B.L., 1990, Biostratigraphic, tectonic, and paleogeographic ties between upper Paleozoic volcanic and basinal rocks in the northern Sierra terrane, California, and the Havallah sequence, Nevada, *in* Harwood, D.S., and Miller, M.M., eds., Paleozoic and early Mesozoic paleogeographic relations; Sierra Nevada, Klamath Mountains, and related terranes: Geological Society of America Special Paper 255, p. 157–173.
Harwood, D.S., Yount, J.C., and Seiders, V.M., 1991, Upper Devonian and Lower Mississippian island-arc and back-arc deposits in the northern Sierra Nevada, California, *in* Cooper, J.D., and Stevens, C.H., eds., Paleozoic paleogeography of the western United States—II: Pacific Section, SEPM (Society for Sedimentary Geology) book 67, p. 717–733.
Harwood, D.S., Fisher, G.R., Jr., and Waugh, B.J., 1995, Geologic map of the Duncan Peak and southern part of the Cisco Grove 7-1/2′ quadrangles, Placer and Nevada counties, California: U.S. Geological Survey Miscelleneous Investigations Series Map I-2341, 12 p., scale 1:24 000.
Hietanen, A., 1981, Extension of Sierra Nevada-Klamath suture system into eastern Oregon and western Idaho: U.S. Geological Survey Proessional Paper 1226-C, 11 p.
Hotz, P.E., 1973, Blueschist metamorphism in the Yreka–Fort Jones area, Klamath Mountains, California: U.S. Geological Survey Journal of Research, v. 1, p. 53–61.
Hotz, P.E., 1977, Geology of the Yreka quadrangle, Siskiyou County, California: U.S. Geological Survey Bulletin 1436, 72 p.
Hotz, P.E., Lanphere, M.A., and Swanson, D.A., 1977, Triassic blueschist from northern California and north-central Oregon: Geology, v. 5, p. 659–663.
Hsu, K.J., 1968, Principles of mélanges and their bearing on the Franciscan-Knoxville problem: Geological Society of America Bulletin, v. 79, p. 1063–1074.
Ingersoll, R.V., 1997, Phanerozoic tectonic evolution of central California and environs: International Geology Review, v. 39, p. 957–972.
Irwin, W.P., 1977, Review of Paleozoic rocks of the Klamath Mountains, *in* Stewart, J.H., et al., eds., Paleozoic paleogeography of the western United States: Pacific Section, Society of Economic Paleontologists and Mineralogists Pacific Coast Paleogeography Symposium 1, p. 441–454.
Irwin, W.P., 1981, Tectonic accretion of the Klamath Mountains, *in* Ernst, W.G., ed., The geotectonic development of California (Rubey Volume I): Englewood Cliffs, New Jersey, Prentice-Hall, p. 29–49.
Irwin, W.P., 1994, Geologic map of the Klamath Mountains, California and Oregon: U.S. Geological Survey Miscellaneous Investigations Series Map I-2148, scale 1:500 000.
Irwin, W.P., Jones, D.L., and Pessagno, E.A., Jr., 1977, Significance of Mesozoic radiolarians from the pre-Nevadan rocks of the southern Klamath Mountains, California: Geology, v. 5, p. 557–562.
Irwin, W.P., Jones, D.L., and Kaplan, T.A., 1978, Radiolarians from the pre-Nevadan rocks of the Klamath Mountains, California and Oregon, *in* Howell, D.G., and McDougall, K.A., eds., Mesozoic paleogeography of the western United States: Pacific Section, Society of Economic Paleontologists and Mineralogists Pacific Coast Paleogeography Symposium 2, p. 303–310.

Irwin, W.P., Jones, D.L., and Blome, C.D., 1982, Map showing sampled radiolarian localities in the western Paleozoic and Triassic belt, Klamath Mountains, California: U.S. Geological Survey Miscellaneous Field Studies Map MF-1399, scale 1:250 000.

Irwin, W.P., Wardlaw, B.R., and Kaplan, T.A., 1983, Conodonts of the western Paleozoic and Triassic belt, Klamath Mountains, California and Oregon: Journal of Paleontology, v. 57, p. 1030–1039.

Jacobsen, S.B., Quick, J.E., and Wasserburg, G.E., 1984, A Nd and Sr isotopic study of the Trinity peridotite—Implications for mantle evolution: Earth and Planetary Science Letters, v. 68, p. 361–378.

Jayko, A.S., 1990, Stratigraphy and tectonics of Paleozoic arc-related rocks of the northernmost Sierra Nevada, California: The eastern Klamath and northern Sierra terranes, *in* Harwood, D.S., and Miller, M.M., eds., Paleozoic and early Mesozoic paleogeographic relations; Sierra Nevada, Klamath Mountains, and related terranes: Geological Society of America Special Paper 255, p. 307–323.

Johnson, J.G., and Pendergast, A., 1981, Timing and mode of emplacement of the Roberts Mountains allochthon, Antler orogeny: Geological Society of America Bulletin, v. 92, p. 648–658.

Jones, A.E., 1990, Geology and tectonic significance of terranes near Quinn River Crossing, Nevada, *in* Harwood, D.S., and Miller, M.M., eds., Paleozoic and early Mesozoic paleogeographic relations; Sierra Nevada, Klamath Mountains, and related terranes: Geological Society of America Special Paper 255, p. 239–252.

Jones, A.E., 1991, Sedimentary rocks of the Golconda terrane: Provenance and paleogeographic implications, *in* Cooper, J.D., and Stevens, C.H., eds., Paleozoic paleogeography of the western United States—II: Pacific Section, SEPM (Society for Sedimentary Geology) book 67, p. 783–800.

Jones, A.E., and Jones, D.L., 1991, Paleogeographic significance of subterranes of the Golconda allochthon, northern Nevada, *in* Raines, G.L., et al., eds., Geology and ore deposits of the Great Basin: Reno, Geological Society of Nevada, p. 21–23.

Jones, D.L., Irwin, W.P., and Ovenshine, A.T., 1972, Southeastern Alaska—A displaced continental fragment?: U.S. Geological Survey Professional Paper 800-B, p. B211-B217.

Keller, J.V.A., Minelli, G., and Pialli, G., 1994, Anatomy of late orogenic extension: The northern Apennines case: Tectonophysics, v. 238, p. 275–294.

Kent, D.V., and Gradstein, F.M., 1985, A Cretaceous and Jurassic chronology: Geological Society of America Bulletin, v. 96, p. 1419–1427.

Kistler, R.W., McKee, E.H., Futa, K., Peterman, Z.E., and Zartman, R.E., 1985, A reconnaissance Rb-Sr, Sm-Nd, U-Pb, and K-Ar study of some host rocks and ore minerals in the West Shasta Cu-Zn district, California: Economic Geology, v. 80, p. 2128–2135.

Lahren, M.M., 1989, Proterozoic and Lower Cambrian miogeoclinal rocks of Snow Lake pendant, Yosemite-Emigrant Wilderness, Sierra Nevada, California: Evidence for major Early Cretaceous dextral translation: Geology, v. 17, p. 156–160.

Lahren, M.M., and Schweickert, R.A., 1994, Sachse Monument pendant, central Sierra Nevada, California: Eugeoclinal metasedimentary rocks near the axis of the Sierra Nevada batholith: Geological Society of America Bulletin, v. 106, p. 186–194.

Lahren, M.M., Schweickert, R.A., Mattinson, J.M., and Walker, J.D., 1990, Evidence of uppermost Proterozoic to Lower Cambrian miogeoclinal rocks and the Mojave–Snow Lake fault: Snow Lake pendant, central Sierra Nevada, California: Tectonics, v. 9, p. 1585–1608.

Landing, E., Bowring, S.A., Davidek, K.L., Westrop, S.R., Geyer, G., and Heldmaier, W., 1998, Duration of the Early Cambrian: U-Pb ages of volcanic ashes from Avalon and Gondwana: Canadian Journal of Earth Sciences, v. 35, p. 329–338.

Lanphere, M.A., Irwin, W.P., and Hotz, P.E., 1968, Isotopic age of the Nevadan orogeny and older plutonic and metamorphic events in the Klamath Mountains, California: Geological Society of America Bulletin, v. 79, p. 1027–1052.

Lapierre, H., 1983, Andésites riches en magnésium, témoins d'un arc insulaire dévonien dans les Klamaths orientales (N. California), États-Unis: Paris, Académie des Sciences Comptes Rendus (Sér. II), v. 296, p. 287–290.

Lapierre, H., and Cabanis, B., 1985, Caractérisation d'une série tholeiitique d'arc d'âge paléozoique (Klamaths orientales–Nord de la Californie, U.S.A.) à l'aide des clinopyroxénes et des éléments en traces: Société Géologique de France Bulletin, ser. 8, v. 1, p. 541–552.

Lapierre, H., Albarede, F., Albers, J., Cabanis, B., and Coulon, C., 1985a, Early Devonian volcanism in the eastern Klamath Mountains, California: Evidence for an immature island arc: Canadian Journal of Earth Sciences, v. 22, p. 214–227.

Lapierre, H., Cabanis, B., Coulon, C., Brouxel, M., and Albarede, F., 1985b, Geodynamic setting of Early Devonian Kuroko-type sulfide deposits in the eastern Klamath Mountains (northern California) inferred by the petrological and geochemical characteristics of the associated island-arc volcanic rocks: Economic Geology, v. 80, p. 2100–2113.

Lapierre, H., Brouxel, M., Martin, P., Coulon, C., Mascle, G., and Cabanis, B., 1986, The Paleozoic and Mesozoic geodynamic evolution of the eastern Klamath Mountains (north California) inferred from the geochemical characteristics of its magmatism: Société Géologique de France Bulletin, ser. 8, v. 2, p. 969–980.

Lapierre, H., Brouxel, M., Albarede, F., Coulon, C., Lecuyer, C., Martin, P., Mascle, G., and Rouer, O., 1987, Paleozoic and lower Mesozoic magmas from the eastern Klamath Mountains (north California) and the geodynamic evolution of northwestern America: Tectonophysics, v. 140, p. 155–177.

Lapierre, H., Tardy, M., Coulon, C., Ortiz Hernandez, E., Bourdier, J.-L., Martínez-Reyes, J., and Freydier, C., 1992, Caractérisation, genèse et évolution géodynamique du terrain de Guerrero (Mexique occidental): Canadian Journal of Earth Sciences, v. 29, p. 2478–2489.

Lindsley-Griffin, N., 1977a, The Trinity ophiolite, Klamath Mountains, California, *in* Coleman, R.G., and Irwin, W.P., eds., North American ophiolites: Oregon Department of Geology and Mineral Industries Bulletin 95, p. 107–120.

Lindsley-Griffin, N., 1977b, Paleogeographic implications of ophiolites: The Ordovician Trinity complex, Klamath Mountains, California, *in* Stewart, J.H., et al., eds., Paleozoic paleogeography of the western United States: Pacific Section, Society of Economic Paleontologists and Mineralogists Pacific Coast Paleogeography Symposium 1, p. 409–420.

Lindsley-Griffin, N., 1991, The Trinity Complex: A polygenetic ophiolitic assemblage, *in* Cooper, J.D., and Stevens, C.H., eds., Paleozoic paleogeography of the western United States—II: Pacific Section, SEPM (Society for Sedimentary Geology) book 67, p. 589–607.

Lindsley-Griffin, N., and Griffin, J.R., 1983, The Trinity terrane: An early Paleozoic microplate assemblage, *in* Stevens, C.H., ed., Pre-Jurassic rocks in western North American suspect terranes: Pacific Section, Society of Economic Paleontologists and Mineralogists, p. 63–75.

Lindsley-Griffin, N., Griffin, J.R., and Wallin, E.T., 1991, Redefinition of the Gazelle Formation of the Yreka terrane, Klamath Mountains, California: Paleogeographic implications, *in* Cooper, J.D., and Stevens, C.H., eds., Paleozoic paleogeography of the western United States—II: Pacific Section, SEPM (Society for Sedimentary Geology) book 67, p. 609–624.

Lucas, S.G., 1996, Correlation and tectonic significance of Lower Jurassic conglomerates in Sonora, Mexico, *in* Morales, M., ed., The continental Jurassic: Museum of Northern Arizona Bulletin 60, p. 497–501.

Lucas, S.G., Kues, B.S., Estep, J.W., and González-León, C.M., 1997, Permian-Triassic boundary at El Antimonio, Sonora, México: Revista Mexicana de Ciencias Geológicas, v. 14, p. 149–154.

Malinverno, A., and Ryan, W.B.F., 1986, Extension in the Tyrrhenian Sea and shortening in the Apennines as result of arc migration driven by sinking of the lithosphere: Tectonics, v. 5, p. 227–245.

Mankinen, E.A., 1984, Implications of paleomagnetism for the tectonic history of the Eastern Klamath and related terranes in California and Oregon, *in* Nilsen, T.H., ed., Geology of the Upper Cretaceous Hornbrook Formation, Oregon and California: Pacific Section, Society of Economic Paleontologists and Mineralogists book 42, p. 221–229.

Mankinen, E.A., and Irwin, W.P., 1990, Review of paleomagnetic data from the Klamath Mountains, Blue Mountains, and Sierra Nevada: Implications for paleogeographic reconstructions, *in* Harwood, D.S., and Miller, M.M., eds., Paleozoic and early Mesozoic paleogeographic relations; Sierra Nevada, Klamath Mountains, and related terranes: Geological Society of America Special Paper 255, p. 397–409.

Mankinen, E.A., Irwin, W.P., and Grommé, C.S., 1989, Paleomagnetic study of the Eastern Klamath terrane, California, and implications for the tectonic history of the Klamath Mountains province: Journal of Geophysical Research, v. 94, p. 10444–10472.

Martin, M.W., and Walker, J.D., 1992, Extending the western North American Proterozoic and Paleozoic continental crust through the Mojave Desert: Geology, v. 20, p. 753–756.

Martin, P., Lapierre, H., and Rocci, G., 1984, Présence d'un arc insulaire permien dans les Klamaths orientales (N. Californie), États-Unis: Paris, Académie des Sciences Comptes Rendus, (Sér. II), v. 298, p. 223–228.

Merguerian, C., and Schweickert, R.A., 1987, Paleozoic gneissic granitoids in the Shoo Fly complex, central Sierra Nevada, California: Geological Society of America Bulletin, v. 99, p. 699–717.

Metcalf, R.V., Wallin, E.T., and Willse, K., 1998, Devonian volcanic rocks of the eastern Klamath Mountains: An island-arc or volcanic cover of the Trinity ophiolite?: Geological Society of America Abstracts with Programs, v. 30, no. 6, p. 54.

Miller, E.L., Bateson, J., Dinter, D., Dyer, J.R., Harbaugh, D., and Jones, D.L., 1981, Thrust emplacement of the Schoonover sequence, northern Independence Mountains, Nevada: Geological Society of America Bulletin, v. 92, p. 730–737.

Miller, E.L., Kanter, L.R., Larue, D.K., Turner, R.J., Murchey, B., and Jones, D.L., 1982, Structural fabric of the Paleozoic Golconda allochthon, Antler Peak Quadrangle, Nevada: Progressive deformation of an oceanic sedimentary assemblage: Journal of Geophysical Research, v. 97, p. 3795–3804.

Miller, E.L., Holdsworth, B.K., Whiteford, W.B., and Rodgers, D., 1984, Stratigraphy and structure of the Schoonover sequence, northeastern Nevada: Implications for Paleozoic plate-margin tectonics: Geological Society of America Bulletin, v. 95, p. 1063–1076.

Miller, E.L., Miller, M.M., Stevens, C.H., Wright, J.E., and Madrid, R., 1992, Late Paleozoic paleogeographic and tectonic evolution of the western U.S. Cordillera, *in* Burchfiel, B.C., et al., eds., The Cordilleran orogen: Conterminous U.S.: Boulder, Colorado, Geological Society of America, Geology of North America, v. G-3, p. 57–106.

Miller, J.S., Glazner, A.F., Walker, J.D., and Martin, M.W., 1995, Geochronologic and isotopic evidence for Triassic-Jurassic emplacement of the eugeoclinal allochthon in the Mojave Desert region, California: Geological Society of America Bulletin, v. 107, p. 1441–1457.

Miller, M.M., 1987, Dispersed remnants of a northeast Pacific fringing arc: Upper Paleozoic terranes of Permian McCloud faunal affinity, western U.S.: Tectonics, v. 6, p. 807–830.

Miller, M.M., 1989, Intra-arc sedimentation and tectonism: Late Paleozoic evolution of the eastern Klamath terrane, California: Geological Society of America Bulletin, v. 101, p. 170–187.

Miller, M.M., and Cui, B., 1989, Submarine-fan characteristics and dual sediment provenance, Lower Carboniferous Bragdon Formation, eastern Klamath terrane, California: Canadian Journal of Earth Sciences, v. 26, p. 927–940.

Miller, M.M., and Harwood, D.S., 1990, Paleogeographic setting of upper Paleozoic rocks in the northern Sierra and eastern Klamath terranes, northern California, *in* Harwood, D.S., and Miller, M.M., eds., Paleozoic and early Mesozoic paleogeographic relations; Sierra Nevada, Klamath Mountains, and related terranes: Geological Society of America Special Paper 255, p. 175–192.

Miller, M.M., and Saleeby, J.B., 1991, Permian and Triassic paleogeography of the eastern Klamath arc and Eastern Hayfork subduction complex, Klamath Mountains, California, *in* Cooper, J.D., and Stevens, C.H., eds., Paleozoic paleogeography of the western United States—II: Pacific Section, SEPM (Society for Sedimentary Geology) book 67, p. 643–652.

Miller, M.M., and Wright, J.E., 1987, Paleogeographic implications of Permian Tethyan corals from the Klamath Mountains: Geology, v. 15, p. 266–269.

Moores, E., 1970, Ultramafics and orogeny, with models of the US Cordillera and the Tethys: Nature, v. 228, p. 837–842.

Moores, E.M., 1998, Ophiolites, the Sierra Nevada, "Cordilleria", and orogeny along the Pacific and Caribbean margins of North and South America: International Geology Review, v. 40, p. 40–54.

Murchey, B.L., 1990, Age and depositional setting of siliceous sediments in the upper Paleozoic Havallah sequence near Battle Mountain, Nevada: Implications for the paleogeography and structural evolution of the western margin of North America, *in* Harwood, D.S., and Miller, M.M., eds., Paleozoic and early Mesozoic paleogeographic relations; Sierra Nevada, Klamath Mountains, and related terranes: Geological Society of America Special Paper 255, p. 137–155.

Nichols, K.M., and Silberling, N.J., 1977, Stratigraphy and depositional history of the Star Peak Group (Triassic), northwestern Nevada: Geological Society of America Special Paper 178, 73 p.

Oldow, J.S., 1983, Tectonic implications of a late Mesozoic fold and thrust belt in northwestern Nevada: Geology, v. 11, p. 542–546.

Oldow, J.S., 1984, Evolution of a late Mesozoic back-arc fold and thrust belt, northwestern Great Basin, U.S.A.: Tectonophysics, v. 102, p. 245–274.

Oldow, J.S., and Bartel, R.L., 1987, Early to Middle(?) Jurassic extensional tectonism in the western Great Basin: Growth faulting and synorogenic deposition of the Dunlap Formation: Geology, v. 15, p. 740–743.

Oldow, J.S., Bally, A.W., Ave Lallemant, H.G., and Leeman, W.P., 1989, Phanerozoic evolution of the North American Cordillera; United States and Canada, *in* Bally, A.W., and Palmer, A.R., eds., The geology of North America—An overview: Boulder, Colorado, Geological Society of America, Geology of North America, v. A, p. 139–232.

Oldow, J.S., Bartel, R.L., and Gelber, A.W., 1990, Depositional setting and regional relationships of basinal assemblages: Pershing Ridge Group and Fencemaker Canyon sequence in northwestern Nevada: Geological Society of America Bulletin, v. 102, p. 193–222.

Oldow, J.S., Satterfield, J.I., and Silberling, N.J., 1993, Jurassic to Cretaceous transpressional deformation in the Mesozoic marine province of the northwestern Great Basin, *in* Lahren, M.M., et al., eds., Crustal evolution of the Great Basin and Sierra Nevada: Reno, Nevada, Department of Geological Sciences, Mackay School of Mines, p. 129–166.

Ortega-Guriérrez, F., Sedlock, R.L., and Speed, R.C., 1994, Phanerozoic tectonic evolution of Mexico, *in* Speed, R.C., ed., Phanerozoic evolution of North American continent-ocean transitions: Boulder, Colorado, Geological Society of America, DNAG Continent-Ocean Transect Volume, p. 265–306.

Palmer, A.R., 1983, Decade of North American Geology geologic time scale: Geology, v. 11, p. 503–504.

Peacock, S.M., and Norris, P.J., 1989, Metamorphic evolution of the Central Metamorphic belt, Klamath province, California: An inverted metamorphic gradient beneath the Trinity peridotite: Journal of Metamorphic Geology, v. 7, p. 191–209.

Peiffer-Rangin, F., 1977, Les zones isopiques du Paléozoique inférieur du nord-ouest mexicain, témoins du relais entre les Appalaches et la cordillère ouest-américaine: Paris, Académie des Sciences Comptes Rendus (Sér. II), v. 288, p. 1517–1519.

Poole, F.G., 1974, Flysch deposits of Antler foreland basin, western United States, *in* Dickinson, W.R., ed., Tectonics and sedimentation: Society of Economic Paleontologists and Mineralogists Special Publication 23, p. 58–82.

Poole, F.G., and Sandburg, C.A., 1991, Mississippian paleogeography and conodont biostratigraphy of the western United States, *in* Cooper, J.D., and Stevens, C.H., 1991, Paleozoic paleogeography of the western United States—II: Pacific Section, SEPM (Society for Sedimentary Geology) book 67, p. 107–136.

Poole, F.G., Stewart, J.H., Palmer, A.R., Sandberg, C.A., Madrid, R.J., Ross, R.J., Jr., Hintze, L.H., Miller, M.M., and Wrucke, C.T., 1992, Latest Precambrian to latest Devonian time: Development of a continental margin, *in* Burchfiel, B.C., et al., eds., The Cordilleran orogen: Conterminous U.S.: Boulder, Colorado, Geological Society of America, Geology of North America, v. G-3, p. 9–56.

Potter, A.W., Hotz, P.E., and Rohr, D.M., 1977, Stratigraphy and inferred tectonic framework of lower Paleozoic rocks in the eastern Klamath Mountains, northern California, *in* Stewart, J.H., et al., eds., Paleozoic paleogeography of the western United States: Pacific Section, Society of Economic Paleontologists and Mineralogists Pacific Coast Paleogeography Symposium 1, p. 421–440.

Potter, A.W., Boucot, A.J., Bergstrom, S.M., Blodgett, R.B., Dean, W.T., Flory, R.A., Ormiston, A. R., Pedder, A.E.H., Rigby, J.K., Rohr, D.M., and Savage, N.M., 1990, Early Paleozoic stratigraphic, paleogeographic, and biogeographic relations of the eastern Klamath belt, northern California, *in* Harwood, D.S., and Miller, M.M., eds., Paleozoic and early Mesozoic paleogeographic relations; Sierra Nevada, Klamath Mountains, and related terranes: Geological Society of America Special Paper 255, p. 57–74.

Quick, J.E., 1981, Petrology and petrogenesis of the Trinity peridotite, an upper mantle diapir in the eastern Klamath Mountains, northern California: Journal of Geophysical Research, v. 86, p. 11837–11863.

Renne, P.R., and Scott, G.R., 1988, Structural chronology, oroclinal deformation, and tectonic evolution of the southeastern Klamath Mountains, California: Tectonics, v. 7, p. 1223–1242.

Rouer, O., and Lapierre, H., 1989, Comparison between two Palaeozoic island-arc terranes in northern California (eastern Klamath and northern Sierra Nevada): Geodynamic constraints: Tectonophysics, v. 169, p. 341–349.

Rouer, O., Lapierre, H., and Coulon, C., 1988, La série calco-alcaline permienne d'arc du N de la Sierra Nevada (Caifornie, U.S.A.): Ultime étape magmatique dans l'évolution de la bordure de la microplaque Sonomia: Paris, Académie des Sciences Comptes Rendus, (Sér. II), v. 307, p. 57–62.

Rouer, O., Lapierre, H., Coulon, C., and Michard, A., 1989, New petrological and geochemical data on mid-Paleozoic island-arc volcanics of northern Sierra Nevada, California: Evidence for a continent-based island arc: Canadian Journal of Earth Sciences, v. 26, p. 2465–2478.

Rubin, C.M., Miller, M.M., and Smith, G.M., 1990, Tectonic development of Cordilleran mid-Paleozoic volcano-plutonic complexes: Evidence for convergent margin tectonism, *in* Harwood, D.S., and Miller, M.M., eds., Paleozoic and early Mesozoic paleogeographic relations; Sierra Nevada, Klamath Mountains, and related terranes: Geological Society of America Special Paper 255. p. 1–41.

Saleeby, J., 1977, Fracture zone tectonics, continental margin fragmentation, and emplacement of the Kings-Kaweah ophiolite belt, southwest Sierra Nevada, California, *in* Coleman, R.G., and Irwin, W.P., eds., North American ophiolites: Oregon Department of Geology and Mineral Industries Bulletin 95, p. 141–159.

Saleeby, J., 1981, Ocean floor accretion and volcanoplutonic arc evolution of the Mesozoic Sierra Nevada, *in* Ernst, W.G., ed., The geotectonic development of California (Rubey Volume I): Englewood Cliffs, New Jersey, Prentice-Hall, p. 132–181.

Saleeby, J.B., 1982, Polygenetic ophiolite belt of the California Sierra Nevada: Geochronological and tectonostratigraphic development: Journal of Geophysical Research, v. 87, p. 1803–1824.

Saleeby, J.B., 1990, Geochronological and tectonostratigraphic framework of Sierran-Klamath ophiolitic assemblages, *in* Harwood, D.S., and Miller, M.M., eds., Paleozoic and early Mesozoic paleogeographic relations; Sierra Nevada, Klamath Mountains, and related terranes: Geological Society of America Special Paper 255, p. 93–114.

Saleeby, J.B., 1992, Petrotectonic and paleogeographic settings of U.S. Cordilleran ophiolites, *in* Burchfiel, B.C., et al., eds., The Cordilleran orogen: Conterminous U.S.: Boulder, Colorado, Geological Society of America, Geology of North America, v. G-3, p. 653–682.

Saleeby, J. B., and Busby-Spera, C., 1992, Early Mesozoic tectonic evolution of the western U.S. Cordillera, *in* Burchfiel, B.C., et al., eds., The Cordilleran orogen: Conterminous U.S.: Boulder, Colorado, Geological Society of America, Geology of North America, v. G-3, p. 107–168.

Saleeby, J.B., and Busby, C., 1993, Paleogeographic and tectonic setting of axial and western metamorphic framework rocks of the southern Sierra Nevada, California, *in* Dunne, G.C., and McDougall, K.A., eds., Mesozoic paleogeography of the western United States—II: Pacific Section, SEPM (Society for Sedimentary Geology) book 71, p. 197–225.

Saleeby, J., Hannah, J.L., and Varga, R.J., 1987, Isotopic age constraints on middle Paleozoic deformation in the northern Sierra Nevada, California: Geology, v. 15, p. 757–760.

Saleeby, J.B., Shaw, H.F., Niemeyer, S., Moores, E.M., and Edelman, S.H., 1989, U/Pb, Sm/Nd, and Rb/Sr geochronological and isotopic study of northern Sierra Nevada ophiolitic assemblages, California: Contributions to Mineralogy and Petrology, v. 102, p. 205–220.

Sanborn, A.F., 1960, Geology and paleontology of the southwest quarter of the Big Bend quadrangle, Shasta County, California: California Division of Mines Special Report 63, 26 p.

Savage, N.M., 1977, Lower Devonian conodonts from the Gazelle Formation, Klamath Mountains, California: Journal of Paleontology, v. 51, p. 57–62.

Savelli, C., 1988, Late Oligocene to recent episodes of magmatism in and around the Tyrrhenian Sea: Implications for the processes of opening in a young inter-arc basin of intra-orogenic (Mediterranean) type: Tectonophysics, v. 146, p. 163–182.

Savelli, D., and Wezel, F.C., 1980, Morphologic map of the Tyrrhenian Sea: Rome, Consiglio Nazionale della Richerche (Italia), scale 1:250 000.

Savelli, C., Beccaluva, L., Deriu, M., Macciotta, G., and Maccioni, L., 1979, K/Ar geochronology and evolution of the Tertiary "calc-alkalic" volcanism of Sardinia (Italy): Journal of Volcanology and Geothermal Research, v. 5, p. 257–269.

Schweickert, R.A., 1976, Early Mesozoic rifting and fragmentation of the Cordilleran orogen in the western USA: Nature, v. 260, p. 586–591.

Schweickert, R.A., 1978, Triassic and Jurassic paleogeography of the Sierra Nevada and adjacent regions, California and western Nevada, *in* Howell, D.G., and McDougall, K.A., eds., Mesozoic paleogeography of the western United States: Pacific Section, Society of Economic Paleontologists and Mineralogists Pacific Coast Paleogeography Symposium 2, p. 361–384.

Schweickert, R.A., 1981, Tectonic evolution of the Sierra Nevada Range, *in* Ernst, W.G., ed., The geotectonic development of California (Rubey Volume I): Englewood Cliffs, New Jersey, Prentice-Hall, p. 87–131.

Schweickert, R.A., 1996, Tectonic setting of the Triassic to Jurassic magmatic arc and its basement, *in* Girty, G.H., et al., eds., The Northern Sierra terrane and associated Mesozoic magmatic units: Implications for the tectonic history of the western Cordillera: Pacific Section, SEPM (Society for Sedimentary Geology) book 81, p. 57–75.

Schweickert, R.A., and Lahren, M.M., 1987, Continuation of Antler and Sonoma orogenic belts to the eastern Sierra Nevada, California, and Late Triassic thrusting in a compressional arc: Geology, v. 15, p. 270–273.

Schweickert, R.A., and Lahren, M.M., 1990, Speculative reconstruction of the Mojave–Snow Lake fault: Implications for Paleozoic and Mesozoic orogenesis in the western United States: Tectonics, v. 9, p. 1609–1629.

Schweickert, R.A., and Lahren, M.M., 1991, Age and tectonic significance of metamorphic rocks along the axis of the Sierra Nevada batholith: A critical reappraisal, *in* Cooper, J.D., and Stevens, C.H., eds., Paleozoic paleogeography of the western United States—II: Pacific Section, SEPM (Society for Sedimentary Geology) book 67, p. 653–676.

Schweickert, R.A., and Lahren, M.M., 1993a, Triassic-Jurassic magmatic arc in eastern California and western Nevada: Arc evolution, cryptic tectonic breaks, and significance of the Mojave-Snow Lake fault, *in* Dunne, G.C., and McDougall, K.A., eds., Mesozoic paleogeography of the western United States—II: Pacific Section, SEPM (Society for Sedimentary Geology) book 71, p. 227–246.

Schweickert, R.A., and Lahren, M.M., 1993b, Tectonics of the east-central Sierra Nevada—Saddlebag Lake and northern Ritter Range pendants, *in* Lahren, M.M., et al., eds., Crustal evolution of the Great Basin and Sierra Nevada: Reno, Nevada, Department of Geological Sciences, Mackay School of Mines, p. 313–351.

Schweickert, R.A., and Snyder, W.S., 1981, Paleozoic plate tectonics of the Sierra Nevada and adjacent regions, *in* Ernst, W.G., ed., The geotectonic development of California (Rubey Volume I): Englewood Cliffs, New Jersey, Prentice-Hall, p. 182–201.

Schweickert, R.A., Armstrong, R.L., and Harakal, J.E., 1980, Lawsonite blueschist in the northern Sierra Nevada, California: Geology, v. 8, p. 27–31.

Schweickert, R.A., Harwood, D.S., Girty, G.H., and Hanson, R.E., 1984, Tectonic development of the northern Sierra terrane: An accreted late Paleozoic island arc and its basement, *in* Lintz, J., ed., Western geological excursions, Volume 4: Reno, Nevada, Department of Geological Sciences, Mackay School of Mines, p. 1–65.

Sedlock, R.L., Ortega-Gutiérrez, F., and Speed, R.C., 1993, Tectonostratigraphic terranes and tectonic evolution of Mexico: Geological Society of America Special Paper 278, 153 p.

Silberling, N.J., 1973, Geologic events during Permian-Triassic time along the Pacific margin of the United States, *in* Logan, A., and Hills, L.V., eds., The Permian and Triassic systems and their mutual boundary: Calgary, Alberta Society of Petroleum Geologists, p. 345–362.

Silberling, N.J., and Wallace, R.E., 1969, Stratigraphy of the Star Peak Group (Triassic) and overlying lower Mesozoic rocks, Humboldt Range, Nevada: U.S. Geological Survey Professional Paper 592, 50 p.

Silberling, N.J., Jones, D.L., Blake, M.C., Jr., and Howell, D.G., 1987, Lithotectonic terrane map of the western conterminous United States: U.S. Geological Survey Miscellaneous Field Studies Map MF-1874-C, scale 1:2 500 000.

Silberling, N.J., Jones, D.L., Monger, J.W.H., and Coney, P.J., 1992, Lithotectonic terrane map of the North American Cordillera: U.S. Geological Survey Miscellaneous Investigations Series Map I-2176, scale 1:5 000 000.

Silver, L.T., and Anderson, T.H., 1974, Possible left-lateral early to middle Mesozoic disruption of the southwestern North American craton margin: Geological Society of America Abstracts with Programs, v. 6, p. 955.

Snow, J.K., 1992, Large-magnitude Permian shortening and continental-margin tectonics in the southern Cordillera: Geological Society of America Bulletin, v. 104, p. 80–105.

Snyder, W.S., and Brueckner, H.K., 1983, Tectonic evolution of the Golconda allochthon, Nevada: Problems and perspectives, *in* Stevens, C.H., ed., Pre-Jurassic rocks in North American suspect terranes: Pacific Section, Society of Economic Paleontologists and Mineralogists, p. 102–123.

Soja, C.M., 1988, Lower Devonian platform carbonates from Kasaan Island, southeastern Alaska, Alexander terrane: Canadian Journal of Earth Sciences, v. 25, p. 639–656.

Soja, C.M., 1990, Island arc carbonates from the Silurian Heceta Formation of southeastern Alaska (Alexander terrane): Journal of Secimentary Petrology, v. 60, p. 235–249.

Soja, C.M., 1993, Carbonate platform evolution in a Silurian oceanic island: A case study from Alaska's Alexander terrane: Journal of Sedimentary Petrology, v. 63, p. 1078–1088.

Speed, R.C., 1977, Island-arc and other paleogeographic terranes of late Paleozoic age in the western Great Basin, *in* Stewart, J.H., et al., eds., Paleozoic paleogeography of the western United States: Pacific Section, Society of Economic Paleontologists and Mineralogists Pacific Coast Paleogeography Symposium 1, p. 349–362.

Speed, R.C., 1978a, Basinal terrane of the early Mesozoic marine province of the western Great Basin, *in* Howell, D.G., and McDougall, K.A., eds., Mesozoic paleogeography of the western United States: Pacific Section, Society of Economic Paleontologists and Mineralogists Pacific Coast Paleogeography Symposium 2, p. 237–252.

Speed, R.C., 1978b, Paleogeographic and plate tectonic evolution of the early Mesozoic marine province of the western Great Basin, *in* Howell, D.G., and McDougall, K.A., eds., Mesozoic paleogeography of the western United States: Pacific Section, Society of Economic Paleontologists and Mineralogists Pacific Coast Paleogeography Symposium 2, p. 253–270.

Speed, R.C., 1979, Collided Paleozoic microplate in the western United States: Journal of Geology, v. 87, p. 279–292.

Speed, R.C., 1984, Paleozoic and Mesozoic continental margin collision zone features: Mina to Candelaria, NV, traverse, *in* Lintz, J., Jr., ed., Western geological excursions, Volume 4: Reno, Nevada, Department of Geological Sciences, Mackay School of Mines, p. 66–80.

Speed, R.C., and Jones, T.A., 1969, Synorogenic quartz sandstone in the Jurassic mobile belt of western Nevada: Boyer Ranch Formation: Geological Society of America Bulletin, v. 80, p. 2551–2584.

Speed, R.C., and Sleep, N.H., 1982, Antler orogeny and foreland basin: A model: Geological Society of America Bulletin, v. 93, p. 815–828.

Speed, R.C., Elison, M.W., and Heck, R.F., 1988, Phanerozoic tectonic evolution of the Great Basin, *in* Ernst, W.G., ed., Metamorphism and crustal evolution of the western United States (Rubey Volume VII): Englewood Cliffs, New Jersey, Prentice-Hall, p. 572–605.

Stanley, G.D., Jr., and González-León, C.M., 1995, Paleogeographic and tectonic implications of Triassic fossils and strata from the Antimonio Formation, northwestern Sonora, *in* Jacques-Ayala, C., et al., eds., Studies on the Mesozoic of Sonora and adjacent areas: Geological Society of America Special Paper 301, p. 1–16.

Stevens, C.H., Stone, P., and Kistler, R.W., 1992, A speculative reconstruction of the middle Paleozoic continental margin of southwestern North America: Tectonics, v. 11, p. 405–419.

Stevens, C.H., Stone, P., Dunne, G.C., Greene, D.C., Walker, J.D., and Swanson, B.J., 1997, Paleozoic and Mesozoic evolution of east-central California: International Geology Review, v. 39, p. 788–829.

Stewart, J.H., 1980, Geology of Nevada: Nevada Bureau of Mines and Geology Special Publication 4, 136 p.

Stewart, J.H., 1988, Latest Proterozoic and Paleozoic southern margin of North America and the accretion of Mexico: Geology, v. 16, p. 186–189.

Stewart, J.H., and Poole, F.G., 1974, Lower Paleozoic and uppermost Precambrian Cordilleran miogeocline, Great Basin, western United States, *in* Dickinson, W.R., ed., Tectonics and sedimentation: Society of Economic Paleontologists and Mineralogists Special Publication 22, p. 28–57.

Stewart, J.H., McMenamin, M.A.S., and Morales-Ramirez, J.M., 1984, Upper Proterozoic and Cambrian rocks in the Caborca region, Sonora, Mexico—Physical stratigraphy, biostratigraphy, paleocurrent studies, and regional relations: U.S. Geological Survey Professional Paper 1309, 36 p.

Stewart, J.H., Murchey, B., Jones, D.L., and Wardlaw, B.R., 1986, Paleontologic evidence for complex tectonic interlaying of Mississippian to Permian deep-water rocks of the Golconda allochthon in Tobin Range, north-central Nevada: Geological Society of America Bulletin, v. 97, p. 1122–1132.

Stewart, J.H., Poole, F.G., Ketner, K.B., Madrid, R.J., Roldán-Quintana, J., and Amaya-Martínez, R., 1990, Tectonics and stratigraphy of the Paleozoic and Triassic southern margin of North America, Sonora, Mexico, *in* Gehrels, G.E., and Spencer, J.E., eds., Geologic excursions through the Sonoran Desert region, Arizona and Sonora: Arizona Geological Survey Special Paper 7, p. 183–202.

Stewart, J. H., Amaya-Martínez, R., Stamm, R.G., Wardlaw, B.R., Stanley, G.D., Jr., and Stevens, C.H., 1997, Stratigraphy and regional significance of Mississippian to Jurassic rocks in Sierra Santa Teresa, Sonora, Mexico: Revista Mexicana de Ciencias Geológicas, v. 14, p. 115–135.

Stone, P., and Stevens, C.H., 1988, Pennsylvanian and Early Permian paleogeography of east-central California: Implications for the shape of the continental margin and the timing of continental truncation: Geology, v. 16, p. 330–333.

Tardy, M., Lapierre, H., Bourdier, J.-L., Coulon, C., Ortiz-Hernández, L.E., and Yta, M., 1992, Intraoceanic setting of the western Mexico Guerrero terrane—Implications for the Pacific-Tethys geodynamic relationships during the Cretaceous: Universidad Nacional Autónoma de México, Instituto de Geología, Revista, v. 10, p. 118–128.

Tardy, M., Lapierre, H., Freydier, C., Coulon, C., Gill, J.-B., Mercier de Lepinay, B., Beck, C., Martinez Reyes, J., Talavera M., O., Ortiz H.E., Stein, G., Bourdier, J.-L., and Yta, M., 1994, The Guerrero suspect terrane (western Mexico) and coeval arc terranes (the Greater Antilles and the Western Cordillera of Colombia): A late Mesozoic intra-oceanic arc accreted to cratonal America during the Cretaceous: Tectonophysics, v. 230, p. 49–73.

Taylor, D.G., Smith, P.L., Laws, R.A., and Guex, J., 1983, The stratigraphy and biofacies trends of the lower Mesozoic Gabbs and Sunrise Formations, west-central Nevada: Canadian Journal of Earth Sciences, v. 20, p. 1598–1608.

Tomlinson, A.J., Miller, E.L., Holdsworth, B.K., Whiteford, W.B., Snyder, W.S., and Brueckner, H.K., 1987, Structure of the Havallah sequence, Golconda allochthon, Nevada: Evidence for prolonged evolution in an accretionary prism: Discussion and reply: Geological Society of America Bulletin, v. 98, p. 615–617.

Torres V., R., Ruiz, J., Patchett, P.J., and Grajales, J.M., 2000, A Permo-Triassic continental arc in eastern Mexico: Tectonic implications for reconstructions of southern North America, *in* Bartolini, C., et al., eds., Mesozoic sedimentary and tectonic history of north-central Mexico: Geological Society of America Special Paper 340, p. 191–196.

Tucker, R.D., and McKerrow, W.S., 1995, Early Paleozoic chronology: A review in light of new U-Pb zircon ages from Newfoundland and Britain: Canadian Journal of Earth Sciences, v. 32, p. 368–379.

Tucker, R. D., Bradley, D.C., Ver Straeten, C.A., Harris, A.G., Ebert, J.R., and McCutcheon, S.R., 1998, New U-Pb zircon ages and the duration and division of Devonian time: Earth and Planetary Science Letters, v. 158, p. 175–186.

Turner, R.J.W., Madrid, R.J., and Miller, E.L., 1989, Roberts Mountains allochthon: Stratigraphic comparison with lower Paleozoic outer continental margin strata of the northern Canadian Cordillera: Geology, v. 17, p. 341–344.

Varga, R.J., and Moores, E.M., 1981, Age, origin, and significance of an unconformity that predates island-arc volcanism in the northern Sierra Nevada: Geology, v. 9, p. 512–518.

Walker, J.D., 1987, Permian to Middle Triassic rocks of the Mojave Desert, *in* Dickinson, W.R., and Klute, M.A., eds., Mesozoic rocks of southern Arizona and adjacent areas: Arizona Geological Society Digest, v. 18, p. 1–14.

Walker, J.D., 1988, Permian and Triassic rocks of the Mojave Desert and their implications for timing and mechanisms of continental truncation: Tectonics, v. 7, p. 685–709.

Wallin, E.T., and Metcalf, R.V., 1998, Supra-subduction zone ophiolite formed in an extensional forearc: Trinity terrane, Klamath Mountains, California: Journal of Geology, v. 106, p. 591–608.

Wallin, E.T., and Trabert, D.W., 1994, Eruption-controlled epiclastic sedimentation in a Devonian trench-slope basin: Evidence from sandstone petrofacies, Klamath Mountains, California: Journal of Sedimentary Research, v. A64, p. 373–385.

Wallin, E.T., Mattinson, J.M., and Potter, A.W., 1988, Early Paleozoic magmatic events in the eastern Klamath Mountains, northern California: Geology, v. 16, p. 144–148.

Wallin, E.T., Lindsley-Griffin, N., and Griffin, J.R., 1991, Overview of early Paleozoic magmatism in the eastern Klamath Mountains, California: An isotopic perspective, *in* Cooper, J.D., and Stevens, C.H., eds., Paleozoic paleogeography of the western United States—II: Pacific Section, SEPM (Society for Sedimentary Geology) book 67, p. 581–588.

Wallin, E.T., Coleman, D.S., Lindsley-Griffin, N., and Potter, A.W., 1995, Silurian plutonism in the Trinity terrane (Neoproterozoic and Ordovician), Klamath Mountains, California, United States: Tectonics, v. 14, p. 1007–1013.

Watkins, R., 1973, Carboniferous faunal associations and stratigraphy, Shasta County, California: American Association of Petroleum Geologists Bulletin, v. 57, p. 1743–1764.

Watkins, R., 1985, Volcaniclastic and carbonate sedimentation in late Paleozoic island-arc deposits, eastern Klamath Mountains, California: Geology, v. 13, p. 709–713.

Watkins, R., 1986, Late Devonian to Early Carboniferous turbidite facies and basinal development of the eastern Klamath Mountains, California: Sedimentary Geology, v. 49, p. 51–71.

Watkins, R., 1990, Carboniferous and Permian island-arc deposits of the eastern Klamath terrane, California, *in* Harwood, D.S., and Miller, M.M., eds., Paleozoic and early Mesozoic paleogeographic relations; Sierra Nevada, Klamath Mountains, and related terranes: Geological Society of America Special Paper 255, p. 193–200.

Watkins, R., 1993a, Permian carbonate platform development in an island-arc setting, eastern Klamath terrane, California: Journal of Geology, v. 101, p. 659–666.

Watkins, R., 1993b, Carbonate bank sedimentation in a volcanic arc setting: Lower Carboniferous limestones of the eastern Klamath terrane, California: Journal of Sedimentary Petrology, v. 63, p. 955–973.

Watkins, R., and Flory, R.A., 1986, Island arc sedimentation in the Middle Devonian Kennett Formation, eastern Klamath Mountains, California: Journal of Geology, v. 94, p. 753–761.

Whiteford, W.B., 1990, Paleogeographic setting of the Schoonover sequence, Nevada, and implications for the late Paleozoic margin of western North America, *in* Harwood, D.S., and Miller, M.M., eds., Paleozoic and early Mesozoic paleogeographic relations; Sierra Nevada, Klamath Mountains, and related terranes: Geological Society of America Special Paper 255, p. 115–136.

Wyld, S. J., 1990, Paleozoic and Mesozoic rocks of the Pine Forest Range, northwest Nevada, and their relation to volcanic arc assemblages of the western U.S. Cordillera, *in* Harwood, D.S., and Miller, M.M., eds., Paleozoic and early Mesozoic paleogeographic relations; Sierra Nevada, Klamath Mountains, and related terranes: Geological Society of America Special Paper 255, p. 219–237.

Wyld, S.J., 1991, Permo-Triassic tectonism in volcanic arc sequences of the western U. S. Cordillera and implications for the Sonoma orogeny: Tectonics, v. 10, p. 1007–1017.

Wyld, S.J., and Wright, J.E., 1993, Mesozoic stratigraphy and structural history of the southern Pine Nut Range, west-central Nevada, *in* Dunne, G.C., and McDougall, K.A., eds., Mesozoic paleogeography of the western United States—II: Pacific Section, SEPM (Society for Sedimentary Geology) book 71, p. 289–321.

Zucca, J.J., Fuis, G.S., Milkereit, B., Mooney, W.D., and Catchings, R.D., 1986, Crustal structure of northeastern California: Journal of Geophysical Research, v. 91, p. 7359–7382.

MANUSCRIPT ACCEPTED BY THE SOCIETY JANUARY 24, 2000

Printed in U.S.A.

# Index

[Italic page numbers indicate major references]

## Z